Library
State University of New York
Agricultural and Technical College
Canton, New York 13617

Statistics

Xerox College Publishing LEXINGTON, MASSACHUSETTS / TORONTO

STATISTICS

An Applied Approach

Neil R. Ullman COUNTY COLLEGE OF MORRIS, NEW JERSEY

Copyright © 1972 by Xerox Corporation.

All rights reserved. Permission in writing must be obtained from the publisher before any part of this publication may be reproduced or transmitted in any form or by any means, electronic or mechanical, including photocopy, recording, or any information storage or retrieval system.

ISB Number: 0/536/00702/0

Library of Congress Catalog Card Number: 77/171918

Printed in the United States of America.

Preface

I have long been aware of the difficulty that many students as well as generally educated people have in understanding the very basic principles that underly the broad field of statistics. It is to this large audience that I address this text, with the hope of providing them with an opportunity to gain some insight into this challenging subject. The more mathematically prepared student may also find the text interesting and should be able to travel at a rapid pace, exploring deeper meaning and working with more extensive techniques.

Rather than work downward from the "standard" probability and statistics textbooks, I have worked up from a layman's approach. For this reason I have found that many portions of this text have been easily read by such diverse groups of people as police officers in a law enforcement program, secretaries, and faculty members attending graduate courses in statistics. I have not tried to "ease the pain" of a junior level probability and statistics course but, rather, have tried to gather together a collection of material that would be relevant, fundamental, interesting, and understandable. To this end I hope I have been successful.

The text is very flexible. There is no set sequence, as a large number of chapters require little or no prior preparation. Essentially all of Chapters 1 through 8, 11, and the Appendices, are self-contained, except for the examples in the beginning of Chapter 2, the elementary notions of histograms in Chapter 3, and mean and standard deviation in Chapter 4. These last items are basic ideas that are referred to throughout the text. In addition, most of Chapters 9, 10, and 14, as well as parts of 12, 15, 19, and 20 may be read essentially independent of the rest of the text. References may be made in these later chapters to various terms and examples discussed in earlier chapters. However, in all cases, the reader should be able to grasp the general concepts being presented even if he has not covered or fully understood all of the preliminary material.

Listed below are some of the important features of this text.

More than enough material is available for a one semester or in some cases a two semester course in basic statistics, with a number of possible options for treating the subject matter.

The first chapter provides a thorough introduction to the purpose, necessity, and approach of statistical analysis. This is then followed by an extensive discussion of the kinds of problems and questions one must consider when working with all types of data—whether it be yours or someone else's.

An attempt is made throughout to introduce basic concepts in a nonmathematical way. Thus, the concept of combining several distributions to yield a broad distribution is introduced in several different ways as a foundation for future discussions of the Analysis of Variance, and an entire chapter is devoted to a nonmathematical description of hypothesis testing.

Nonparametric and parametric tests are integrated according to the function of the test. The emphasis is always on discussing the underlying reason for a test and then to consider the type of test.

A separate chapter discusses the role of calculators and computers with illustrations of time-sharing library programs.

A thorough collection of tables is provided. Each of these tables is accompanied by unique detailed diagrams and examples to help the reader avoid errors in reading the tables and to eliminate the need to search the text for explanations.

Numerous projects that any reader can perform are presented throughout the text. It is this type of "doing" statistics and working with real data that provides meaning to the entire subject of statistics.

In summary, I have attempted to provide a foundation, both for the casual reader for whom this text may act as a survey and for the more serious student who desires a carefully and slowly developed treatment of the major concepts of statistics and statistical analysis.

I should mention that except where specifically stated to the contrary all examples, problems, and data are fictitious.

I would like to express my appreciation to my wife, numerous students, colleagues, and friends, who through encouragement and patience helped this text to materialize. I am particularly grateful to Dr. Edward Blum of the New York City Rand Institute for his ideas and comments which sparked much of the organization of the early part of this book and to Dr. and Mrs. Thomas Boardman for their valuable reviews of the manuscript.

I am indebted to the Literary Executor of the late Sir Ronald A. Fisher, F.R.S., to Dr. Frank Yates, F.R.S., amd to Oliver & Boyd, Edinburgh, for permission to reprint Table III and VII from their book *Statistical Tables for Biological, Agricultural and Medical Research* (the current edition of their work is the sixth of 1963).

Finally, I would like to thank all the students who have suffered through partially written material, the various secretaries at the County College of Morris who helped type numerous portions of the manuscript and last, but certainly not least, Dana Andrus whose patient and hard work as manuscript editor helped make this book a reality.

Spring 1972   NEIL R. ULLMAN

Contents

Introduction   xv

PART I   What is statistics?   1

1   *An Overview*   3

1.1 Nonmathematical Descriptions   *3*
1.2 Measurements   *7*
1.3 Statistics   *10*
1.4 What Now?   *20*
    REVIEW   *21*
    PROJECTS   *21*
    PROBLEMS   *22*

2   *Questions to Ask*   25

2.1 Some Examples of Data   *25*
2.2 Questions about Reliability   *29*
2.3 Questions about the Data   *35*
2.4 Common Sense   *45*
    REVIEW   *45*
    PROJECTS   *45*
    PROBLEMS   *46*

PART II Descriptive methods — 49

3 *Pictures of a Distribution: Histograms* — 51

3.1 Descriptions *51*
3.2 Graphical Techniques *53*
3.3 Histograms *54*
3.4 Areas *59*
3.5 Grouping *60*
3.6 Measurement Data *64*
3.7 Cumulative Frequency Diagrams: Distributions *67*
3.8 Percentiles *71*
3.9 Shape of the Distribution *73*
3.10 Describing a Distribution *74*
3.11 Normal Probability Paper *79*
   REVIEW *81*
   PROJECTS *82*
   PROBLEMS *82*

4 *Mathematical Descriptions* — 87

4.1 General Characteristics *87*
4.2 Central Values or Averages *88*
4.3 Coding to Find Means *97*
4.4 Which Average to Choose *101*
4.5 Spread *105*
4.6 Variance and Standard Deviation *105*
4.7 Coding to Simplify Calculating Variances *112*
4.8 Interpretation of Standard Deviation and Variance *116*
   REVIEW *122*
   PROJECTS *122*
   PROBLEMS *122*

5 *Computers and Computation* — 129

5.1 Computational Equipment *129*
5.2 Calculators *131*
5.3 Computers and Library Programs *136*
5.4 Examples of Library Programs *137*
   REVIEW *148*
   PROBLEMS *148*

6 *Graphing I: Correlation and Regression* — 151

6.1 Descriptions of Pairs of Quantities *151*
6.2 Correlation *153*
6.3 Correlation Coefficient *156*

6.4   Interpretation of Correlation and the Correlation Coefficient   *159*
6.5   Cause and Effect   *161*
6.6   Regression   *162*
6.7   Correlation or Regression?   *167*
6.8   Scales: Change in Size   *168*
6.9   Range of Prediction   *169*
6.10 Warning: Plot the Data!   *169*
6.11 Other Scales and Graph Paper   *173*
    REVIEW   *176*
    PROJECTS   *177*
    PROBLEMS   *178*

7 *Graphing II: Control Charts*   **185**

7.1   Time-Dependency   *185*
7.2   Variability in Different Time Spans   *188*
7.3   Control   *190*
7.4   Control Charts   *192*
7.5   Nonrandom Variation   *203*
7.6   A Type of Experimental Design   *206*
    REVIEW   *209*
    PROJECTS   *210*
    PROBLEMS   *211*

PART III   Distributions   **215**

8 *Probability and Distributions*   **217**

8.1   Probability and Statistics   *217*
8.2   Replacement   *224*
8.3   Discrete and Continuous Distributions   *228*
8.4   Random Numbers   *229*
    REVIEW   *231*
    PROBLEMS   *232*

9 *The Binomial Distribution*   **235**

9.1   The Need for a Theoretical Distribution   *235*
9.2   Binomial Distribution   *236*
9.3   Means and Standard Deviations   *244*
9.4   Approximation to the Binomial Distribution   *247*
    REVIEW   *248*
    PROJECT   *248*
    PROBLEMS   *248*

10 The Normal or Gaussian Distribution — 253

10.1 The Nature of the Normal Distribution   253
10.2 Using the Tables   264
10.3 The Normal Approximation of the Binomial Distribution   273
10.4 An Interpretation of the Normal Distribution   275
REVIEW   276
PROBLEMS   276

PART IV   Inference and hypothesis — 279

11 A Nonmathematical Approach to Hypotheses — 281

11.1 What Is a Hypothesis?   281
11.2 Kinds of Errors: Decision Problems   282
11.3 The Null Hypothesis   291
REVIEW   292
PROBLEMS   292

12 Sampling — 295

12.1 What Is Sampling?   295
12.2 Distributions of Sample Values   297
12.3 Central Limit Theorem   304
12.4 Selecting Estimators and Efficiency   310
12.5 The Sampling Distributions: t, χ^2, F 315
REVIEW   322
PROJECTS   323
PROBLEMS   323

13 Estimation and Hypothesis — 327

13.1 Point Estimation   327
13.2 Interval Estimates or Confidence Intervals   328
13.3 Hypothesis Testing   333
13.4 Introducing β Error   335
REVIEW   343
PROBLEMS   343

PART V   Statistical tests — 347

14 Univariate Tests: General Discussion and Tests of General Assumptions — 349

14.1 Statistics and Significance   350
14.2 One-sided and Two-sided Tests   350

14.3 Parametric and Nonparametric Tests   *352*
14.4 Univariate Tests   *353*
14.5 A Runs Test for Randomness   *354*
14.6 A Test for Extreme Values   *358*
14.7 Some Other Assumptions   *361*
     REVIEW   *362*
     PROBLEMS   *362*

15 *Univariate Tests of a Categorical Nature*   *365*

15.1 Tests to Compare Data to a Theoretical Distribution   *365*
15.2 Tests of Proportions   *374*
     REVIEW   *378*
     PROJECTS   *378*
     PROBLEMS   *379*

16 *Univariate Tests Involving Interval or Ratio Measurements and an Assumption of Normality*   *385*

16.1 Comparing the Spread of a Sample to a Theoretical Spread   *385*
16.2 Comparing the Mean of a Sample to a Theoretical Population Mean   *389*
     REVIEW   *397*
     PROBLEMS   *397*

17 *Comparing Two Independent Samples*   *403*

17.1 Statistical Tests for Comparing Two Independent Samples   *403*
17.2 χ^2 Test for Independence of Two Sets of Data   *404*
17.3 Difference in Proportions Test   *410*
17.4 Mann-Whitney U-Test   *413*
17.5 Difference in Variances: F-Test   *417*
17.6 Difference in Two Means: z- and t-Tests   *419*
     REVIEW   *422*
     PROJECTS   *423*
     PROBLEMS   *423*

18 *Comparison of Two Related Samples*   *431*

18.1 Statistical Tests for Comparing Two Related Samples   *431*
18.2 Sign Test   *433*
18.3 Wilcoxon Matched-Pairs Test   *435*
18.4 Paired t-Test   *437*
18.5 Comparison of These Three Tests: Power   *438*
     REVIEW   *439*
     PROBLEMS   *439*

19 Multiple Sample Cases: Analysis of Variance 443

19.1 Investigating Multiple Sets of Data   *443*
19.2 The Basic Principle of Analysis of Variance ANOVA   *445*
19.3 One-way or Single-factor ANOVA   *449*
19.4 Comparisons for Significant Differences   *455*
19.5 Two-way Design and Interaction   *457*
19.6 General Kinds of Design Problems   *461*
REVIEW   *464*
PROBLEMS   *465*

20 Correlation and Regression Continued 469

20.1 Correlation: Nonparamagnetic Tests   *469*
20.2 How Good Is the Regression Line?   *471*
20.3 Confidence Intervals for the Regression Line   *477*
20.4 Multiple Correlation   *480*
20.5 Multiple Regression   *482*
20.6 Nonlinear Regression   *483*
REVIEW   *486*
PROBLEMS   *487*

Appendices 491

Appendix A Appendix Tables   *495*

Appendix B Summation and Subscript Notation   *560*

Appendix C Probability   *571*

Bibliography 581

Answers to Selected Problems and Portions of Problems 587

Index 600

Introduction

You are about to embark upon the study of one of the most widely misunderstood and misused subjects known to man—statistics. It is a topic that seems to rank number one in its ability to frighten away or turn off more prospective students than virtually any other college subject. And yet, in ignorance, it is worshiped by these same individuals. Statistics, whatever they are, are quoted continuously as proving this thing or that thing or something else. And of course, if statistics say so, it must be so. Thus, we accept as having valid meaning, statements such as the following:

"Four our of every five dentists questioned said . . ."
"A poll showed that 57% were in favor of . . ."
"Crime statistics indicate an increase of 129% . . ."

and so on. Worst of all, we accept these statements as significant or consequential since numbers have been quoted and numbers do not lie.

Unfortunately, numbers can lie. It may happen because information is not disclosed or because the numbers are distorted, or not complete. This is possible because most of us have not developed an awareness or a feeling toward numbers or data. We are not trained to realize what is necessary to evaluate data. Yet they are so much a part of our lives that to remain oblivious to some of the concepts behind proper interpretation of data is almost a sin. Today, virtually every field of study has found itself caught up in the rush to do statistical analysis. The manner in which the statistics are used, the types of questions asked, and the techniques and tests employed may vary from discipline to discipline.

An engineer may be able to actually control all aspects of his design or his test and then observe the results that are a direct consequence of the one thing that he varies. On the other hand, a psychologist or a teacher doing research must use a less controlled specimen—perhaps you. He has a more difficult time isolating the effects of something he has done. The nature of the data he actually obtained may be different from the engineer's and the way he interprets it will depend largely on his subject matter. This text can hardly hope to explore all of the areas or all of the techniques you might need to know in order to evaluate data. Rather, I hope that you will gain some basic understanding of the concepts underlying uncertainty and variability—the two critical areas in which statistics come into play. Later, you should discover the ways of performing detailed analyses in courses related to your chosen field of work. The material you will encounter in this text will form a foundation for your future studies. It favors no discipline but should be useful in all disciplines.

The first chapter will attempt to give an overview of the kinds of questions and problems we will explore and gain an understanding of through the use of statistics. This becomes the first of my answers to the inevitable question asked by nearly every student, "Why should I study statistics? What good is it?"

You must first be aware of the types of inquiry with which most disciplines are concerned. It is important for you to realize that nearly every subject and nearly every profession is involved in some way with data or numbers. Wherever there are numbers, there are people who attempt to interpret them and to grasp meaning from them. The purpose of statistics is primarily to *assist* people in their analysis and interpretation of their data. It can help you gather more information with less effort and testing. It is a way of determining just what your chance of being wrong might be whenever you make a statement. It is necessary if you are to understand the articles in many of the technical journals. It is vital if you are to comprehend the consequences of research work in such fields as the social sciences, the biological sciences, education, anthropology, economics, or business and management.

A more immediate interest for you would be developing an awareness of the problems of grading and the inexactness of a method which says 90–100 is an A, 80–89 is a B, and so forth. You should also be able to tell whether the results of a poll have any meaning. You should acquire an inquisitive nature regarding statements you read which give conclusions based on "statistics."

You may start this book and this course with some apprehension. This text is written for the person with little prior mathematical background, skill, or interest as well as for the student who may have a more extensive mathematical experience. I may dwell at length on some topics. In teaching this subject I have found that a subtle concept is difficult to understand at first encounter. It is for this reason that I have introduced something once and then tried to say it again another way. I probably will reintroduce it at a later time with the hope that the second time around it may make a lot more sense. If you happen to be quick enough to comprehend the concepts, then you should find the reading fairly rapid and, I hope, interesting.

PART 1

What is Statistics?

An Overview

I will begin by asking you what may seem to be a trivial question. I will ask you to describe an egg. After we struggle for a short time with a single egg, we will expand our discussion and attempt to look at a group of eggs. Then we will try to ask some questions that may be of interest to biologists, farmers, marketing people, or perhaps, even consumers.

1.1 Nonmathematical Descriptions

Words

You might start by saying that an egg is a smooth, oval or spheroidal-shaped object. Here, we are describing the object with a collection of words that are merely symbolic representations or condensations of the actual object. *Provided you and I both understand the words in the same way*, we can convey to each other an accurate image of the real thing. The set of words thus serves as a *substitute* for the object. The substitute never fully replaces the object, but often it is the only adequate way of transmitting information regarding the object.

You may claim that the above description is not complete and you would be correct. Words alone can never fully replace the object. We can expand the definition from four words to 40 or even 400 words by enlarging upon details and by including descriptions of additional aspects of an egg. Perhaps we would introduce the fact that it consists of an outer layer or shell and inside this shell is contained a fluid of varying composition. Of course, this description could still be elaborated upon and it would still be inadequate as a *complete* description. Let me restate the fact that we will never completely *replace* the object with one or more words. We will, however, have a very useful substitute, one that can be carried around completely independent of the real object, one that will not break and make a mess, one that can fit in my head as well as in a book.

Although words are useful tools, the immediate problem with them is in the area of interpretation of the words. Different people derive different meaning from the same words. We must be aware not only of the incompleteness of the description, but also of the *different ways different people interpret individual words*. The complete set of words can combine to give quite a diverse set of images different for each person. Remember, you just have the description "a smooth, oval or spheroidal-shaped object, having an outer layer or shell and containing a fluid of varying composition." From this description you have been asked to construct an image of an object. Will it be an egg? Maybe you are thinking of a plastic bag full of mucous? Try to improve upon the description.

Comparisons

From the preceding we see that the direct use of words to replace an object is limited. Try to keep these limitations in mind as you read. They will recur later in a different setting—with numbers.

We can add another dimension to our description if we establish some standards, some item we will mutually agree upon as a *basis for comparison*. We can then compare our object to the chosen standard. For instance we may select as standards two objects with which both of us are equally familiar. The two objects selected may be a marble and a 100 watt light bulb. In fact, if we stopped at the first part of our description, we could possibly have considered these objects as the ones under consideration. We shall accept these as our standards. If we were to further state that we will select a marble of a particular size and a light bulb of a brand name and catalog number, we will establish more exact standards. However, for our purposes we might find the less specific values sufficient.

In our future work we will find that we will need to have standards for comparison and that they will serve as important aids in describing many problems. We will also encounter the notion of exactness in setting these standards. We will state at times that the standard is a theoretical value. For now, however, let us see how these two standards might be used.

The marble is much smaller than the egg. The light bulb is much larger. We observe that a marble is a perfect sphere. What about the egg? The egg may be spherical on one end, but it is usually elongated. It is generally longer in one direction than in another, so it is not nearly as spherical as the marble. If we now look at the light bulb, ignoring the thread on the small end, we could state that the light bulb has a much odder shape than an egg. Both have an elongated shape with a broad end and somewhat smaller end. As we go from the large end of the light bulb to the narrow end we have a concave region. This is not present on the egg. The egg has an outward bulge throughout.

Let us stop and look at what I have attempted to do. I have attempted to make some direct comparison of the shape of several objects by using words only. It has been a rather poor attempt. I ask you to improve upon it. Remember that the person you are talking to does not know what an egg looks like. I think you will experience the same kind of difficulty that I have encountered. Once again, we have found words limited in their ability to perform certain operations. We have observed that word descriptions may not be sufficient for complete comparisons.

Pictures

There is an old adage, which you have undoubtedly heard, that goes "a picture is worth a thousand words." So now let us resort to the picture to gain some further insight or description of our egg.

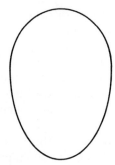

Figure 1.1   A picture of an egg.

Figure 1.1 is a drawing of an egg; with no additional information the general shape of the egg is fairly obvious. The shape is more easily conveyed with a picture than with the use of the words. This picture, however, is pretty much limited to showing us the general shape or outline and outer surface characteristics. It does not tell us much about the structure or the variation of composition that we know exists.

If we refine our drawing or change the type of drawing, we may find that we can describe even more than we did before. If we make a cross-sectional drawing, we can illustrate the shell thickness and some of the internal structure. In Figure 1.2, we have a better image of how the various compositions of liquid are distributed inside the egg, but we have lost some of the information about surface.

So we now see that *no picture is perfect* and *different pictures can be made for the same object*. The different pictures may give different kinds of insight, answer different questions, or provide different descriptions for the object. Thus, we might want a whole collection of pictures, each of which may serve a different purpose or may elaborate upon a different facet of the object.

6 What is Statistics?

Figure 1.2 A cross section of an egg showing the shell and some internal structure.

What about comparisons? Can pictures be used to make comparisons? Of course. In fact, the difficulty we ran into before with an egg and a light bulb is almost nonexistent. In Figure 1.3 we have a picture of a marble, an egg, and a light bulb. With this picture the general size and shape relations are easily communicated. You can quickly see certain similarities and differences among these three objects. But again, the picture is limited both in the total scope of information it can possibly convey and in the nature of the questions it can answer.

An Observation: At this point we may be able to stop. We have been asked to provide a déscription of an object. An attempt was made to explain the characteristics of that object by providing first a verbal or word description, and second, a pictorial description. It *is* possible that these methods of communication are sufficient. If so, then the subject of statistics is not applicable to the problem and we need not go any further. The key word is *possible*. A

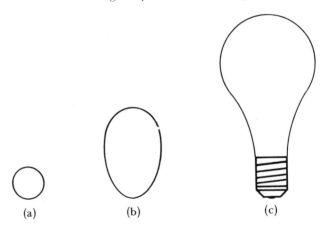

Figure 1.3 Pictorial comparison of (a) a marble (b) an egg (c) a light bulb (120 watt).

certain amount of description, comparison, and even inference, can be made through the use of the technique we have been discussing. However, our capacity for comprehensive examination of even just one egg is extremely limited. The more detailed and more exact descriptions involved in most of today's world generally require measurements of some type. We use numerical and mathematical quantities as better ways of characterizing or describing an object.

1.2 Measurements

Our first step requires us to identify those characteristics we feel are important for our description. A chemist would describe the egg by giving you its chemical composition. An investor or an economist might tell you what it costs. A dietician would tell you its protein value. A packaging engineer would tell you its impact strength. Depending upon the purpose of our investigation or discussion we could arrive at a large collection of *different kinds of measurements* for the *single egg*. A few of the possible types of characteristics we might observe could be:

cost	diameter (at largest round section)
weight	shape
color	height
shell thickness	protein, fat, etc. composition
shell breaking strength	yolk size
volume	flavor
opaqueness	(fertilized or not fertilized)
incubation period	roughness
mass	

For virtually all of these characteristics we can arrive at some form of numerical value. We will call these numbers by the name *data*. Throughout the text I will consider this term as referring to *any set of useful numerical information*. Hopefully, the purpose of having the data is for future evaluation and analysis. However, regardless of what it is to be used for, the data comes from some *source*.

I claim *the principal source of all data is via a measuring device*.

Not so, you say—the data is the thing itself that you are measuring.

But is that so? The numbers are *not a property* of the thing we are observing, they are values we obtain with some kind of instrument.

The physicist tells us that mass is a property of matter. Our particular egg has a precise, absolute, definite amount of mass.

Now determine its mass.

The egg doesn't reveal its mass by announcing it to you. Rather you must use some tool to measure or weight that blob. Depending on the quality of the tool—namely, its precision and accuracy (two words we will consider later)—you may get pretty close, but you will never get an *exact* result. You must always remember that your values are only as good as the measuring devices you use.

I shall use the words "instrument, tool, or measuring device" in their most general sense. I would consider a thermometer, an oil pressure gauge, an intelligence (I.Q.) test, a presidential poll, a census, all as different types of measuring devices. In whatever way you might collect numbers there is something between the actual object or attribute you measure and the final set of numbers you obtain. It is this in-between thing that I consider as the measuring device that is always present and is the source of our data.

Now comes one of the problems that is the plague of statistics. As long as we must have some sort of instrument to measure a quality or a quantity, there exists (1) the possibility of an error in measurement and (2) the presence of variation in measurements.

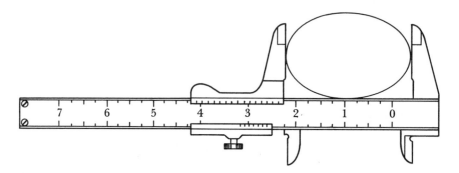

Figure 1.4 Measuring the longest dimension of an egg using a vernier caliper. This instrument can read dimensions to the nearest 0.001 inches.

Let us consider the measurement of "height" or longest dimension of our egg. If possible, do this yourself. Pick an egg from the refrigerator and using a "standard" twelve-inch ruler, measure it to the nearest $\frac{1}{16}$ of an inch. On my first try with my egg I measured $2\frac{7}{16}$ inches for the longest dimension. Was that a good value? We really have no way of saying yes or no yet. It might be true that the egg is exactly $2\frac{7}{16}$ inches, but that is quite doubtful. First, any true value from $2\frac{13}{32}$ to $2\frac{15}{32}$ would be called $2\frac{7}{16}$. Here we see that there exists a completely unavoidable error because we cannot measure close enough to the actual value. If in fact the egg was $2\frac{13}{32}$, we would be off by *at least* a thirty-second of an inch. For other actual exact sizes there would be other errors that would necessarily be present because the instrument is not *precise* enough.

If I provided you with another instrument, one which is more precise, we still would have the problem of possible error. Consider a vernier caliper (see Figure 1.4), which can measure dimensions to 0.001 inches. I now find my egg measures 2.421 inches. This is different than the $2\frac{7}{16}$ obtained with the ruler, since $2\frac{7}{16}$ is about 2.438 inches. We are now able to get a much finer reading. But if I now say this is the exact size, I am probably wrong. The exact value could be 2.4213 and we would therefore still have an unavoidable error. It would be a smaller error, but, nevertheless, it is still an error.

What I have been saying is that with many measuring devices we have a scale of some sort. We then gauge the thing we are measuring against this particular scale. However, we can almost always magnify the scale and find spaces that suggest our readings are not actually identical to the thing we are observing.* We have made an *error* in measuring the object.

A second cause of discrepancy in reading, which might also be called error, I classify as variability. We will encounter variability in several different ways and most of the ways are present with our one egg.

If we remeasure we may have variability from reading to reading. Why?

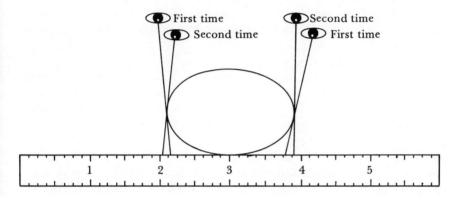

Figure 1.5 Trying to measure an egg by eye with a ruler. Two times that you look will probably give two different readings.

If I look at the egg a second time with the ruler, my eyes will probably *not line up exactly the same way as the first time*. Look at Figure 1.5. The first arrow on the right shows the direction I looked the first time. The next time I made an observation my eye was toward the side and the reading was different.

With either instrument (ruler or vernier caliper), if another person made the next measurement he would quite likely have a slightly different way of seeing and of interpreting the values. The vernier caliper requires a certain amount of feel or touch and judgement in closing against the egg. The second time I might press a little harder and compress the egg a little, thus getting a smaller reading. Or I might not press as much and get a larger reading. I might read the instrument a little differently and be off by a small amount. Thus, there would be a *variation* in measurements we made.

Another, perhaps more obvious notion, is that the egg might not be perfect.

★ This concept should also be recognized in such instruments as, for instance, a questionnaire measuring attitude toward cafeteria food. Perhaps two answers (Good, Bad) are provided. You can see this is imprecise. If five answers are permitted, say (Excellent, Good, Average, Poor, Terrible) we have a more precise instrument. However, there still is no way of saying "slightly below average" because you would have to pick "average".

If we chose two slightly different points on the egg as our points to measure between, we quite probably would have different actual, true values. If the instrument is precise enough, it would be affected by this variation. Then repeated measurements would be different, due in part to the *variation in the object* we are measuring.★

The main point for you to remember is that every time you gather data, or read about it, or look at it, or evaluate it, you must keep in mind the presence of error and variation. If neither of these were present, then one of the main reasons for the existence of statistics would be eliminated. But as long as they are present, then we generally need help in working around the difficulty presented by these uncertainties. Statistics is really just a set of tools that can assist us in making decisions or answering questions. They take into account that there is in fact a certain degree of uncertainty associated with our information.

AN EXPLANATION. Let us see what kinds of jobs these tools are prepared to do for us. Just as a tool chest contains a variety of general purpose hammers, pliers, wrenches, screwdriver, and so forth, some specialty tools for limited jobs, and perhaps some multipurpose power tools, so does the statistical tool chest. The rest of this chapter will describe some of the categories of tools we will use, some of the handicraft that can be produced, a few dangers that may be involved, but no detailed instructions. I want you to begin to realize for what purposes this tool chest has been constructed. Then you can begin to learn how to build something with the tools.

1.3 Statistics

So you've got some numbers that describe at least one egg. *Now what are you going to do with the numbers?*

This question is paramount. To begin, I offer two basic possibilities:

1. You will attempt to *describe* the egg or eggs.
2. You will attempt to *infer* something about the egg or eggs.

At first you may say these two statements seem to be the same. They are not.

Descriptive Statistics

We searched for ways to describe our egg and we came up with some measurements. If we were to gather a collection of measurements that were made with

---

★ As for a questionnaire type of measurement, if you were to answer the questionnaire a second time, you might decide the cafeteria food is "just better than poor." You would probably then say *poor*— a different result than before. Thus, the questionnaire is subject to some amount of variability associated with the individual evaluation. Here I am assuming the variability has to do with the object, the attitude in this case. It might however, be construed to be associated with the measuring scale. In either case the effect is the same—variability in the final reading and, therefore, in the data.

a ruler with subdivision at every 0.05 inches (20 lines to an inch), we might have obtained the values in Table 1.1. The 10 numbers serve as the *raw data*—rather unappealing as presented.

Table 1.1 Heights of an egg as measured with a ruler (10 times) to the nearest 0.05 inches

2.45, 2.40, 2.45, 2.50, 2.40, 2.40, 2.35, 2.45, 2.45, 2.35

In order to more readily convey the information contained in the data, we employ a pair of methods to condense the data. The methods I am referring to are ways of describing the data itself, which, in turn, are ways of describing the object. They fall into the same two categories of descriptive methods we previously discussed, namely, *words* and *pictures*. The difference, however, is that the words are now *mathematics* and the pictures are *graphs*.

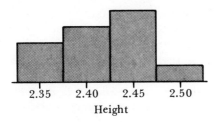

Figure 1.6 A pictorial way of representing data. A histogram or frequency diagram for the number of times that different measurement values have been observed.

Thus, we describe the set of values by using mathematical "words" and "sentences" *to summarize* the data. Later we will see how such descriptions are derived. Now we state that

Mean = $\bar{X}$ = 2.42
Standard deviation = σ = 0.048
Range = $X_{max} - X_{min}$ = 0.15
Median = $\tilde{X}$ = 2.45

These are called *statistics* because they are *derived from the original data*. They are calculated with the use of some or all of the individual values but they are not the actual values. They are called *descriptive statistics* because they are used to describe the set of numbers.

First, they tell us approximately where the *center of the original numbers are*. Second, they tell us to what extent the numbers are *distributed or spread out around that center*.

These two characteristics are the primary ones we consider important about a collection of data.

The second way we describe the data is through the use of mathematical pictures called *graphs*. One of the most commonly used graphs for picturing the nature of a single set of values is a histogram or frequency diagram. In Figure 1.6 we have a histogram for the measurements of our egg. Here we have a way of visualizing where the concentration of measurements lie and to what extent they are distributed. You probably don't know exactly how to interpret this "picture" yet. We will spend a chapter's time working with graphs of this type.

The descriptive facet of statistics is very important. Much of the statistics we read about in newspapers, and most of the statistics included in such sources as the Federal Government's various volumes of statistics and statistical abstracts, fall under the category of descriptive statistics. They are merely mathematical summaries of multitudes of numbers. However, most of the other statistical methods rely upon such an initial condensation of the data.

Inferential Statistics

A second and perhaps more important type of statistical study involves the making of *inferences*. Here we begin to extend our work into more general, broader statements, judgements, or decisions.

We look at the ten measurements and realize that there is variation between them. *Where there is variation, there is uncertainty.* We cannot be absolutely positive what the *true* size of this egg is. We will *infer* or *estimate* that it is *most likely* to be 2.42 inches. It may actually be a little larger or a little smaller but the best guess we can make is 2.42.

The next plausible question you might ask could be:

"How much larger or how much smaller?"

To answer this I would introduce the idea of establishing a pair of limits around that *estimated* value. I choose a pair of values that are believed to be reasonable when compared to the variation in the different measurements we took. I will declare that *the true size of the egg is between 2.39 inches and 2.45 inches*.

You should now ask, where did these values come from? They are based on the use of certain statistical methods and certain assumptions that will be discussed later in the text. They are not just pulled out of the air, they are computed by a particular procedure and based on a specific theory.

In addition to limits our work will give us another important piece of information. The use of the statistical procedure I have just mentioned will allow us to say what is the *chance*, the *probability*, or the *odds** that the true size of the egg is *not* between the limits I have. Sometimes I will by chance make a mistake. I might be giving you the wrong limits and you would like to know just what is the possibility that I could be in error. Thus, by the use of these methods I can state that there is—(1) less than a 5% chance, or (2) a probability of 0.05, or (3) 1 to 19 odds—*that I am wrong*. The possibility of error is quite slim, but I admit its possibility and can declare what it might be. Thus:

---

* I am assuming that you have an intuitive feel for at least one of these three terms. You may refer to Appendix B for a fuller explanation of these concepts.

1. *I recognize the chance of error, the chance my statement might be wrong*, and
2. *I am provided with a means of determining just how large or small that chance of error might be.*

The idea of making inferences from our data is really where our main study of *statistical analysis* will begin. The descriptive part of statistics is necessary because in many cases we actually work with the condensed values it provides. We will often manipulate these condensed values instead of the actual data. Thus, we will arrive at newly computed quantities that we also call statistics. Based on these new statistics we will make statements about the original data and, perhaps even more important, about the *source* of the data.

Populations

To introduce you to the field of statistical analysis, we should now go beyond our single egg. I will ask you two very pertinent, practical questions:

"What *population* are you really interested in?"

"Are you just interested in *this* egg?"

If the second question is true then we can stop and deal with the concept of measurement error only, in fact, with measurement error associated with the very simple population of one. But this is not usually the case. The more realistic situation involves a desire to extend your information. You generally will want to answer a question more like:

What should I expect *eggs* in general to measure?

Now you must consider this egg as merely a typical one out of a collection of many eggs. Thus, although you measure this egg and arrive at a value for one of its characteristics, you really want to be making statements—inferences—about an entire population of eggs. Think about the situation. You measured one egg. Is it typical of all eggs? How reliable an estimate do you think you can make of the size of another egg?

Before you really can proceed, you will need to know something about variability of eggs. You will need to look at a sample of eggs. But there still will remain an amount of uncertainty unless you examine the entire population of eggs. What about this population? I ask the question again,

"What population are you interested in?"

You might be concerned with:

The dozen eggs in the box you bought today at the local supermarket.
The two cases of eggs delivered today to the local supermarket.
The eggs delivered by a particular farm.
The eggs laid today by the hen Sally.
The eggs laid this month by the hen Sally.
The eggs laid today by the hens in coop A.
The eggs laid this season by the hens in coop A.
The eggs laid today by a particular variety of hen.
The eggs laid by any type of hen.

The eggs laid by any type of poultry.
The eggs sold in New Jersey.

Any of the above, or, for that matter, any of a thousand other descriptions could yield satisfactory populations.

Statistics will not help you select a population. You must understand what population you wish to deal with at the point when you decide it is important to obtain data. The main consideration is that whenever we make statements we want them to be about the chosen population. This leads to one of the most difficult problems of statistics—*making certain the samples chosen are representative of the population we are interested in*.

An improper method of selecting samples often can lead to erroneous conclusions. We generally will want to have what are called *random* samples. In a random sample each and every individual from the population has an equal chance of being chosen to be in the sample. When this principle is violated then definite biases are possible and your results may be subject to question—they may not be representative of the population. We will discuss some of the problems of sampling at various points throughout the text.

The next step is to select a question of interest. You decide it is useful to know the height of large eggs delivered by the Hawk Chicken Farm. Perhaps you make egg boxes and it is important to know the typical egg size before you design the box. Maybe you are going to make the egg trays for a refrigerator storage compartment. Or possibly you buy eggs and you would just like to know how big a large egg should be. Whatever the reason, you begin by formulating a question.

Hypothesis

The question usually will be put in the form of an hypothesis. Perhaps the standard size of a large egg is known (by some previous measurements) to be 2.45 inches. We may question if Hawk's large eggs are smaller. So we hypothesize that they are *not* smaller. We then proceed to select a sample of eggs and measure them.

We first decide upon the size of the sample to take. Too large a sample takes too much effort at measuring. Too small a sample will not allow us to detect a "statistically significant" difference in the values. The problem is that what we are looking at is only a sample. With a sample *we will undoubtedly find a difference*. The average size of the eggs we look at will almost always be different than the 2.45—but remember, it is only a sample. Since we are really interested in the *population*, I now pose the actual question: "*What is the chance that this sample could have come from a collection of eggs that did average 2.45?*"

A population of eggs might have a distribution something like that in Figure 1.7a. This is a pictorial histogram, which can give you a feel for the proportion of differently sized eggs that are in the population. Each egg in the diagram would be 2% of the population (since 50 eggs are shown). So 28% of the eggs are 2.45 inches. But remember, you don't know what that distribution actually is. You may surmise it but you don't know it. A sample like that in Figure 1.7b may be selected from the population—at least we must say that we have little reason to deny that it might have come from the proposed population. The sample in Figure 1.7c, however, would cause me to be highly suspicious. It would be a very rare case for me to pick, *at random*, only very small eggs.

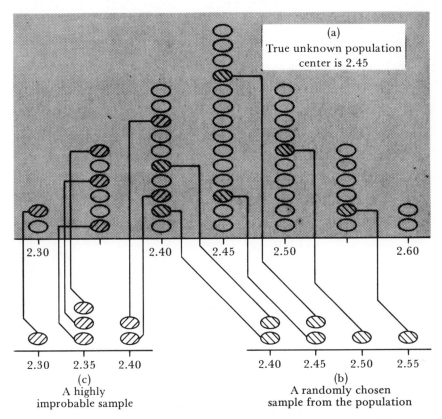

Figure 1.7 Taking a sample from a population of eggs which average 2.45. (a) The true population (unknown to you). (b) A random sample from the population. Eggs are chosen from *all* sections of the population. (c) A very unlikely sample. All eggs are from the bottom section of the distribution. I would not expect such a sample to have come from the distribution in (a).

So we look at the sample that was randomly picked from the large collection of eggs delivered by the Hawk Chicken Farm. We compute some statistics (in this case it is something called a z-value, which we will discuss later as the subject of several chapters) and we compare them to some quantities in a table. If the calculated values are greater (or smaller in some cases) than some theoretical limits that we decide upon, then we say that it *is unlikely* that this sample could have come from the given population. In this case, we say there is less than a

5% chance that these eggs belong to a population which averages 2.45 inches and, therefore, we should say the *entire shipment* is not "large eggs."

That was a strong statement to make. We are accusing that chicken farm of mislabelling eggs and we want to be sure we are correct. But since we only took a sample we can never be certain we are absolutely correct. There is a chance of error—they may be shipping large eggs. What we have gained, though, is a gauge of our chance of being right: a set of odds to back us up.★

An alternate approach involves the setting up of an interval around the average of the sample. We call the interval a *confidence interval*. It is so defined because we will be creating the interval in such a way that enables us to state that we are willing to give perhaps 19 to 1 odds that the population is centered within that interval. For the sample in Table 1.2 we would have the following statements to make:

1. Average value of sample = 2.36.
2. 95% confidence interval is from 2.33 to 2.39 or there is a 95% chance, 19 to 1 odds, that the population has an average between 2.33 and 2.39.

Table 1.2 Heights of six different eggs from Hawk Chicken Farm; each egg was measured once

2.35, 2.35, 2.35, 2.40, 2.40, 2.30

Let me review and emphasize a few points. Almost everything we do that involves a measurement or a number has associated with it a certain degree of error or imprecision. In addition, most data is not complete; it is a sample. In this sense we can consider the *taking of an individual measurement* also as a sample of several possible readings. Thus, there is always uncertainty regarding our knowledge. But with each illustration of a statistical analysis concept we have also described something we do to try to compensate for this degree of uncertainty. *We have been assigning* a numerical value, a probability, a chance, some odds, to the uncertainty. You have been given an indication of how reliable the estimate might be, what the chances are that the range of possible values might be valid, or what the probability is of being wrong.

Two Sample Cases

Another type of problem is that of comparison. We already encountered an example when we discussed word descriptions. There I posed the question of comparing a marble and an egg. We only looked at one of each and we declared that the egg was larger than the marble. Now I want you to think about how reliable the comparison is if instead of "this marble" and "this egg" I say "marbles" and "eggs." I want you to confront the question as though you have

---

★ It is also possible to make another error. The eggs might be small, but your sample shows up as "large" and so you say "the eggs are large."

no previous knowledge about typical marbles or typical eggs. Then if you conclude there is a significant difference between these two objects, I say you are on shaky ground.

The difficulty is that you have no measure of how variable the sizes of eggs and marbles are. Without an indication of how different in size another egg might be or another marble might be, we have no valid reason to suspect that they both might have come from a collection of items all with the same average size.

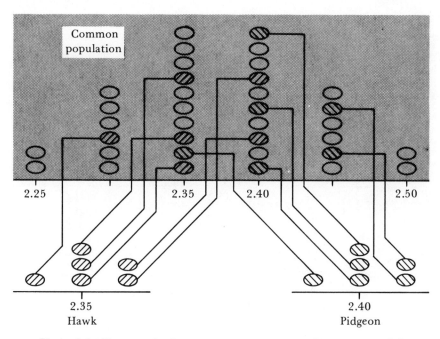

Figure 1.8 Two samples from a common population. The averages of the samples are different but the source is the same.

Another example might serve as a better illustration of what I mean. We have already examined a sample of eggs delivered by the Hawk Chicken Farm. Let's say we also can buy eggs from another company, Pigeon Egg Corporation. Our question now is:

"Are large eggs from Hawk a different size than large eggs from Pigeon Egg Corporation?"

I think you can see that this question can be applied just as easily to the two brands of eggs in your local supermarket.

Your approach is to take a sample of the eggs delivered by Pigeon and then compare these eggs to the ones we measured from Hawk. We will compare not all the values but rather the averages of these two samples. *We will undoubtedly find a difference in the average sizes of the two samples.*

We may find that the eggs from Hawk have an average that is 0.05 inches smaller. The real question must be: *Could the two samples have come from the same population?* Is it unreasonable or improbable that two samples taken from a common, single source would be as different as our samples?

In Figure 1.8 we have the case where two samples are taken from a common population. The samples differ. What we hope is that when we look at the difference we can conclude that the difference is not excessively large. We will declare that there is not a significantly large difference in our egg samples—that these are two typical samples from a single population. Thus, Hawk and Pigeon supply eggs of equal size.

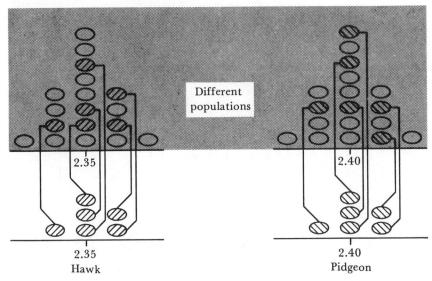

Figure 1.9 Two samples from two different populations. The averages are different but observe that you might possibly have chosen two samples that would be identical even though the populations were different.

The alternate kind of decision that we can make is that there *is* a difference. This difference is *not* that of the *samples, but* rather of *the populations*. We want to determine if we really have two different population sources from which we are taking our samples. For example, Figure 1.9 shows two possible populations. Perhaps one is the source of Pigeon Egg Corporation eggs and the other is the source of Hawk Farm eggs. The two distributions are different. However, the difficulty is that the samples might end up being very similar. The result would be that we would conclude there is no difference—an error on our part.

I should repeat the main theme of this section. We are looking at samples. The conclusion we make concerns the population we believe produced our sample. It is in this extending beyond the sample for conclusions that our uncertainty lies. And again with statistical methods we will be able to say what chance we have of making a wrong statement.

Table 1.3 Heights of six eggs from Pigeon Egg Corporation

| 2.45, | 2.40, | 2.35, | 2.40, | 2.45, | 2.40 |

In Table 1.3 I have listed the sizes of a sample of eggs from Pigeon Egg Corporation. If we carried out a statistical analysis we would conclude that there is less than a 5% chance (1 chance in 20) that the two farms are supplying eggs of the same size. You would now know where to buy your eggs.

Many Sample Cases

So you aren't satisfied with looking at just two items. You want to look at several at the same time. Now we enter into the realm of what is called *design of experiments*. This very broad field, which will only be introduced briefly in this text, is one of the powerful techniques available to researchers.

For instance, you may have six suppliers of eggs and you want to know if there are significant differences between suppliers. You take a sample from each. If we applied the previous method of looking at pairs of samples, we would have fifteen separate pairs to compare. (If you don't believe there are fifteen, try listing them.) However, we can substitute for the fifteen separate individual tests a single slightly more complicated test procedure, which will be discussed later.

Another situation occurs when you desire to consider more than one kind of comparison. Perhaps we buy brown and white eggs from the two original suppliers. By merely adjusting the samples we choose, we can compare *simultaneously* brown and white eggs and Hawk Farms vs. Pigeon Corp. We do not have to look at a larger total sample this time even though we are now considering two factors. We will introduce some additional chance of error, but often we will be able to gain additional knowledge. We will be able to observe if there is some interrelation between egg color and egg supplier. If Hawk supplied larger white eggs and Pigeon had larger brown eggs we would have a difficult time identifying what was happening if we limited our examination to the two individual comparisons (brown vs. white and Hawk vs. Pigeon).

I do not want to get too involved at this point, but I do want you to be aware of the ultimate type of problem we can study. If you were operating a farm you might be able to identify five different things to consider (or adjust). You might look at categories consisting of: eggs laid by different kinds of hens; hens eating different feeds; hens being allowed to wander or being confined; eggs laid in morning or evening; hens given different vitamins. A single quantity is measured for each egg that is sampled. After the analysis we are able to identify which of these items seems to be a significant factor in final egg size. The information that can be obtained from this technique can be very substantial.

Relationships

Another fairly broad area of statistics involves comparing two different kinds of measurements of the same item and then trying to determine if they are related to each other and if you can predict one from the other.

CORRELATION. The first of the two important concepts in such comparisons involves what is called correlation. Here we will merely attempt to determine if two quantities vary in a similar manner. So we look at height and weight of a sample of eggs. We find that most of the tall eggs are also the heavy ones, and the short ones are also the lighter ones. We do not imply that the weight *causes* the height or vice versa. We merely observe that the two quantities vary in a similar manner. We might conclude that the mechanism that creates tall eggs also apparently creates heavy ones. A statistic can be computed (called a correlation coefficient) that indicates how close a relationship we have.

REGRESSION. The second type of relationship we deal with does involve cause and effect. Here we assume that we are able to control one factor by accurately varying it. We can independently set its value. We might administer different amounts of a vitamin to different hens. We can adjust and accurately determine the amount we give each hen. We then measure a result that we believe depends upon the thing we have performed: the size of the eggs. This is called a dependent variable (it depends upon other factors).

What we want to know is what is the best mathematical equation relating the thing we can manipulate to the result. This equation will allow us to make the best prediction of outcome.

In some cases regression analysis and certain types of design of experiments are closely related. Some of the same statistical methods are used for both, although their purposes and their interpretations may vary.

1.4 What Now?

A number of other important aspects of statistics have not been discussed. Some will be introduced in the course of this text. Others will remain for you to encounter in future work. The main purpose of this chapter was to answer your question—what good is the course? You have now had a rapid capsule summary of a good portion of this book. But you have not been told the hows. That will follow. I do not expect you to understand the implication of all that was discussed. That too will be a major topic. What I do hope is that some of the mystery of where we are headed has been cleared up. We want to come out of the fog. I want you to lose the fear of statistics. Remember, it is only a tool. It should help you to think, evaluate, appraise your knowledge, and give you a means to make critical judgments.

REVIEW

Here are a number of keywords and concepts to study and review

Sample · Population · Data · Measurement · Graph · Statistic · Descriptive statistic · Inference statistic · Error · Hypothesis

PROJECTS

Several small "Projects" are to be introduced here and expanded upon as we continue through the course of the text. Choose at least one of these as your own to follow. You may find that more than one will interest you and they should not be so difficult that you cannot do several. Conscientiously gather the data and work out the problems as we progress. The realities of using live data can be intriguing and will give meaning to the course.

Two questions should be answered for each problem at this point:

How are you going to assure that your sample is random?
What might be the purpose of the experiment?

1. Put a thermometer in your refrigerator (you can purchase an inexpensive one for about 59¢ or else try to scrounge up something—perhaps from a neighbor or through a lab). The thermometer should be capable of reading to 1 or 2 degrees. Take a collection of readings at different times and carefully record the values. Do you have any observations to make? Are all the readings the same? What do you think might influence the values or cause them to vary?
2. Record your gas mileage over a period of time. What can influence the measurements?
3. Obtain a collection of sugar packages. You should be able to find them in the cafeteria or a local restaurant. Weigh the sugar in each packet. Try to use an accurate scale such as would be available in the chemistry lab. Also try weighing them on a scale such as a postage scale. How does the different scale affect the values you obtain? Are all the values the same? If not, why might this be expected?
4. Go to a local hospital and find out the birth weights (or lengths) of some babies born at that hospital. Are they all the same? Why not?
5. Record the number of songs per hour on a particular radio station.
6. Record the number of commercials per hour on a TV channel.
7. Drop a tennis ball about 15 times. Measure the height it bounces to each time. Are the values all the same? How easy was it to measure the height? What do you think would happen if you used several different tennis balls?
8. Using a spring scale (if available) measure the tear strength of cotton thread. Try a number of different pieces.

9. Measure your foot about 10 times with a twelve-inch ruler. Does your foot vary?
10. As an alternative to Project 3 scoop out a teaspoon full of sugar and weigh it. Do this for a number of spoonfuls.

PROBLEMS

Section 1.2

1.1 Try to describe what is good coffee. If you were going to try to determine what you should do to make good perked coffee, what are some of the different things that you might consider (for example—different brands)? How might you go about devising an experiment to investigate this question?

1.2 Do the same as Problem 1.1 but consider the quality of a razor blade, the selection of a brand of razor blade (or an electric razor) to use, and so forth.

1.3 For each of the following items describe several different *types* of measurements that can be taken. What kinds of instruments are used to perform the measurement and who might perform each of the kinds of measurements?

   a. Reading (5th graders).
   b. Eyeglasses.
   c. Crime.
   d. Ice cream (chocolate chip).
   e. Stamps.

1.4 A bug spray is to be tested. What might be some ways of measuring its effectiveness? If you sprayed several sets of bugs why might you get different results?

1.5 Light bulbs are labeled according to long-life, according to wattage, according to voltage, and according to amount of light. Under what circumstances would you be interested in different qualities (don't forget about street lights or those bulbs in out of the way places)? How would your interest lead to different types of measurements?

Section 1.3

1.6 Look up the definition of the word "statistics" in several unabridged dictionaries. How do the definitions differ from the idea of inference statistics discussed in this chapter?

1.7 The ability to recognize the population you are dealing with or are interested in is extremely important. Provide a list of at least 5 populations from which each of the following could be samples. In each case also show how these populations might then be included in other larger populations.

a. College Board scores.
b. Number of traffic tickets.
c. Number of divorces or age of persons getting divorces.
d. Mail.
e. Bug spray (or the results of using the bug spray).

1.8 One way of showing the relationship between populations that fall within each other is called a tree diagram. With this we can show how sub-populations branch out of larger populations. For example,

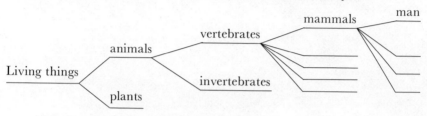

Using the same technique show how different populations branch out of the following types of overall populations:

a. The people who live in the United States (consider various types of regions or ethnic groups).
b. Blood (include the question of blood samples).
c. Crimes.
d. Automobile repairs.
e. Light bulbs.

1.9 Refer to Problem 1.3. What kinds of hypotheses can be formulated? In what way might you write an hypothesis to involve a known population and in what way might you write one to involve a comparison of two possible populations?

2

Questions to Ask

In this chapter I will introduce some more examples of data and make some observations about them. I will then present several sets of questions you must answer about your data. The first will deal with problems regarding the source of the data that you must investigate and clear up before you can even think about using statistics. The second set will include those questions you should begin to realize will be important in order to make the most elementary kinds of analyses.

2.1 Some Examples of Data

The following set of example problems, with the associated data, will act as a source for analysis and evaluation throughout the text. We shall find out in later chapters that these same sets of numbers can reveal much information when analyzed in different ways, or when additional information about the problems is divulged. Of course, these will not be the only examples we will use, but they will keep reappearing.

EXAMPLE 2.1

Over the course of several weeks the number of speeding tickets issued at radar trap locations was recorded. A total of 112 separate quantities was listed and presented in the form of Table 2.1. We could describe this set of numbers as raw data.

COMMENT ON EXAMPLE 2.1. Because of the way that these values are presented we are unable to say very much about the data except for very gross approximations. We can observe that all the numbers are not the same. By this statement we recognize that there exists a distribution. If we are very dilligent and carefully examine the data, we might also make a reasonable guess about the nature of the distribution. Later we will discuss some formal ways of reducing this mass of data into some small set of easy-to-use numbers and some meaningful graphs.

Table 2.1   Traffic tickets issued

14	17	18	14	12	15	12	10	11	2	4	9	14	13
11	9	12	8	10	13	11	8	6	3	3	7	12	11
13	15	11	11	14	13	19	10	11	4	2	12	13	11
9	10	11	9	9	11	9	6	9	1	2	8	9	10
15	15	9	10	10	12	14	10	12	10	8	13	14	12
11	11	5	8	10	13	14	7	9	9	9	12	10	10
11	8	12	10	9	15	13	9	12	12	10	12	11	17
8	10	8	10	8	9	6	3	6	7	8	9	9	7

EXAMPLE 2.2

A debate arose as to the time it takes to get to school. A route was chosen and one trial was made. The first value was 16.5 minutes. What can you conclude? Additional trials were made and all were recorded in Table 2.2.

Table 2.2   Time (in minutes) to get to school

16.5	16.5	15.1	16.4	15.8
16.1	14.1	15.7	14.3	15.1
13.6	15.0	15.9	16.4	15.6
16.0	16.7	15.2	14.5	15.5

COMMENT ON EXAMPLE 2.2. What should you have replied to the question presented above? From one piece of data, one individual recording, you should realize that not much can be stated about the next trial run or about the typical length of time it should take. You do know a lot more than you did before you made that reading. However, after you obtained one value, you still do not know how much confidence to place in it. Depending on the car, the weather conditions, the driver, the driving conditions (city or highway) and so forth, we can end up with all types of different values on subsequent tests. Even under identical conditions random variations will dispense any additional values.

When we look at the collection of data that was accumulated, we should try to observe patterns and a distribution. Hence, collections of data convey information; individuals just give a point of departure.

EXAMPLE 2.3

Loaves of white bread are being baked in a factory. A hundred of these loaves were carefully measured to determine their actual weights. The data is presented in Table 2.3.

Table 2.3 Weights of bread (in ounces)

20.055	20.040	20.019	20.025	20.018
20.052	20.024	19.993	19.984	19.982
20.046	20.036	19.968	20.013	19.982
19.994	20.008	19.987	19.984	20.006
20.033	20.041	19.997	20.003	20.037
19.999	19.975	19.985	19.968	20.022
20.022	20.011	20.002	19.975	19.988
20.002	20.026	20.009	19.960	19.997
19.976	19.976	20.000	19.988	20.006
19.994	19.989	19.970	19.958	20.019
19.981	19.987	20.013	20.018	19.994
19.990	19.968	20.013	20.009	19.986
19.965	19.970	20.053	20.026	20.018
19.977	19.975	20.005	20.024	20.002
19.986	19.969	20.004	20.003	20.024
20.052	20.017	19.986	19.972	20.006
19.982	20.017	20.004	20.030	19.983
20.032	19.955	20.019	20.013	20.001
20.007	20.011	19.983	19.962	20.035
20.025	19.999	20.010	20.000	19.993

COMMENT ON EXAMPLE 2.3. Since the breads are produced one at a time by a machine, a time-dependency situation is actually established. A similar situation might have arisen in the previous examples, and we will also be considering these problems from a time-dependency or a time-sequencing point of view. This is a typical type of problem that might also be considered in the category of quality control.

The more "standard" or parochial idea is that quality control deals with the determination of whether or not the produced items are being maintained with acceptable characteristics. Are the dimensions within the tolerances established? Are the tolerances realistic in light of the capabilities of the production machinery? We will tend to broaden this definition and observe time-dependency where it exists in many non-manufacturing and non-traditional quality control areas.

EXAMPLE 2.4

Cups are to be filled with beer. We are to check the sizes of twenty of them. The volume is determined by measuring the diameter and the depth and performing a simple calculation. All three values are listed for twenty cups in Table 2.4.

COMMENT ON EXAMPLE 2.4. What other way of determining the volume could have been used that may have been superior?*

---

* ANSWER: We could have filled it with a liquid and used a standardized set of volume measuring devices such as graduated beakers.

Table 2.4 Dimensions of twenty cups (in inches)

Diameter	Height	Volume
2.48	4.39	21.195
2.49	4.41	21.464
2.50	4.36	21.391
2.53	4.45	22.360
2.49	4.40	21.415
2.48	4.44	21.437
2.47	4.37	20.929
2.53	4.35	21.857
2.53	4.43	22.259
2.47	4.41	21.120
2.51	4.43	21.909
2.54	4.43	22.436
2.47	4.45	21.312
2.49	4.42	21.512
2.48	4.39	21.195
2.47	4.41	21.120
2.42	4.42	20.320
2.49	4.40	21.415
2.48	4.44	21.437
2.51	4.39	21.711

Here we should become aware of a mathematical manipulation that we are performing with the actual measurement data. Quite frequently we obtain numbers which we then add, multiply, or convert in some way into other numbers. Sometimes this conversion or calculation is performed for us by some physical piece of equipment such as a thermometer, radar trap, clock, transducer, optical gage, extensometer, and so forth. This conversion or transformation may alter the nature of the distribution and our way of interpreting it.

EXAMPLE 2.5

The number of absences per day in my last semester statistics class are recorded in Table 2.5. There were twenty-five students originally registered in the course and a total of thirty class periods to the semester.

Table 2.5 Absences from statistics class out of twenty-five students

0	1	2	1	0
1	2	0	1	0
0	0	0	2	1
1	1	1	2	2
3	2	4	4	2
6	3	2	2	2

COMMENT ON EXAMPLE 2.5. There is something different about the types of numbers listed here. Stop, go back, and try to find *another* way to write down the numbers. As a hint, there should be a *word* following each number.

What did you choose (if any)?

You should have used the word *percent*. If you didn't, then go back again and try to see

1. how the word percent can be applied, and
2. why this makes the numbers different than any we have looked at so far.

We call this kind of data by the name *attribute* or *enumerative* data. These are nothing more than tallies or counts of the number of times that certain individual characteristics are observed or encountered. Other examples of the kinds of problems that are counting problems might be the following:

1. Polls with numbers in favor and numbers against.
2. Number of bad items and number of good items.
3. Number of one piece, two piece, and three piece bathing suits sold in a week.
4. Number of students admitted from each of five high schools.*
5. Number of runs by several baseball teams.

Each of these examples can be written in terms of percents:

1. Percent of people in favor versus percent against.
2. Percent of bad items (usually called defectives).
3. Percent of one-piece, two-piece, and three-piece bathing suits.
4. Percent of admitted students from each of five high schools.
5. Percent of runs by each of several teams.

This use of percentages introduces our first question related to the type of measurement or the type of scale to use. We will return to this problem later in the chapter.

Two large sets of questions will now be presented. The first will deal with the reliability or reasonableness of data presented to you. The second set concern themselves more directly with the nature of the data.

2.2 Questions about Reliability

Before we go headlong into any statistical work, I want to introduce some of the assumptions that should be investigated and verified. Whenever you look at data for the purpose of extracting some valuable meaning, you must be conscious of where the data came from. You want to judge if it is worth your while to even read it, let alone use it. This judgment must first be concerned with the reliability of the source and then with the reasonableness of the data.

---

* If we knew how many graduates there are from each school then we would have five percentages, each of which might be considered a measurement.

I will outline the questions that relate to the nature of the source of the data. We will then carefully examine each question to see how it is important and what effect a yes or a no to the question may have on the interpretation of the data. By no means should you consider this list as absolute or exhaustive. In some cases several questions may seem identical and indeed they are very similar. They do, however, take on a different point of view. Taken together, the entire set of questions should begin to set your inquisitiveness into motion. I want you to get used to challenging things you read and hear about.

Our main topic here is *reliability of the data*. In this section we are *not* dealing with mathematical reliability. Mathematical or statistical reliability would pertain to such notions as accuracy, precision, or magnitude of error of the data. I am, instead, first concerned with the concept of *usefulness* of the data, *reliability* or confidence you can place in the source, and actual *reasonableness* of the data. These constitute problems that may create enough suspicion about the data's lack of validity that we need not go any further. Thus, I present the following set of questions:

How reliable is the data?

I. How useful is the data?
   A. Is it meaningful?
   B. Is it discoverable?
   C. Is it truly relevant (to context)?
   D. Has there been a change (externally)?
II. How is it reported?
   A. Who gathered it?
      Is the source biased?
   B. How was the data gathered?
      What type of plan was used, if any?
   C. Do you have the raw data?
      1. Is it condensed or filtered?
      2. Is information lacking?
III. Is the data reasonable?
   A. Would you expect such values?
   B. Are there any wild or erratic values?

I. *How useful is the data?* By usefulness I am asking you to try to determine whether the quantities you have or intend to obtain bear any relation to the real thing. Are you able to justify that the unit you work with, the data gathered, or the manner of gathering it, can be validly associated with the subject being investigated?
   A. *Is it meaningful?* We may hear that the streets are 80% clean. Does this mean 80% of the time the streets are clean? Does it mean that 20% of all the streets are dirty—1 out of every 5 streets? Or does it mean that only 80% of my street is clean and 80% of your street is clean? What is the meaning of the figure given?

How about the following? "Cost of living rose 4%." Does that mean it now actually costs 4% more than at the previous time listed? In other words, if it used to cost you $125.00 a week, does it now cost 4% more or $5.00 more—therefore, $130.00? Or is the 4% a rise in the price index which uses 1959 as a base of 100%? Is the rise then from, say, 125% to 129% of the 1959 income level? In this latter case it would mean that the $125.00 was 125% of the cost of living in 1959. Then in 1959 the same goods cost $100.00. With the 4% rise, we are now up to 129% of the 1959 value or $129.00. The rise is only $4.00 as opposed to the $5.00 rise we had the other way. Which is meant by the quotation? We don't know and, therefore, this data or information is meaningless.

B. *Is it discoverable?* The following data was presented in the *New York Times* (my italics).*

> An "alarming" increase in venereal diseases in the city was reported yesterday by Health Commissioner Mary C. McLaughlin, who estimated that there were *200,000 new cases a year—"most of them undetected."*

If most of the cases were undetected (and according to the rest of the article over 160,000 were implied to be undetected) then how does the Commissioner "estimate" how many new cases were contracted?

The main difficulty with a poll is that a person may not tell you the truth. He may say he is for one candidate and even campaign for him, but in the secrecy of the voting booth he may change his mind and switch to the other candidate. A chief complaint of the famous Kinsey Report centered about the question of how much could actually be believed. Often premarital and extramarital relations have a very good chance of being the result of fantasy and bragging of certain individuals, while other individuals will not discuss these relationships regardless of the supposed anonymity of the report.

Can an author ever really know if the reader understands what he is saying? Can he know if the reader even cares?

C. *Is it truly relevant?* I am really asking if the thing measured can reflect upon or be considered associated with the thing we are inquiring about. In the previous chapter, I introduced you to a measuring device. I am now asking if you have any reason to believe that what that device is measuring bears any relation to your concern.

Following the 1968 elections, seven United States Senators reported to the Senate that it cost them *nothing* to get elected. Five others said campaign costs were less than $5000.00. If you were to rely on those figures, which were in accordance with Federal law on disclosure of campaign spending, your conclusion would be slightly out of whack.

---

* *New York Times*, 9 February 1970, p. 42. © 1970 by the New York Times Company. Reprinted by permission.

It should be noted that under the Federal Law the candidates for the thirty-four Senate seats at stake in 1968 reported to the Senate a combined spending of $2,714,464.00. As a contrast, the candidates for the Senate seat of California reported to the State of California expenditures of $4,451,849.00. Would you consider the spending reports to the Senate as relevant to the question of campaign spending?

Educators often use the difficulty of a textbook as a measure of how good or poor a course may be. The harder the textbook the better the course. The fallacy is that we don't know if the student has read the book, or even bought it. What difference does it make if we prescribe a hard book, but nobody opens it?

School libraries are evaluated according to how many books they have on their shelves. That is fine if you are checking on theft rate or on utilization of funds, but if you really want to know about how much of the knowledge stored in those books has transferred to the heads of some number of students, then the number of books on the shelf will get you nowhere. A dusty, unopened book may be part of the library but really has no relation to learning.

The hourly earnings of a bricklayer may seem to be excessively high, but what is his yearly income? Which is the more relevant measure when it comes to deciding if an income is satisfactory or if wages are inflationary?

D. *Has there been a change?* One of the best illustrations of external changes that influence data involves crime statistics. For example the FBI uniform crime report of 1968 showed a 17% rise over the previous year, and from 1960 to 1968 a rate of crime that was climbing eleven times faster than the population. What happened? What has changed? Of the total number of crimes 28% of them involve "larceny of $50.00 and over." But inflation during the period of 1960 to 1968 pushed many items from petty larceny to the next level of seriousness. What had been included as a petty crime in earlier reports now becomes a major crime because everything costs more. A large number of auto thefts (also included in the crime index) include kids taking a joyride. Reporting methods change. Precincts may be encouraged for a period of time to be more thorough in recording all cases and events.

II. *How is it reported—where did the data come from?* I propose two approaches to thinking about the source of the data. The first involves questions about the procedure utilized in taking data. The second consists of questions about the stability, independence, and randomness of the data and/or source. We will deal with the second set of ideas in the next section. Right now, I want to ask about the *individual or individuals* reporting the data.

A. *Who gathered the data? Is the source biased?* Does the person or group have something to prove? Is a reputation at stake? When the toothpaste manufacturer or the aspirin company declares that the results of a certain study showed its product to be more effective than some other product, you must be critical. Does the other company also have a study showing the reverse to be true? You must challenge the authority that is presenting the data or the conclusions. Inquire if there might be an ulterior motive for studying the subject other than for the sake of knowledge.

B. *How was the data gathered—was there any plan?* Two important possibilities exist:
1. Was the data from a planned, designed experiment?
2. Was the data from an uncontrolled, unplanned, just gathered source?

To previously plan or design an experiment implies the complete preplanning of an experiment prior to carrying it out. This includes determining the number of samples, the sequences or order of samples, and an anticipation of the future statistical analysis and evaluation to be performed. Most of our goals will be directed toward this first category, since preplanning can save considerable time, effort, and money, and can often lead to answers for questions that could not be legitimately asked without planning.

In the category of unplanned sources are those types of data presented in many final and summary reports, newspapers, and collections of past experimental data. They generally lack adequate explanation or information. These sources may present limited opportunity for exploratory work, but should not be relied upon. Be prepared to encounter much of this type of data. Also expect to encounter evaluation by inspection. I will try to give you some insight and some tools for these types of problems, but the best way is: don't just run some tests—first plan.

C. *Do you have the raw data?*
1. *Has it been condensed?* If you do not have the raw data then it has been condensed. By raw data, I mean the actual individual measurements and the details on how they were obtained. If you do not have the complete collection of values, then there may be serious doubt about the interpretation of the data. If somebody else has already made some analysis you must be conscious of the possibility of misinterpretation. Values may have been ignored or dropped because they "seemed" bad. Statistics that favored a certain viewpoint may be reported while some neutralizing observations or opposing statistics may be ignored. We will see this effect when we study four different "averages" and find that they may all give *different* values. You may be able to select the one you like to "prove" your point of view.
2. *Is information lacking?* Are you just given an average? As stated in (1) we will later find that there is more than one way to arrive at

what may be called the average. The numerical result also will generally be different. As a consequence, if you are not told what kind of average is being discussed you are unable to adequately interpret the "average."

During 1969 a report was released showing a large number of color TV sets emitted X-rays. An article in the *New York Times* stated:

> "In a survey of 5000 Long Island homes, covering the sets of 37 manufacturers, at least 1 color receiver of each brand was found to be giving off radiation . . ." (of an excessive amount).*

Question 1: Does it mean 5000 color sets were measured? We don't know.

Question 2: Could it be that some manufacturers had only 1 set in the sample and others 1000 or more? If so, could it be that all of the sets of some manufacturers were defective while only one or a few of others were defective?

We don't know because we don't have enough information.

"Four of five dentists polled, recommended. . . ." Were only 5 dentists polled? If so, then to attempt to infer that about 80% of all dentists would make the same recommendation is pure folly. But, if the 4 out of 5 was based on a poll of 1000 dentists, then you can expect that pretty close to 80% of the population of dentists will agree on the recommendation.

III. *Is the data reasonable?* I cannot emphasize enough the importance of being honest. Report the data and use the data exactly as it is obtained. If numbers should appear to be in error, it does not mean that they are actually in error. Great discoveries have been made because researchers obtained data which appeared to be wrong when it really was the theory that was wrong.

Often the equipment we are using is too inaccurate to produce the kind of results we think we should obtain or which we hypothesize to be correct, and only through repeated measurements can we make sense out of the experimental results. Even if you believe a value is way off and you feel you should "discard" it, don't throw anything away or modify it until you have studied all the data.

Most students who have had a laboratory course would remember trying to make "minor" changes in numbers that were read on the various instruments just so the data could come "close" to what the equation predicted. Don't do it. In most cases you do not usually have any basis for stating what is reasonable for an experimental value until you have studied the process fairly carefully.

---

* *New York Times*, 8 April 1969, p. 39. © 1969 by the New York Times Company. Reprinted by permission.

A. *Would you expect such values?* A student was assigned the task of taking several recordings of the temperature in his refrigerator. He came back with readings that were around 31°. What do you think?

If the temperature was really 31° then I would suspect some of the food would be freezing and I would question the accuracy of the thermometer.

If you were suddenly getting 4 miles to a gallon or 40 miles to a gallon, I would be suspicious. Perhaps the odometer just broke or the gas tank wasn't filled up.

Sometimes the wrong units are used. A fluid ounce and a weight ounce are two different things and yet they get confused. Dollars per ounce may be quite different depending upon which ounce you use (for example with ice cream). Be conscious of what seems totally unreasonable.

B. *Are there any wild or erratic values?* I may note that the miles per gallon had suddenly dropped. Nothing had changed except that the gas station I went to was on a down slope, on a hill. When the tank was filled this time it was filled more than ever before according to the gauge. To consider this value together with the previous data could lead to an error in interpretation.

Any extreme values of a set must be considered and questioned. They must be handled with care. If they actually are valid readings then tossing them out will cause distortion. Of course, keeping them in if they are invalid will also cause distortion.

Was the instrument banged or dropped? Was it hotter or colder than before? Is there any reason to be suspicious that the particular value was a consequence of some external influence? Such knowledge may in fact lead to new insight and further inquiry. So, in some cases, a wild value may actually be a blessing, not a disaster.

2.3 Questions About the Data

At this point I assume that we will not question the sincerity of the reporting. I expect you to accept the data as being useful, meaningful and of sufficient calibre to bother attempting to perform some analysis. Now we have to consider some of the assumptions and questions we must pursue as we look toward making evaluations. Certain assumptions we do or do not make may severely limit what would be an acceptable procedure. I also want you to begin to recognize certain types of classifications that data may fall into.

The set of questions that will now be considered are as follows:

I. What are the assumptions related to the actual source of the data?
   A. What concerns are there about the sample and source?
      1. Is the source stable?
      2. Is size of sample adequate?
      3. Is the sample random?

4. Are all values independent of previous values?
   5. Is data missing?
  B. Do we have any prior knowledge?
   1. Do we expect a particular type of distribution?
   2. Do we know any parameters?
II. What kinds of measurements are involved?
  A. Is it only category or attribute data? (Nominal scale)
  B. Is order present? (Ordinal scale)
  C. Do we have equal intervals? (Interval scale)
  D. Do we have a true zero? (Ratio scale)
III. What types of comparisons are to be considered?
  A. How many sets of values are included?
   1. Is the data *univariate*—*one* set of values?
   2. Is it a comparison of 2 sets of values?
   3. Does it involve many (more than 2) sets of values?
  B. What kind of comparisons?
   1. Are we concerned with spread?
   2. Are we concerned with central values?
   3. Are we concerned with the type of shape of the distribution?
  C. How are the comparisons being made?
   1. Do we have a standard that we are comparing to?
   2. Are we merely comparing sets of values?
   3. Do we have relationships to consider?
   4. Is there any possible time-dependency?
  D. What is the direction of comparisons?
   1. Are we only concerned if something is larger?
   2. Are we only concerned if something is smaller?
   3. Are we concerned if there is a difference either way?

I. *What are the assumptions related to the actual source of the data?*
  A. *What concerns are there about the sample and the source?*
   1. *Is the source stable?* If there has been a sequential nature to the accumulation of the data, then there may well be a time effect. Things occur over a period of time. A car goes out of adjustment. A tool wears, causing a change in the size of a manufactured part.
      Many processes involve a form of learning. The first pizza a pizza maker makes will probably be very poor. As he makes more pizzas, an improvement (we hope) will occur.
      Sudden, abrupt changes in the environment may be present. Vibration and drafts (cold air) are two common problems. People themselves have changing reactions and attitudes, sometimes occurring very rapidly.
   2. *Is the size of the sample adequate?* During the controversy about the safety of birth control pills one prominent specialist in research and treatment of illness related to the female reproductive system

estimated that "it would take a *four-year follow-up study of 20,000 pill-users* in the 20- to 30-year-old age bracket to have a reasonable chance of detecting a *two-fold increase* in the incidence of breast cancer."*

What should be noted is not that it is impossible to find a smaller sample that would show a two-fold increase. You could look at a sample of 100. If you were normally to expect 1% of the population to have this form of cancer and a sample of a hundred contained 2 cancer cases, you would have a two-fold increase. But, with this size sample (100), two cancer cases is *not* unusual (there is about a 25% chance of finding two cancer cases). So, the results with this size sample would be questionable.

An even further extreme was our "four out of five dentists polled recommended...." If only five dentists were polled you can trust almost nothing about the poll.

3. *Is the sample random?* Nonrandomness in selecting the individuals for a study significantly alters our ability to *generalize any results.* Our concern is not for the few individuals we have observed, but rather for the implications of any conclusions that we make in the overall estimations and decisions about the population.

When we conduct a poll we need to use proper selection procedures to be able to generalize about an entire population. We need a representative sample with each person having an equal chance of being selected.★

When we measure breads or beer cups we assume every bread or cup has an equal chance of being chosen. If each is picked in an arbitrary way then there may be good reason to say random selection has taken place. If samples are drawn by grabbing a bunch of bananas, or several ears of corn from a stalk, or all breads from one side of the conveyor, then we may not be getting an overall sample —we may be introducing some sort of bias.

It is not always easy to assure that the selection of a sample has been purely random. The draft lottery conducted in December 1969 was an excellent case. Shortly after the lottery was conducted, statisticians at several universities observed evidence of non-

---

* *New York Times*, 16 January 1970, p. 26. © 1970 by the New York Times Company. Reprinted by permission.

★ Actually with polls we often use what is called stratified sampling. If the population can be broken into true groups that are not equal in size, such as numbers in different disciplines in a school, we first see what the proportions of the different areas are to the total. We then sample in proportion to the total. Thus, if we have 100 English majors, 50 science majors, 25 engineers, and 25 psychology students, a sample of 24 would be broken into a ratio of 100:50:25:25 or 4:2:1:1. Then 12 English majors, 6 science majors, 3 engineers, and 3 psychology students would be polled. But among the 100 English majors the 12 would be *randomly selected.*

randomness. One approach that was given national publicity and used as a basis for a legal challenge to the lottery, involved adding up the lottery numbers for all the dates of each month and dividing by the number of days in the month. If the system were random, each month would be expected to have an average of about 183 or 184. The averages of each of the first 6 months were above this and the last six months were all below it. In Figure 2.1 we have a graph illustrating the change by month. Such a pattern has an extremely small chance of occurring by pure random selection. (The odds are at least 1000 to 1 against such an occurrence.)

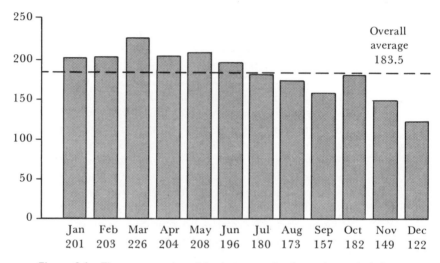

Figure 2.1 The average value of the lottery number for each month during the first draft lottery (in 1969). If the system were completely random we would have about 183 or 184 as the average number each month. The lower the number the better the chance of being drafted.

It was later revealed that the capsules for January were prepared first, placed in a wooden box, and pushed to one side with a cardboard divider, leaving part of the box empty. The February capsules were then prepared and dropped in the empty side of the box. These were then pushed into the January capsules and mixed. The process was continued for each month's worth of capsules. Thus the December capsules were only mixed once (since they were placed last) while some capsules were partially mixed together as many as eleven times (January and February). They were all then poured into a large bowl. *But*, most of the December capsules apparently ended up on top, not randomly mixed, and they tended to be selected first.

The moral—beware of how the sample is chosen.

4. *Are the values independent of each other?* Does a large bread now mean that the next bread is also large? If samples or individuals are independent, then because one bread was large, there is no reason to believe the next will be large. Most of the tests we deal with depend upon the assumption that no value or sample will influence the next one or will have been influenced by a previous one. Certain processes can be considered as producing dependent sequences of data. The weather this evening is quite definitely influenced by yesterday's or this morning's weather. The temperature tomorrow is generally dependent on today's temperature (at least the chance that it will be a certain temperature tomorrow will be affected by today's temperature).

In a fight or a war the effectiveness of each hit or shot often depends on what has previously occurred. The chance of hitting a target will depend on how many targets are left, which depends on the result of the last shot.

The quality of the painting during one stroke is perhaps dependent on what has happened to the brush and paint during the previous stroke.

We must carefully consider our problem to determine if there is any reason to believe that any outcome is affected by past history.

5. *Is data missing?* Are there any gaps in the data? Was a value lost? When you do a follow-up study on graduates, it may be the ones that do not answer that are crucial to a proper evaluation. If I am measuring change in attitude towards statistics between the first day in class and the day after the final exam, how do I include the four students that dropped the course before the exam?

The biologist who is measuring growth as affected by certain vitamins has a problem when one of his mice dies. Measuring tire wear is complicated when one of the tires goes over a nail and has a blowout.

B. *Do we have prior knowledge?*
1. *Do we expect a particular distribution?* Data always comes from some distribution. Sometimes we have a *valid* reason to believe that the distribution is of a particular shape or type. Many statistical tests depend heavily upon knowing the type of distribution that the data came from. If the assumption about the distribution is wrong, then the results may be quite invalid. A considerable amount of the text will be spent discussing distributions.

2. *Do we know any parameters?* A parameter is a quantity that describes or distinguishes distributions. The arithmetic average is one parameter that describes a set of data. Do we have any reason to assume what the average is?

Do we have previous data to suggest that the variability of the population is a certain value? Has the machine been calibrated?

Do we have a history of data and information to suggest or provide values for certain parameters? We will find that if we have reasonable estimates of certain parameters then we can make more reliable judgments and decisions about other factors.

II. *What kinds of measurements are involved?* Several types of scales are used in taking measurements. We have not yet described and defined these different levels of measurements. Certain statistical tests are not applicable to data that does not have at least an interval scale. Categorical or merely ordered values may not yield the types of interpretation that the more sophisticated types of scales will permit.

Let us look at some examples:

A. *Categorical or attribute* scale: this type of measurement was introduced in Example 2.5. In each of the examples no real scale and no particular order is present. We just classify the data into several possible categories.

EXAMPLE 2.6

Grading in a course can be recorded as follows: pass, fail, withdraw (there is no scale or order).

EXAMPLE 2.7

An automobile race can be classified as follows: finished; didn't finish.

EXAMPLE 2.8

Loaves of bread can be classified as follows: above minimum standard weight, below minimum standard weight.

EXAMPLE 2.9

Men are classified as follows: six footers or less than six footers.

EXAMPLE 2.10

Temperature is recorded as follows: above or below average.

B. *Ordinal or ordered scale:* here the data can be ranked in some way.

EXAMPLE 2.6 (Cont.)

Grading can be put on an A, B, C, D, F scale or students can be ranked as 1st, 2nd, 3rd, 4th, 5th and so forth.

EXAMPLE 2.7 (Cont.)

The contestants in an automobile race can be ranked according to order in which they crossed the finish line.

EXAMPLE 2.8 (Cont.)

Loaves of bread can be ranked according to weight by observing which of two is heavier and which is the lighter. Additional breads are possibly placed between these two.

EXAMPLE 2.9 (Cont.)

We can line up the men with the shortest on one end and the tallest on the other end.

EXAMPLE 2.10 (Cont.)

Temperature is ranked as highest today, next highest two days ago, coolest five days ago.

C. *Interval scale:* we now add a *numerical scale* which provides a magnitude to the differences. We can now say not only that the first, second, and third differ, but by how much they differ.

EXAMPLE 2.6 (Cont.)

Now we shift grading to a numerical scale of 50, 55, 60, 65, 70, 75, 80, 85, 90, 95, 100.

EXAMPLE 2.7 (Cont.)

After the winner crosses the finish line we start counting the number of seconds for each of the next racers to cross the line. So the winner is 0 seconds, the next is 2 seconds, the third is 6 seconds, and so forth.

EXAMPLE 2.8 (Cont.)

Loaves are recorded by how many ounces more or less than the allowable minimum they weigh.

EXAMPLE 2.9 (Cont.)

The men are classed as so many inches taller than John Jones.

EXAMPLE 2.10 (Cont.)

The temperature is recorded in degrees Fahrenheit.

D. *Ratio scale:* the "best" type of scale is one that preserves the concept of ratio. Thus a quantity which is measured as a 100 is *twice* as large as a value of 50. This requires that a true absolute zero be present, not an arbitrary zero.

EXAMPLE 2.6 (Cont.)

If it is knowledge acquired that we want to measure, it is almost impossible to find a valid test which can judge that a 100 is twice as good as a 50 and a 90 is twice as good as a 45. For most tests of the educational or psychological type an arbitrary value or a norm is assumed and used as a center. Hence, no zero is present in the College Board scores or I.Q. tests. Instead centers of 500 and 100 respectively are chosen with deviations around that *center* as the main measurement.

EXAMPLE 2.7 (Cont.)

We report the number of seconds after the start. This has a true zero point. The person completing in 1 minute is three times as fast as one who finishes in 3 minutes.

EXAMPLE 2.8 (Cont.)

If loaves are recorded in actual ounces, a 24 ounce loaf will weigh twice as much as a 12 ounce loaf.

EXAMPLE 2.9 (Cont.)

Actual heights are recorded and we can say that the ratio of a 6 footer to a 5 foot 6 inch person is the same as a 5 foot 6 inch person to a 5 foot $\frac{1}{2}$ inch individual.

EXAMPLE 2.10 (Cont.)

We must change thermometers or scales and use the Rankine scale, which has an absolute zero of $-459.7°$ if we are to use a ratio scale. Recall that on the Fahrenheit scale boiling water is 212° or 6.6 times as high as the freezing point. But another scale, centigrade, can be used. It records the boiling temperature of water at 100°, and the freezing point of 0°. Now is the boiling point still 6.6 times as high as the freezing point? No, it is infinitely larger (100°/0° is defined as infinity). If, however, you use the true absolute zero for centigrade, $-273.2°$, you will find that either new scale will work in ratio. In the Rankine scale $32°F = 491.7°R$ and $212°F = 671.7°R$. The ratio of boiling to freezing is now 671.7/491.7 or 1.37. For the centigrade scale and its counterpart called Kelvin, we have $0°C = 273.2°K$ and $100°C = 373.2°K$ with the ratio 373.2/273.2 or 1.37.

III. *What types of comparisons are to be considered?*
  A. *How many sets of values are included?* This question is generally self-evident. Bear in mind that we are asking about sets of data. Thus, univariate data usually pertains to some single distribution and usually we are mainly concerned with estimating the nature of this *distribution*. What is its shape, how does it compare to some norm? About the only serious point of confusion with univariate data is when we encounter some of the distributions of attribute data. One-piece, two-piece, and three-piece bathing suits form a single distribution consisting of three categories.

   Much of the time we will be more interested if there is difference of some sort between two groups or kinds of categories. In these cases we will obtain samples from both categories and compare the samples in some way. We will then be asking if they differ or are related in some way.

   The extension of one- and two-sample cases to the many category

problem involves the very broad area of design of experiments. Here we ask if differences among several different conditions occur.
B. *What kind of comparisons?* What is the basic characteristic we will be investigating?
  1. *Spread:* In certain cases the variability is critical. Until a large variability is brought under control we often cannot regulate the level or bring the individuals within the tolerances we would desire. Many tests that we will use require that the variations of the different populations under consideration be equal.

     Other times we just want to compare the spreads of some populations. We ask if the voltage at the plug varies more in my house or yours, if the sizes of tomatoes vary more with one species than another, or if the strength of steel wire varies more than the strength of aluminium wire.

     An excellent example of the consequences of a large variability was revealed in a study of hallucinogenic drug users. The usual dosage of LSD should be around 100 micrograms (according to a Department of Justice pharmacologist). A laboratory analysis of samples by the Department of Justice showed that some 14 of 15 LSD samples supplied by 6 donors had dosages that varied from *50 to 200* micrograms.* Such variances can result in the situation where a user gets little effect out of the first pill because it may have been a low dosage pill. So, he takes another, which may now be a higher than normal dosage. The combined effect may give him his final trip.†
  2. *Central value:* More frequently we ask about central values (although sometimes we will have to go back and ask about variability first). We acknowledge that the samples will undoubtedly differ, but the question is if they differ sufficiently to say that there is a *statistically significant* difference, as well as a *practical* difference (the fact that there may be a statistically significant difference of 1 degree in average temperature between July and August may be an absurd result or at least of questionable value).

     Since so much of our work will deal with comparisons of averages, I will not elaborate any further here.
  3. *Type of shape of the distribution:* In some cases we will actually be more interested in what shape the distribution has. Sometimes, we will do this because we want to use certain statistical tests that assume that the population has a particular shape. Then we would want to ascertain what the shape might be before we proceed with analyzing the data.

---

* That 15th pill is another question. From the data reported in the news article there is *no* indication of what that pill was.
† *New York Times*, 14 December 1969, p. 62. © 1969 by the New York Times Company. Data reprinted by permission.

C. *How are the comparisons being made?*
  1. *Do we have a standard given that we are using for comparison?* If we are to make a direct comparison to a given standard, our job will be fairly straightforward. We often have some kind of *previously determined* norm or specification that is given. I do not mean a control group, which will come up later; I mean an established basis. We have a "standard clock" that is a hundred times more accurate than any watch. We can check a set of watches against the standard to determine if these watches "significantly" differ from the standard. We have standards that we can use to verify most physical measurements.

     Psychological and educational tests work with a norm or standard to which future tests can be compared.
  2. *Are we merely comparing sets of values?* Most of the problems we encounter involve direct comparisons between two or more groups of individuals. We want to know if one toothpaste is different from another, if aluminum pots cook better than steel, if the color of one brand of TV is better than another, if whipping students makes them learn statistics better than pleading with them.
  3. *Are relationships being considered?* Several kinds of questions can be asked here:
     a. Do we merely ask for a yes or no (significance test)?
     b. Do we want to determine the degree to which they are related (correlation)?
     c. Are we looking for prediction equations (regression)?

     There is a definite hierarchy here that will depend both on the answers we are looking for as well as the kind of measurements we are working with (true prediction equations cannot be developed for categorical and ordered data). Observations taken from before and after studies, with matched pairs, or by making simultaneous measurements of different quantities may lead to relationships.
  4. *Is there any time dependency?* Since stability and time dependence are so closely related, perhaps this is a redundant question. However, this still remains a fundamental question that must be posed. We will discuss several ways of evaluating time dependencies.
D. *What is the direction of comparison?* The question of *larger, smaller,* or *both* will not influence our choice of test. This is mainly to enable us to identify limits that must be specified. This topic will reappear several times in the course of our studies.

   Perhaps we do not want to change from regular gas to high test unless there is a significant increase in gas mileage. We would design the statistical test to assume we are only interested in making a decision if there is strong evidence of a positive change. We do not care if mileage decreases because we will not make any change from regular to high test if a significant decrease in mileage occurs.

Other times we may only worry about a decrease. There is a minimum requirement of 400 USP units of vitamin D per quart of milk. If a significant decrease has taken place, an adjustment (and perhaps a fine) will be necessary.

Frequently we ask if a shift in either direction has happened. The dimensions of a particular part for a machine are required to be produced within a pair of limits, and if the size departs in either direction the piece may not function properly. Hence, we set up two limits.

2.4 Common Sense

We have by no means exhausted all of the possible questions that may be asked about a set of data. But, I hope I have begun to intrigue you with some of the difficulties that you can encounter when you try to make sense out of data. Much of what I have been stating is nothing more than good common sense and so is much of the analysis of data. However, common sense is something that also must be learned. You need to be trained to be aware of the many pitfalls into which you can stumble.

REVIEW

Keywords

Attribute data · Time dependency · Usefulness of data

Parameter · Randomness · Independence · Comparisons

Kinds of measurements

Categorical · Ordinal · Interval · Ratio

Spread · Central value

PROJECTS

1. Look through a daily newspaper (I would suggest the *New York Times*—your library probably has it). Find examples of statistics quoted or statements made using statistics. Comment about them.
2. Go to the library and count the number of books on several shelves. Find the arithmetic average of the number of books per shelf. Count the number of shelves in the library. If you multiply these two values together (the average number per shelf and the number of shelves) you have determined how many books there are in your library. Check with the librarian to find out how many books the library actually has. Why are the values different? How close is your estimate? If you were to count more shelves why would you expect that your estimate would improve?
3. The cost of living is always a great topic of discussion, especially between people who lived during the great depression and are alive now.

    Look up in the government statistical abstracts or an almanac (the *New*

York Times Encyclopedic Almanac) the average U.S. retail food prices for several years (say 1929, 1939, 1969). What has happened to the price of eggs, round steak, coffee, and butter?

Then look up either the number of pounds (or dozens) that can be purchased with one hour of labor, or the number of minutes and hours that it takes to earn enough to purchase one pound (or dozen) of the items. How do these values compare to the retail cost statistics? Why is the pattern different? Can you now "prove" that things are worse now than they used to be, as well as better than they used to be?

Go to the library and look up on microfilm the food advertisements for one year from a local newspaper. Find the cost at different stores, in different months, for the same items. Comment on average retail price.

PROBLEMS

2.1 Statistics deals with all fields of study that somehow accumulate or use data. A collection of problems follow that individuals working in different fields might pursue. For each statement discuss the following:
1. Who might be concerned and why?
2. What type of data and/or measurements might be gathered? (Specify if they are categorical, ordered, etc.)
3. What would be the source or population associated with the data?
4. What difficulties might distort the data, cause it to be in error, or lead to misinterpretation?
5. What could be the effect of an error in interpreting the data?
6. What problems are there in assuring that the following studies are random?
    a. The effect of a stimulant and a depressant on learning is studied.
    b. The noise level around a rock group is suspected of leading to deafness.
    c. Thickness of tire tread seems to be related to accidents and to the time that a person will purchase a new tire.
    d. A psychologist has determined how many of the college students in a particular college are well adjusted, satisfactorily adjusted, or poorly adjusted.
    e. A new car dealer has been keeping a record of the kinds of defects that customers reported after picking up their new car.
    f. A survey is taken to determine how many people watched Channel 2, 3, 4, and so forth.
    g. The number of strawberries in a pint container is counted for several different containers.
    h. The number of tomatoes per tomato plant is recorded.

2.2 Comment on the following statements.
    a. In a leading hospital study 2 ——— pills relieved pain faster than 4 aspirin tablets.

b. 120,000 people watched last night's spectacular on TV.
c. More people have purchased foreign automobiles this year than ever before.
d. Deaths per million automobile miles driven on the Parkway are less than on any other major toll road in the U.S.
e. I flipped a coin twice. It came up heads both times. Therefore, it must be a biased coin or a two-headed coin.

2.3 A study was to be conducted regarding the change in attitude of students after taking an English literature course. At the beginning of the course students were asked if they planned to take an advanced literature course. After the course records were checked to see how many of the students then signed up for an advanced literature course.

Conclusions were then drawn about the number of students who changed their attitude (by taking a course or not taking a course).

What is wrong with the conclusion and the approach?

2.4 The average cost for college is estimated as $1200.00 per year. How useful is this figure if you are contemplating attending Harvard? Commuting to a community college? Going to a state-supported college?

2.5 The average cost for all hardback books was $8.47. How does the average cost of your textbooks this semester compare? How about last semester? Comment.

2.6 Statements are often made about how large a percentage of the eligible population voted or did not vote in a particular election. How does this depend on the accuracy of the estimate of how many people are eligible to vote? What can influence the official census or cause it to be unreliable? What about its reliability in the 1972 or 1976 elections?

2.7 Describe what difficulties we have in knowing *precisely* what the true values of the following are. Are they discoverable? Why or why not?
a. The number of missiles in the U.S.
b. The number of missiles in the U.S.S.R.
c. Number of square feet of sandy beach in Florida.
d. Proportion of some population contemplating suicide.
e. Number of people driving through red lights.

2.8 Explain how the following items are measured (for example, eggs are classed as small, medium, large, extra large, and jumbo)

a. Olives
b. Toothpaste
c. Cars
d. Soap
e. Shoes (both length and width)

PART II

Descriptive Methods

3

Pictures of a Distribution: Histograms

You have been introduced to some sets of data. Our first task is to try to make some sense out of the mass, or rather mess, before us. Somehow there must be ways of reducing it to some reasonable collection of descriptive terms.

3.1 Descriptions

Let me briefly review part of Chapter 1 to get us back on the track. Often we observe an object and we are asked to convey to another person a description of what we see. The description may be verbal, consisting of several words. The definitions of these words are agreed upon ahead of time so that both of us understand what is meant by the words. Frequently we lose much information in describing the object by words, but this may not necessarily be serious; it may be the only practical way of dealing with or discussing the object. If you were asked to come to a classroom to compare an elephant and a giraffe, you would dread the prospect of bringing both animals into the classroom. Instead, you bring with you a set of words that, for your purpose, adequately describes the animals. You say that they are both mammals, are four-legged, and are grass eaters. The elephant has a long nose (trunk) and the giraffe has a long neck.

In cases of collections of data, we also would like to use a small set of descriptive words. Refer to Figure 3.1a and ask yourself, "Would I like to describe or compare those sets of numbers when they are presented in this way—complete and haphazard?" I suspect and hope you would answer no.

Well, in the fields of mathematics and statistics we also have descriptive "words." They will usually end up being in the same language as the data—they will be in the form of numbers with some symbols or mathematical short-

52 *Descriptive Methods*

OBJECTS FOR COMPARISON	DATA FOR COMPARISON
A living elephant (you provide)	175 195 207 178 209 192 139 168 149 201
A living giraffe (you provide)	165 155 172 131 184 137 201 135 142 157

(a)

Both are large mammals and plant eaters. The elephant has a long trunk, 4 teeth, 2 ivory tusks and large ears. The giraffe has a long neck and long legs and grows to 18–19 feet tall.	$\bar{X}_1 = 181.3$ $S_1^2 = 575.3$ $\bar{X}_2 = 157.9$ $S_2^2 = 526.1$ $T = 2.12$ Signif at .05 $F = 1.09$

(b)

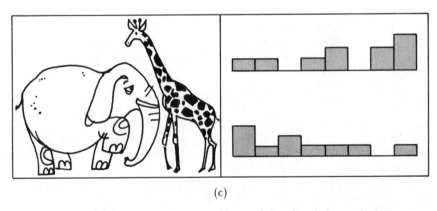

(c)

Figure 3.1 Analogy between object and data descriptive methods.

hand around them. You will have to learn the meaning of those terms so that you will understand them and be able to convey the proper descriptions to other people. We shall spend much time in discussing the calculations and interpretations of the important descriptive "words" of statistics.

We also discussed in Chapter 1 a second very important way of describing an object that does not require seeing the actual object itself. We can provide a picture. We let the eye observe a condensation, a smoothing out, or an approximation of the real thing. Often this conveys a large enough amount of information, particularly if we are not too demanding in our needs for precision. With data too, pictures may be invaluable in giving us a clear understanding of the nature of the data and its consequences. They are called graphical descriptions.

In general, however, we will want to couple together the two types of descriptions (word and picture) with our data. With the animals we could show the picture, but would probably have difficulty completely depicting all of the characteristics of the animals by pictures alone, and we would want to resort to a combination of both types of descriptive methods. This argument for simultaneous graphical (picture) and mathematical (word) presentation of data holds equally well.

Now let us examine some of these methods that enable us to make more sense out of the data. In this chapter we shall cover some important graphical techniques and in the next chapter we will deal with the mathematical descriptions.

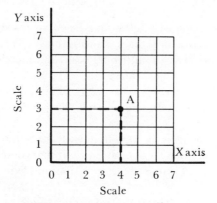

Figure 3.2 A graph with an X-axis and a Y-axis. A point on the graph has an X-value and a Y-value found by following a straight line to the axis. Point A has an X-value of 4 and a Y-value of 3.

3.2 Graphical Techniques

There are several methods of displaying a collection of data. The ones we will be concerned with at this time are those that deal only with distributions of a *single set of data*. In Chapters 6 and 7 we will look at some other types of pictorial or graphical techniques.

Two general approaches will be dealt with in detail:

1. Those involving frequencies or numbers of times that particular individual values appear—histograms.
2. Those involving accumulated frequencies or the total number of individuals less than or equal to some particular value—cumulative frequency diagrams.

In addition, we may speak of the following:

1. Actual frequencies—actual number of times different values appear.

2. Relative frequencies—the ratio of times the value occurs out of all the occurrences.

We are going to start talking about graphs. You should first recall that a graph has two axes that are usually labeled the X-axis (the horizontal or abscissa) and the Y-axis (the vertical or ordinate) as shown in Figure 3.2. There are scales along these two axes. A point that is above the X-axis is said to have an X-value associated with a location on the X-axis directly below it. So point A has an X-value of 4. The same point has a Y-value associated with it, in this case 3.

3.3 Histograms

We will start by working with sets of numbers having *only one measured characteristic*. Only one axis can contain a measurement or a classification. (By classification I am referring to attribute data, and therefore I mean that the measurements are actually just categories.) The second axis will be *the number of times that the particular measurement occurs*. For univariate data (only one variable or thing measured) this becomes the only graphical form that is meaningful.

What I am introducing is the most frequently used graphical method, called a *histogram*. The histogram is essentially a chart that contains bars showing the frequencies or number of times that the various individual values appear.

Table 3.1 Traffic tickets issued

14	17	18	14	12	15	12	10	11	2	4	9	14	13
11	9	12	8	10	13	11	8	6	3	3	7	12	11
13	15	11	11	14	13	19	10	11	4	2	12	13	11
9	10	11	9	9	11	9	6	9	1	2	8	9	10
15	15	9	10	10	12	14	10	12	10	8	13	14	12
11	11	5	8	10	13	14	7	9	9	9	12	10	10
11	8	12	10	9	15	13	9	12	12	10	12	11	17
8	10	8	10	8	9	6	3	6	7	8	9	9	7

EXAMPLE 2.1 (Cont.)

Let us refer to Example 2.1 (the traffic tickets problem), to examine this idea. Table 3.1 shows the data for that problem again. With numbers as seemingly haphazard as these are, your first step should be to do a little organization. The old tally sheet that you probably used for keeping score as far back as your third grade baseball games should come to mind. All you did then was to set up a place for each category (Home and Visitors) and then place a little slash line next to the value or category every time you observed the particular category.

When you kept score your categories were teams and each time a run was made you put a check mark down next to the team that got the run (Figure 3.3).

RUNS

Home	/
Visitors	⌢⌢⌢⌢⌢ ⌢⌢⌢⌢⌢ ////

Figure 3.3 A tally sheet for a sad but well remembered baseball game of my past.

You may have extended this problem a little bit more when you enlarged the number of classifications to include innings as well as teams (Figure 3.4). I think that little picture should convey a lot of meaning. Dwell a little more on the data presented by those slash marks. This is attribute data since for each category we have only a measure of the times that something occurred in that category (the number of runs in a particular inning by a team).

Of course if the occurrences we are tallying are different, the tally sheet would also be different (see Figure 3.5). Here we are looking at errors not runs, and the picture is completely altered. The basic idea has not changed. There are only two X-values: errors by Home; and errors by Visitors. The Y-value corresponding to each of these categories is the number of errors, or 18 and 1 respectively.

If we now go back again to our ticket problem we obtain a tally sheet that looks something like Figure 3.6 with a separate place for each possible number of tickets per day. This approach is a convenient and simple recording device that one usually employs as a first step in simplifying data.

A *histogram* resembles the tally sheet. It has the same measurement "axis" with provision for identifying each value. Alongside each measurement value or category is recorded the number of times that the particular value occurred. This time, however, instead of making slashes we introduce the formal second axis and second scale. This second axis is the frequency axis. The height or length along this second axis indicates the *number of times the associated value occurred*. The scale, which we will call the *frequency scale*, generally has equal subdivisions. It is a linear scale. The histogram for Example 2.1 is shown in Figure 3.7.

We sometimes encounter some other similar types of graphs, such as dot charts (Figure 3.8a) where dots are placed along the frequency axis instead of

RUNS

Innings	1	2	3	4	5
Home	/	/		/	///
Visitors	⌢⌢⌢⌢⌢ ⌢⌢⌢⌢⌢ //	⌢⌢⌢⌢⌢ ⌢⌢⌢⌢⌢ ⌢⌢⌢⌢⌢	⌢⌢⌢⌢⌢ ///	⌢⌢⌢⌢⌢ ⌢⌢⌢⌢⌢ /	/

Figure 3.4 An extension of the previous game. This time I have included several innings from that game.

ERRORS

Home	~~IIII~~ ~~IIII~~ IIII
Visitors	/

Figure 3.5 A different tally sheet for that same baseball game. This time looking at errors instead of runs.

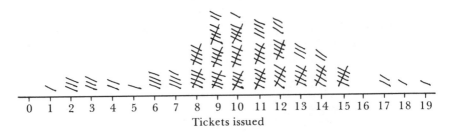

Figure 3.6 Tally sheet for Example 2.1—tickets issued per day.

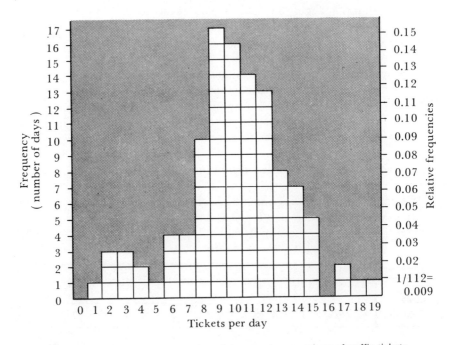

Figure 3.7 Histogram for number of times various numbers of traffic tickets were issued.

the bars, or line charts (Figure 3.8b), where lines are drawn from the X-axis up to the necessary height. Sometimes the dots of a dot chart are connected and we obtain a diagram like Figure 3.8c.

Even though all of these different techniques are used someplace by someone, we will only employ the histogram. It is the most commonly used method and will help in our future understanding of distributions. The main reason I will devote so much time to the histogram is that the histogram may be considered about the best picture of a distribution. If you can begin to internalize a feeling of the relationship between this picture and the real distributions then

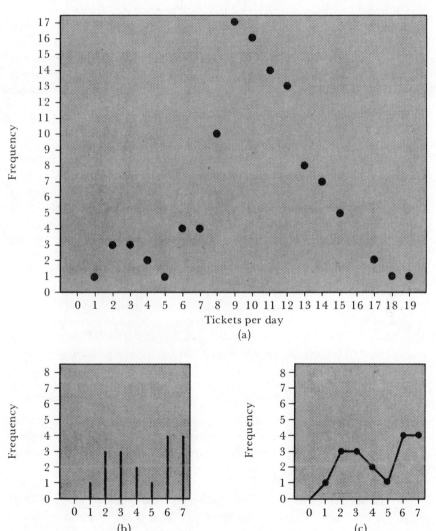

Figure 3.8 Other types of graphs for univariable data. (a) Dot chart, (b) line chart, (c) frequency polygon.

much of the future work will be substantially easier. You might picture in your mind a day from a calendar to be placed in each of the little boxes of the histogram. After all, each box supposedly represents a day during which the particular number of tickets were issued.

In constructing a histogram you must first determine the categories and the scales to be used. I will give you a few basic principles to aid in selecting categories. They follow the rules of common sense, but should be stated.

RULE 1   *A location must be provided for each and every possible category or measurement.*

Sometimes this will appear to be unwieldy. For example, maybe on one day forty-four tickets were issued (perhaps a car rally was being held or the radar was on the fritz). We would need twenty-five more positions on our scale just to satisfy the one additional value. Some type of simple adjustment should be available for this situation. We can provide a single additional category of perhaps "twenty-to-fifty". *But*, there must be extreme care in interpreting this category.*

RULE 2   *An individual value must fit into one and only one category.*

This means that we must provide definite *limits* to the interval that any location or category on the measurement axis represents. Here the values are discrete, there is no such thing as half a ticket. However, you will probably encounter more situations that involve continuous measurement scales. In those cases there is always an intermediate, inbetween possible value and we must be aware of the number of decimal places to which we will be measuring. We will dwell at length on the idea of limits.

RULE 3   *Each category has the same number of possible values.*

This may seem obvious to you. I already indicated a possible exception while discussing Statement 1 above, but I also caution you on making interpretations from histograms which have unequal-sized intervals. It should be noted that commonly 2, 5, 10, 25, 50, or 100 is used. These usually provide convenient interval sizes to work with.

RULE 4   *There should be at least six and no more than about twenty-five categories.*

This is the most flexible rule and is truly one of common sense. Too small a number of categories leads to a complete loss of information. Too many categories (particularly with measurement data) can lead to an enormously long histogram with many gaps.

The second axis contains the frequency scale. The numbers along this axis would be the *actual frequencies*, the number of times that the specific quantity occurred. Thus we observe that on ten occasions eight tickets were issued. The

---

* We will consider the proper adjustment in the section on grouping.

measurement or category is the number of tickets (8) and this happened with a frequency of 10. How many times were 14 tickets issued?*

A very prevalent alternate to the actual frequency scale is one called a *relative frequency* scale. For this type of scale we divided each of the actual frequencies by the total number of individuals included in the collection. In this example the total number of individuals is 112 (you will see this expressed as $n = 112$). The relative frequency for a particular value is simply

$$\text{Relative frequency} = \frac{\text{Actual frequency}}{\text{Total number}}. \quad (3.1)$$

For our example, the relative frequency of eight tickets being issued is

$$\text{Relative frequency} = \frac{10}{112} = 0.089.$$

What is the relative frequency for 14?†

Look back to Figure 3.7 and observe that there are two scales for the Y-axis, or frequency axis. The left scale has already been discussed and is merely the actual frequency. The right scale is a *relative frequency scale*. The first point on that scale corresponds to an actual frequency of 1 and is therefore a relative frequency of 1/112 or 0.009. In a similar manner other points are located to establish the entire scale.

Two important observations should be made about relative frequencies. First, they are easy to convert to percents. *Multiply the relative frequency by 100 and you have percent.* So our officer can state that 8 tickets are issued 8.9% of the time since

$$\text{Relative frequency} \times 100 = \text{percent}$$

or
$$0.089 \times 100 = 8.9\%.$$

The second observation is that no relative frequency can be greater than 1.00, and the sum of all the relative frequencies will be 1.00. This is a useful check on your computations.

3.4 Areas

When it comes to drawing the histogram, it is best if the intervals that we use are equal in size. If this is done, and if the frequency scale is also linear, then any histogram consists of collections of little rectangles. For a histogram with an "actual frequency" scale we can think of the little rectangles as being one unit along the frequency scale and one category wide (see Figure 3.7). If relative frequencies are used, then the height of the little rectangle would be $1/n$ units. In either case, areas of the histogram can be equated to the proportion of

---

* ANSWER: 7.
† ANSWER: $7/112 = 0.062$.

the time that the corresponding values occur. It is in this way that the picture presented by a histogram is most important.

If you think of the little boxes as weights, then the more frequently a value occurs the more weight it has. Another way of thinking about this is, the higher the column for a value, the more weight or area the particular value has. The more area associated with the X-value, the more important that value is out of the whole collection of values or the entire distribution.

3.5 Grouping

Quite often we find it desirable to compress a large number of intervals into a smaller set of categories. For instance, it might be desirable for presentation purposes to use only seven or eight categories for our current example. In some cases it is necessary (our example in the next section will illustrate the necessity situation). The procedure is called *grouping*.

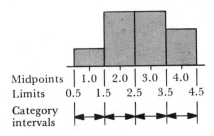

Figure 3.9 Midpoints, limits, and intervals for several categories of Figure 3.7.

To perform grouping we are merely going to use a different set of categories for the same set of data. You should realize that each category has a midpoint and a pair of limits. If you look at Figure 3.9 you will find the first four of the categories of the complete histogram of traffic tickets (Figure 3.7).

The categories are the sections separated by vertical lines. These lines constitute our *interval limits* and a category (or class, as it is sometimes referred to) is considered as encompassing the complete set of possible values between these limits.★ A midpoint is the center of an interval or category and is the value you consider that category to represent. So the first interval is from 0.5 to 1.5 and the midpoint is 1.0. You therefore consider any value in that category as though it were a 1.0. This will become obvious as we continue. In grouping we must choose a new set of class or category intervals or a new set of central points. If

---

★ In the cases where we are dealing with attribute data the category interval really has no meaning. Even in this case, since we have been dealing with discrete units, there has been no need to worry about intervals. However, as we begin to look at decimal measurement groups there will be concern about the limits.

we decide on intervals, then we automatically establish the centers and if we provide centers, the ends or limits are also established.

Let us pick a set of intervals that appears reasonable. I have chosen as a first possibility the following set of intervals (remember—the rules for selecting categories still hold):

Set 3.1
$$
\begin{aligned}
&0.5–\ 3.5\\
&3.5–\ 6.5\\
&6.5–\ 9.5\\
&9.5–12.5\\
&12.5–15.5\\
&15.5–18.5\\
&18.5–21.5
\end{aligned}
$$

The histogram for the interval Set 3.1 is shown in Figure 3.10.

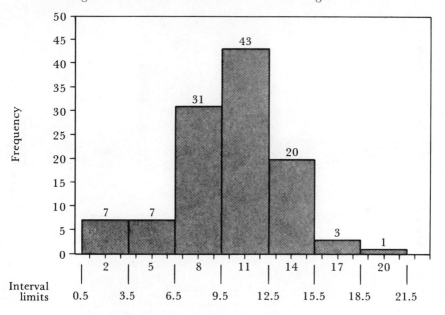

Figure 3.10 Grouped data for Example 2.1 using intervals of
0.5— 3.5   12.5—15.5
3.5— 6.5   15.5—18.5
6.5— 9.5   18.5—21.5
9.5—12.5

Each category contains three integer values since the only kinds of values in this problem are integers. As long as each interval contains the same number of integers there is no confusion. Note that the centers of the intervals have also been established. The first one is 2. What are the remaining midpoints?*

---

* ANSWER: 5, 8, 11, 14, 17, 20.

Set 3.2

You might say why not write the intervals as

1– 3
4– 6
7– 9
10–12
13–15
16–18
19–21

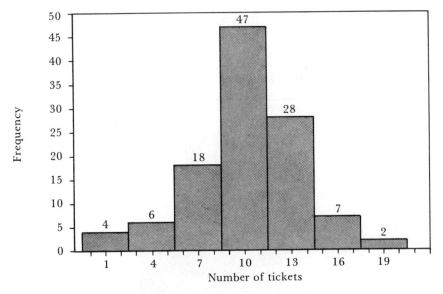

Figure 3.11 Grouped data for Example 2.1 using new centers.

For this problem it is permissible. However, the end points of two adjoining intervals normally should not have any space between them. Later, when we look at cases where decimal values are expected to occur it will be important that no spaces are present. With decimal measurements we always have the possibility that an individual belongs in the space (has a value which corresponds to one in the space). It is best at this point to have rules that apply to a more general problem and then add special conditions or modifications later.

Observe that our condensed version of the distribution shown in Figure 3.10 is very similar in shape to the original histogram we drew (Figure 3.7). I will admit that our new histogram is not as accurate as the original, since much of the variation has been eliminated. *Whenever we group we lose accuracy.* However, in most cases the loss is not great enough to be concerned with, provided we do not reduce the histogram to only a couple of categories. Common sense is the best rule here. Obviously, if you see just two categories *almost nothing* can be learned from the data. In other cases, if, say, thousands of intervals are possible, then by size alone the histogram is also meaningless.

Let's try another histogram, this time selecting midpoints. How about midpoints of

$$1, 4, 7, 10, 13, 16, 19.$$

What are the intervals?*

The histogram using these centers is shown in Figure 3.11. The two grouped histograms, Figures 3.10 and 3.11, look different. This is to be expected. If a third set of intervals having different widths or different central points were used, it too would probably be different from these two. After all, we have been approximating and condensing data and each approximation is somewhat different.

One last concept should be explored. I mentioned that sometimes unequal intervals may be employed (see Rule 1, page 58). In Table 3.2 I present a set of *unequal categories* (unequal in width). In Figure 3.12 you can see two incorrect

Table 3.2 Unequal grouping of data for Example 2.1

Interval	Midpoint	Values included	Frequency
0.5– 7.5	4	1, 2, 3, 4, 5, 6, 7	18
7.5– 8.5	8	8	10
8.5– 9.5	9	9	17
9.5–10.5	10	10	16
10.5–11.5	11	11	14
11.5–12.5	12	12	13
12.5–19.5	16	13, 14, 15, 16, 17, 18, 19	24

ways of plotting the histogram for these categories. The first histogram uses *unequal* intervals along the measurement axis (X-axis). This tends to distort the impression of the relative spans of the intervals (relative coverage along the X-axis) and therefore is not recommended.

The second histogram (Figure 3.12b) has a linear scale. However, the frequencies of 18 for the first category and 24 for the last category are plotted in such a way as to give an extremely distorted image of their importance, since the areas of these categories are so much larger than all the others. The first category (frequency of *18*) really should not be much larger than the third category (frequency of *17*).

In Figure 3.12c the scale is linear again but the vertical scale is no longer of similar value. Normally the heights correspond to the frequency. With this modified scale the heights of the two end categories are not the same as the frequency for the category. Observe that the category whose midpoint is 4 covers seven possible values. The next five categories have only one possible value. So the 18 observations in the first category should really be averaged

---

* ANSWER: 0.5–2.5, 2.5–5.5, 5.5–8.5, and so forth.

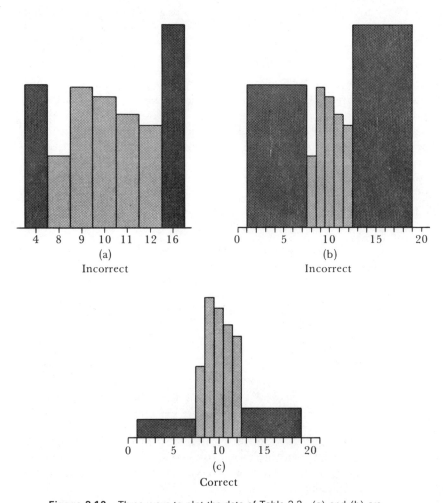

Figure 3.12 Three ways to plot the data of Table 3.2. (a) and (b) are incorrect: In part (a) the scale is not linear, therefore widths are not comparable; in part (b) the scale is linear, but the height is out of proportion. (c) is correct: In part (c) the scale is linear and the heights have been adjusted in accordance with different widths of intervals.

out over the seven locations on the scale, giving an *overall typical frequency* of 2.5 for the interval from 0.5 to 7.5.

3.6 Measurement Data

EXAMPLE 2.2 (Cont.)

Let us draw a histogram for the data in Example 2.2. Now we have to be more careful when considering the rules for intervals. If I can assume that the time was read to a definite one-tenth of a minute, then the histogram in Figure 3.13 would be the proper one for showing all the data.

The main reason I ask about the assumption is that with measurements we are always approximating. The approximation is usually established by the instrument used. With a more accurate device we could always measure to 1 or 2 more decimal places. Thus, 16.5 is an approximation and does not necessarily mean 16.50 or 16.500 or 16.5000 but instead includes a range of possible values, one of which may be 16.50. We may consider that we are grouping all of the ten values from 16.45 to 16.54, or all the 100 values from 16.450 to 16.549, etc, into one point which we identify as 16.5. We are rounding off in this case and obviously again we loose some accuracy. In most cases the loss is not enough to be concerned with but there is always some loss. The important thing to keep in mind is that the value 16.5 is really a *grouping of many different possible values* and that every point on our measurement axis is a grouping of finer, smaller points.

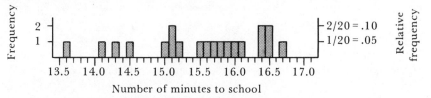

Figure 3.13 Histogram for Example 2.2.

Now, if 16.45 minutes is a possible value, it must be placed in a category and there must be no question about what category it belongs to. This problem belongs to the general concept of grouping data. As a measurement, 16.45 must fit into a category. Perhaps the category is 16.5. How many other 2 decimal values would fit into the category 16.5?

For the values we have worked with up to now, we have already stated that the point 16.5 was really the center of an interval that extends from 16.450 to 16.549 (the interval depending on the number of decimal values you are able to work with). If, in addition to 16.45, there are the 9 values from 16.45 to 16.54, then 10 different possible 2-decimal cases fit in the class interval or the category we called 16.5. If we continued along the X-axis, our next midpoint would be 16.6 and would also contain 10 different 2 decimal individuals, 16.55 to 16.64. All other values would follow similarly.★

★ You should be questioning something here. I am fudging a little bit by stating that 16.5 is the center of the interval 16.45 to 16.54. I suggest that you draw a line, subdivide it, and see what I mean (see Figure 3.14). You will find that the actual center is 16.495, which is generally close enough to consider it as 16.5. Note that if these intervals 16.450–16.549 are to be used, the actual center would still be slightly offset since there is a gap, but the space would be one-tenth of the previous space and the actual center is now 16.4995. You should realize, though, that this is only valid if you actually are measuring values to 3 decimals and you can actually distinguish 16.549 from 16.550. If you cannot make this distinction then you cannot consider that the true center of the interval is 16.4995.

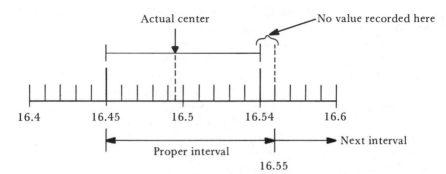

Figure 3.14 Line showing interval subdivisions. If the interval units to be used are 16.45–16.54 then the actual center is 16.495 not 16.5.

I would now like to group this data into, say, seven or eight categories. I will select a set of midpoints. One reasonable set of points would be:

13.5, 14.0, 14.5, 15.0, 15.5, 16.0, 16.5

They are equally spaced and include five values per category. Now, what would be the interval limits? First let us look at a magnified portion of the *X*-axis scale.

As you can see, the value 14.0 contains all of the 1 decimal values from 13.8 to 14.2. However, to provide a more continuous line, we should designate the interval as 13.75 to 14.25. Thus the intervals would be

Set 3.3
13.25–13.75
13.75–14.25
14.25–14.75
14.75–15.25
15.25–15.75
15.75–16.25
16.25–16.75

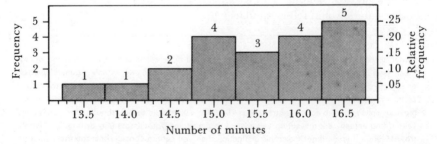

Figure 3.15 Grouped histogram for Example 2.2 using centers.

A histogram for this set of points has also been drawn (Figure 3.15). Have you made an important observation? The *intervals have been given to one decimal place more than to what the measurements are recorded*. This has been done so that there is no question of where to place any quantity. It also provides assurance that we will have a closer approximation to a continuous line for our measurement axis (the X-scale). If there exists the possibility that a value would be recorded to more than one decimal place then you must carry the interval limits one more decimal place. What if the possibility of a 14.75 could arise? In such a case the limits for the two possible categories it could fit into would need to be modified, perhaps to 14.250–14.749 and 14.750–15.249.★

So far in this problem we have been dealing with actual frequencies. If you look at the right hand columns of Figure 3.15 you will see that I have also included *relative frequencies*. These indicate (in Figure 3.15) that out of the total of 20 observations $\frac{4}{20}$ or $\frac{1}{5}$ of them had an approximate value of 15.0 (fell within the limits of 14.75 to 15.25). We can also state it as 20% of the data had an approximate value of 15.0. What percent (or relative proportion) of the original histogram (Figure 3.13) had a value of 15.5?*

Relative frequencies will generally be more useful and more informative for all types of data. This is especially true for continuous measurements and for large collections of data. In addition, when we encounter theoretical and infinitely large distributions, relative frequencies will be our only way of identifying frequencies and for drawing histograms.

3.7 Cumulative Frequency Diagrams

Up to now we have dealt with graphs of individual values. Another technique that conveys a considerable amount of meaning involves adding up all the numbers of times values less than or equal to a designated value occur. This is called a cumulative frequency diagram.

For instance we may find it necessary to talk about the proportion or the probability of a bread weighing *less than* 20.00 ounces. We might want to know what our chance is of taking *more than* 18 minutes to get to school. Our ball player wants to know how frequently his team gets *more than* two runs. All these involve cumulative frequencies as opposed to frequencies or occurrences of exact, individual, or particular values.

In addition, some types of distributions will only be discussed from a standpoint of cumulative frequencies. When we encounter certain theoretical distributions, the only way to consider the histogram will be by using areas, which in turn will involve the use of cumulative frequencies.

---

* Answer: 5% since 1 out of the 20 had an "exact" value of 15.0.

★ The previous situation of not having the true center of the interval as the midpoint arises if we work with these very fine interval limits. However, the effect will be negligible. We are now covering 500 different values with each interval. The difference between the true and the working center is half a value. As a result, our loss of accuracy is in the neighborhood of a hundredth of a percent.

EXAMPLE 2.1 (Cont.)

Let us start with the value of zero. We find no times (0 times) that the quantity zero (or less than zero) occurred. Our officer never gave up without giving a ticket. Therefore, the cumulative frequency for the number 0 is 0. The smallest number of tickets issued was 1 and it occurred with a frequency of 1. So the cumulative frequency for 1 ticket (less than or equal to 1 ticket) is 1.

For 2 tickets issued we have a frequency of 3. The sum of all previous frequencies plus this case is 4, (1 + 3). So the cumulative frequency for 2 tickets is 4. By the time we reach 5 tickets we have accumulated a total of $0 + 1 + 3 + 3 + 2 + 1 = 10$. The cumulative frequency for 5 (up to and including 5 tickets), is therefore 10.

A complete table showing the frequencies and cumulative frequencies for each number of tickets issued are shown in Table 3.3.

Notice that the cumulative frequencies are found by simply adding the actual frequency of the particular value to the previous cumulative frequency. This is the simplest way to create the table of values. It also works in reverse (by

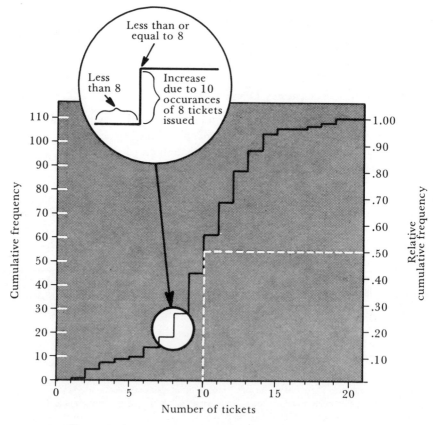

Figure 3.16 Cumulative frequency diagram for Example 2.1.

subtracting) if you want to determine the individual frequencies from the cumulative frequencies—try it.

I have drawn the complete cumulative frequency diagram for Example 2.1 in Figure 3.16. Two scales have been provided along the frequency axis. One is actual cumulative frequencies and the other is relative cumulative frequencies. In the previous section you should have noted that relative frequencies or percents are easier to make interpretations with. From the relative frequency you can immediately determine percentages. With cumulative diagrams this is even more valuable. For example, you can tell where the bottom 50% of the numbers are by finding the value corresponding to a *relative cumulative frequency* of 0.50. For this problem it would be 10 (see the dotted lines).

One note on construction and interpretation. The diagram I presented contains abrupt jumps as each new value is accumulated (see the insert of Figure 3.16). You should interpret the cumulative frequency for a particular value as the upper level at that point. Thus for 8 use 0.25, not 0.16, as the relative

Table 3.3 Frequency measures for tickets issued

Number of tickets	Frequency	Relative frequency	Cumulative frequency	Relative cumulative frequency	Cumulative percent
20	0	0.000	112	1.000	100
19	1	0.008	112	1.000	100
18	1	0.008	111	0.991	99.1
17	2	0.018	110	0.982	98.2
16	0	0.000	108	0.964	96.4
15	5	0.045	108	0.964	96.4
14	7	0.063	103	0.920	92.0
13	8	0.071	96	0.857	85.7
12	13	0.116	88	0.786	78.6
11	14	0.125	75	0.670	67.0
10	16	0.143	61	0.545	54.5
9	17	0.152	45	0.402	40.2
8	10	0.089	28	0.250	25.0
7	4	0.036	18	0.161	16.1
6	4	0.036	14	0.125	12.5
5	1	0.008	10	0.089	8.9
4	2	0.018	9	0.080	8.0
3	3	0.027	7	0.063	6.3
2	3	0.027	4	0.036	3.6
1	1	0.008	1	0.008	0.8
0	0	0.000	0	0.000	0.0
SUM	112	0.998			

Table 3.4 Frequency measures for time to get to school (Example 2.2)

X-value	Frequency	Cumulative frequency	Relative frequency	Relative cumulative frequency	Relative cumulative frequency —modified
16.7	1	20	1/20	20/20 = 1.00	20/21 = 0.95
16.6	0	19	0	19/20 = 0.95	19/21 = 0.90
16.5	2	19	2/20	19/20 = 0.95	19/21 = 0.90
16.4	2	17	2/20	17/20 = 0.85	17/21 = 0.81
16.3	0	15	0	15/20 = 0.75	15/21 = 0.71
16.2	0	15	0	15/20 = 0.75	15/21 = 0.71
16.1	1	15	1/20	15/20 = 0.75	15/21 = 0.71
16.0	1	14	1/20	14/20 = 0.70	14/21 = 0.67
15.9	1	13	1/20	13/20 = 0.65	13/21 = 0.62
15.8	1	12	1/20	12/20 = 0.60	12/21 = 0.57
15.7	1	11	1/20	11/20 = 0.55	11/21 = 0.52
15.6	1	10	1/20	10/20 = 0.50	10/21 = 0.48
15.5	1	9	1/20	9/20 = 0.45	9/21 = 0.43
15.4	0	8	0	8/20 = 0.40	8/21 = 0.38
15.3	0	8	0	8/20 = 0.40	8/21 = 0.38
15.2	1	8	1/20	8/20 = 0.40	8/21 = 0.38
15.1	2	7	2/20	7/20 = 0.35	7/21 = 0.33
15.0	1	5	1/20	5/20 = 0.25	5/21 = 0.24
14.9	0	4	0	4/20 = 0.20	4/21 = 0.19
14.8	0	4	0	4/20 = 0.20	4/21 = 0.19
14.7	0	4	0	4/20 = 0.20	4/21 = 0.19
14.6	0	4	0	4/20 = 0.20	4/21 = 0.19
14.5	1	4	1/20	4/20 = 0.20	4/21 = 0.19
14.4	0	3	0	3/20 = 0.15	3/21 = 0.14
14.3	1	3	1/20	3/20 = 0.15	3/21 = 0.14
14.2	0	2	0	2/21 = 0.10	2/21 = 0.14
14.1	1	2	1/20	2/21 = 0.10	2/21 = 0.14
14.0	0	1	0	1/21 = 0.05	1/21 = 0.10
13.9	0	1	0	1/21 = 0.05	1/21 = 0.10
13.8	0	1	0	1/20 = 0.05	1/21 = 0.05
13.7	0	1	0	1/20 = 0.05	1/21 = 0.05
13.6	1	1	1/20	1/20 = 0.05	1/21 = 0.05
13.5	0	0	0	0	0/21 = 0.00
13.4	0	0	0	0	0/21 = 0.00

cumulative frequency, since the term cumulative implies up to *and* including the frequencies of the particular value.

Sometimes the centers of the horizontal lines are connected, to give an approximation of a smooth curve (Figure 3.17).

There is one disadvantage with drawing a cumulative frequency diagram in the manner just covered. Refer to the percent scale in Table 3.3; realize that we have not exhausted *all* possible samples (have not looked at all possible days that tickets have been issued). I am not ready to say definitely that 100% of all cases are below 19 or that 0% of all times no tickets were issued. Of course I haven't said that all cases are in the range we obtained, but most people tend to use the curve to make interpretations of all possible cases.

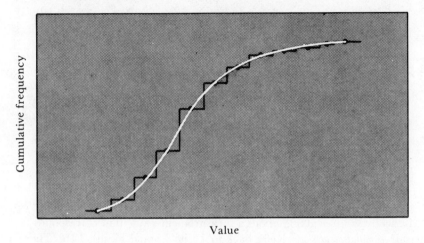

Figure 3.17 Cumulative frequency diagram drawn as an approximate smooth curve.

One simple way to alleviate this difficulty is to arbitrarily add one more to the total number of individuals. Thus, use 113 instead of 112 for computing the frequency values. As a result, there will not be a 100% value. If you do not connect the first value to the bottom axis, there will not be a zero percent either. This is a more realistic type of distribution.

EXAMPLE 2.2 (Cont.)

Table 3.4 shows the various frequency calculations for times to school. Cumulative frequencies are presented for the "standard" method first discussed and for the modified method just presented. The differences here are at most about 5%. Figure 3.18 shows a cumulative frequency diagram using this modified method. You would tend to use this type of diagram to assist you in deciding your relative chance of taking different lengths of time to get to school. Note that we do *not* deny the chance of taking less than 13.6 minutes or longer than 16.7 minutes.

3.8 Percentiles

Quite frequently we talk about percentiles. You hear that you are at the 90th percentile with a certain College Board score. Or those below the 25th percentile are excluded from a certain activity. Or the middle 50 percent are expected to go to a certain type of college.

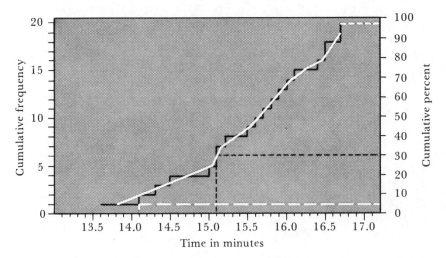

Figure 3.18 *Modified* cumulative frequency diagram for Example 2.1.

Basically we are talking about a measurement value that corresponds to a certain *cumulative percent*. Refer back to Table 3.3 and Figure 3.16. There a cumulative percent of 20% corresponds to 8 tickets per day. Hence, the 20th percentile was 8 because 20% were below the value of 8. This is frequently written as $P_{20} = 8$ where the P stands for percentile, the 20 represents the

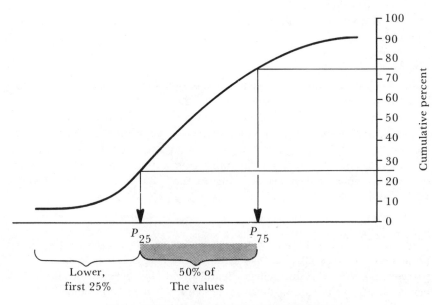

Figure 3.19 How to find the middle 50 per cent.

percent we are dealing with and the 8 is the corresponding value (and is read as P sub 20). Similarly, $P_{75} = 12$.

What about the middle 50%? You must find 2 different values since it must be bounded on two sides (see Figure 3.19). Here you find P_{75} and P_{25}. The amount between them is 50% of the total. Thus the limits would be 12 and 8.

Of course, these values are approximate. If we refer to the cumulative frequency diagram for Example 2.2 (Figure 3.18), we can have a fraction of a value. Here with a smooth curve as an approximation to the relative cumulative frequency diagram we can pick off intermediate values. Then, for example, the 30th percentile, P_{30}, would be about 15.15 (following the dotted line).

3.9 Shape of the Distribution

The histogram is a picture. It is a way of showing *how* our values are distributed and it has a very important property—namely, shape. We have looked at a couple of distributions and you should have observed first, how irregular they were and second, the fact that the shape may be affected by slight changes in the method of grouping. In addition, as you look at a sample or the histogram of a sample, you should have more on your mind than just that set of values.

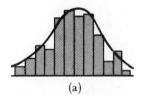
(a)
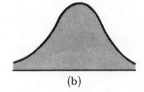
(b)

Figure 3.20 (a) A smooth curve drawn over a histogram. The curve is only an approximation of what we believe the actual population is like. (b) The smooth curve by itself.

You should be thinking about the population from which that sample came. The sample is a means to a more significant or general end—an understanding of the population. As a consequence, we should feel free to approximate a shape and smooth out the histogram of a sample. You will be looking at the jagged peaks and jumps of your histogram, but I hope you will be seeing the overall picture and the smooth curve. We do this by drawing a continuous, smooth line through the peak areas and over the gaps as in Figure 3.20a. After we obtain the smooth curve we can erase from our minds the sample distribution and strictly work with the smooth curve (Figure 3.20b).

Next let me turn the situation around. Quite frequently I will talk about "a distribution" and I will show a smooth curve to represent that distribution. That smooth curve, whatever its shape, is an *ideal* curve. It is not the curve of a particular sample but rather an estimation of what the histogram for the infinite population might be. So when drawing the smooth curve remember that an actual set of data would not be exactly like it.

3.10 Describing a Distribution

There are two ways that distributions are distinguished. The first involves a truly descriptive approach for which I will discuss three characteristic shapes a distribution may possess: symmetry; skewness; and bimodality. The second approach involves naming a hypothetical model or distribution that approximates the shape of the distribution (or histogram) we are evaluating; the main case here is the introduction of what is called the *normal distribution*.

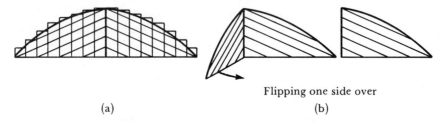

(a)   Flipping one side over
                              (b)

Figure 3.21 A symmetrical distribution.

Characteristic Shapes

A curve might be *symmetrical*. A symmetrical distribution is one where each half of the distribution is a mirror image of the other half (Figure 3.21a). In a symmetrical distribution one side, say the left one, could be flipped around the center and matched up with the other side, the right side (Figure 3.21b). Symmetry is a very desirable characteristic.

Most distributions that are not symmetrical (nonsymmetrical) can be described as *skewed*. A skewed distribution has a tail on one side and a concentration on the other. If this tail is on the right, it is called positively skewed (Figure 3.22); if the tail is on the left, it is a negatively skewed distribution.

Skewness is very important in many of the types of statistical tests we will perform. If a distribution we will be working with is very skewed, certain techniques should not be used. Unfortunately, however, there is no very good measure of skewness that can be used for reliable comparisons of distributions.

In a continuous, smooth distribution a peak is called a *mode* of the distribution. The mode is actually the X-value corresponding to the peak (see Figure 3.23a). In some cases a noticeable second peak or mode is present (Figure 3.23b). This may be a situation to be very concerned about. Many statistical

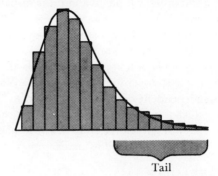

Figure 3.22 A positively skewed distribution. The tail or extended section is toward the right.

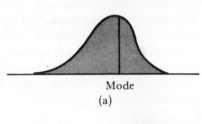

Figure 3.23 Distributions showing (a) one mode and (b) showing two modes or bimodal.

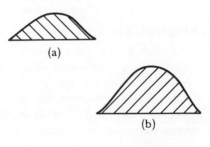

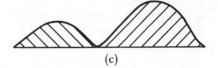

Figure 3.24 Creation of a bimodal distribution from two distributions having single modes.

techniques will be invalid if the distribution is bimodal. This is often a more serious condition than skewness.

When we observe a bimodal distribution, one of the primary reasons for the concern I have expressed is that *you probably do not have a single distribution.* You may not have a single source for the data. Other factors appear to be contributing to the overall distribution. Look at Figure 3.24. We have two completely independent distributions, a and b. When we combine them we obtain c, which is bimodal. When we look at c we should be wary. It is very infrequent that a single source will yield this shape.

Consider the following examples.

EXAMPLE 3.1

A ball is tossed 10 times to a person and the number of times he catches it with his right hand is recorded. Figure 3.25 shows a histogram for a sample of 124 different persons to whom a ball was tossed (10 times).

Here we have two different groups of individuals, right-handed and left-handed. The distribution is strongly influenced by the duality of categories present. We are able to attribute the shape to an expected dual cause system. However, if we did not know about right- and left-handedness, we would indeed have an interesting problem.

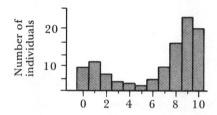

Figure 3.25 Number of catches with right hand out of 10 tosses for 124 different people. A bimodal distribution.

EXAMPLE 2.1 (Cont.)

Look carefully at the histogram in Figure 3.7. Does this show any possibility of containing a second mode? If so, can you arrive at any possible explanation? We will return to this case soon.

Often two distributions will have different locations or spreads and a combination of them will not be bimodal. The combined distribution will just be spread out much more than either of the individual distributions. This is illustrated in Figure 3.26. A and B are distributions with different central values. The combined distribution is much broader than either of the individual distributions, but does not indicate that two different sources may be present. This kind of situation will form the basis of much of our work in hypothesis testing and experimental design. We will attempt to determine if it is reasonable to state that the data is coming from more than one source.

Pictures of a Distribution: Histograms   77

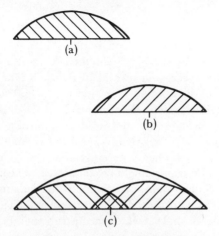

Figure 3.26 Two differently located distributions (a) and (b) summed to yield a third (c) more variable distribution.

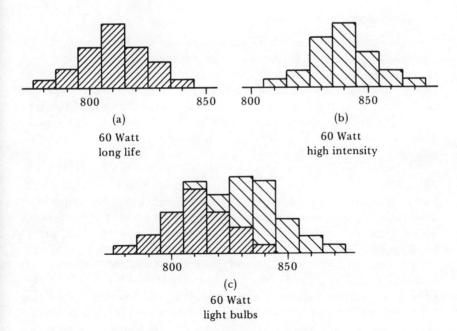

Figure 3.27 Intensity of light bulbs: (a) and (b) have different means; (c) disregards different types of light bulbs.

EXAMPLE 3.2

The distribution of light intensity (in lumens) of two different brands of 60 watt light bulbs is shown in Figure 3.27a and b. If the bulbs are all lumped together and not identified as to brand of light bulb, then the resulting distribution is the much broader, more spread out one, as in Figure 3.27c. However, by just looking at the overall distribution we would not have reason to question the width of the overall distribution. We must also have the data broken down into subdivisions. If we never subdivide the data according to any additional classifications, then we never can determine if we have more than one factor at work. We cannot discover other important influences. In most real situations as many as 5 to 50 major factors might be considered. The intensity of the light bulb might be dependent upon such things as the manufacturer, the actual voltage at the wall, the age of the bulb, the instrument used to measure the bulb, the color of the wall, the angle at which the instrument is held, or the temperature of the room. Similarly, for Example 2.1 we can question day of week, time of day, weather, equipment used, officer involved, age of driver, and so forth.

In most practical situations many different factors will influence the final distribution. Some of these factors will create additional modes, others will just spread out or distort the distribution. The net result is that one rarely encounters simple distributions. Even less frequently will you find one which would also be theoretically expected.

Models of Distribution

There are a number of different distributions that can be represented by a mathematical equation. These are called *models* and we try to determine if our data can be substituted by one of these models. We ask if the model and the data are essentially the same, or if a model can accurately duplicate the actual histogram. When this substitution can be made, we can use the model in place of the data, and we often can make many of the useful statements about the data such as probabilities connected to hypothesis and confidence intervals that were described in Chapter 1.

One of the most commonly referred-to models in statistics is called the *normal* distribution or the *Gaussian* distribution. The name *normal* should not be interpreted as meaning that any other distribution is abnormal. We will be using the term non-normal but that should merely convey the fact that the data does not fall on a normal distribution curve. It does not imply that there is anything especially peculiar about the data or the distribution.

The normal distribution is just one of many possible kinds of distributions. It is symmetrical and often described as bell-shaped (see Figure 3.28). It is particularly important since numerous kinds of data fall into distributions that can essentially be called normal distributions. Many statistical tests *assume* that the source of the data can be considered a normal distribution. Thus, we will continually refer to this kind of distribution and Chapter 10 will be entirely devoted to the normal distribution.

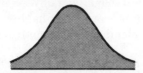

Figure 3.28 The normal distribution.

A few other basic shapes include the *flat-rectangular* or uniform distribution (Figure 3.29a), the J-shaped (Figure 3.29b), and the U-shaped distribution (Figure 3.29c). Other distributions have names associated with mathematical terms such as the exponential (Figure 3.29d), binomial, Poisson, and gamma distributions. These serve mainly as additional ways of describing distributions that we sometimes observe. However, we should be wary about the source when we see unusual shapes as our distributions.*

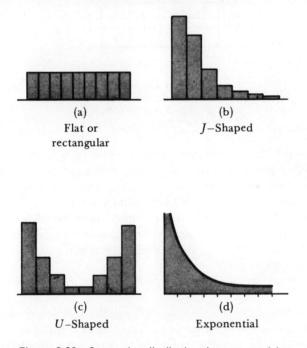

Figure 3.29 Some other distribution shapes or models.

3.11 Normal Probability Paper

The normal distribution described above is a very important distribution. I have stated that the distribution has a particular shape. It is often difficult,

---

* I have not provided diagrams for the last three since the shape varies according to the values of certain parameters or constants of the particular distribution.

however, to try to determine if your histogram is essentially identical to a normal distribution. A special type of graph paper has been developed to assist in our assessing how well a particular distribution approaches a normal distribution.

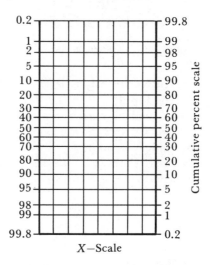

Figure 3.30 A sample of normal probability graph paper. Observe how the X-axis has a uniform scale *but* the Y-axis is distorted.

You must calculate the relative *cumulative frequencies* or cumulative percents for your distribution. You then plot these on a graph paper called normal probability paper. Figure 3.30 shows a small piece of this graph paper. The bottom is a linear (equally divided) scale that you use as the X-axis. You fill in the scale to be used for the particular problem or particular distribution. The vertical axis is provided for you and corresponds to the cumulative percent. The right side gives the up-to-and-including percent points and the left scale gives the alternate, or percents above that value. So if 20% are below a certain value (20.0 on the right side) then 80% are above that value (see the 80.0 on the left scale).

The main difference between this graph and the ones we previously drew is in the construction of the percent scale. Notice there is no 0% or 100%, so this is somewhat like our modified approach to cumulative distributions. Even more important, however, the spaces between the percents varies for different percents. The spaces between adjacent extreme values are very large, so the space between 1 and 2% or 98 and 99% is very large. These spaces are as large as the space between 50 and 60%. It is as though the scale we had originally used was stretched out in both directions. The effect is to straighten out the curve that would be drawn. In Figure 3.31a I have smoothed out the cumulative frequency diagram for tickets issued (Example 2.1, Figure 3.16). Observe the curved shape. In Figure 3.31b I have plotted the same values on the normal probability paper. Notice how the curved portions are straightened out. It is not a perfect straight line but it is fairly close (the shape of the histogram is not that perfectly bell-shaped and the cycling of the points or failure to fall on a straight line seems to confirm that).

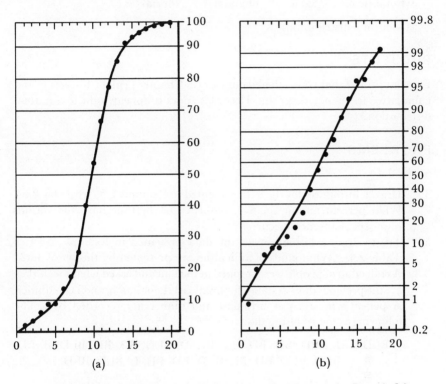

Figure 3.31 Comparison of cumulative frequency diagrams for Example 2.1: (a) On linear graph paper—this yields an S-shaped curve, (b) on normal probability paper—this approaches a straight line.

You should observe that if the distribution is in fact normal, then only two points need to be plotted. Thus, if only the cumulative percent corresponding to the value of 7 (16.1) and the cumulative percent for 14 (92.0) were plotted, virtually the same line would have been drawn. So we really wouldn't have to figure out a very large number of points in order to arrive at a plot on normal probability paper.

REVIEW

Keywords

Histogram · Category · Midpoints · Intervals · Frequency · Relative frequency · Area · Grouping · Measurements ·

Cumulative frequency · Relative cumulative frequency · Percentiles

Shape of a Distribution

Symmetrical · Skew · Bimodal · Normal ·

Model (of a distribution)

PROJECTS

For your project that you gathered data for (in Chapter 1) plot a histogram and cumulative frequency diagram. How close does it appear to fit to a normal distribution?

PROBLEMS

Section 3.4

3.1 Draw a histogram for the data presented in Example 2.5 (see Table 2.5). What percent and/or relative frequency of the time have the various numbers of absences occurred?

3.2 Try to draw a histogram for the data presented in Example 2.3 (see Table 2.3). What kinds of difficulties are presented by this set of data?

3.3 A collection of people were stopped in the hall and asked what meals they ate the previous day. B stands for breakfast, L for lunch, and D for dinner. A person who ate just breakfast and dinner are recorded BD. The following is the resulting data:

BLD, LD, BLD, BD, BD, LD, BL, D, BLD, BD, BD, BLD, B, LD, BLD, BLD, BD, BL, BLD, LD, BD, L, BLD, BLD.

Draw a histogram showing how many people had which meals. What kind of data is this?

3.4 Referring to the data in Problem 3.3 draw a histogram showing the number of people who ate breakfast, lunch, and dinner. What are the relative frequencies?

3.5 A poll of students found that 25% of the students favored the food in the cafeteria, 40% would never eat it, 15% tolerated it, 10% didn't know there was a cafeteria, and 10% brought their own lunch.
   a. What type of data do we have?
   b. What are the categories?
   c. What goes on the frequency axis?
   d. Draw the histogram.

3.6 The number of customers on each day of the week are found to be

127, 56, 32, 41, 37, 67, 167.

What are the X-values? Draw a histogram.

Section 3.6

3.7 Draw another histogram for Example 2.3. This time use about 12 categories. What is the difference between this histogram and the one in Problem 3.2?

3.8 Refer to Example 2.1. If midpoints of

$$0, 2.5, 5. 7.5, 10, 12.5, 15, 17.5, 20$$

were selected,
   a. How many values belong to each category?
   b. What rule is violated?
   c. What effect does the violation have on the histogram and the interpretation of the histogram?
   d. Draw a histogram.
   e. How do you interpret a center of 2.5 as a center of a category?

3.9 If you wanted 10 categories for the histogram for Example 2.1, what would you use for limits and midpoints?

3.10 For the data in Example 2.4 (Table 2.4) draw three histograms—one for diameter, one for height, and one for volume. How do they compare to each other?

Section 3.8

3.11 Draw a cumulative frequency diagram for Example 2.3. Below what value are 40% of the breads? What is the 60th percentile? What is the chance of a bread weighing more than 20 ounces?

3.12 Referring to Example 2.5 (Table 2.5):
   a. What percent of the time should I expect at least 2 absences?
   b. What is the chance of 20% or more of the class being absent?
   c. Discuss the following statements:
      "Since on no days were there 5 absences, the chance of 5 absences must be zero."
      "Since on no days were there 7 or more absences I would never expect to have 7 students absent."

3.13 Look at the data in Problem 3.6. In what different ways might a cumulative frequency diagram be drawn?

3.14 Referring to Example 2.4 (Table 2.4), answer the following:
   a. Draw relative cumulative frequency diagrams for diameter, height, and volume of the beer cups. Use the same scale and compare the curves.
   b. What is the chance you will not get 22 ounces in a full cup?
   c. The beer dispenser is 4.40 inches from the bottom of the table. How many cups will fit without having to be tilted?
   d. Find the middle 50% of the amount of beer that you can expect a full cup to contain.

3.15 Percentiles refer to the X-value for which a particular percent or hundredth of the population lie below. Two additional subdivisions are also

frequently used: quartiles and deciles. As the words imply, quartiles (symbolized by Q) are quarters (thus splitting the population into 4 equal parts) and deciles (symbolized by D) are tenths (splitting the population into 10 parts). Subscripts designate the particular quartile or decile (with the 1st the smallest X-value).
- a. What percentile would Q_1 be?
- b. What percentile would the third quartile be and what would its subscript be?
- c. What quartile is the median?
- d. What decile is the median?
- e. What percentile is D_3?
- f. What percentile is D_6?
- g. What would the upper decile be?
- h. How can you express the middle 50% of the population in terms of quartiles?

3.16 Refer to Example 2.1.
- a. What value of number of tickets is the bottom or lowest quartile?
- b. What value of number of tickets is the upper decile?
- c. What is the value of D_3?
- d. Twelve falls in what decile?
- e. Twelve falls in what quartile?

Section 3.11

3.17 A collection of different types of tests are to be given to different students. What would you expect the distribution of grades to look like in each of the following cases? Why?
- a. An addition test is given to high school freshmen.
- b. A college physics test is given to high school freshmen.
- c. An addition test is given to two groups of students—50 second graders and 50 high school freshmen.
- d. A college physics test is given to 50 second graders and 50 high school freshmen.

3.18 How would you describe the distribution of Example 2.5 (see Problem 3.1)?

3.19 Plot the cumulative frequency distribution for Example 2.3 (see Problem 3.11) on normal probability paper. How closely does this plot resemble a straight line?

4

Mathematical Descriptions

I have already mentioned several times that we will employ mathematics to describe data. Since measurements and data are generally expressed in the form of numbers, the kinds of words we will find most useful to assist us in discussing this data, or to aid us in condensing the data, should, naturally, also be in the form of numbers. The importance of being able to describe the data and thereby find a useful way to relate to the data should be obvious. As to the necessity of condensing the data I merely refer you back to Table 2.1 (Example 2.1) and ask you to "tell me about it!" We will now explore some of the kinds of mathematical descriptions.

4.1 General Characteristics

In the last chapter you drew a type of picture. It told you that the data was somehow distributed over several different values, with some being used more frequently than others. If you now look carefully at the distribution (and this will be true of any distribution) you should observe that the pattern has at least *two* very important characteristics. They are:

1. A *center* or central value, and
2. A *spread* around that center.

It will be necessary to find mathematical values to describe these two characteristics for virtually all of the sets of data you encounter. Of course these aren't the only characteristics, but they are the most important and about the most useful. Some other types of descriptive characteristics (from a mathematical approach) will be mentioned and perhaps even described, but we will generally avoid them in our work.

We are now about to engage in some mathematics. There will be some basic algebra used to tell you about what is going on and what has to be done. This simplifies the writing. The actual manipulation of the data is arithmetic (mostly addition, multiplication, and squaring). The problems are designed so that if you must work by hand without the aid of *any* computational equipment, there are enough problems available so you can learn and apply what I'm talking about without exhausting yourself doing arithmetic. I will also give you some hints and shortcuts to use, since everyone who uses a computer at some time is faced with a broken one, and must resort to old-fashioned techniques. For those individuals having access to calculators or computers I have provided some additional and more realistic sets of data for you to work with.

4.2 Central Values or Averages

You have already encountered at some point in time the term "average." You probably have computed a baseball "average." You certainly have read about "average" salaries or "average" costs of living. Well, "average" and central value are one and the same. The difficulty is that you have been deceived with the word "average." There is not just one exclusive quantity that can serve as *the* measure of the center of a set of numbers. There are several kinds of averages and they may have different values *for the same set of data*. I referred to this problem in the second chapter as something to be wary of in somebody else's conclusions.

We will work with four kinds of averages: *midrange, mode, median, mean*. Each plays its role, although one, the mean, will be used more than any of the others. Some will be easier to calculate than others, several may have the same value for some data. Sometimes one (the mode) may not even exist for some data. I will introduce them in the above order, which is somewhat in order of increasing difficulty.

Midrange

The midrange is perhaps the simplest to calculate. It is halfway between the two extreme values.

$$\text{Midrange} = \frac{\text{Smallest} + \text{Largest}}{2}. \qquad (4.1)$$

EXAMPLE 2.1 (Cont.)

Our officer gave a maximum of 19 tickets, therefore the largest *value* (not the largest *frequency*) was 19. The smallest value was the issuance of 1 ticket. Thus the midrange is

$$\text{Midrange} = \frac{1 + 19}{2} = \frac{20}{2} = 10.$$

EXAMPLE 2.2 (Cont.)

The smallest value is the fastest time it takes to get to school or 13.6 minutes. The largest value was 16.7 minutes. Thus

$$\text{Midrange} = \frac{13.6 + 16.7}{2} = \frac{30.3}{2} = 15.15.$$

Mode

The mode is generally the next easiest to find. It is the number that appears most frequently. The main difficulty is that there may not be just one single mode, there may be several—as we have already seen in the last chapter. I have already introduced you to the idea that the peak of a distribution or histogram is called a mode (recall Figure 3.23). We have also explored the situation where two modes may exist—a bimodal distribution. When discussing data it is often not the actual mode that is important, but rather the number of modes that are present.

EXAMPLE 2.1 (Cont.)

Nine occurs seventeen times. No other value occurs as frequently as this and therefore 9 is the mode.

EXAMPLE 2.2 (Cont.)

Here there is no single mode since 15.1, 16.4, and 16.5 all occur with the same frequency—2. This, as I have said before, may not be unusual. We merely state that there is no single mode.* The condensed histogram of Figure 3.15 (page 66) does have a single mode of 16.50. What are the modes for the grouped data for Example 2.1?†

Median

The median is a very simple value to calculate *if* the values have been placed in order of largest to smallest (or smallest to largest) or *if* a histogram has been constructed (a tally sheet will serve equally well). The median is simply the middle value *when the numbers have been placed in order*. *Half* of the numbers are *greater than the median* and *half* of the numbers are *smaller than the median*.

If there are an *odd amount of numbers*, say the five numbers 1, 3, 9, 2, and 6; *one* of them is the median. Arranging them in order of increasing value we would have 1, 2, 3, 6, 9. Then, two of the numbers are less than 3 and two are greater than 3, so 3 is the median. The median is *not* the number 9. It should be obvious that if the numbers have just been randomly written down then they could be rewritten again and the middle one in that random set could be any value. But

---

* With small sets of data, say less than 40 values, the mode is often not very useful for any purpose.
† Answer: Figure 3.10, Mode = 11; Fig. 3.11, Mode = 10.

there can *only be one median* to any set of numbers. The way to assure that there is one value is always to place the numbers in numerical order.

If there are 11 numbers, the median is the 6th one when the numbers are placed in order. So you would have 5 numbers, then the median, then 5 more numbers. A quick formula for figuring out which is the median for odd amounts of numbers is simply

$$\text{median} = [(n + 1)/2]\text{th value},$$

where n is the number of numbers. (Throughout this text and virtually everywhere data is used, n stands for the *number* of values or the size of the sample.)

So for 11 numbers the median is the

$$\frac{11 + 1}{2} = \frac{12}{2} = 6\text{th value}.$$

For *even* sets of numbers, the median does not have to be one of the values. Given the 6 numbers 1, 3, 8, 5, 1, 7, again you *first* must put them in numerical order—or 1, 1, 3, 5, 7, 8. Split them into two groups: 1, 1, 3 and 5, 7, 8. The median is halfway between the two middle numbers—or $(3 + 5)/2 = 4$. If the two middle numbers are the same then obviously the median is that common value.

EXAMPLE 2.1 (Cont.)

Since there are 112 numbers, they would be split into 2 groups of 56 values each. The median would then fall between the 56th and the 57th number. The easiest way to find these numbers is to refer to the tally sheet, frequency table, or even the histogram (Figure 3.6 or 3.7). Start on one end, add up the frequencies until you reach the 56th observation. Record that value and then record the next one (the 57th). If they are identical then that value is the median. If they are different, then the median is halfway between them. Here, both are 10's (in fact the 46th through the 61st are all 10's), so the median is 10.

EXAMPLE 2.2 (Cont.)

We have 20 numbers. The median would be between the 10th and the 11th ones. Looking back at the histogram of Figure 3.13 and counting until we reach the tenth and eleventh numbers I obtain: 10th = 15.6, and 11th = 15.7. Therefore the median is 15.65.

COMMENT. If the cumulative distribution is available, then we should even have less difficulty in finding the median. The median is the value corresponding to a Relative Cumulative Frequency of 0.50 (the 50th percentile). Refer back to the cumulative tables and diagrams of Chapter 3 and note the ease of finding the median for the two examples we have just worked on.

If you have a histogram that is drawn with a smooth curve, the median is the value for which half of the *area* is below it and half of the *area* is above it (see

Figure 4.1). However, we will not be particularly interested in this interpretation.

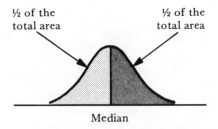

Figure 4.1 Location of the median for a continuous distribution.

Mean

We now come to what is generally regarded as the most important and most commonly used measure of central value. You have probably encountered it by the name of arithmetic average. This is the one you generally think of when you hear the word "average." To compute it you merely add up all the values and divide by n (the number of values). To write this instruction down mathematically I will symbolize the mean by $\bar{X}$ (read this as "X bar"). If we call each of the individual numbers X's we must be able to distinguish them apart. We do this by calling them X_1, X_2, X_3 and so on. You read X_1 as "X sub 1" (the 1 is a subscript) and you consider this as the first one of the set of numbers. X_2 is the second, X_3 is the third, and X_n is the nth number (whatever the n might be). These are merely substitutes for the actual numbers you have in a set of data. We may also use a little i or a little j as a subscript to stand for any one of the individuals.
So
$$\text{Mean} = \bar{X} = \frac{X_1 + X_2 + X_3 + X_4 + \cdots + X_n}{n}. \tag{4.1}$$

The $+ \cdots +$ is used to indicate that we may be adding many more than just the five X-values. If we were adding up 112 values (as we will do in Example 2.1) then n would be 112 and I would have to write down 112 different X's. But you should get the meaning of what you are to do from just the few terms written down.

We will use the operation of adding up a set of numbers quite frequently. Again, for the sake of making the writing easier and simplifying the equations we will encounter, mathematicians use a symbol $\sum$ (a Greek capital sigma) to stand for "Take the sum of a certain set of numbers."

If we let X_i stand for those X's in Equation (4.1), then we can rewrite the sum (the numerator) as
$$\sum_{i=1}^{n} X_i = X_1 + X_2 + X_3 + X_4 + \cdots + X_n. \tag{4.2}$$

The left-hand part is interpreted as follows: The $\sum$ says you will add up a group of numbers; the types of numbers are those next to the $\sum$, in this case the X's; the $i = 1$ means the first X you will start with is X_1; the n on top of the $\sum$ says you will keep on adding X's until you reach the one called X_n. So if n is 112 you will start with X_1 and add up all the X's up to and including X_n or X_{112}. We can now write the mean as simply

$$\text{Mean} = \bar{X} = \frac{\sum_{i=1}^{n} X_i}{n}. \qquad (4.3)$$

If you need more work with the subscripts and the summation ($\sum$) notation refer to Appendix B—now. (The next few examples may help but the work in the appendix may help much more.)

EXAMPLE 2.1 (Cont.)

Refer again to Table 2.1 for the set of data. n is 112. If we start from the upper left corner and go to the right then X_1 is 14, X_2 is 17, X_3 is 18, ..., X_{15} is 11, X_{16} is 9, ..., X_{110} is 9, X_{111} is 9, X_{112} is 7. What is X_{21}, X_{35}, X_{100}?*
So

$$\sum_{i=1}^{112} X_i = X_1 + X_2 + X_3 + \cdots + X_{110} + X_{111} + X_{112}$$

$$= 14 + 17 + 18 + \cdots + 9 + 9 + 7$$

$$= 1132,$$

and

$$\text{Mean} = \bar{X} = \frac{\sum_{i=1}^{n} X_i}{n} = \frac{14 + 17 + 18 + \cdots + 9 + 9 + 7}{112} = \frac{1132}{112} = 10.11.$$

Often you will find the $\sum$ written without the $i = 1$ below it or the n above it. I will also leave off these symbols. You should realize that if you are adding up a set of numbers it must begin with one value and end with another value. The only times I will include the extra instructions is when they are necessary to avoid confusion (it is possible that you could be starting with $j = 5$, and end with $m = 12$). My analogy would be the instruction "start the car." If I assume it has an automatic transmission then most cars will start with the shift lever in neutral or in park and I do not have to include in my statement where the shift lever must be placed. But my car *might* be an exception and I would then be forced to say what the particular exception is: for example, put the lever in park, pull the lever toward you until you hear a click, then turn on the ignition.

---

* Answer: $X_{21} = 11$, $X_{35} = 19$, $X_{100} = 10$.

EXAMPLE 2.2 (Cont.)

Here $n = 20$. Refer to Table 2.2 for the individual values. Then

$$\text{Mean} = \bar{X} = \frac{\sum X_i}{n} = \frac{16.5 + 16.5 + \cdots + 14.5 + 15.5}{20} = \frac{310.0}{20} = 15.5.$$

Means for Grouped Data

The means we have calculated for Examples 2.1 and 2.2 were done directly from the original data. Quite often, however, the data has been put into categories, as in a table, tally sheet, or histogram. In these cases we will work with the categories and the frequencies associated with those categories. We spent a considerable amount of time discussing categories in the last chapter. The important idea now is that all the individuals in a particular category are treated *as though they all have the same value*—the X-value of the center of the category or interval to which they belong.

In Table 4.1 I have listed the individual X-values and their frequencies for each of the X-values of Example 2.1. Each of the X-values represents a separate category and there are 21 of them. If we take the sum of the frequencies by adding up Column 2, we obtain the quantity 112. This is also the total number of observations (the n). Thus if I use the symbol j to stand for *category* and I have c categories, I can write this as

$$\text{Sum of the frequencies} = \sum_{j=1}^{c} f_j = n$$

or simply

$$\text{Sum of the frequencies} = \sum f = n.$$

Before, we added up a quantity every time that it appeared. Since 9 appeared seventeen times we made seventeen separate additions of 9 in our summation. But I can also multiply them together to obtain the same 153 (17 times 9 is the same as $9 + 9 + 9 + \cdots + 9 + 9$) and I will have eliminated a large number of additions. Thus, I will multiply each X-value by the number of times it appears (its frequency) and I will add up all of these products. This sum will be the same as the sum of all of the individual values. We write this as

Sum of all of (frequency of a category) times (X-value of the category)

$$= \sum_{j=1}^{c} (f_j \cdot X_j).$$

(We are adding up the *products* and that is why I am using the parenthesis around the $f \cdot X$.) Then, finally,

$$\text{Mean} = \frac{\sum_{j=1}^{c} (f_j \cdot X_j)}{\sum_{j=1}^{c} f_j} = \frac{\sum (f_j \cdot X_j)}{n}. \qquad (4.5)$$

Table 4.1 Calculation of the mean for the number of tickets issued (Example 2.1)

Column	1	2	3
Quantity	X-value	f frequency	f · X frequency times X-value
Data	20	0	0
	19	1	19
	18	1	18
	17	2	34
	16	0	0
	15	5	75
	14	7	98
	13	8	104
	12	13	156
	11	14	154
	10	16	160
	9	17	153
	8	10	80
	7	4	28
	6	4	24
	5	1	5
	4	2	8
	3	3	9
	2	3	6
	1	1	1
	0	0	0
Sum		112	1132

n = Sum of frequencies = 112

$$\text{Mean} = \bar{X} = \frac{\text{Sum of (frequency times } X)}{\text{Sum of frequency}} = \frac{\Sigma (f \cdot X)}{\Sigma f}$$

$$= \frac{\text{Sum of Column 3}}{\text{Sum of Column 2}} = \frac{1132}{112} = 10.11.$$

EXAMPLE 2.1 (Cont.)

Refer to Table 4.1. Here Column 3 contains the products of $f \cdot X$. Then

$$\text{Mean} = \frac{(0 \cdot 0) + (1 \cdot 1) + (3 \cdot 2) + (3 \cdot 3) + (2 \cdot 4) + (1 \cdot 5) + \cdots + (1 \cdot 19) + (0 \cdot 20)}{0 + 1 + 3 + 3 + 2 + 1 + \cdots + 1 + 0}$$

$$= \frac{0 + 1 + 6 + 9 + 8 + 5 + \cdots + 19 + 0}{0 + 1 + 3 + 3 + 2 + 1 + \cdots + 1 + 0} = \frac{1132}{112} = 10.11.$$

Table 4.2 Mean for times to school (Example 2.2)

Column	1	2	3
Quantity	X-value	Frequency	Frequency times X-value
Data	16.7	1	16.7
	16.6	0	0
	16.5	2	33.0
	16.4	2	32.8
	16.3	0	0
	16.2	0	0
	16.1	1	16.1
	16.0	1	16.0
	15.9	1	15.9
	15.8	1	15.8
	15.7	1	15.7
	15.6	1	15.6
	15.5	1	15.5
	15.4	0	0
	15.3	0	0
	15.2	1	15.2
	15.1	2	30.2
	15.0	1	15.0
	14.9	0	0
	14.8	0	0
	14.7	0	0
	14.6	0	0
	14.5	1	14.5
	14.4	0	0
	14.3	1	14.3
	14.2	0	0
	14.1	1	14.1
	14.0	0	0
	13.9	0	0
	13.8	0	0
	13.7	0	0
	13.6	1	13.6
Sum		20	310.0

$n = $ Sum of frequency $= 20$

$$\text{Mean} = \frac{\sum (f \cdot X)}{\sum f} = \frac{\text{Sum of Column 3}}{\text{Sum of Column 2}}$$

$$= \frac{310.0}{20} = 15.5.$$

Observe that there were 21 categories (0 to 20) although only 18 had frequencies of 1 or more. This points out that the total number of categories is somewhat arbitrary and really doesn't matter as long as all of the values are counted.

EXAMPLE 2.2 (Cont.)

Table 4.2 contains the data for Example 2.2.

$$\text{Mean} = \frac{\sum_{j=1}^{32}(\text{frequencies} \cdot X_j)}{\sum_{j=1}^{32}(\text{frequencies})} = \frac{\sum (f_j X_j)}{\sum f_j}$$

$$= \frac{(1 \cdot 13.6) + (1 \cdot 14.1) + \cdots + (2 \cdot 16.5) + (1 \cdot 16.7)}{1 + 1 + \cdots + 2 + 1}$$

$$= \frac{13.6 + 14.1 + \cdots + 33.0 + 16.7}{1 + 1 + \cdots + 2 + 1} = \frac{310.0}{20} = 15.5.$$

(Note that I have omitted all products that equal zero.)

When data has been grouped into wider intervals we can still follow the same procedure.

EXAMPLE 2.1 (Cont.)

In column 1 of Table 4.3 I have listed a group of intervals (which we previously called Set 3.1—see page 61). In column 2 I give the center of the intervals that we will use as the *X-values* for the various intervals or categories.

Table 4.3 Computing the mean number of tickets using grouped data and given intervals (Example 2.1)

Column	1	2	3	4
Quantity	Interval	Center or X-value	Frequency	Frequency times X-value
	18.5–21.5	20	1	20
	15.5–18.5	17	3	51
	12.5–15.5	14	20	280
Data	9.5–12.5	11	43	473
	6.5–9.5	8	31	248
	3.5–6.5	5	7	35
	0.5–3.5	2	7	14
Sum			112	1121

$$\text{Mean} = \bar{X} = \frac{\sum (\text{frequency} \cdot X)}{\sum (\text{frequency})} = \frac{\text{Sum of Column 4}}{\text{Sum of Column 3}} = \frac{1121}{112} = 10.03.$$

Column 3 contains the frequencies for each interval, and Column 4 has the products of the frequencies times the center or X-values. Then

$$\text{Mean} = \bar{X} = \frac{\sum (\text{frequencies} \cdot X\text{-values})}{\sum (\text{frequencies})}$$

$$= \frac{(7 \cdot 2) + (7 \cdot 5) + \cdots + (3 \cdot 17) + (1 \cdot 20)}{7 + 7 + \cdots + 3 + 1}$$

$$= \frac{14 + 35 + \cdots + 51 + 20}{112} = \frac{1121}{112} = 10.03.$$

This mean is slightly different than the original 10.11 that we obtained by using all of the numbers. However, the calculations here were simpler and you should realize that there was only a 0.08 difference between the two means, or an error of less than 1%. Most of the time a small error of this type is not objectionable since the data itself is not as precise as the mean we are able to calculate.★

4.3 Coding to Find Means

We will now attempt to make some observations regarding the use of a scale, and the relationship between a scale (the X-axis) and the mean. We will find that there exists a technique called *coding*, which is useful for somewhat simplifying our calculations. If you have access to calculating equipment or a computer then you will not really need to concern yourself with this section—at least not now. But invariably the time will come when you really want an answer in a hurry and either you are miles away from a calculator (like the Sunday night before a report is due or a test is imminent) or the computer is down (that's computerese for it's on the fritz, which can be more often than not). So some ideas on how to ease up the computations should be welcomed by anyone.

RULE 1 *If you subtract a single constant from each and every individual of a set of numbers, then the same constant has been subtracted from the mean.*

Often we can find a number to subtract from our collection of X-values that makes the multiplications and additions much easier to perform. We subtract this constant from *each* X-value to obtain what we call "coded X-values" or perhaps X'-values (X' is read as "X prime"). We then calculate the *average* (the mean) of these new X'-values *using the same frequencies* as before. If we then add the constant back to this new mean, we have the true mean of the original uncoded set of values. Thus,

Actual mean = Mean of coded data + Constant.

★ Here we have individual values accurate to at most two digits. The mean we calculated has four digits, two of which must be regarded a little bit skeptically. Generally, we calculate the mean to *one* more significant place than the data is recorded. However, I have gone an extra place mainly to illustrate how good the approximations are.

If the "constant" is called C and the "mean of the coded data" is $\bar{X}'$, then
$$\text{Mean} = \bar{X} = \bar{X}' + C. \tag{4.6}$$

EXAMPLE 2.1 (Cont.)

In Table 4.4 Columns 1 and 2 list the original set of numbers and their frequencies. A quick glance would indicate that 10 seems to be pretty close to the center (although 11 or 9 would be just as good a guess). If we subtract 10 from each and every one of these X-values we obtain a set of "coded" X-values or X'-values which I have included in Column 3. We thereby have established that the constant C is 10.

Table 4.4 Table for calculating the mean number of tickets using coded values (Example 2.1)

Column	1	2	3	4
Quantity	X-value	f frequency	X' $X - 10$ coded X	$f \cdot X'$ frequency times coded X
Data	20	0	10	0
	19	1	9	9
	18	1	8	8
	17	2	7	14
	16	0	6	0
	15	5	5	25
	14	7	4	28
	13	8	3	24
	12	13	2	26
	11	14	1	14
	10	16	0	0
	9	17	-1	-17
	8	10	-2	-20
	7	4	-3	-12
	6	4	-4	-16
	5	1	-5	-5
	4	2	-6	-12
	3	3	-7	-21
	2	3	-8	-24
	1	1	-9	-9
	0	0	-10	0
Sum		112		$+12$

Constant $= C = 10$

Mean of coded data $=$ Mean about 10

$$\bar{X}' = \frac{\sum (\text{frequency} \cdot \text{coded } X)}{\sum (\text{frequency})} = \frac{\text{Sum of Column 4}}{\text{Sum of Column 2}} = \frac{12}{112} = 0.11$$

Then $\bar{X} = \bar{X}' + C = 0.11 + 10.00 = 10.11$.

We next calculate a mean for the X'-values, a mean which might also be called the mean around the number 10. This is calculated as

$$\bar{X}' = \text{Mean of coded } X\text{-values} = \frac{\sum (f_j \cdot X'_j)}{\sum f_j}$$

$$= \frac{[1 \cdot (-9)] + [3 \cdot (-8)] + \cdots + (2 \cdot 7) + (1 \cdot 8) + (1 \cdot 9)}{1 + 3 + \cdots + 2 + 1 + 1}$$

$$= \frac{-9 - 24 + \cdots + 14 + 8 + 9}{112} = \frac{+12}{112} = 0.11.$$

Then, for this problem,

$$X = \text{Mean of coded } X\text{-values} + \text{constant}$$

or

$$\bar{X} = \bar{X}' + C = 0.11 + 10 = 10.11.$$

Basically, all we have done is to change the scale of the data or the histogram and we work with what is effectively just a smaller scale. Finally we convert back to the original scale (see Figure 4.2). The main point, however, is that the multiplications in Table 4.4 are much simpler (and as a result should lead to fewer errors) than those in Table 4.1. Similarly, the additions are easier.

Coding may be generalized one step further. Often the intervals are large, as perhaps with grouped data. You might have numbers that jump in 5's or 25's or 100's so that even after subtracting a constant you have something like:

0, 25, 50, 75, 100, 125, 150, 175.

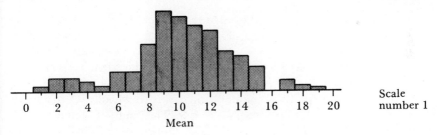

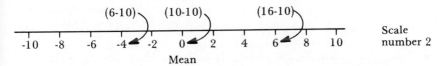

Figure 4.2 A histogram for Example 2.1 with the original scale which I designate as *Scale number 1*. The mean is 10.11—observe its relative location in the distribution. *Scale number 2* was drawn by subtracting 10 from *all* the values. The relative location of the mean is the same but its value is now only 0.11. The same amount has been subtracted from the mean as from the individual values.

Multiplying these numbers can also be quite unwieldy. So we come to:

RULE 2 *If each and every individual in a set of numbers is divided by a constant, then the mean has also been divided by that same constant.*

This rule says, in effect, that I can replace that row of numbers

$$0, 25, 50, 75, 100, 125, 150, 175$$

by

$$0, 1, 2, 3, 4, 5, 6, 7$$

simply by dividing each of the numbers in the first set by 25.

We will combine Rules 1 and 2 in the following way. If

L = width of the interval or the number each value is divided by
C = constant subtracted from each value
$\bar{X}'$ = mean of fully coded values
$\bar{X}$ = true mean

then we

1. Subtract C from each X-value.
2. Divide each of these new quantities by L (*now* call them X').
3. Compute the mean of the new coded X'-values to obtain $\bar{X}'$.
4. Calculate

$$\bar{X} = (L \cdot \bar{X}') + C. \tag{4.7}$$

EXAMPLE 2.1 (Cont.)

Table 4.3 showed grouped categories for Example 2.1, from which we computed a mean. If we decided to code, then what is a good choice of center? It looks like 11 would be pretty close to the center. So we should subtract 11 from each value (then $C = 11$). I have reproduced the original center values and frequencies in Table 4.5. In Column 3 of this table I have subtracted 11 from each value (partially coding them). Next we divide each of the new values by the width of the interval, which is 3. The result is Column 4, which *contains a series of integers going from* -3 *to* $+3$. In fact, we can go directly from Column 1 to Column 4 by just selecting a spot we would like to call 0 (which was the old 11), then counting up and down from that point (where up is positive and down is negative).* We then compute $\bar{X}'$ as before and substitute into Equation (4.7) to obtain the mean of 10.01. This slightly different result (compared to the 10.03 in Table 4.3) is 2/1000 or a 0.2% difference and is the consequence of performing different divisions and of rounding off. Again, do not be alarmed with very small errors when different calculating procedures are followed.

---

* Of course this only works when all the intervals are the same size. If the intervals vary then we cannot fully code.

Table 4.5 Grouped data for Example 2.1, fully coded

Column	1	2	3	4	5
Quantity	Center or X-value	f frequency	Partially coded X $X - 11$	Fully coded X $X' = \dfrac{X - 11}{3}$	$f \cdot X'$
Data	20	1	9	3	3
	17	3	6	2	6
	14	20	3	1	20
	11	43	0	0	0
	8	31	-3	-1	-31
	5	7	-6	-2	-14
	2	7	-9	-3	-21
Sum		112			-37

$C = 11, L = 3$

$$\bar{X}' = \frac{\sum f \cdot X'}{\sum f} = \frac{\text{Sum of Column 5}}{\text{Sum of Column 2}} = \frac{-37}{112} = -0.33$$

Then

$$\bar{X} = L \cdot \bar{X}' + C = 3(-0.33) + 11 = -0.99 + 11.00 = 10.01.$$

I hope that you can see the advantage of coding when a calculator is not available.

4.4 Which Average to Choose

I have introduced you to four different numbers that can be used to state what the "average" or typical value for a set of numbers could be. We have seen that they were not necessarily the same. Table 4.6 summarizes the averages that we have computed for two examples. But how do we decide which *one* to use if only one is to be quoted? This is a very important question that I already mentioned in Chapter 2 and is one that constantly pops up. Let me begin to answer this by first exploring the question of how to interpret the different "averages."

Table 4.6 Summary of averages for two examples

Example	2.1	2.2
Midrange	10	15.2
Mode	9	No single mode
Median	10	15.65
Mean	10.11	15.5

You have already seen that the mode can be pictured as the *peak* of a distribution. The median was shown to be a point that splits the distribution into two *equal areas*. Finally, you might observe that the midrange cuts the *length* of the distribution into two equal parts (see Figure 4.3). What about the mean?

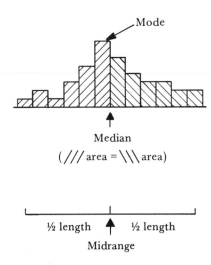

Figure 4.3 Geometrical interpretation of the median, midrange and mode.

Let us look at the histogram of a distribution. In the last chapter I asked you to think of the areas as weights, with the greater heights having greater weight and lower heights being less heavy. Carrying this analogy one step further, consider the *X*-axis as a support for these weights. Then the mean is the one value that can act as a balancing point for the entire distribution. So if we place the distribution on a seesaw (Figure 4.4), the spot that allows it to be in a perfect balance, what the physicists call the fulcrum, is also called the *mean*. Now what significance does this have with respect to our way of interpreting or thinking about the mean? Think about the seesaw. A single point two units away from the fulcrum can counter the effect of two points (or

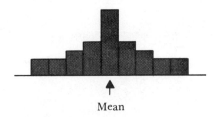

Figure 4.4 If a distribution is placed on a seesaw, the mean is the point which allows a perfect balance.

Figure 4.5 A seesaw: The fulcrum is the balancing point (for a distribution this was the mean). Two weights one unit from the fulcrum are balanced by *only* one weight placed two units from the fulcrum on the other side.

weights) that are only half as far away from the fulcrum (or one unit away). Thus a single point that is some distance from the center of a distribution can significantly alter the value of the *mean* by offsetting the effect of *several* points that are close to the center (see Figure 4.5). Of course the mean is not the only average strongly influenced by a single very large or very small value—the midrange depends entirely upon the extreme values. However, because the midrange is so seriously affected by an individual wild value, it is not regarded as a very important measure of average. This is particularly true with large sets of data. The median is not affected by wild or peculiar individuals or even small sets of wild individuals. For this reason it is usually a good indicator of where the center of a mass of data is located.

Statistical theory tells us that for the important application called estimation, the best estimator (the value which best describes the population center or the place from which all these values come from) is the *mean*. But it *isn't*

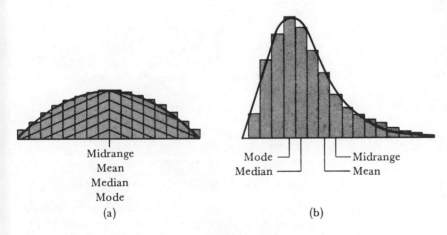

Figure 4.6 Relationship of the various "averages": (a) For a symmetrical distribution, (b) for a skewed distribution.

always right and it *isn't always* the closest to the true center. It often does not even act as the best description of the center. But, it is the only one that uses all of the numbers, and generally is the one most often quoted and used in statistical calculations.

The most common problem with averages is that encountered with skewed distributions, and skewed distributions, unfortunately, are far from uncommon. If the distribution is symmetric then all four averages are identical (Figure 4.6a). However, when the distribution is skewed the four averages fall into a pattern usually like that in Figure 4.6b, having an order going from *mode* to *median* to *mean* to *midrange*. As a result the word "average" alone can provide misleading information. A good illustration would be the following example:

EXAMPLE 4.1

A company which employs ten people reported the average income as $14,500. Is this a good figure? The actual incomes were also listed (Table 4.7 and Figure 4.7).

Table 4.7 Salaries of personnel in dollars

4000	5000
4300	5100
4600	5500
4600	7000
4900	100,000

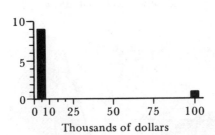

Figure 4.7 Grouped histogram for the distribution of salaries of Table 4.7.

The average that was reported was the mean. What are the other averages?

Median = 4950—quite a bit lower, but do you agree that it is a much more typical value, and presents a much better single description of the collection?

Mode = 4600.

Midrange = 52,000 = this is really high. Ninety percent of the people make less than this average. Is this a "good" value?

The main question always is whether one particular average is better than another as a *description of the data*. Do *not* fall into the easy trap of saying that you like the mean or the midrange *"because they are bigger* than the median." The criterion for saying that you like a particular average in any case should be "because it tells it like it *is*," not because it looks better.

4.5 Spread

The second "word" that is extremely valuable for describing a set of numbers is the measure of *spread*. Since a distribution has been defined as a collection of numbers that are not identical, we need some way of saying just how different they are. I will present three different measures of spread: *range, variance*, and *standard deviation*. A fourth called the *mean deviation* will be introduced in the problems, but it is not nearly as important as these three.

Range

This is about the simplest of all descriptive statistics to calculate. It is the total spread of the numbers and is just the difference between the two extreme values. So:

$$\text{Range} = \text{Largest value} - \text{Smallest value}. \tag{4.8}$$

EXAMPLE 2.1 (Cont.)

$$\text{Range} = 19 - 1 = 18.$$

EXAMPLE 2.2 (Cont.)

$$\text{Range} = 16.7 - 13.6 = 3.1.$$

The range will be particularly useful when working with very small sets of numbers. When there are only four or five values the range is a very good description of spread. But observe that it is extremely susceptible to individual wild values. Thus, since large sets of numbers tend to have *some* extreme individuals we frequently ignore the range when we deal with samples greater than about 20 to 25. In addition, the range disregards the middle of the distribution, and thereby treats a U-shaped distribution no differently than one that is highly concentrated at the center.

4.6 Variance and Standard Deviation

The most common and generally most useful measures of the variability or spread of a set of numbers are the *variance* and the *standard deviation*. We will use them interchangeably during our discussion because the *variance is just the square of the standard deviation.* If one is calculated, I will assume you will be able to easily convert to the other.

$$\text{Variance} = (\text{Standard deviation})^2$$

or

$$\text{Standard deviation} = \sqrt{\text{Variance}}. \tag{4.9}$$

We will always calculate the variance. Then to find the standard deviation we will take the square root of the variance. A table of squares and square roots is provided in Appendix Table A.17.

The definition of variance that we will use will be

DEFINITION *The variance is an average of the squares of the differences from the mean.*

Mathematically it is

$$\text{Variance} = s^2 = \frac{\sum_{i=1}^{n}(X_i - \bar{X})^2}{n-1}. \tag{4.10}$$

You will see both s^2 and σ^2 (read as sigma squared) used as symbols for the variance at different times. You will also see the two quantities $n-1$ and n in the denominator. The differences have to do with the notions of sample and population. If we are dealing with a sample, which is what most data is, then we will use the s and the $n-1$. The few times we are dealing with a complete universe or a theoretical population, we will use σ and n. We will also use μ, the Greek letter mu, to stand for the *population* average or mean. When n is very large (say 100 or more) there is essentially no difference between the two.

So the variance may be written in two ways:

Sample

$$s^2 = \frac{\sum(X - \bar{X})^2}{n-1}$$

Population

$$\sigma^2 = \frac{\sum(X - \mu)^2}{n}.$$

We will almost always be working with the sample variance or s^2.

The mathematical definition, Equation (4.10), is not usually the best one to follow when we must actually calculate a variance. There are several other modifications of the formula that can be used that are often easier (though not necessarily easy). I will work with two formulas now. In the next section we will deal with coding and further reduce the difficulty of calculating variance. Two equations follow. *If you are not grouping the numbers* (even to the extent of using a tally sheet), then just *ignore the f_i-values*, since they will all be 1's anyhow.

$$\frac{\sum f_i(X_i - \bar{X})^2}{n-1}. \tag{4.11}$$

Equation (4.11) is the same as the definition, and generally it is very tedious to use.

$$\frac{\sum f_i(X_i^2) - (\sum f_i X_i)^2/n}{n-1}. \qquad (4.12)$$

Equation (4.12) is fine for a calculator. The numbers are usually very large and you must carry out the calculations to an excessive number of decimal places. If *the numbers are not organized* this is what you would generally use.

Now let's go through a very simple illustration to see what all those different summations are and what kind of calculations are involved.

EXAMPLE 4.2

I have observed the number of students awake and alert in class at 8:00 A.M. on six consecutive Monday mornings as follows:

$$6, 9, 8, 8, 12, 9.$$

In categories they would be

Category	Value	Frequency
1	6	1
2	8	2
3	9	2
4	12	1

The mean would be

$$\bar{X} = \frac{\sum_{i=1}^{4} f_i \cdot X_i}{\sum f_i} = \frac{(1 \cdot 6) + (2 \cdot 8) + (2 \cdot 9) + (1 \cdot 12)}{1 + 2 + 2 + 1} = \frac{52}{6} = 8.67$$

$$\sum f_i = n = 6.$$

Using Equation (4.11) the variance would be calculated as

$$s^2 = \frac{\sum_{i=1}^{4} f_i(X_i - \bar{X})^2}{n-1}$$

$$= \frac{f_1 \cdot (X_1 - \bar{X})^2 + f_2 \cdot (X_2 - \bar{X})^2 + f_3 \cdot (X_3 - \bar{X})^2 + f_4 \cdot (X_4 - \bar{X})^2}{n-1}$$

$$= \frac{1 \cdot (6 - 8.67)^2 + 2 \cdot (8 - 8.67)^2 + 2 \cdot (9 - 8.67)^2 + 1 \cdot (12 - 8.67)^2}{6 - 1}$$

$$= \frac{1(-2.67)^2 + 2(-0.67)^2 + 2(0.33)^2 + 1(3.33)^2}{5}$$

$$= \frac{1(7.129) + 2(0.449) + 2(0.109) + 1(11.089)}{5}$$

$$= \frac{19.33}{5} = 3.87.$$

Observe that each term that is added together contains both the product of a difference that has already been squared and a frequency.

With Equation (4.12) we have two separate summations to perform. The first is

$$\sum f_i \cdot X_i^2 = f_1 \cdot X_1^2 + f_2 \cdot X_2^2 + f_3 \cdot X_3^2 + f_4 \cdot X_4^2$$
$$= 1 \cdot (6)^2 + 2 \cdot (8)^2 + 2 \cdot (9)^2 + 1 \cdot (12)^2$$
$$= (1 \cdot 36) + (2 \cdot 64) + (2 \cdot 81) + (1 \cdot 144)$$
$$= 470,$$

the second is

$$\sum f_i X_i = f_1 X_1 + f_2 X_2 + f_3 X_3 + f_4 X_4$$
$$= 1 \cdot 6 + 2 \cdot 8 + 2 \cdot 9 + 1 \cdot 12 = 52.$$

Then observe that

$$\left(\sum f_i X_i\right)^2 = (52)^2.$$

This is because the squaring is done *after* the summation—the square is outside the parenthesis. The previous summation had the square around the X itself, so that *each term had a square*.
Thus

$$s^2 = \frac{\sum f_i X_i^2 - (\sum f_i X_i)^2 / n}{n - 1} = \frac{470 - 52^2/6}{6 - 1} = \frac{470 - 450.67}{5}$$

$$= \frac{19.33}{5} = 3.87.$$

So several different formulas will all yield the same result, even though the calculations may be slightly different. I should point out two additional useful things to be aware of:

1. If the numbers were *not grouped* but just listed, then Equation (4.12) would have been easiest to work with (with a few additional calculations).

Since all the f_i's are 1's then

$$s^2 = \frac{\sum X_i^2 - (\sum X_i)^2/n}{n-1}$$

$$= \frac{6^2 + 9^2 + 8^2 + 8^2 + 12^2 + 9^2 - (6+9+8+8+12+9)^2/6}{6-1}$$

$$= \frac{36 + 81 + 64 + 64 + 144 + 81 - (52)^2/6}{5}$$

$$= \frac{470 - 450.67}{5} = 3.87.$$

2. You will sometimes see Equation (4.12) rewritten as

$$s^2 = \frac{\sum f_i \cdot X_i^2 - n(\bar{X})^2}{n-1}. \qquad (4.13)$$

The only difference is in the right-hand term. If you happen to already know the mean this may be easier to use than Equation (4.12).

For actually carrying out the calculation of a variance you should put the data in a table form. The following formats are particularly suitable for the formulas that I have used. In addition, you should be able to use these same tables to compute the mean.

By Equation (4.11).

Column	1	2	3	4	5	6
Quantity	X	f	$f \cdot X$	$(X - \bar{X})$	$(X - \bar{X})^2$	$f \cdot (X - \bar{X})^2$
Individual values	. . .	. . .	. . .	. . .	. . .	. . .
		Sum	Sum			Sum

$$\text{Mean} = \frac{\sum f \cdot X}{\sum f} = \frac{\text{Sum of Column 3}}{\text{Sum of Column 2}},$$

$$\text{Variance} = \frac{\sum f \cdot (X - \bar{X})^2}{n-1} = \frac{\text{Sum of Column 6}}{n-1},$$

Where the word *Sum* is stated this means add up the column.
Note that here you must compute the mean *before* you can fill in columns 4, 5, and 6.

By Equation (4.12).

Column	1	2	3	4*	5
Quantity	X	f	$f \cdot X$	X^2	$f \cdot X^2$
Individual values	. . . . . .	. . . . . .	. . . . . .	. . . . . .	. . . . . .
		Sum	Sum		Sum

then

$$\text{Mean} = \frac{\text{Sum of Column 3}}{\text{Sum of Column 2}},$$

$$\text{Variance} = \frac{\sum f \cdot X^2 - (\sum f \cdot X)^2/n}{n-1}$$

$$= \frac{\text{Sum of Column 5} - (\text{Sum of Column 3})^2/n}{n-1},$$

EXAMPLE 2.1 (Cont.)

I have calculated and presented in Table 4.8 the variance using the methods of Equations (4.11) and (4.12). The values and frequencies are in part A of the table. The calculations are in parts B and C. Both methods give answers with differences of less than one in a thousand. Considering the precision of the data itself, these differences are rather minute. You should now see the disadvantages of the first two techniques. The first requires you to square and multiply awfully messy numbers. Squaring a number like 7.11 and then multiplying it even by a 3 is not my idea of spending an enjoyable afternoon. The second method requires you to square or multiply some very large numbers. Again, 156 times 13 is not the nicest kind of calculation to do over and over. In addition, you often must carry out the calculations to six or seven places in order to obtain an accurate answer. Coding we will now find will help relieve some of the difficulties that are inherent in these two methods (a computer would even help more and we will discuss the operation and use of a computer and calculator in the next chapter).

---

* This column may be omitted. I find it easier to square each X-value and then multiply by the f-value. After a while I have memorized most of the smaller squares and find it simpler to then multiply by the frequency rather than taking an odd $f \cdot X$-value and multiplying it again by X. It is entirely a personal preference and you can eliminate the square columns and still reach the last column, which after all is the most important consideration.

Table 4.8 Calculation of variance of the number of tickets issued (Example 2.1); two different methods are used

	A. Data		B. By the definition				C. For a calculator		
Column	1	2	3	4	5	6	7	8	9
Quantity	X	f	$f \cdot X$	$X - \bar{X}$	$(X-\bar{X})^2$	$f \cdot (X-\bar{X})^2$	$f \cdot X$	X^2	$f \cdot X^2$
	20	0	0	9.89	97.81	0	0	400	0
	19	1	19	8.89	79.03	79.03	19	361	361
	18	1	18	7.89	62.25	62.25	18	324	324
	17	2	34	6.89	47.47	94.94	34	289	578
	16	0	0	5.89	34.69	0	0	256	0
	15	5	75	4.89	23.91	119.55	75	225	1125
	14	7	98	3.89	15.13	105.91	98	196	1372
	13	8	104	2.89	8.35	66.80	104	169	1352
	12	13	156	1.89	3.57	46.41	156	144	1872
	11	14	154	0.89	0.79	11.06	154	121	1694
	10	16	160	−0.11	0.01	0.16	160	100	1600
Data	9	17	153	−1.11	1.23	20.91	153	81	1377
	8	10	80	−2.11	4.45	44.50	80	64	640
	7	4	28	−3.11	9.67	38.68	28	49	196
	6	4	24	−4.11	16.89	67.56	24	36	144
	5	1	5	−5.11	26.11	26.11	5	25	25
	4	2	8	−6.11	37.33	74.66	8	16	32
	3	3	6	−7.11	50.55	151.65	6	9	27
	2	3	6	−8.11	65.77	197.31	6	4	12
	1	1	1	−9.11	82.99	82.99	1	1	1
	0	0	0	−10.11	102.21	0	0	0	0
Sum		112	1132			1290.53	1132		12,732

$\sum f = n = 112$

$\bar{X} = \dfrac{1132}{112} = 10.11$

Equation (4.11)

$$s^2 = \dfrac{\sum f \cdot (X - \bar{X})^2}{n - 1}$$

$$= \dfrac{\text{Sum of Column 6}}{n - 1}$$

$$= \dfrac{1290.53}{111}$$

$$= 11.626$$

Equation (4.12)

$$s^2 = \dfrac{\sum f \cdot X^2 - (\sum f \cdot X)^2 / n}{n - 1}$$

$$= \dfrac{\text{Sum of Column 9} - (\text{sum 7})^2/n}{n - 1}$$

$$= \dfrac{12{,}732 - (1132)^2/112}{111}$$

$$= \dfrac{12{,}732 - 1{,}281{,}424/112}{111}$$

$$= \dfrac{12{,}732 - 11{,}441.28}{111}$$

$$= 1290.72/111 = 11.628$$

Standard deviation: $s = \sqrt{11.626} = 3.41$; $s = \sqrt{11.628} = 3.41$

4.7 Coding to Simplify Calculating Variances

RULE 3 *Adding or subtracting a constant from every value in a set of values does not change the variance or standard deviation of the values.*

To see why this is true look at Figure 4.8. The upper histogram has a certain spread that can be described mathematically by the variance or the standard deviation. If we are to subtract 10 from *each* of the values in the upper histogram we get the lower histogram (b). The *shape has not changed*. Since the shape is *exactly the same*, the spread is the same, and therefore the variances and standard deviations are the same.

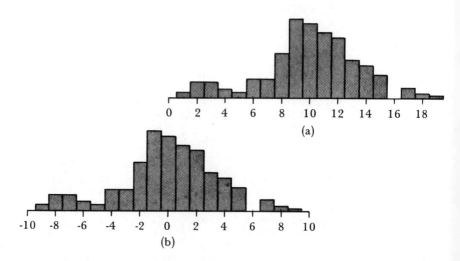

Figure 4.8 (a) Histogram of Example 2.1—observe the spread—the variance, a measure of that spread, is 11.6. (b) The constant 10 has been subtracted from each value in (a). The Spread has *not* been altered, therefore the variance is still 11.6.

Thus, if I code the data by subtracting a constant from every value, I do not have to modify my calculating technique to obtain the variance. I offer you the following formulas:

$$s^2 = \frac{\sum f_i(X_i - C)^2 - (\sum f \cdot (X_i - C))^2/n}{n - 1} \tag{4.14}$$

or

$$s^2 = \frac{\sum f_i(X_i - C)^2 - n(\bar{X} - C)^2}{n - 1}, \tag{4.15}$$

where C is the constant you have subtracted from each value.

Again a table helps in organizing the steps to perform the calculations. Below is a list of headings to help when using Equation (4.14):

Column	1	2	3	4	5	6
Quantity	X	f	$X - C'$	$f \cdot (X - C)$	$(X - C)^2$	$f \cdot (X - C)^2$
Individual values	...	...	...	...	...	...
				Sum	Sum	Sum

$$\bar{X} = \text{mean} = \frac{\sum f \cdot (X - C)}{\sum f} + C = \frac{\text{Sum of Column 4}}{\text{Sum of Column 2}} + C$$

$$\text{Variance} = \frac{\sum f \cdot (X - \dot{C})^2 - (\sum f \cdot (X - C))^2/n}{n - 1}$$

$$= \frac{\text{Sum of Column 5} - (\text{Sum of Column 4})^2/n}{n - 1}.$$

As in the last table of this type, Column 5, $(X - C)^2$, may be omitted.

EXAMPLE 2.1 (Cont.)

In Table 4.9 I have coded the data and provided the calculations of the variance using Equation (4.14). Compare the sizes of the numbers here with those in Table 4.8 to see how much easier the calculations are after coding the data.

When we were coding data to calculate the mean we found that sometimes even after subtracting a quantity from all of the values we still might have large numbers to work with. In some of these cases I said you could divide each value by some convenient amount. In order to get the true mean you then merely had to multiply the resulting mean by that same amount. With a slight change, a similar situation occurs with the variance.

RULE 4   *If you divide each and every number in a set of numbers by a single constant, you can then multiply the resulting variance by the square of that constant and the standard deviation by the constant to obtain the true variance and standard deviation.*

If

L is the *constant* we divide each value by
$(s')^2$ is the variance of the *CODED* data
s' is the standard deviation of the *CODED* data
s^2 is the true variance
s is the true standard deviation

Table 4.9 Calculating the variance and standard deviation of Example 2.1 by fully coding the data

Column	1	2	3	4	5	6
Quantity	X	f	X' $(X-C)$	$f \cdot X'$ $f \cdot (X-C)$	$(X')^2$ $(X-C)^2$	$f \cdot (X')^2$ $f \cdot (X-C)^2$
Data	20	0	10	0	0	0
	19	1	9	9	81	81
	18	1	8	8	64	64
	17	2	7	14	49	98
	16	0	6	0	36	0
	15	5	5	25	25	125
	14	7	4	28	16	112
	13	8	3	24	9	72
	12	13	2	26	4	52
	11	14	1	14	1	14
	10	16	0	0	0	0
	9	17	−1	−17	1	17
	8	10	−2	−20	4	40
	7	4	−3	−12	9	36
	6	4	−4	−16	16	64
	5	1	−5	−5	25	25
	4	2	−6	−12	36	72
	3	3	−7	−21	49	147
	2	3	−8	−24	64	192
	1	1	−9	−9	81	81
	0	0	−10	0	100	0
Sum		112		12		1292

$$C = 10$$

$$\bar{X} = \frac{\sum f \cdot (X-C)}{f} + C = \frac{\text{Sum of Column 4}}{\text{Sum of Column 2}} + C$$

$$= \frac{12}{112} + 10 = 10.11$$

By Equation (4.14):

$$s^2 = \frac{\sum f \cdot (X-C)^2 - (\sum f \cdot (X-C))^2/n}{n-1}$$

$$= \frac{\text{Sum of Column 6} - (\text{Sum of Column 4})^2/n}{n-1}$$

$$= \frac{1292 - (12)^2/112}{111} = \frac{1292 - 1.285}{111} = \frac{1290.715}{111} = 11.628$$

Standard deviation $= \sqrt{11.628} = 3.41$

then
$$s^2 = L^2(s')^2 \qquad (4.16)$$
and
$$s = Ls'. \qquad (4.17)$$

These formulas can be developed mathematically. You can get some idea why they work by looking at Figure 4.9, where a change in the shape of a histogram is caused by dividing each value by 10. The reason for multiplying the variance by L^2 is because the variance basically is an average of squares. By dividing before squaring you have avoided squaring the divisor, but it must be done later (consider the following: 50 squared is 2500, 50 divided by 10 yields 5. Square it and you get 25. Then 25 times 10 is only 250 but 25 times 10^2 is 2500).

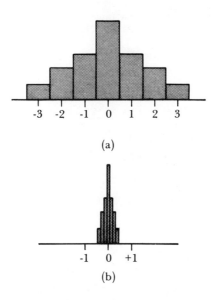

Figure 4.9 (a) A histogram (hypothetical data) (b) The values in the histogram of (a) have been divided by 10. Observe the effect on the spread of the distribution.

EXAMPLE 2.1 (Cont.)

Refer to the data in Table 4.5. It has the calculations for the mean of the grouped data for this example. Column 4 contains the fully coded values where L is 3 and C is 11. In Table 4.10 I have expanded Table 4.5 to include an $f \cdot (X')^2$ column and have completed the calculations of the variance and standard deviation as fully coded data.

Table 4.10 Example 2.1 (cont.). Calculation of the variance using grouped data. An example of fully coded data

Quantity		1	2	3	4	5	6
		X	f	$X - C$ partially coded	$(X - C)/L = X'$ fully coded	$f \cdot X'$	$f \cdot (X')^2$
Data		20	1	9	3	3	9
		17	3	6	2	6	12
		14	20	3	1	20	20
		11	43	0	0	0	0
		8	31	−3	−1	−31	31
		5	7	−6	−2	−14	28
		2	7	−9	−3	−21	63
Sum			112			−37	163

$$L = 3; \quad L^2 = 9$$

$$s'^2 = \frac{\sum f \cdot (X')^2 - (\sum f \cdot X')^2/n}{n - 1} = \frac{163 - (-37)^2/112}{111}$$

$$= \frac{163 - 12.2}{111} = \frac{150.8}{111} = 1.358.$$

Standard deviation of the coded data $= \sqrt{1.358} = 1.168 = s'$

Variance $= s^2 = L^2(s')^2 = 9(1.358) = 12.25$

Standard deviation $= s = Ls' = 3(1.168) = 3.50$

4.8 Interpretation of Standard Deviation and Variance

Now that you have learned how to go through the drudgery of calculating a variance and standard deviation, you probably are saying, "So what?" Well, initially we set out to find some kind of numerical or mathematical way of expressing how much a set of numbers was spread out. We have done just that. Your problem now is to find some way of figuring out if your number is large or small. I recognize that you have no built-in mechanism that tells you how to make sense out of the numerical size of a variance or a standard deviation. I will now try to give you a way of relating the standard deviation to a distribution. This discussion will deal principally with the standard deviation, since it is virtually impossible to give a simple illustration which will provide a meaningful idea of how the variance is related to the distribution or the data.

Interpretation of Standard Deviation

Over the years you have used many different units to measure different things. You have paid for a hamburger with so many units of cents (perhaps

45); you have filled the gas tank with units of gallons (maybe 12.6); your height has been measured in units of inches (say 69). If you took a number of measurements of one of them, you would end up with one of our distributions and you could compute a mean and a standard deviation. Remember now that the mean and the standard deviation would be in the *same* units as the measurements were. So the mean cost of 10 different hamburgers might be 50 *cents* and the standard deviation 4.3 *cents*. Similarly, you might have filled the gas tank 5 times with an average (mean) of 10.8 *gallons* and with a standard deviation of 3.5 *gallons*. The height of one person measured on 12 days might have a mean of 69 *inches* and a standard deviation of 0.29 *inches*.

From now on we will freely convert from the units of the distribution (such as the cents, gallons, and inches) to units of standard deviations. We will say that two numbers in a particular distribution are a certain number of standard deviations apart. If we calculate the difference between the two numbers we then merely divide this difference by the standard deviation of the distribution to convert the units. In general

$$\text{Number of standard deviations} = \frac{X_2 - X_1}{s} = k, \quad (4.18)$$

where X_2 and X_1 are the two different values, s is the standard deviation of the distribution, and k is the number of standard deviations. We can also express the difference in terms of s, or as

$$X_2 - X_1 = ks. \quad (4.19)$$

So we have two ways of talking about differences—the second being in terms of standard deviations.

Often you are pretty certain that you are talking about a population or you are told that the standard deviation you have is the population standard deviation. In these cases the "number of standard deviations" is expressed as $k\sigma$ (or k sigma). So you frequently will hear differences stated as "so many sigma." These merely are the number of standard deviations the two numbers are apart. To convert back to the actual units you merely multiply the quoted number (k) by the σ of the population (or s of the sample). I will generally talk about sigma and not s and we will use Equation (4.18) or (4.19) with σ in place of s.

Now let's look at a few examples (each refers to data just mentioned).

EXAMPLE 4.3

Two hamburgers that cost 48 cents and 54 cents have a cost difference of 6 cents. Since $\sigma = 4.3$ cents they are also

$$\frac{X_2 - X_1}{\sigma} = \frac{54 - 48}{4.3} = \frac{6}{4.3} = 1.4$$

standard deviations apart. Therefore, they are 1.4 σ (1.4 sigma) apart.

EXAMPLE 4.4

The two times I filled the car this week I got 15.5 gallons and 9.5 gallons. This difference was 6 gallons, or since $\sigma = 3.5$, it is also

$$\frac{15.5 - 9.5}{3.5} = \frac{6}{3.5} = 1.72$$

standard deviations apart. Therefore they are 1.72σ (or 1.72 sigma) apart.

EXAMPLE 4.5

Twice I was measured and the readings were 69 and 69.5 inches. The difference is 0.5 inches or since $\sigma = 0.29$ it is also

$$\frac{69.5 - 69}{0.29} = \frac{0.5}{0.29} = 1.72$$

standard deviations apart or 1.72σ (1.72 sigma) apart.

I would like to point out two important observations: (1) Examples 4.3 and 4.4 both had differences of the same actual magnitude, 6 units, but because the standard deviations were different the numbers of sigma were also different. Thus, with different distributions, the same amount of spread may not be the same in terms of standard deviations; (2) Examples 4.4 and 4.5 both have differences that are of different magnitudes (6 and 0.5 respectively) but the differences are both 1.72 sigma. Thus, in terms of sigma units *they are the same* and we will find that there is more similarity between these two values than the previous pair we looked at. We will see that expressing certain differences in terms of units of sigma will permit you to make comparisons between different distributions, assist in making a sensible judgement on how the distribution is spread out, and help you determine what would be an excessive or an unreasonable individual value.

For the most part we will talk only about one type of difference—those where one of the X-values is the *mean*. For the previous examples, if we wanted to convert our first measurements into sigma units from the mean, it would have been done as follows:

EXAMPLE 4.3 (Cont.)

If
$$\bar{X} = 50, \quad X_1 = 45, \quad \sigma = 4.3,$$

then the number X_1 has a difference from the mean which can be expressed as

$$\frac{X_1 - \bar{X}}{\sigma} = \text{number of standard deviations}$$

$$= \frac{45 - 50}{4.3} = \frac{-5}{4.3} = -1.16 \text{ sigma}.$$

(The minus sign says that the X-value is less than the mean, and 45 *is* less than 50.)

EXAMPLE 4.4 (Cont.)

If
$$\bar{X} = 10.8, \quad X_1 = 12.6, \quad \sigma = 3.5,$$
then
$$\text{Number of standard deviations} = \frac{X_1 - \bar{X}}{\sigma} = \frac{12.6 - 10.8}{3.5} = \frac{1.8}{3.5} = 0.51.$$

Thus, 12.6 is just about $\frac{1}{2}$ a standard deviation greater than the mean.

EXAMPLE 4.5 (Cont.)

If
$$\bar{X} = 69, \quad X_1 = 69, \quad \sigma = 0.29$$
then
$$\frac{X_1 - \bar{X}}{\sigma} = \frac{69 - 69}{0.29} = 0.$$

The value we got was the mean itself, so it is 0 sigma from the mean.

Okay, you can express values in terms of sigma or in terms of "so many sigma units from the mean." Now I offer what is often called the *empirical rule*.

EMPIRICAL RULE   If the *distribution can be regarded as a normal distribution* (or closely resembles the bell-shaped distribution) *then approximately*
68% of the individuals are between minus 1 sigma and plus 1 sigma from the mean;
95.5% of the individuals are between minus 2 sigma and plus 2 sigma from the mean;
99.73% of the individuals are between minus 3 sigma and plus 3 sigma from the mean.

Let me use Example 2.1 to explain what I have just stated.

EXAMPLE 2.1 (Cont.)

Recall that the mean was 10.11 and the standard deviation was 3.41. Then

$$\begin{aligned}
\text{Mean plus 1 sigma} &= 10.11 + 1(3.41) = 13.52 \\
\text{Mean plus 2 sigma} &= 10.11 + 2(3.41) = 16.93 \\
\text{Mean plus 3 sigma} &= 10.11 + 3(3.41) = 20.34 \\
\text{Mean minus 1 sigma} &= 10.11 - 1(3.41) = 6.70 \\
\text{Mean minus 2 sigma} &= 10.11 - 2(3.41) = 3.29 \\
\text{Mean minus 3 sigma} &= 10.11 - 3(3.41) = 0.12.
\end{aligned}$$

In Figure 4.10 I have drawn the histogram and I have labeled these points as $+1\sigma$, $+2\sigma$, $+3\sigma$, -1σ, -2σ, and -3σ respectively. These then can be used interchangeably with the actual values of 13.52, 16.93, and so forth.

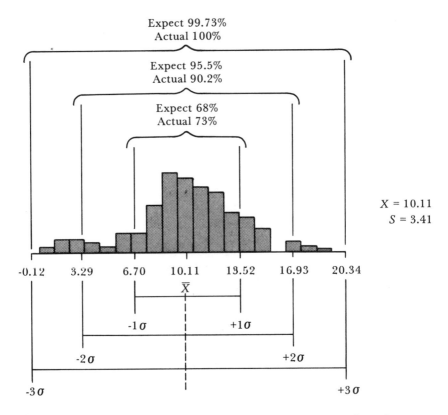

Figure 4.10 The plus and minus 1, 2, and 3 sigma limits are shown for Example 2.1.

The space between -1σ and $+1\sigma$ is the distance between 6.70 and 13.52 and it includes 82 individual values. Then

$$82/112 \times 100 = 73.2\%$$

of the distribution lies between 6.70 and 13.52 or from -1σ to $+1\sigma$. This compares well with the empirical approximation of 68%.

Between -2σ and $+2\sigma$ (from 3.29 to 16.93) are 101 values, or

$$101/112 \times 100 = 90.2\%$$

are between $\pm 2\sigma$.* We expect approximately 95.5% to be between $\pm 2\sigma$.

Between $\pm 3\sigma$ (from -0.12 to 20.34) are all 112 values so

$$112/112 \times 100 = 100\%$$

are included. We expect approximately 99.73% of the values here.

---

* I will now use $\pm$ to mean plus *and* minus. It will mean that you should consider *both* the plus 2σ and the minus 2σ value. It can be used with any quantity, so that ± 5.2 means $+5.2$ and -5.2.

As you can see we are quite close in our approximation—the most we are off is about 5%, which is quite acceptable. Here has been the most valuable use of the standard deviation. *With it we can essentially reconstruct the distribution.* Knowing that the mean is 10.11 and the standard deviation is 3.41, I could draw a distribution that practically perfectly fits the distribution we actually had. I can also tell you that rarely should the number of tickets have fallen below 4 or above 16, just about never should the number be less than 0 (which would be absurd anyhow) or more than 20, and the bulk of the time, about two-thirds of the time, there would be between 7 and 13 tickets. And I would be pretty close to what actually happened—without all the numbers or the monstrous tables. Hence, we have accomplished a monumental task, we have essentially reduced the entire distribution to *two* numbers (the mean and the standard deviation).

Interpretation of Variance

You might now be asking, "Okay, I understand standard deviation, but what's this variance business?" I offer three reasons for studying the variance.

1. We calculate it because we must calculate it before we obtain the standard deviation.
2. There will be several statistical tests that will use it directly. (I won't bother you with them yet.)
3. There is a reason I will try to give you a little insight into now (which you may encounter in more advanced work)—*variances can be added, but standard deviations cannot.*

To see how this third concept is useful, recall Example 2.2, which listed the length of time it took to get to school. You should have found the variance of a one-way trip to be 0.770 minutes and the standard deviation to be 0.877 minutes. But what about a round trip? This consists of 2 trips. I declare that the sum of the variances (or $0.77 + 0.77 = 1.54$) is equal to the variance of a round trip (the sum of the trips). Thus, the standard deviation of the round trip would be $\sqrt{1.54}$ or 1.24. If you just added the standard deviations for both trips you would have gotten $0.877 + 0.877$ or 1.754, a much larger and less valid result.★

This concept arises in many measurement problems where you may have to measure an object in several steps, such as the measuring of a room with a 12″ ruler or measuring a wall with several windows (up to a window, across the window, etc.). In each case your overall dimension is subject to the variation of each measurement you make and you are then involved in multiple errors.

★ This is reminiscent of an old friend from geometry, the Pythagorean theorem. It stated that the sums of squares of the lengths of two sides added up to the square of the length of the third side, but the lengths themselves did *not* add up. Think about it.

REVIEW

Keywords and key formulas

Central value or Average · Midrange · Mode · Median · Mean · Summation notation · Coding · Spread · Range · Standard deviation = $\sqrt{\text{Variance}}$ · s versus σ · Sigmas · Empirical rule

$$\bar{X} = \frac{\sum fX}{\sum f} = \frac{\sum fX}{n} \qquad \bar{X} = L\bar{X}' + C$$

$$\text{Variance} = s^2 = \frac{\sum f(X - \bar{X})^2}{n - 1} = \frac{\sum fX^2 - (\sum fX)^2/n}{n - 1}$$

$$= \frac{\sum f(X - C)^2 - (\sum f(X - C))^2/n}{n - 1}$$

$$s^2 = L^2(s')^2 \qquad s = Ls'$$

$$k = \frac{X_2 - X_1}{s}.$$

Table for calculations of mean and standard deviation

	X	$X - C$	f	$f(X - C)$	$f(X - C)^2$
Data	. . .	. . .	. . .	. . .	. . .
	. . .	. . .	. . .	. . .	. . .
	. . .	. . .	. . .	. . .	. . .
Sum			x	x	x

PROJECTS

For the projects in Chapter 1 compute the mean, median, midrange, mode, range, and standard deviation of the data. Do you have any additional comments to make?

PROBLEMS

Section 4.2

4.1 The price of one style of new homes in a development sold during the first 3 months of its opening were as follows:

29,990; 28,850; 27,900; 30,280; 28,300; 30,220; 29,340; 30,790; 28,500; 29,200; 30,000; 30,400; 28,650; 30,270.

Find the mean, median, mode, and midrange prices.

4.2 Twelve women attended "Weightwatchers" and agreed to eat the identical sets of food. After four weeks they reported the following weight loss in ounces (a negative value means that they increased weight):

$$6, 12, -3, 5, 8, 7, 4, 1, 1, 3, 7, 9.$$

Compute the mean, median, mode, and midrange weight loss for the group of women. How do the averages compare to each other?

4.3 Refer to the data in Problem 3.3. To calculate the different "average" number of meals that the people interviewed have eaten, what must be done to the data that was presented? Calculate the four "averages." Which average is easiest to calculate?

4.4 For the data in Problem 3.6, what are the "average" numbers of customers? Which is easiest to calculate? Which seems most realistic?

4.5 For Example 2.5 (see Table 2.5) what are the "average" numbers of absences? Why are there such differences?

4.6 a. For the data in Example 2.4 (Table 2.4) calculate the various "average" dimensions (diameter, height, and volume) of the cups.
   b. For each kind of average (mean, median, midrange, mode) perform the following calculations:

$$\text{Volume} = \pi \frac{(\text{average diameter})^2}{4} (\text{average height}).$$

Do these calculated values of volume equal the "average" volumes from the Table?

★c. Try to demonstrate mathematically why you got the results in part (b). (*Hint:* Try the problem with just a couple of products instead of the twenty in this problem.)

4.7 Refer to Problem 3.5. Can you calculate any types of average? Comment.

4.8 For the data in Example 2.3 (Table 2.3) compute the median, midrange, and mode. Consider calculating the mean. Comment on the relative difficulty you had in arriving at values for the different "averages." How difficult do you feel the mean would be to calculate?

Section 4.3

4.9 Calculate the mean number of absences in Example 2.5 by grouping the data. You may refer back to your histogram of Problem 3.1. Is there any difference between this calculated mean and the one you calculated in Problem 4.5? Explain.

4.10 Group the set of sale prices for the various homes listed in Problem 4.1. What intervals should you use? Calculate the "averages" as before. Are there any differences between the averages obtained from the grouped values and the ones calculated in Problem 4.1?

4.11 In Problem 3.7 twelve categories were used as groups for Example 2.3. Using those categories calculate the "averages." Compare these averages with the ones calculated in Problem 4.8. How much error have you made by grouping? How much difference is there in the effort to calculate the "averages"?

4.12 In Problem 3.10 you drew a grouped histogram for Example 2.4. Now redo Problem 4.6 using the grouped data.

4.13 a. Look at Table 4.3 and find the median. How close is this to the actual median?

★b. When we encounter grouped data there is one modification to the method of calculating the median that often gives us a value closer to the actual median than we have just used. This method involves determining just how far across the category interval our median should be. In this problem the 46th value is associated with 9.5 and the 98th value with 12.5. We want to know what would go with the 56th value. It would be found by substituting into the formula

$$\text{Median} = X_L + \frac{d}{f_m} L. \qquad (4.20)$$

where X_L is the lower value for the group containing the median (the 9.5 here)
L is the width of the interval (2.5 in this case)
f_m is the frequency associated with the group that the median belongs to (43)
d is the difference or number of values away from the bottom value that the median is ($56 - 46 = 10 = d$)

Now recalculate the median for Example 2.1 from Table 4.3. How does it compare to the value obtained in part (a)? How does it compare to the actual value? Discuss just what is happening and why we should often get a better value.

4.14 Refer back to the data in Problem 4.1. Code the data by subtracting 29,000 from each value. Then calculate each average. Is there any advantage to also divide by some other constant?

4.15 For Example 2.3, refer to your histogram in Problem 3.7 (or the data in Problem 4.11). Fully code the grouped data (by subtraction and division) and recalculate the mean.

4.16 Sales of gasoline were recorded for twenty customers. They were (in dollars):

2.75; 1.50; 2.00; 3.25; 2.50; 2.75; 3.50; 4.50; 6.25; 1.00; 2.50; 3.50; 3.25; 5.50; 4.00; 1.75; 2.25; 3.25; 4.00; 1.00.

a. If you were to fully code, what would you choose for L and C?
b. Calculate the coded X'-values.
c. Find the mean, median, mode, and midrange.

4.17 For Example 2.4 (Table 2.4) code the values of diameter and height. Then calculate the means. How do the computations compare to the previous attempts (Problems 4.6 and 4.12)?

4.18 Quite often we must calculate an average of several sets of values when they carry different weights or degrees of importance. If you want to average together the means of several different-sized samples, you would not regard them as equally important. For example, twenty flashlight batteries had an average life of 150 hours. Another sample of 15 batteries had an average life of 200 hours and a third sample of 5 batteries had an average life of 250 hours. What is the average life of these batteries? We must calculate the average $\bar{X}$ as

$$\bar{X} = \frac{\sum wX}{\sum w} \qquad (4.20)$$

where w represents the weight (these are essentially frequencies). Here the weights are 20, 15, and 5 respectively, and the X are 150, 200, and 250. Thus,

$$\bar{X} = \frac{20 \cdot 150 + 15 \cdot 200 + 5 \cdot 250}{20 + 15 + 5} = \frac{7250}{40} = 181.2 \text{ hours.}$$

If an additional sample of 20 batteries averaged 210 hours, what would be the new overall average life of the batteries?

4.19 A company employs machinists earning $3.00 per hour, assemblers earning $2.00 per hour, and packers at $2.50 per hour. If there are 25 machinists, 50 assemblers, and 25 packers, what is the average earning of workers in this company?

4.20 A teacher considers his four tests as equally important and the final exam equivalent to three tests. A student obtains the following grades on the tests, 61, 73, 67, 71, and a 76 on the final exam. What average will the instructor calculate as his average?

Section 4.4

4.21 Comment on this statement: "Something is wrong with the way they are teaching our children. Half of our children scored below the National Average! Let's get better teachers."

4.22 The following donations (in dollars) were made to a local charity:

4, 10, 6, 2, 1, 3, 1, 150, 3, 5, 6, 2, 8, 2, 12.

Calculate the various averages. Which is the most representative or the "best" average? Explain.

Section 4.5

For Problems 4.20 and 4.25 calculate the range. Calculate the variance and standard deviation by using the definitions of Equations (4.10), (4.11), (4.12), or (4.13).

4.23 Refer to the data in Problem 4.2.
4.24 Refer to Example 2.5 (use the grouped data in Problem 4.9).
4.25 Refer to the data in Problem 4.3.
4.26 Refer to the data in Problem 4.4.

126 Descriptive Methods

4.27 Refer to the data in Example 2.4 (try diameters and heights only).

4.28 Refer to the data in Problem 4.1.

4.29 Another measure of variability that uses all of the data is called the *average deviation*. It is found by computing the mean of the data $(\bar{X})$. You then find the absolute value of the difference between each quantity and the $\bar{X}$. If you compute the mean of these absolute differences you now have the average or mean deviation. If you refer to the data on page 107, the mean was 8.67. The deviations or differences would be: $6 - 8.67$, $9 - 8.67$, $8 - 8.67$, $8 - 8.67$, $12 - 8.67$, and $9 - 8.67$. Therefore they are $-2.67, 0.33, -0.67, -0.67, 3.33,$ and 0.33. The absolute values of these values are $2.67, 0.33, 0.67, 0.67, 3.33,$ and 0.33. The mean of these last quantities is $8.00/6 = 1.33$. Thus the average deviation is 1.33.

The formula for the average deviation is

$$\text{Average deviation} = \text{Mean deviation} = \frac{\sum |X - \bar{X}|}{n}. \quad (4.21)$$

Now do the same for Problem 4.2.

4.30 Find the average deviation for the data in Example 2.5. How does this compare to the standard deviation? Should it? Comment on the relative difficulty in computation.

Section 4.6

4.31 Redo Problem 4.23, this time selecting an integer to subtract from each value and using Equation (4.14) or (4.15).

4.32 Redo Problem 4.24 first selecting an integer to subtract from each value.

4.33 Redo Problem 4.25 first selecting an integer to subtract from each value.

4.34 Redo Problem 4.26 first selecting an integer to subtract from each value.

4.35 Calculate the variance and standard deviation of the heights, diameters, and volumes of the cups in Example 2.4 (refer to Problem 4.17). First choose the appropriate formula.

4.36 Redo Problem 4.27. Fully code the data (see Problem 4.14). Now calculate the variance and compare it to the value in Problem 4.27.

4.37 Refer to the data in Problem 4.16. Fully code the data and recalculate the standard deviation.

4.38 Using the grouped data in Problem 4.15 calculate the variance and standard deviation of the bread weights of Example 2.3.

4.39 What value of C will make Equation (4.14) the same as

a. Equation (4.11).
b. Equation (4.12).

Discuss how this affects your interpretation of Equation (4.14).

Section 4.7

4.40 The weights of a group of students have a mean of 147 and a variance of 25. How many standard deviations from the means is 164? 135? 147?

Are they unusual? What weight is 2 standard deviations from the mean? -3.1 standard deviations? 0.3σ?

4.41 Refer to the data of Example 2.2.
   a. What are the 2 and 3 sigma limits? (See Problem 4.27.)
   b. Are you willing to accept the necessary assumption about a "bell-shaped" distribution to apply the empirical rule to this data? Explain.
   c. Compare the percentages within 1, 2, and 3 sigma limits to those predicted by the empirical rule. Comment.
   d. Would you be surprised if it took you twelve minutes to get to school? How about 17?

4.42 Between what pair of weights do you expect to find about 95% of all the breads listed in Example 2.3? First look at the histogram (Problem 3.7) to decide if you should apply the empirical rule.

4.43 Calculate the standard deviation of the donations listed in Problem 4.22. The 3σ limits should include about 99.75% of the values. Do they? What change would occur if that $150 donation were not included? Discuss.

4.44 If the data does not fall into a distribution that resembles a normal or bell-shaped form, then the empirical rule is no longer valid. There is a rule called Tchebychev's inequality (also spelled Chebyshev), which states that *no matter what the shape of the distribution at least* $(1 - 1/k^2)$ *100%* *of the data must fall within* $\pm k$ *standard deviations of the mean.* Thus, within ± 2 standard deviations, or $\pm 2\sigma$, there *must be* at least

$$(1 - 1/2^2)100 = (1 - 1/4)100 = (0.75)(100) = 75\%.$$

How do the percents within the 1 and 3 sigma limits compare to the empirical rule? Can you create a distribution that contains the minimum possible number of points inside 1 sigma limits?

★4.45 If we have found the range (R) and the standard deviation (σ) of a set of data, an interesting relationship that is sometimes useful to compute is $R/2\sigma$. It is particularly interesting since it is bounded—it has a largest possible value and a smallest possible value for any given sample size n. Try to compute the maximum and minimum possible $R/2\sigma$ for various values of n. (*Hint:* For minimum $R/2\sigma$ you need to consider n being even and n being odd in separate ways. In all cases try to visualize first just what way the data must be organized to give you various kinds of ratios.)

5

Computers and Computation

Applied statistics cannot really be dealt with without also considering the problems of doing the computations involved in every analysis. To omit some discussion of the way we *usually* arrive at our computed values is like teaching one to drive and not providing an automobile.

Today's world is in a technological whirlwind with new inventions, developments, and techniques being produced at an ever rapid pace. One of the most dynamically changing fields is that involving calculating equipment and computers. This chapter will be somewhat outdated even at the time of printing, but the lack will be because of omission of the most recent innovations rather than the inclusion of invalid material.

5.1 Computational Equipment

Undoubtedly, one of the primary reasons for the universal acceptance of statistical methods is the increasing availability of computational equipment. Much of the drudgery usually associated with the enormous calculations of descriptive statistics and statistical tests are able to be eliminated, or at least reduced to minimal proportions. Not only is the equipment in existence, but it is rapidly becoming inexpensive enough so that most colleges and most industry have some form of advanced machines (called hardware) available for general use by its students or employees.

Until the last few years, the field of computational equipment was simply divided into two easily defined groups. There existed a variety of desk calculators that could perform an array of operations. Some units were capable of doing automatic multiplication and division and could even store or hold some intermediate sums. Thus, one could accumulate the sum of a collection of products just by multiplying the pairs of values in a particular way. Most of these machines were mechanical and slow and very noisy (it was not uncommon for them to "walk" around on the table while grinding away a long division problem—but mechanical calculators have improved over the years).

The second area was at the other extreme, the full scale computer. It was large, expensive, and required an additional knowledge—that of computer programing. The computer is electronic and it is fast. It is capable of storing many intermediate values as well as the data, and of performing many different kinds of calculations quickly and accurately. But its biggest asset is also one of its biggest drawbacks. It is able to be given a set of instructions that it stores, and then refers to as it carries out the solution of a problem. This set of instructions is called a program. Generally, once a program is written it may be used over and over for many different sets of data, as long as the problem is of the same type. The difficulty is in writing the program, and it can be an enormous undertaking.

To write the program for a sizable or a reasonable problem may involve a hundred times more effort than the actual hand computation of the problem. So a trade-off must be made. One way to alleviate some of the programing difficulties of communicating with the computer is through intermediate languages. Certain languages that the computer can easily translate and use have been invented. Thus you will hear of FORTRAN, BASIC, ALGOL, COBAL, PL/1, and so forth. These are designed so that you can converse in something that is more like the application that you are involved in, rather than the words the computer uses. The language of the computer is called a machine language (and is a terrible thing to have to work with). You just have numbers or small sets of letters that are used to tell the computer the minute steps it must go through (a typical instruction might be 15 12500 13000). To tell it to add together two numbers might take four or five steps. To tell it to compute $Y = AX^2 + BX + C$ could take twenty small steps and require you to keep track of a dozen locations in the computer. A reasonably sophisticated problem would require hundreds or thousands of these minute steps. However, in Fortran the equation $Y = AX^2 + BX + C$ would simply be written as $Y = A*X**2 + B*X + C$ and the computer can understand this. The similarity between Fortran and algebra is unmistakable and intentional. The ease with which one can identify and work mathematical problems with languages such as Fortran is simply due to the close connection between it and the mathematical language. However, in order to use any of the computer languages we require a moderately large computer.

In the past few years two major developments of extreme importance have occurred in the computer world—the use of time-sharing and electronic calculators (some even programable). I will talk about each of them and then we will expand the discussion to a fuller, but by no means exhaustive, description of computational equipment.

Time-sharing is a technique whereby many people at many locations can share the same large computer. Each person works with a teletype unit that resembles a typewriter and couples together with the telephone. You just dial in to the computer and start computing. Problems may be quickly solved, modified, corrected, and rerun in short order. Perhaps even more important is that the computer has stored internally for your immediate call a large library of previously written programs. For many problems you don't have to write the difficult program, you just need the instructions on how to use the ones that are already there. There are drawbacks though. Not all problems may have programs available in every time-sharing system. Some of these programs may be on only a very few, but many of the basic ones are widely available. The other possible disadvantage is that for some problems time-sharing is slow, and time is equal to money here. For most basic problems, however, this will not be the case and we will find time-sharing quite satisfactory for our problems.

The second development mentioned, that of relatively inexpensive programable calculators, decreased much of the extra handling involved in many tests and analysis. By extra handling I refer to the need to enter each value once, in order to calculate perhaps the sum of the X and then a second time for the sum of the X^2. After these sums are calculated, other computations often must be performed, possibly requiring the re-entry of these sums as well as other numbers. A programable calculator alleviates the entry and re-entry problem by providing a number of storage locations where intermediate results may be retained as they are computed. In addition, the steps involved in the final solution may also be stored in the calculator.

5.2 Calculators

Calculators today are primarily electronic (and I will therefore ignore discussing mechanical calculators even though the only calculator you may have might be mechanical). This means that they have no moving parts (unless they also print). They are therefore quiet (so silent that they sometimes scare you) and they are generally quite fast. There is a whole spectrum of calculating equipment and with luck you will have something in this group available to you to do your multiplications and additions. I highly recommend the use of any form of desk calculator. I would strongly suggest, though, that you master the basic operations involved in calculating by hand first, so that you learn what the actual operations are, and then in subsequent problems freely use whatever computational means are provided for your use.

The following descriptions are general in nature. Specific makes and models will differ in the exact manner of operation and you must read the instruction booklets that accompany whatever machines you might be able to use. In addition, there will be omissions of certain types of units. Again, the field is very large and the main concern here is not to make you an expert on calculators.

The most basic unit involves a set of keys with individual digits from 0 to 9, perhaps a . which represents a decimal point, and separate + and − keys so you can do addition and subtraction (see Figure 5.1 for an example). You merely enter the digits of the number to be added by pressing the keys in the same order as the digits are in (entering from left to right). The number 12.65

would be entered by hitting first the 1, then the 2, then the *decimal point*, the 6 and the 5. Striking the + key stores that number someplace in the machine so that you may add to it additional numbers (a button called a clear button is usually also available to erase all the previous values stored in the machine and you generally use it before anything else). You may add to the previous value by now entering another number, say 4.8. This would be done by striking the 4, then the decimal point and then the 8. Hitting the + key will enter the number and *add* it to the previous one (the 12.65). Now if you strike the = key you will obtain the sum.

Figure 5.1 A ten key adding machine. The numbers to be added and the total are printed on paper tape.

I should mention *how* the sums will appear. Two different techniques are usually used for displaying the numbers. In one case the numbers are printed on a roll of paper (this is mostly with mechanical units), and you will see

$$12.65 +$$
$$4.80 +$$
$$17.45 =$$

printed out on the paper (the sum being next to the = sign). The nice part here is that you have a *permanent record* of the *data and the answer*. If something doesn't seem to be correct you can check back to see if the numbers were all correctly entered (a very, very common problem).

The second method is truly just display and involves a collection of bulbs that light up showing the digits, and a small light that shows the decimal point. One complete number (say 12.65) would appear. A few types of machines have a large screen that resembles a TV screen and several different numbers (say all 3 of the above) might be displayed. In either case *you* must record the sums after they appear on the screen.

Next comes multiplication. Provision for multiplication costs more. This usually involves another key with an × on it, and you merely enter the first number using the + key, and then the second number, this time followed by the × key. If the = key is then hit the answer is immediately displayed. Again, this may vary from model to model.

Sometimes to do squaring you do not even have to re-enter the number a second time. Many machines have a sequence like:

1. Enter the number,
2. Press the "+" key,
3. Press the "×" key,
4. Press the "=" key,

and the answer will be the square of the original value (this may sound complicated, but it really is very rapid).

Many machines that do multiplication also have a routine or procedure to do division (again it may be a little more complex machine).

The next major step in sophistication and of particular value in statistical work is the introduction of storage, memory, or accumulators. The simplest type of unit is one which permits you to multiply a pair of numbers together and then store the product. Let us say you want to calculate the mean of the grouped data in Table 5.1. You first calculate $12 \cdot 2 = 24$ and store it. Now you do *not*

Table 5.1 Some grouped data

X	f
12	2
9	4
8	4
6	2

have to write down the 24, but can store it in a location in the machine. Next the $9 \cdot 4 = 36$ is calculated, and if there is a storage, accumulation, or summation part of the machine then this 36 will enter that location *along with the 24 that is already there* and they will be added together to obtain an intermediate result of 60. The 60 probably will not appear, only the 36. Then you calculate $8 \cdot 4 = 32$ and that too is added to the 60 to give an accumulation of 92. Calculating $6 \cdot 2 = 12$ brings the accumulator to a sum of 108. By pressing an appropriate key the value of 108 will appear. You did *not* have to add all the products separately. You can now go on to divide the 108 by 12, the sum of the frequencies, to obtain a mean of 9. Similarly, we can quickly add squares of numbers (and sometimes by only entering the number once).

One storage or memory area allows you only to accumulate one kind of number, but we might also like to automatically add frequencies as well as frequencies times X. So another location would be desirable to serve as a place to store the sum of the frequencies, and these should be added each time we enter the frequency portion of the products. Finally, we might desire a third quantity, frequency times X^2, to be added together as we go along. In this way we would quickly find the three sums we need to compute the mean and the variance (see page 107 under Equation 4.12), and by entering each X-value and each frequency only one time. And we do *not* have to write down any products or intermediate values. Of course, the price now begins to mount. Figure 5.2 shows a small electronic unit with one memory, which costs around

Figure 5.2 A small electronic calculator. This unit can be held in the palm of your hand. It can perform multiplication and division and has a storage or memory location. The results of the calculations are displayed with lights.

five hundred dollars. (Prices are for the 1971 market. Prices are merely to provide you with some rough idea of cost. There may be considerable variation depending on model, manufacturer, and the prevailing economic situation.) Most basic calculators stop at two memories (although there are some with 3 or more) and then the big jump takes place both in sophistication and in cost.

The next major development involves the ability of the calculator to store the steps that you take in calculating a problem. Thus, if you were to do a complicated calculation (like finding the standard deviation) you run through all of the steps only once. The next time you merely need to enter the new sets of data. In some units you may insert a card (the unit in Figure 5.3 is such a type) or a magnetic cartridge, or a tape. These contain the steps "written" on them in some form of a code that the calculator recognizes. Again all you need to do is enter the data, except this time you do not even have to enter the instructions.

Figure 5.3 About the most sophisticated type of desk calculator—a programmable calculator. A card can hold predetermined steps for the calculator to carry out. Trig functions, logs, square roots, and numerous other mathematical steps can be carried out.

Of course, the unit must have many storage places to hold the instructions and it usually contains a considerable number of working storage or memory areas. The price rises again, this time to anywhere from 2 to 5 thousand dollars. In some cases (usually the more expensive ones) there are added bonuses like trigonometric functions, logarithms, decision keys (which can allow you to truly do some elementary programing), and other capabilities that are valuable for many different types of mathematical computations. These units are usually called *programable calculators* and are essentially the missing link between the desk calculator and the computer.

5.3 Computers and Library Programs

Some of you have access to computers or time-sharing computer terminals. I will not talk about programing a computer, nor will I even allude to the idea that you should learn how to program at this time. But computers are a way of life today, especially in the real world, and people who have computers available to them normally take advantage of them. One aspect of computer solution to statistics problems even you will be able to handle involves the use of *library programs*.

Library programs are prewritten computer programs that are stored inside the computer. In order to use the program you merely ask the computer to ready the program you want, maybe give it a few simple instructions, type in the data, and tell the computer to RUN. The computer will do all the necessary computations and then type out the results. The particular programs that might be available to you would depend on the time-sharing service you might be able to use (such as General Electric, LEASCO, CSI, and so forth) or the programs that are stored in your computer. In either case you must refer to an instruction booklet or some prepared instructions for the details on how to use the particular program that you will work with.

I will now give a couple of examples of the use of several library programs. I will use the General Electric time-sharing service for my examples. I claim no advertising intent nor any endorsement of this service but merely use it for illustration purposes. I will not explain all of the output since you often get a lot more than you need. I offer some pages reprinted from the instruction booklet and then a copy of the printout for an example. The printout includes the part *I typed in*, the part the *computer typed out*, and some *comments*. The parts I circled are all that I had to type in. All the rest of the typed portions and all the computations are done by the computer.

One last note before we start. The computer may use any of about four major languages (ALGOL, FORTRAN, BASIC, PL/1). You need to know nothing about these languages except which of the languages the program you want to use is written in. This is sort of like having a friend who knows French, another who knows German, and a third who knows Russian. If you know none of these languages but you want a passage translated from German you must call for your friend who knows German. You then ask him to translate the passage and give you the results in English. You do not care about how he goes about translating, nor how the language is structured, you are just interested in the results of the translation—something you can read. Similarly you need not concern yourself at this point about how the computer translates your instructions, or goes about doing the calculations. Only if you do large amounts of perhaps critical analyses will you be so concerned. Such would be the case when you are doing literary interpretation of the German passage, but are not versed in the German language. The nuances and special meanings will be hidden from you. If in the future you do considerable work with a computer, I recommend you learn about programing and the way a computer calculates. Only in that way will you understand the problems and possible errors involved in computer calculations.

5.4 Examples of Library Programs

Here I am starting from scratch and will use a program called DESTAT***, which is written in the language called ALGOL.

The following two pages contain the instructions and a part of the sample problems provided in the information booklet supplied by the time-sharing service. Observe that if you still need more instructions, you merely type DESTEX*** and you will be given additional assistance by the computer.

Category MATHEMATICAL
Name   DESTAT * * *

Language ALGOL
Character 4800

PURPOSE

To analyze a set of observations on one variable.

INSTRUCTIONS

To use this program enter input data using the following format:

```
200  X (1), X (2),....., X(N)  (THE VARIATES TO BE DESCRIBED)
         (DATA MAY BE CONTINUED ON LINES 201-398)
400  L, U      (A TYPICAL CLASS, L=LOWER LIMIT, U=UPPER)
500  P1, P2, .....  (ADDITIONAL PERCENTILES DESIRED)
RUN
```

The program output consists of summary statistics. At run time you may also select the following:

    Order Statistics
    Frequency Distribution
    Cumulative Distribution
    Data arranged in ascending order

More detailed instructions may be obtained by listing the program DESTEX * * *

SOURCE: Instructions reprinted by permission from MARK I Quick-Use Programs User's Guide, 1966, 1967, 1968, 1970 by the General Electric Company.

SAMPLE PROBLEM

Determine the statistical characteristics of the following data points.

261.4	252.1	255.5	258.3	253.2
270.8	268.3	249.6	256.3	266.4
265.4	250.3	280.9	259.3	
261.4	272.3	270.3	270.1	
258.1	262.8	263.2	259.3	

The commas after the last values in lines 200 and 201 were necessary since the following lines contained additional values.

This is the same data as used in STATAN * * * and MANDSD * * *

SAMPLE SOLUTION

These Sample Solutions are copies of the printout that will appear at your terminal. Supply all underlined information.

```
200 261.4,270.8, 265.4,261.4,258.1,252.1,268.3,250.3,
201 272.3, 262.8,255.5,249.6,280.9,270.3,263.2,258.3,256.3,
202 259.3,270.1, 259.3,253.2,266.4
RUN
WAIT.
DESTAT

DARTMOUTH ALGOL.

     S U M M A R Y    S T A T I S T I C S

         NUMBER OF VARIATES = 22
            ARITHMETIC MEAN = 262.059
         STANDARD DEVIATION = 7.78374
                   VARIANCE = 60.5865
                   SKEWNESS = .385134

     TYPE '7' FOR MORE OUTPUT, OR '3' TO STOP NOW.  WHICH? 7
     O R D E R    S T A T I S T I C S

              SMALLEST VARIATE = 249.6
                 LOWER DECILE = 250.84
                FIRST QUARTILE = 256.1
                       MEDIAN = 261.4
                THIRD QUARTILE = 268.75
                 UPPER DECILE = 271.85
               LARGEST VARIATE = 280.9

                  TOTAL RANGE = 31.3
                 DECILE RANGE = 21.01
             SEMI-INT-Q RANGE = 6.325

     F R E Q U E N C Y    D I S T R I B U T I O N

                     UP TO BUT                           PERCENT
          FROM      NOT INCLUDING     FREQUENCY         FREQUENCY

           240           245               0               0
           245           250               1               4.54545
           250           255               3               13.6364
           255           260               6               27.2727
           260           265               4               18.1818
           265           270               3               13.6364
           270           275               4               18.1818
           275           280               0               0
           280           285               1               4.54545
           285           290               0               0
```

```
CUMULATIVE  DISTRIBUTION

            NUMBER LESS     PERCENT LESS    VARIATE SUM - PCT
VALUE       THAN VALUE      THAN VALUE      LESS THAN VALUE
245         0               0               0
250         1               4.54545         4.32935
255         4               18.1818         17.4353
260         10              45.4545         44.2648
265         14              63.6364         62.4564
270         17              77.2727         76.3343
275         21              95.4545         95.1277
280         21              95.4545         95.1277
285         22              100             100.

ORDERED  ARRAY.
249.6       258.1           262.8           270.1
250.3       258.3           263.2           270.3
252.1       259.3           265.4           270.8
253.2       259.3           266.4           272.3
255.5       261.4           268.3           280.9
256.3       261.4
```

EXAMPLE 2.1 (Cont.)

Below is the actual run that I performed on our terminal for Example 2.1.

HELLO ← *This tells the computer that I am ready*

GE TIME-SHARING SERVICE

ON AT   9:51   09 TUE 08/24/71   TTY 10

USER NUMBER--OLD ← *The code number of the school*
PROJECT ID--NRD ← *The code for this project or perhaps your class*
SYSTEM--ALGOL ← *The computer language needed for the program (see instructions)*
NEW OR OLD--OLD ← *Old, because I am using a program already written*
OLD FILE NAME--DESTAT*** ← *The name of the program*

READY. ← *The computer says ready for me to enter data*

```
200  14,17,18,14,12,15,12,10,11,2,4,9,14,13,
201  11,9,12,8,10,13,11,8,6,3,3,7,12,11,
202  13,15,11,11,14,13,19,10,11,4,2,12,13,11,
203  9,10,11,9,9,11,9,6,9,1,2,8,9,10,
204  15,15,9,10,10,12,14,10,12,10,8,13,14,12,
205  11,11,5,8,10,13,14,7,9,9,9,12,10,10,
206  11,8,12,10,9,15,13,9,12,12,10,12,11,17,
207  8,10,8,10,8,9,6,3,6,7,8,9,9,7
```
← *The data: The 200, 201, 202, etc., are line numbers and are necessary for this program*

RUN ← *The data is complete. This tells the computer to go ahead and compute*

DESTAT   9:57   09 TUE 08/24/71
MODIFIED 11/1/68

```
    SUMMARY   STATISTICS

        NUMBER OF VARIATES  =  112   ← This is n
            ARITHMETIC MEAN  =  10.1071  ← x̄
        STANDARD DEVIATION  =  3.39474  ← s
                  VARIANCE  =  11.5242  ← s²
                  SKEWNESS  = -.31792  ← One measure of skewness
```

TYPE '7' FOR MORE OUTPUT, OR '3' TO STOP NOW. WHICH? 7   *We have a choice to stop here or go on. I said go on by typing 7. If you want to stop, you type 3.*

```
    ORDER   STATISTICS

        SMALLEST VARIATE  =  1       Smallest number
           LOWER DECILE  =  6       10% are below this number
         FIRST QUARTILE  =  8.25    25% are below this
                MEDIAN  =  10
         THIRD QUARTILE  =  12      75% are below this — upper 25%
           UPPER DECILE  =  14      10% are above this — upper 10%
         LARGEST VARIATE  =  19     Largest number

            TOTAL RANGE  =  18
           DECILE RANGE  =  8       Range of the middle 90% (14-6=8)
         SEMI-INT-Q RANGE  =  1.875  Range of the middle 50% divided by 2  (12-8.25)/2=1.875
```

The rest is self-explanatory:

```
FREQUENCY   DISTRIBUTION

            UP TO BUT                   PERCENT
FROM        NOT INCLUDING   FREQUENCY   FREQUENCY

-2.         0               0           0
 0.         2.              1            .892857
 2.         4.              6           5.35714
 4.         6.              3           2.67857
 6.         8.              8           7.14286
 8.        10.             27          24.1071
10.        12.             30          26.7857
12.        14.             21          18.75
14.        16.             12          10.7143
16.        18.              2           1.78571
18.        20.              2           1.78571
20.        22.              0           0

CUMULATIVE   DISTRIBUTION

            NUMBER LESS   PERCENT LESS   VARIATE SUM - PCT
VALUE       THAN VALUE    THAN VALUE     LESS THAN VALUE

 0.           0              0             0
 2.           1               .892857      8.83392 $-2
 4.           7              6.25          1.41343
 6.          10              8.92857       2.56184
 8.          18             16.0714        7.15548
10.          45             40.1786       27.7385
12.          75             66.9643       55.477
14.          96             85.7143       78.4452
16.         108             96.4286       93.7279
18.         110             98.2143       96.7314
20.         112            100           100

ORDERED    ARRAY

 1          9       10       12
 2          9       10       12
 2          9       10       12
 2          9       10       12
 3          9       10       13
 3          ..      11       13
 3          9       11       13
 4          9       11       13
 4          9       11       13
 5          9       11       13
 6          9       11       13
 6          9       11       13
 6          9       11       14
 6          9       11       14
 7          9       11       14
 7          9       11       14
 7          9       11       14
 7         10       11       14
 8         10       11       14
 8         10       12       15
 8         10       12       15
 8         10       12       15
 8         10       12       15
 8         10       12       15
 8         10       12       17
 8         10       12       17
 8         10       12       18
 8         10       12       19
```

(USED   23.83 UNITS) *This tells how much "time" I used on the computer.*

EXAMPLE 2.3 (Cont.)

I decided to use a different program for this example. This one is called UNISTA*** and is written in the language called BASIC. I have not included the Program Library example, just the initial information.

Category MATHEMATICAL
Name   UNISTA * * *

Language BASIC
Character 4600

PURPOSE

To provide a description of univariate data with up to 300 observations on one variable.

INSTRUCTIONS

Enter data on lines numbered 0-699.

Additional instructions may be obtained by listing the program STADES * * *.

SAMPLE PROBLEM

Determine statistical characteristics for the following data points:

261.4	252.1	255.5	258.3	253.2
270.8	268.3	249.6	256.3	266.4
265.4	250.3	280.9	259.3	
261.4	272.3	270.3	270.1	
258.1	262.8	263.2	259.3	

The next two pages include the computer runs for Example 2.3.

SOURCE: Instructions reprinted by permission from MARK I Quick-Use Programs User's Guide, 1966, 1967, 1968, 1970 by the General Electric Company.

I want to change programs and the new program is written in a different language or system.
SYSTEM
NEW SYSTEM NAME-BASIC *The new language is Basic.*
READY.

FILE *I typed the wrong word and the computer didn't understand.*
WHAT? *It typed "what?"*
OLD *I want the computer to get an old program ready.*
OLD FILE NAME-N←UNISTA*** *The program is UNISTA***. Note that I started by typing the letter N. To erase that letter I type the arrow ←. The computer then ignores the N.*
READY.

```
100 DATA 20.055, 20.040, 20.019, 20.025, 20.018
101 DATA 20.052, 20.024, 19.993, 19.984, 19.982
102 DATA 20.046, 20.036, 19.968, 20.013, 19.982
103 DATA 19.994, 20.008, 19.987, 19.984, 20.006
104 DATA 20.033, 20.041, 19.997, 20.003, 20.037
105 DATA 19.999, 19.975, 19.985, 19.968, 20.022
106 DATA 20.022, 20.011, 20.002, 19.975, 19.988
107 DATA 20.002, 20.026, 20.009, 19.960, 19.997
108 DATA19←← 19.976, 19.976, 20.000, 19.988, 20.006
109 DATA 19.994, 19.989, 19.970, 19.958, 20.019,
110 DATA 19.981, 19.987, 20.013, 20.018, 19.994
111 DATA 19.990, 19.968, 20.013, 20.009, 19.986
112 DATA 19.965, 19.970, 20.053, 20.026, 20.018
113 DATA 19.977, 19.975, 20.005, 20.024, 20.002
114 DATA 19.968,←
114 DATA L9.←
114 DATA 19.986, 19.969, 20.004, 20.003, 20.024
115 DATA 20.052, 20.017, 19.986, 19.972, 20.006
116 DATA 19.982 , 20.017, 20.004, 20.030, 19.983
117 DATA 20.032, 19.955, 20.019, 20.013, 20.001
118 DATA 20.007, 20.011, 19.983, 19.962, 20.035
119 DATA 20.025, 19.999 ,20.010, 20.000, 19.993
RUN
```

The extra space doesn't matter.
I didn't leave a space after DATA so I typed two arrows to erase the 19.
Here I copied the number.
I started the line over but typed an L for a 1, so I started over again.
By typing 114, with this line I replace all the previous 114 lines. I can redo any line by simply retyping the line number.

UNISTA   10:42   09 TUE 08/24/71

MODIFIED 12/27/1967

PLEASE SPECIFY A TYPICAL CLASS INTERVAL
FOR FREQUENCY DISTRIBUTIONS: L,U = ? 20.000,20.010 *I have to choose the interval widths.*

S U M M A R Y   S T A T I S T I C S

      NUMBER OF VARIATES = 100 n
       ARITHMETIC MEAN = 20.002 $\bar{x}$ *This is 2.37×10^{-2} or $2.37 \times (0.01)$ or 0.0237106.*
      STANDARD DEVIATION = 2.37106E-02
             VARIANCE = 5.62191E-04
    COEFF OF VAR (PCT) = .119
    STANDARD SKEWNESS = -13.736 } *Don't worry about these.*
     STANDARD EXCESS = 12356.3

O R D E R   S T A T I S T I C S

        SMALLEST VARIATE = 19.955
          LOWER DECILE = 19.97
        FIRST QUARTILE = 19.984
               MEDIAN = 20.002
        THIRD QUARTILE = 20.0188   } *See Example 2.1.*
          UPPER DECILE = 20.0348
        LARGEST VARIATE = 20.055

            TOTAL RANGE = .1
           DECILE RANGE = .0648
    SEMI-QUARTILE RANGE = .017375
       BOWLEY'S SKEWNESS = -.036
      PEARSON SKEWNESS = -.003

FREQUENCY DISTRIBUTION

FROM	UP TO BUT NOT INCLUDING	FREQUENCY	PERCENT FREQUENCY
19.95	19.96	3	3.
19.96	19.97	8	8.
19.97	19.98	7	7.
19.98	19.99	18	18.
19.99	20	9	9.
20	20.01	18	18.
20.01	20.02	15	15.
20.02	20.03	9	9.
20.03	20.04	6	6.
20.04	20.05	3	3.
20.05	20.06	4	4.

CUMULATIVE DISTRIBUTION

VALUE	NUMBER LESS THAN VALUE	PERCENT LESS THAN VALUE	VARIATE SUM - PCT LESS THAN VALUE
19.96	3	3.	2.993
19.97	11	11.	10.98
19.98	18	18.	17.97
19.99	36	36.	35.955
20	45	45.	44.952
20.01	63	63.	62.954
20.02	78	78.	77.964
20.03	87	87.	86.974
20.04	93	93.	92.984
20.05	96	96.	95.99
20.06	100	100.	100.

ORDERED ARRAY

19.955	19.984	20.002	20.019
19.958	19.985	20.003	20.019
19.96	19.986	20.003	20.019
19.962	19.986	20.004	20.022
19.965	19.986	20.004	20.022
19.968	19.987	20.005	20.024
19.968	19.987	20.006	20.024
19.968	19.988	20.006	20.024
19.969	19.988	20.006	20.025
19.97	19.989	20.007	20.025
19.97	19.99	20.008	20.026
19.972	19.993	20.009	20.026
19.975	19.993	20.009	20.03
19.975	19.994	20.01	20.032
19.975	19.994	20.011	20.033
19.976	19.994	20.011	20.035
19.976	19.997	20.013	20.036
19.977	19.997	20.013	20.037
19.981	19.999	20.013	20.04
19.982	19.999	20.013	20.041
19.982	20	20.017	20.046
19.982	20	20.017	20.052
19.983	20.001	20.018	20.052
19.983	20.002	20.018	20.053
19.984	20.002	20.018	20.055

USED   36.00 UNITS.

EXAMPLE 2.2 (Cont.)

As my third example I will use a program called STATAN***. This program provides an extensive collection of different calculations or statistics about a single set of data. At this point it is not necessary to discuss them but observe the ease of obtaining them. As in the last example I will only provide the cover page of information about the program.

Category MATHEMATICAL
Name   STATAN * * *

Language BASIC
Character 5400

PURPOSE

To perform a statistical analysis on data for one variable. It computes 34 different measures for an array of weighted (as with frequencies) or unweighted values of the variable. It also gives a 10-class frequency distribution summary, and a recapitulation of the input data in terms of deviations from the mean and as an ordered array.

INSTRUCTIONS

To use this program supply data in either of these two formats as determined by the problem

    1)   FOR UNWEIGHTED VALUES:

           10 DATA 0, X(1), X(2), X(3),......

    2)   FOR DATA WITH WEIGHTS OR FREQUENCIES:

           10 DATA 1, X(1), F(1), X(2), F(2), X(3), F(3),......

Where the initial 0 or 1 signals the presence or the absence of weights. Lines 11 through 99 are available for additional input data.

NOTE: This program produces output corresponding to the listing in the National Bureau of Standards Handbook No. 101.

    There can be no more than 99 values entered. If the values are weighted the maximum amount of data cannot exceed 99 values and 99 weights.

Additional instructions may be obtained by listing the program STATEX***.

SOURCE: Instructions reprinted by permission from MARK I Quick-Use Programs User's Guide, 1966, 1967, 1968, 1970 by the General Electric Company.

OLD — *I just want to change the program. This is also in the language*
OLD FILE NAME--STATAN*** *Basic so I do not have to tell the computer anything about "system."*
READY.

10 DATA O, 16.5, 16.5, 15.1, 16.4, 15.8, 16.1, 14.1
11 DATA 15.7, 14.3, 15.1, 13.6, 15.0, 15.9, 16.4, 15.6 — *Data for Example 2.2.*
12 DATA 16.0, 16.7, 15.2, 14.5, 15.5
RUN

STATAN   10:52   09 TUE 08/24/71

INCORRECT FORMAT   IN   10   *The computer is telling me something is wrongly written in line 10. You can't tell by looking, but the O after the word DATA is the letter O not the number zero.*

USED   6.00 UNITS.

10 DATA 0, 16.5, 16.5, 15.1, 16.4, 15.8, 16.1, 14.1 — *I just have to retype the one line that was incorrectly written.*
RUN

STATAN   10:54   09 TUE 08/24/71

MODIFIED 1/23/1970

COMPUTATIONS ON THE DATA ARRAY:

NUMBER OF VALUES = 20 n
NUMBER OF NONZERO WEIGHTS = 20
SUM OF WEIGHTS = 20
SUM OF VALUES = 310.
WEIGHTED MEAN = 15.5
UNWEIGHTED MEAN = 15.5
MINIMUM VALUE= 13.6
MAXIMUM VALUE= 16.7
RANGE = 3.1
WEIGHTED SUM OF SQUARES = 4819.64
VARIANCE = .770526
STANDARD DEVIATION = .877796
STANDARD ERROR OF MEAN = .196281   *We will use this later.*
COEFFICIENT OF VARIATION = 5.6632
STUDENT'S T = 78.9683
MEAN SQUARE SUCCESSIVE DIFFERENCES = 1.24
(MEAN SQ SUCC DIFF)/(VARIANCE) = 1.60929
MEDIAN = 15.65
NUMBER OF RUNS UP AND DOWN = 14
EXPECTED NUMBER OF RUNS = 13
STD ERROR OF NUMBER OF RUNS = 1.79815
(ACTUAL RUNS - EXP RUNS)/(STD ERR) = .556128
FREQUENCY DISTRIBUTION (TEN EQUAL CLASSES):   *You have a simple histogram here.*
   1   1   2   0   3   1   3   3   1   5

COMPUTATIONS ON DEVIATIONS FROM MEAN:

```
NUMBER OF + SIGNS IN DEVIATIONS = 12
NUMBER OF - SIGNS IN DEVIATIONS = 8
NUMBER OF RUNS (SIGN CHANGES + 1) = 9
EXPECTED NUMBER OF RUNS = 10.6
STD ERROR OF NUMBER OF RUNS = 2.08453
(ACTUAL RUNS - EXP RUNS)/(STD ERR) = .767559
TREND VALUE =-2.42105E-02
STD ERROR OF TREND = 8.13256E-03
(TREND)/(STD ERROR) =-2.97698
BETA ONE = .310896
BETA TWO = 2.39653
MEAN DEVIATION = .71
```
These are differences between the values and the mean.

RECAPITULATION OF INPUT:

VALUE	DEVIATIONS	WEIGHTS	ORDERED ARRAY
16.5	1.	1	13.6
16.5	1.	1	14.1
15.1	-.4	1	14.3
16.4	.9	1	14.5
15.8	.3	1	15
16.1	.6	1	15.1
14.1	-1.4	1	15.1
15.7	.2	1	15.2
14.3	-1.2	1	15.5
15.1	-.4	1	15.6
13.6	-1.9	1	15.7
15	-.5	1	15.8
15.9	.4	1	15.9
16.4	.9	1	16
15.6	.1	1	16.1
16	.5	1	16.4
16.7	1.2	1	16.4
15.2	-.3	1	16.5
14.5	-1.	1	16.5
15.5	7.45058E-08	1	16.7

7.4 × 10⁻⁸ or 0.000000074 or essentially 0.0

```
USED    10.17 UNITS.
```
BYE *This is how I tell the computer I'm finished.*
```
*** OFF AT 10:59    09 TUE 08/24/71

    29 ELAPSED TERMINAL MINUTES

    52.17 TOTAL CRU'S USED
```

S U M M A R Y   S T A T I T

REVIEW

Keywords

Calculator · Computer · Program · Language · Storage · Memory · Time-sharing · Library program

PROBLEMS

5.1 What is the difference between a calculator and a computer? (You may have to do some research.)

5.2 Describe in your own words what a program is.

5.3 Examine the three examples to see the difference in the way the data is entered. Research to find how data is entered for programs written in the FORTRAN language.

5.4 Examine the printout for the STATAN program. How many of the statistics do you understand? Try to figure out what some of the ones not discussed yet might be.

5.5 Very often data or calculations are printed in an "E" or "floating point" format. The number following the E is the power of 10 that the rest of the number is multiplied by. Thus, 2.471E05 is 2.471×10^5 or 247100. Why is this kind of notation useful? Where has it been used? (Also note that most computers can work with numbers from 10^{-49} to 10^{49}.)

5.6 How would the following numbers be written in E format?
   a. 25.74   b. 269108
   c. 0.06873   d. 423.68
   e. 0.0000000563   f. 98560034000000000

5.7 What is the decimal equivalent for the following numbers:
   a. 6.7865E06   b. 2.75E12
   c. 5.982E−04   d. 1.988E−18
   e. 9.99999E+44   f. 7.458986543E − 7

5.8 In all computers only a fixed number of decimal places can be used in the calculations. In certain types of problems this limited number of decimal places means that there may not be sufficient numbers of significant values to reach a correct answer. For example, if a computer only can use 8 significant decimal places and is to calculate a variance using Equation (4.12) (without coding) what would the computer find to be the variance of the following values?

$$5000000; \quad 5000005; \quad 5000010; \quad 5000005.$$

How close is the computer to the actual value?

5.9 A way of sometimes remedying the difficulty in Problem 5.8 is to employ a technique called "double precision arithmetic." Research to determine what double precision is and its limitations. If you have a computer that is available, investigate its capabilities and limitations.

6

Graphing I: Correlation and Regression

In Chapter 3 we looked at a pictorial way of showing the relationship between a measurement of some quantity and the frequency with which that quantity appeared. This constituted one graphical way of visualizing and describing a distribution of values, one which then told us something about a population that we were interested in. This approach, however, was limited to the distribution of a single set of data. Furthermore, that distribution could only deal with a single characteristic. Now we will explore the idea of looking at two different measurements or *kinds* of measurements simultaneously and observing the ways that they may be related to each other.

6.1 Descriptions of Pairs of Quantities

Two sets of data have been provided in Table 6.1. They are measurements of the height and weight of twenty college students. In Figure 6.1 a pair of histograms have been drawn to show the distributions of these two types of measurements.

Table 6.1 Heights and weights of twenty students

Height (feet and inches)				Weight (pounds)			
5′ 9″	5′10″	5′10″	5′10″	178	175	150	185
5′11″	5′ 4″	5′ 5″	5′ 3″	171	128	124	118
5′ 6″	5′ 7″	6′ 1″	5′ 9″	138	138	200	154
5′ 8″	5′ 9″	5′ 6″	5′ 6″	142	172	130	144
5′ 7″	5′ 6″	5′ 8″	5′ 8″	163	155	162	133

Observe the general nature of the distributions. The quantities have been recorded in the same order as the measurements were taken, and from casually looking at the table or the graphs we have no reason to say that heights and weights are not totally independent of each other.

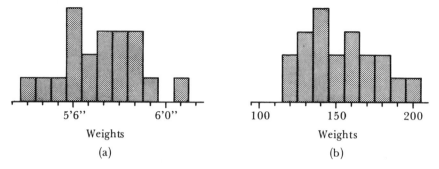

Figure 6.1 The distributions of heights and weights. Can you tell if there is any relationship?

In order to determine if there is any type of relationship between these two quantities we must bring together the pairs of values for each individual by listing which heights go with which weights. We must say that John is 5 feet

Table 6.2 Heights and weights of the twenty students listed in Table 6.1. This time the heights and weights are arranged by student

Student	Height	Weight
1 (John)	5' 9"	178
2 (Frank)	5'11"	171
3 (Mary)	5' 6"	138
4 (Jane)	5' 8"	142
5	5' 7"	163
6	5'10"	175
7	5' 4"	128
8	5' 7"	138
9	5' 9"	172
10	5' 6"	155
11	5'10"	150
12	5' 5"	124
13	6' 1"	200
14	5' 6"	130
15	5' 8"	162
16	5'10"	185
17	5' 3"	118
18	5' 9"	154
19	5' 6"	144
20	5' 8"	133

9 inches *and* 178 pounds, Frank is 5 feet 11 inches *and* 191 pounds, Mary is 5 feet 6 inches *and* 138 pounds, and so forth. Table 6.2 lists the heights and weights for the twenty students; now each height is identified with the weight of the same student.

We will now draw a graph for this data. This time the graph will include measurements along *both* the *X*- *and* the *Y*-axes (remember that the histogram only had measurements recorded on the *X*-axis, the *Y*-axis contained frequencies). We will have to locate the two measurements for each student. Heights are to be recorded along the *X*-axis and weights along the *Y*-axis. Every single point on the graph will correspond to some combination of height and weight and we will place a dot at the point where the height and weight of the individuals we observed coincide. Thus the value for John, a height of 5 feet 9 inches and a weight of 178 pounds, is located at the point where the two lines on Figure 6.2a intersect (a line going up from a height of 5 feet 9 inches and a line going to the right from 178 pounds). The rest of the values in Table 6.2 have been plotted in Figure 6.2b (I have identified the first four points by placing the student's names next to the corresponding points).

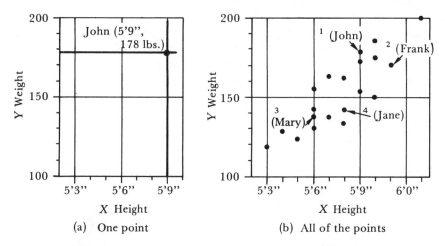

Figure 6.2 Graph of the heights and weights, (a) shows how one point is plotted, (b) contains the points representing all of the 20 students listed in Table 6.2.

6.2 Correlation

Now that we have the graph we must define or describe what it is that we will be attempting to observe or learn from the graph. There are two important things that we will look for, both of which deal with a concept called *correlation*.

1. We will try to see if there is a *relationship* between these two sets of values and
2. We will observe how closely the points seem to fall along or *fit* some straight line.

These two ideas go hand-in-hand, and for the most part can be treated as one concept. However, there are subtle differences in the underlying theory and perhaps in the way you visualize them.

When you look at the graph you need to interpret the *number or concentration of dots* in the various locations or areas of the graph as indicating the frequency with which *those pairs of values have simultaneously* occurred. These points show the *distribution of the pairs of quantities*. Depending upon the degree of random scatter or the concentration into some pattern we can state if there is *no* relation or *some* relation between the two quantities. The pattern we are looking for is usually a straight line.

Look carefully at Figure 6.2b. There are no short people, those under 5 feet 6 inches, who are over 160 pounds, and there are no tall people, over 5 feet 10 inches, who are under 160 pounds. As we look at larger heights, the weights become larger, so there appears to be some relationship between height and weight. If you know someone's height, then looking at the graph will provide you with a narrow band of possible values to which the corresponding weights belong (see Figure 6.3a). By narrow I really mean narrow with reference to the total possible distribution of weights. Compare the distribution of weights for the 5'9" point to the overall distribution of weights plotted along the Y-axis of Figure 6.3a. The same holds true if you want to "predict" a height from a given weight (see Figure 6.3b).

At this point we are not discussing actual predictions, as that falls in the realm of regression analysis, but simply trying to gauge if two kinds of data

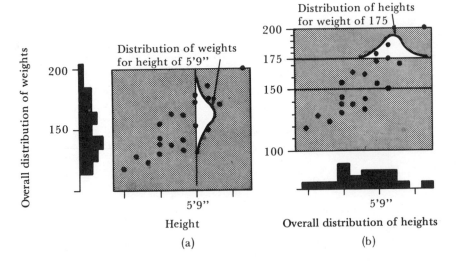

Figure 6.3 Two graphs demonstrating the importance of a correlation. (a) Given one value, the height, and looking for the weight you are able to identify the value as being within a distribution that is smaller than the overall distribution of weights, (b) same as in (a) except going from weight to height.

may vary together. To illustrate further let us look at two extreme cases of possible "relationships."

The first is a graph of heights and the score that each student received on the first statistics test taken, Figure 6.4. Apparently there is no relation

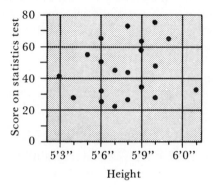

Figure 6.4 Graph of student's height and score on the first statistics test. There is no apparent relationship.

between these two sets of values. If you were to select a height, say 5 feet 8 inches, and look at the distribution of test scores for the students with that particular height, you do not have a distribution that is smaller than the overall distribution of tests. So you do not have any predictive possibilities or discrimination. This is the extreme that implies that there is *no relation* or *no correlation*.

The second graph is that of height measured in feet and inches as before, and height in centimeters. This is merely a simple conversion factor and therefore values should fall on a perfect straight line. Knowing the height as recorded on one scale should quickly and perfectly yield a single value on the

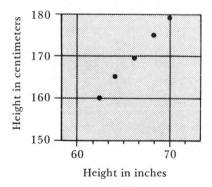

Figure 6.5 Graph of height as measured in inches and in centimeters.

second scale. From a practical standpoint however, if we remeasure the same height we will obtain a different value. In addition, the scales for centimeters will be finer than inches (there are approximately $2\frac{1}{2}$ centimeters for every inch, so the centimeter scale is actually recording to better than the nearest $\frac{1}{2}$ inch). Therefore, there is the possibility of slight variation. Nevertheless, the dots will fall about as close to a straight line as we could expect (see Figure 6.5). Pick a value of height along the X-axis and this corresponds to a distribution that is a mere point, a small location on the broad distribution of possible points on the Y-axis. This is the other extreme, what is essentially a perfect correlation.

6.3 Correlation Coefficient

So far what we have done has been very imprecise, at least as far as the mathematics is concerned. If I gave you a set of different graphs with points scattered all over them, which we call scatter diagrams, and asked you which were more or less correlated, you might have difficulty. The extremes are usually obvious, although the question of no correlation may not always be obvious. However, we will now explore a way of arriving at a numerical value to describe the degree of relation, the degree of closeness to a straight line, or the amount of scatter that exists. This quantity is called a correlation coefficient and is generally given the symbol r.

There are several different types of correlation coefficients. The one we will primarily deal with is called the *Pearson product moment correlation coefficient*. Others, such as the Spearman rank order correlation coefficient will be dealt with in Chapter 20. They may also have other symbols, r_{rho} in this case.

The value of r can range from -1 to $+1$ with the following general interpretations:

1. If $r = 0$ then we have *no* correlation, or more precisely, we have *no linear* correlation. Usually we just have scatter, but there *may* be other ways in which the values are related. The distribution in Figure 6.4 is close to an r of 0.
2. If $r = +1$ or $r = -1$ then we have a *perfect* linear correlation; all of the values fall on a straight line. Figure 6.5 is an excellent case.
3. Values of r between 0 and 1 or between 0 and -1 indicate varying degrees of correlation. Thus, the correlation coefficient for the points in Figure 6.2b would be greater than 0 but less than 1.
4. *Positive* values of r indicate a line going *up* to the right (as X increases so does Y). Figure 6.2b and Figure 6.5 fall into this category.
5. *Negative* values of r indicate a line going *down* to the right (as one of the values increases the other decreases).

Figure 6.6 contains a variety of graphs illustrating the range of correlation and giving the associated correlation coefficients. There are two types of scatter diagrams illustrated. The first set, Figure 6.6a to c, contains dots at

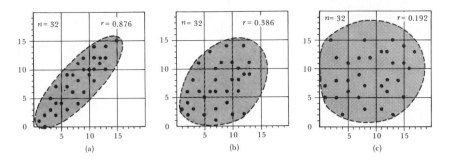

Scatter diagrams using dots at points of intersection

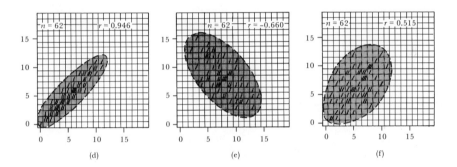

Scatter diagrams containing tally boxes

Figure 6.6 Correlation scatter diagrams illustrating different degrees of correlation and two techniques of construction. For both sets of diagrams dotted lines have been drawn around the points to emphasize the general nature of the distributions. Correlation coefficients and the size of the sample are given for all the distributions.
(a), (b), and (c) Intersection of two lines are used to represent the values. The measurements are located along the lines. (d), (e), and (f) Boxes represent the values with the centers of the boxes (or intervals) corresponding to the measurement values.

the points of intersection of the two values. The second set, Figure 6.6d to f, resemble frequency tally sheets, except that these frequency charts cover two possibilities at one time. The latter group of diagrams also demonstrates a method that lends itself more readily to graphical presentation of repeated occurrences of the same pairs of values. Thus, if two students have the same height and weight this can easily be recorded (rather than superimposing dots on top of each other, making interpretation difficult).

The calculation of r is the next item. It is a fairly involved procedure, particularly when computation equipment is not available. We will define r as

$$r = \frac{\text{Variance of the products of (the } X \text{ times the } Y)}{(\text{Standard deviation of the } X)(\text{Standard deviation of the } Y)}. \quad (6.1)$$

For computational purposes (disregarding multiple frequencies of X and Y combinations) this is written as:

$$r = \frac{[\sum (X \cdot Y) - (\sum X)(\sum Y)/n]/(n-1)}{\sqrt{\frac{\sum X^2 - (\sum X)^2/n}{n-1}} \cdot \sqrt{\frac{\sum Y^2 - (\sum Y)^2/n}{n-1}}}. \quad (6.2)$$

The $(n - 1)$ all divide out, and this then reduces to merely:

$$r = \frac{\sum (X \cdot Y) - (\sum X)(\sum Y)/n}{\sqrt{\sum X^2 - (\sum X)^2/n} \cdot \sqrt{\sum Y^2 - (\sum Y)^2/n}}. \quad (6.3)$$

This formula may look formidable but it really involves little more than we have already been doing in calculating standard deviations. Let us apply the formula to our data and determine r for our sample of heights and weights. Heights were the X and weights were the Y. So

$$r = \frac{\sum (\text{heights})(\text{weights}) - \frac{(\sum \text{heights})(\sum \text{weights})}{\text{number of students}}}{\sqrt{\sum (\text{heights}^2) - \frac{(\sum \text{heights})^2}{\text{students}}} \cdot \sqrt{\sum (\text{weights}^2) - \frac{(\sum \text{weights})^2}{\text{students}}}}. \quad (6.4)$$

In Table 6.3 we have the computations of all of the sums. Substituting the sums into Equation (6.4) we have

$$r = \frac{24{,}608 - (155)(3060)/20}{\sqrt{1317 - (155)^2/20} \sqrt{477{,}674 - (3060)^2/20}}$$

$$= \frac{24{,}608 - 23{,}715}{\sqrt{1317 - 1201.25} \sqrt{477{,}674 - 468{,}180}} = \frac{893}{\sqrt{115.75} \sqrt{9494}}$$

$$= \frac{893}{(10.8)(97.4)} = \frac{893}{1051.9} = 0.849.$$

Our value of r confirms the high degree of correlation between height and weight that was apparent from the graph.

Table 6.3 Computations needed to calculate a correlation coefficient. Data is height and weight of twenty students

Student	Height*	Weight	$(Height)^2$	$(Weight)^2$	$(Height)(Weight)$
1	9	178	81	31,684	1602
2	11	171	121	29,241	1881
3	6	138	36	19,044	828
4	8	142	64	20,164	1136
5	7	163	49	26,569	1141
6	10	175	100	30,625	1750
7	4	128	16	16,384	512
8	7	138	49	19,044	966
9	9	172	81	29,584	1551
10	6	155	36	24,025	930
11	10	150	100	22,500	1500
12	5	124	25	15,376	620
13	13	200	169	40,000	2600
14	6	130	36	16,900	780
15	8	162	64	26,244	1296
16	10	185	100	34,225	1850
17	3	118	9	13,924	354
18	9	154	81	23,716	1386
19	6	144	36	20,736	864
20	8	133	64	17,689	1064
Sum	155	3060	1317	477,674	24,608

* Values of heights have been converted to inches greater than 5 feet. Thus 5'6" is shown as 6, 6'1" is 13 (1 foot 1 inch greater than 5 feet), and so forth.

6.4 Interpretation of Correlation and the Correlation Coefficient

The first thing that you should question now that we have a value for correlation coefficient is: How large must the correlation coefficient become in order to be considered "large"? If we look at only three pairs of values, or have only three points to plot, there is a pretty good chance that they will come close to being on a straight line *regardless of whether there truly is a correlation.* Let's look at a case which would be absurd. Pick three books, a medium size book, a fairly long book, and a very short book. Now count the number of words in the first sentence of each of these books. There is a pretty good chance that you will just randomly end up with the second book having quite a few more words in its first sentence than the first book had, and the third book having less than either of the other two, or vice-versa. In either case (as well as some other possible cases) you will end up with varying degrees of correlation, sometimes as high as 0.7 or 0.8. Yet this "high" value perhaps should not be considered important.

Now if we were to increase the sample by picking up another two dozen books, thus giving us a sample of 27, we should start to be suspicious that a relation may actually exist if we were suddenly to find that the correlation coefficient was as large as 0.7 or 0.8.

So, the significance of a large value of correlation coefficient depends primarily upon the *size of the sample* or the number of pairs of values that we have observed. But you should inquire further; you should want to know just *how large* the coefficient should be for *whatever sample size* you are dealing with. To make this determination we will consult Appendix Table A.7. For convenience in discussing how to use this table I have reproduced some parts of it in Figure 6.7. We have a number of columns in the table. The first is a value related to the size of the sample. The value is labeled df, which stands for degrees of freedom. You need only to know that it is $n - 2$. So, since the values of n that have just been discussed were 3 and 27, the degrees of freedom would be 1 for the first case and 25 for the second case. Both are found in the table and the figure.

$df = n - 2$	Level of significance for two-tailed test	
	0.05 (5%)	0.01 (1%)
1	0.9969	0.9999
18	0.4438	0.5614
25	0.3809	0.4869

Figure 6.7 Part of Appendix Table A.7 showing what are considered significantly high correlation coefficients for several different sample sizes.

In the table in the appendix there are five more columns. For this discussion we will only concern ourselves with two of them: those labeled as 0.05 and 0.01 below the heading of "significance level for two-tailed tests." These are reproduced in part as the second and third columns of Figure 6.7.

These columns list values of correlation coefficient. For an n of 27 we have listed in the second column, the column labeled 0.05 level, the number 0.3809. The 0.3809 is a value of correlation coefficient that is considered unusually large. If the correlation coefficient I obtain is as large as the 0.3809 or larger than 0.3809 then it is sufficiently large enough to say that there is less than a 5% chance that *no* relation exists, or that the odds are about 19 to 1 in favor of some sort of a relation.

The third column also lists correlation coefficients. These are larger values than those in the previous column and for $n = 27$ it is 0.4869. When I obtain a correlation coefficient larger than 0.4869 I know that there is *less than a 1%* chance that *no* relation exists. The odds are increased to almost 100 to 1 that some relation is present, or I say there is 1 chance in 100 of obtaining a correlation of this sort (larger than 0.4869) when the source of the data is *totally uncorrelated*.

Going back now to our example of the correlation between height and weight, we can determine what the critical or significantly large values of correlation coefficient are for this sample. The sample size was 20, since we measured 20 students. Referring to Table A.7 (or Figure 6.7) we find that the critical values of correlation coefficient corresponding to $n = 20$ are 0.4438 (at the 0.05 or 5% level of significance) and 0.5614 (at the 0.01 or 1% level of significance). Our computed coefficient was 0.849, which far exceeds even the upper value of 0.5614 (the 1 in a 100 chance that there is actually no relation). So we can now state that the *correlation is* actually *very significant* and that there is a *significant relationship between height and weight*.

6.5 Cause and Effect

If we do find that there seems to be a relation between two types of measurements this *does not mean that there is or is not a cause and effect relation*. Any of the following possibilities may exist:

1. X *may* cause Y (large height may cause a large weight).
2. Y *may* cause X (large weight may cause a large height).
3. X and Y may be dependent on a *third factor* (large height and large weight may depend on genetic characteristics and both increase together).
4. The correlation may be an accidental random situation.

The last two possibilities are the critical ones to keep in mind when we discuss correlation. Item 4 pertains to those 5% and 1% values just discussed, and is the general reason why we even deal with the subject of statistics. There always exists the possibility that we have a peculiar situation arising in a set of data, one in which the data just happens to fall close to a straight line. The table provides us with a gauge that permits us to limit the chance of accidentally calling two things related when, in fact, there is no relationship. However, the possibility of an accidental relationship still exists whenever we *do declare there is a relation*.

The third item, that the two sets of measurements may be both dependent on some third factor, is the very reason we have used the word correlation. You might read this word as co-relation, in which case we would emphasize that at best we would envision a mutual relationship. Many absurd correlations are possible, e.g. the way in which an inflationary trend in the economy or population growth may affect totally unrelated subjects. If you were to look up the salaries of teachers or ministers and compare them to the tons of aluminum produced that year, or the federal budget, or the price of books, or the cost of mailing a letter, you would find a high correlation. Obviously they are coincidental and occur because they both depend upon a third factor, the economy in this case.

The first two items (X may cause Y, and Y may cause X) involve true cause and effect. However, if the data was obtained merely by the simultaneous recording of both of the measurements, then *we cannot know which is the cause and which is the effect*. In order to identify a true causal relationship we must be capable of *precisely controlling one of the variables*. If, however, we have the ability to control one of the variables then generally this means that we are no longer just concerned with *if* a relation exists. We are more interested in the *nature* of this cause and effect relationship. This will mean finding a mathematical or graphical way of *predicting the effect*, and this leads us to *regression*.

6.6 Regression

Regression is concerned with replacing the data by a *single line* that also represents a mathematical equation. The line that we are looking for is the same line we previously found for the dots in a correlation problem. The purpose for finding this line generally is to predict. It should be stated here that *there is no necessity for a cause and effect relation to exist in order for you to want to predict one value from the other*. For example, it has generally been found that there is a high correlation between I.Q. scores and the future success in school. The psychologist recognizes that the score on the I.Q. test does not cause good grades in school and it should be obvious that a future success or failure could not influence or cause an earlier test score to be high or low. Thus, cause and effect is to be ruled out as a possibility. However, the im-

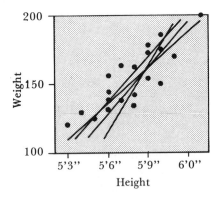

Figure 6.8 Four possible regression lines to predict weight from height. The lines are drawn on the scatter diagram of Figure 6.2b. Can you pick the best one? Is there a better line?

portance of knowing the nature of the relationship and being able to accurately *predict* what an individual will do in the future may be considerable.

We will return to our initial correlation problem and determine the line and formula to be used to predict weight when an individual's height is known. First, look back to Figure 6.2b and see how well you can fit *one* line between the dots. One way is to hold a piece of string tightly over the dots and see if you can sort of balance all of the dots above and below the string. Easy? Do you get *one perfect* line? Did you find it difficult to decide which points should go above or below, or whether to go between certain points? In Figure 6.8 I have drawn 4 possible lines. Which one is the "best"?

In order to decide which line to select as the "best" line, we must have some criterion. With any line that we might select there are differences, deviations, or errors, between the points representing the actual values of the heights and weights of the students and that line that we will be selecting. Figure 6.9 shows *one* such line and the differences or distances between this particular line and

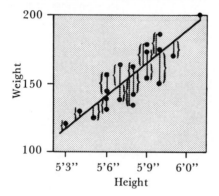

Figure 6.9 Differences between one possible regression line and the points of the scatter diagram. The least-squares regression line is the one which has been adjusted so that the squares of those distances are minimized.

the points for each student. If we were to move that line in any way, some of these distances would decrease and others would increase. The idea is to move the line around until you find the location in which some kind of average distance becomes smallest. The average distance that we try to minimize is the square of the distances (leading to the definition of the method as the "least-squares" method). Without going into great detail, the equation used to estimate or predict values of Y (weights in this case) for any particular X value (height) is as follows:

$$Y_{\text{predicted}} = \bar{Y} + b(X - \bar{X}) \tag{6.5}$$

where

$$b = \frac{\sum (X \cdot Y) - (\sum X)(\sum Y)/n}{\sum X^2 - (\sum X)^2/n}. \qquad (6.6)$$

(Compare Equation (6.6) to Equation (6.3) and note the similarity between b and r.

We need $\bar{X}$ and $\bar{Y}$. For our example they are

$$\bar{X} = \frac{\sum X}{n} = \frac{155}{20} = 7.75 \qquad \text{(an average height of 5 feet } 7\tfrac{3}{4} \text{ inches)}$$

$$\bar{Y} = \frac{\sum Y}{n} = \frac{3060}{20} = 153 \qquad \text{(an average weight of 153 pounds)}$$

Then,

$$b = \frac{\sum (\text{heights})(\text{weights}) - \dfrac{(\sum \text{heights})(\sum \text{weights})}{\text{number of students}}}{\sum (\text{heights}^2) - \dfrac{(\sum \text{heights})^2}{\text{number of students}}}$$

$$= \frac{24{,}608 - (155)(3060)/20}{1317 - (155)^2/20} = \frac{24{,}608 - 23{,}715}{1317 - 1201.25}$$

$$= \frac{893}{115.75} = 7.71.$$

Substituting these values into Equation (6.5) we have

$$Y_{\text{predicted}} = \bar{Y} + b(X - \bar{X})$$
$$= 153 + 7.71(X - 7.75)$$
$$= 153 + 7.71X - 59.8$$
$$= 93.2 + 7.71X. \qquad (6.7)$$

This now is the best *linear or straight line equation* for the data given. If you substitute values for X you will obtain predicted values for Y. For instance, if $X = 10$, meaning a height of 5 feet 10 inches, we estimate or predict the weight to be

$$Y_{\text{predicted}} = 93.2 + 7.71(10) = 93.2 + 77.1 = 170.3.$$

This is not too far from the one actual value (individual number 6) who was 5 feet 10 inches and weighed 175 pounds.

You might also observe that if you calculate the predicted Y value for the point X equal to $\bar{X}$ you will find that the $b(X - \bar{X})$ part of Equation (6.5) becomes 0 and then $Y = \bar{Y}$. Thus, *one point that the line will always go through is the point* $(\bar{X}, \bar{Y})$, in this case the point $(X = 7.75, Y = 153)$.

Now that we have two points, the $(\bar{X}, \bar{Y})$ point and the previous one $(X = 10, Y = 170.3)$ we can draw a straight line on the graph. This line is

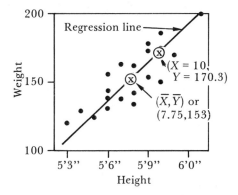

Figure 6.10 The least-squares *regression line* (of *X* on *Y*). Having found two points, the circled *X*, we can draw a straight line. This line provides the best prediction from height to weight.

called the *regression line*, or more specifically the *regression line of X on Y*, or height on weight. In Figure 6.10 this regression line is drawn on the graph. The reason it is stated as the regression of *X* on *Y* is to emphasize that *this line is only for predicting from X-values to Y-values*. You are not to use the line, nor rework the equation, to use it to go from *Y*-values to *X*-values. If you want to *also* predict from *Y* to *X*, knowing weight to predict height, you must recalculate *b* [(with the denominator as $\sum (\text{weights}^2) - (\sum \text{weights})^2 / \text{students})$] and obtain a new equation. This is necessary because if you look

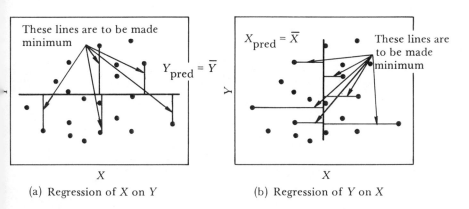

(a) Regression of *X* on *Y*   (b) Regression of *Y* on *X*

Figure 6.11 The best regression lines when the data is uncorrelated. (a) The regression of *X* on *Y*, or going from *X* to *Y*—the best guess or prediction of the *Y* value, regardless of *X*, is $\overline{Y}$. Observe that the lines to be minimized are those parallel to the *Y* axis. (b) The regression of *Y* on *X*, going from *Y* to *X*. Here the best guess or best prediction is $\overline{X}$, regardless of the *Y* value. The lines to be minimized this time are those parallel to the *X* axis. The two prediction lines, *X* to *Y* and *Y* to *X*, are perpendicular to each other.

back to Figure 6.9 you will see that the distances we were working with were vertical distances. In this case we are now working with horizontal distances.

In the most extreme case, where *no* correlation is present, the best line to use to predict Y from X will be a line parallel to the X-axis. This arises because b will essentially be zero and then $Y_{predicted}$ will be $\bar{Y}$ for *all* of the X-values. Thus no matter what you choose for X you end up with the average value of Y (see Figure 6.11a).

The reverse "prediction" would be when we know the Y-value and look for an X-value. If we look at the extreme case again, the uncorrelated one, we would find that this time regardless of the Y-value, the best prediction is always $\bar{X}$. Now we end up with a line parallel to the Y-axis, as in Figure 6.11b. You can now compare the two prediction lines to see how different they are.

Getting back to weights and heights let us see about the equation predicting *height from weight*. This time b would be equal to

$$b = \frac{\sum (\text{heights})(\text{weights}) - \frac{(\sum \text{heights})(\sum \text{weights})}{\text{students}}}{\sum (\text{weights}^2) - \frac{(\sum \text{weights})^2}{\text{students}}}$$

$$= \frac{893}{9494} = 0.094,$$

and the resulting equation is

$$X_{predicted} = \bar{X} + b(Y - \bar{Y})$$

$$= 7.75 + 0.094(Y - 153) = 7.75 + 0.094Y - 14.38$$

$$= -6.63 + 0.094Y. \tag{6.8}$$

This equation looks different than Equation (6.7) and will yield different points. For instance, go back to the point we predicted with the previous equation. At that time we substituted $X = 10$ and obtained $Y = 170.3$. Now let us substitute $Y = 170.3$ (pounds) and see what the prediction of height would be

$$\text{Height} = X_{predicted} = -6.63 + 0.094Y$$

$$= -6.63 + 0.094(170.3)$$

$$= -6.63 + 16.0 = 9.38 \text{ inches},$$

or we would expect him to be 5 feet 9.38 inches (which oddly enough is also 5 feet $9\tfrac{3}{8}$ inches). Compare this X to the 10 above. The resulting line is plotted in Figure 6.12, along with a few of the differences that *this line* attempts to reduce.

Thus we have found two possible lines to serve as predictors. If a true cause and effect relationship exists, though, we really are *only* concerned with the prediction equation *from the cause factor to the effect factor*, not vice-versa. Hence, only one line would ever be drawn in these cases.

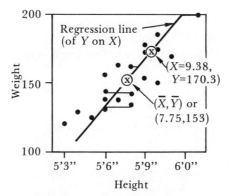

Figure 6.12 The least-squares regression line of Y on X, of weight on height. This is the best line for predicting from weight to height. A few of the types of differences that are being minimized with this line are shown. Compare this figure with the line of Figure 6.10 and the distances minimized in Figure 6.9.

6.7 Correlation or Regression?

Correlation studies and the computation of correlation coefficients are usually done with data gathered in the social sciences and behavioral sciences such as sociology, education, psychology, and anthropology. In these disciplines data is often collected, under conditions in which true control over the things measured is very difficult, and perhaps impossible. On the other hand, the physical and biological sciences and the engineering disciplines are generally capable of closely controlling variables. The data accumulated in these fields is more likely to show cause and effect relationships, and, in fact, most experiments are designed to determine *what the relationship actually is*, rather than *if* it exists at all. In these latter cases regression becomes the principal tool.

Generally, the field you are working in and the purpose behind taking the data will dictate whether a correlation is sufficient, necessary, or useful, or if the regression equation is more valuable. However, as you have just seen, we can *always* calculate a regression line. We can determine an equation any time we see highly correlated values and prediction of one from the other is desirable.

In Chapter 20 we will explore regression and correlation in more depth. The remainder of this chapter will deal with problems of interpretation and some cautions.

6.8 Scales: Change in Size

The actual scales used do not affect the correlation coefficient or the regression equation, but changing a scale may create widely different impressions in graphs. Figure 6.13 shows a collection of graphs of the number of people employed for the 1960–1968 period of time. The only difference between these three plots is the scale that is used. The first graph (Figure 6.13a) gives a true

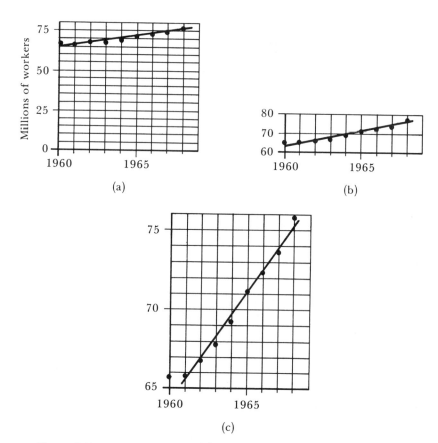

Figure 6.13 Three graphs of the United States labor force from 1960 to 1968. By changing the size of the scales or the amount of scale, different interpretations or illusions, some invalid, may result. (a) The proper graph, (b) the same graph only this time the bottom of the graph has been cut off, (c) the same graph except the Y axis has been stretched so that the scale is 10 times larger than the previous ones.

picture of what has been occurring, as it shows a gradual increase in the work force. However, if the work force is compared to population increase you would find that there is no percentage increase in workers. If we now lop off the bottom of the graph, as in Figure 6.13b, the effect is to place emphasis on the increasing nature of the curve, even though the scale has not been increased.

For additional emphasis we can expand the scale along the Y-axis so that 1 unit, say one million workers, occupies the same space as 10 of the previous units did. As a consequence we no longer see a *gradually* increasing work force, but rather we see an extremely *rapid* increase taking place (Figure 6.13c). The scale has created an illusion and you must be very careful not to be deceived.

6.9 Range of Prediction

A particularly important comment should be made regarding predictions. The regression equation and its graph are *only valid within the range for which the data has been accumulated*. We cannot validly predict above or below the limits for which we have plotted our graph. We have no basis for stating or believing that the rest of the curve will be a straight line. We don't know if a sudden increase or decrease will take place. We should not make predictions about the future. In Figure 6.14 I have shown the net growth rate of the U.S. population for the years from 1935 to 1952. There appears to be a high correlation,

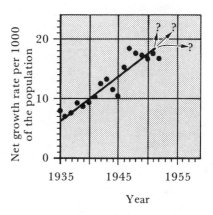

Figure 6.14 Net rate of growth of the United States population for the years 1935 to 1952. A rate of 10 means a growth rate of 0.1% per year. A fairly good straight line implies a continuation in the future years.

and it looks like population should keep increasing. If you extended the curve to 1970 you would expected about a 30 per 1000 or a 3% a year growth rate to continue.

In Figure 6.15 the graph has been extended to 1968. The straight line suddenly became a curve and our prediction would have been extremely wrong. Now, by the looks of this graph we soon should hit a zero growth rate and we should not have to worry about a population explosion. Don't trust it.

6.10 Warning: Plot the Data!

A *zero* correlation may *not* mean *no* relationship.

A *high* correlation may *not* mean *a* relationship.

Let's look at these two odd statements.

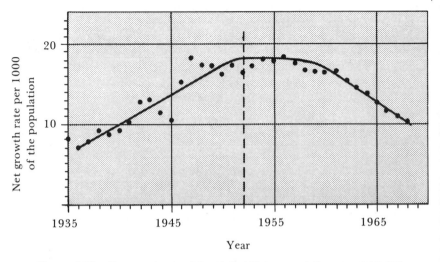

Figure 6.15 Net growth rate of the United States population now extended to the year 1968. Observe the complete change in direction of the curve.

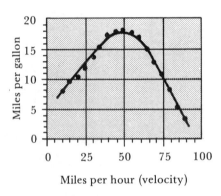

Figure 6.16 A graph of miles per gallon versus velocity of an automobile. There is a straight line relation from about 0 to 40 miles per hour and again from about 60 to 90 miles per hour, but between the graph is a curve.

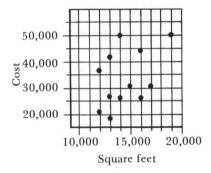

Figure 6.17 Scatter diagram for the cost of a house and the size (in square feet) of the house. The correlation is poor.

A *near zero correlation*, or a poorly fitted straight line, does not necessarily mean that no relation exists. *First*, the relation *may not be linear* to start off with, it may be a curve. The last graph we just looked at was one example of a curve. As another example, Figure 6.16 shows the gas mileage for a car when it is driven at different speeds.

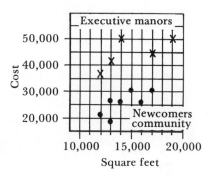

Figure 6.18 The same data as in Figure 6.17 except the houses have been identified according to the development to which they belong. One is high cost and the other relatively low cost. Each have good correlations.

Figure 6.17 is a graph of the cost of a dozen homes versus the size of the homes (in square feet). Observe how poor the overall correlation appears to be. It is not significant ($r = 0.42$). If, however, we now identify the houses as belonging to two different housing developments in different communities, we might obtain a graph more like Figure 6.18. In this second figure the houses have been identified according to the community to which they belong. It is fairly evident that if you separate the groups then you have two fairly good lines. As would be expected, there is an increase in cost as the size of the house gets

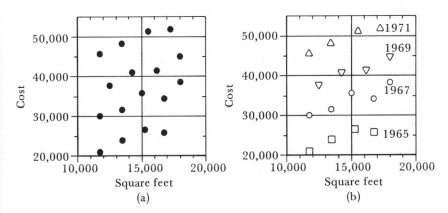

Figure 6.19 Scatter diagram of the cost and size of homes in a single development. (a) The collection of points are not distinguished in any way. A poor correlation. (b) Houses are classified according to year of sale. Observe how good the correlation is if we neglect the increase in cost from year to year.

larger. However, for prediction purposes, or even to identify a correlation, we are required to distinguish between the two sets of houses. The result is two correlations that are both significant.

Sometimes the low correlation is a result of a collection of many groups of values. A common example of this type of problem is when the data is accumulated over a period of years and changes occur from year to year. Figure 6.19a shows another large sample of house costs versus sizes of houses, but this time for the same housing development. Again the correlation seems poor. This time, however, we have costs that are spread over four different years (and a span of six years). If we identify the house according to the year sold, we find a set of parallel lines. If we discount the increase in cost each year, we will obtain an extremely high correlation.

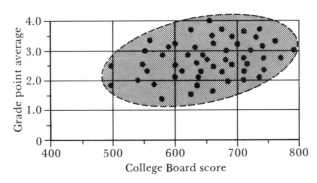

Figure 6.20 Grade point average versus College Board score for a group of students at a selective university. 4.0 is highest, 0 is lowest. There appear to be no correlation.

A *third* circumstance that can lead to low correlation occurs when the range of values is *truncated* or cut short. The argument that says that College Board scores or I.Q. scores are not good predictors of performance in college might be based on the low correlation between College Board scores and grade point average as shown in Figure 6.20. The problem with this contention is that it misses the fact that all those who were accepted were already *selected* at least partly on the basis of College Board scores. If in fact all the students were admitted randomly, then took *all of the same courses*, and also represented a *complete range* of College Board scores, the diagram in Figure 6.21 would be closer to the final result. Here we have a much higher correlation. By cutting off this distribution essentially at the 500 mark you will be able to approximate the distribution of Figure 6.20.

Now let us examine one case where a *high* correlation is actually a deception. We have already looked at cases when two or more sources or conditions that separately had high correlations were lumped together, only to lose any semblance of correlation. The reverse is also possible. We may have two or more clusters of points, each essentially uncorrelated, but located in such a way that they will cause a high overall correlation.

If you consider the starting salaries for a first job you may find that it is correlated to the age of the person taking the job. Why? Is it because the older you get the wiser you become? No, it is because the older job applicants are

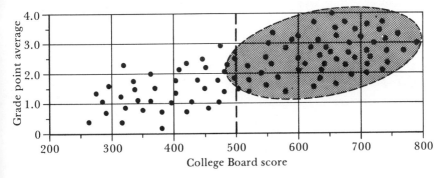

Figure 6.21 Grade point average versus College Board score if the selectivity factor were not considered. If all the points to the left of the dotted line were eliminated you would essentially have Figure 6.20. You now have a fairly high correlation.

older generally because they have completed either a 2 year or a 4 year college program. The first beginning salaries to high school graduates as a group are not correlated to age. Neither are first salaries to college graduates correlated to age. But lump the several groups together and you end up with a high correlation as in Figure 6.22.

6.11 Other Scales and Graph Paper

Previously we observed how many sets of data may show up as curves. Often we obtain these curves because the process does not behave according to a straight line equation. We are no longer considering that the prediction equa-

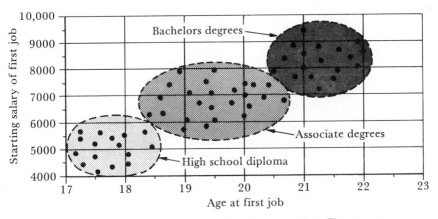

Figure 6.22 Starting salary and age at the time of first job. There appears to be a high correlation. This is actually because of the clustering of graduates of different educational levels who enter their first job at similar ages and the fact that more education generally leads to higher starting salary.

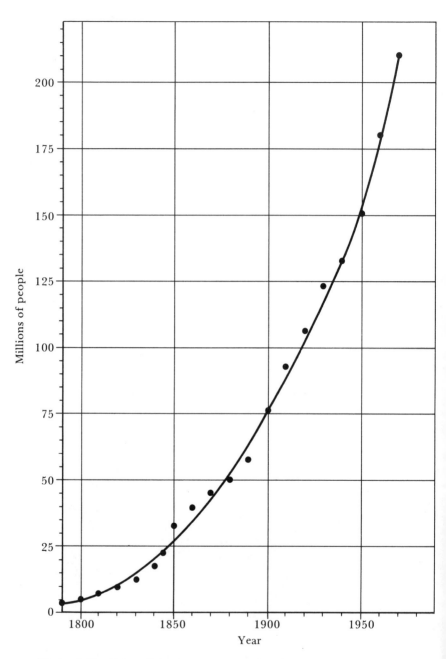

Figure 6.23 United States population from 1790 to 1970. Plotted on linear graph paper. The population looks like it is going to explode.

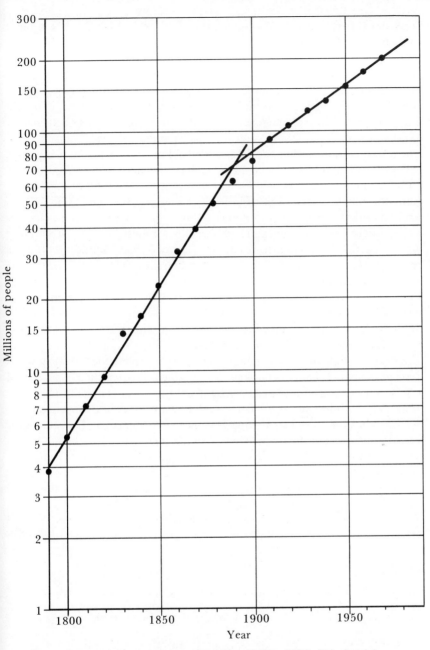

Figure 6.24 United States population from 1790 to 1970. This time it is plotted on semi-logarithmic paper. We now have what appear to be two straight lines, one going from 1790 to about 1870 and a second from about 1900 to the present. These lines indicate that there are two different rates of growth that have taken place over the two different periods of time. We have a much more realistic estimate of how the population "might" continue to grow (if we want to risk predicting into the future).

tion is linear (a type which says perhaps that $Y = 92.7 + 7.77X$), but rather it could have powers of X (maybe it would be $Y = 92.7 + 7.77X + 1.2X^2$) or it could involve many other functions or ways of including X (such as $1/X$, or a^X, or $\log X$). The method whereby a graph can assist us here is essentially by the same approach we took in Chapter 3, when we found that changing the scale of cumulative frequencies could make certain curved graphs essentially straight lines. Sometimes by replacing the linear scales that we have been using throughout this chapter by different scales we *may obtain a straight line instead of a curve*. The interpretation of the line will then depend upon the scales actually used.

Figure 6.23 shows a graph of the United States population from 1790 to 1970. Observe how curved it is. You might interpret from this curve that we are going to explode pretty soon. If instead we now plot the population data on semi-logarithmic graph paper (Figure 6.24) we have what looks much more like a straight line. In fact, if you look carefully it almost looks like two straight lines. Something appears to have happened around 1900. Actually there was a pretty constant 33% increase for each decade from 1790 to about 1860. The increase then dropped to about 25% until 1900 when it again dropped to about 15 to 20%. This lower rate of population increase has been present during most of the 20th century. A property of the semi-logarithmic graph paper is that it yields a straight line when there is a constant *rate* of percent *change*— a constant rate of either increase or decrease.

This recognition of certain changes in conditions is often more evident when using graph paper that does not have linear scales. Frequently the presence of a curve on linear graph paper may be strictly due to a change in the way a process acts at different stages. Thus, fluids flowing slowly have certain properties, but when they flow rapidly they become very turbulent and may have different properties and this change may occur at a very definite speed. Certain types of human growth and personality development make definite changes at the time of puberty and adolescence. The rate at which grass or corn grows will change with a heavy dose of fertilizer. Situations such as these may alter what otherwise might be close to a straight line relation. Or they may just present strong additional reasons for being ready to look at different graphs and to spend some time exploring the possibility that different relationships might be present.

REVIEW

Correlation · Scatter diagram · Relationship ·
Correlation coefficient · r · Prediction · Significance ·
Cause and effect

$$r = \frac{\sum XY - \frac{(\sum X)(\sum Y)}{n}}{\sqrt{\sum X^2 - \frac{(\sum X)^2}{n}} \sqrt{\sum Y^2 - \frac{(\sum Y)^2}{n}}}$$

Regression · Linear relationship · Scales · Range of scale ·
Nonlinear relation

$$Y = \bar{Y} + b(X - \bar{X})$$

$$b = \frac{XY - \dfrac{(\Sigma X)(\Sigma Y)}{n}}{\Sigma X^2 - \dfrac{(\Sigma X)^2}{n}}$$

PROJECTS

1. You can easily duplicate the following experiment. Take an elastic band and attach it to a nail. On the other end of the elastic attach a paper clip as a hook. Pass the paper clip through a small envelope. (See Figure 6.25.)

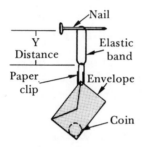

Figure 6.25

Now measure the length of the elastic band. This distance corresponds to a load in the envelope of zero. Place a coin, perhaps a nickel, in the envelope and observe the new length of the elastic band. Continue to put nickels in the envelope and measure the length. Calling the length Y and the weight (number of coins) X you can plot a regression curve. Take several readings with different amounts of coins. Calculate a correlation coefficient and the regression equation. How close is the line to a straight line?

Try the same experiment but use a different length or thickness elastic band. Compare the results. Try it with other elastic bands of the same size.

What happens if you use quarters or pennies? How about the correlation of cents (amount of money) vs. length?

2. Ask a group of people what their weights are. Then weigh them (on the same scale). Prepare a correlation diagram of the weight they claimed they were versus their actual weight.

Do a second correlation. This time compare the "weight they claimed they were" versus "the difference between the actual and claimed weight." How do the two correlations compare?

Do the same for height or foot size (or some other similar characteristic).

3. If you have been recording temperature of your refrigerator (Project 1, Chapter 1) you can now expand the project. Each refrigerator has a control dial that usually reads from 1 to about 6. Change the dial setting and measure the temperature (you had better leave it for several hours to let the refrigerator reach the proper temperature). Try all of the readings and then plot a correlation curve and a regression line. How close is the line to a straight line?
4. Drop a tennis ball from several different heights (see Project 7, Chapter 1). Measure the distance it bounces up. Is there a correlation? What is the regression line?

PROBLEMS

Section 6.2

6.1 A group of secretaries were given a typing test. The speed and number of errors were recorded as follows:

Speed	65	60	55	75	100	35	45	60	70	50
Errors	3	4	2	5	5	2	3	3	2	3

Plot a scatter diagram and calculate the correlation coefficient.

6.2 A study was done for the number of staples (out of 25) that stapled properly through different numbers of pages of paper. The results were as follows:

Number of staples	23	23	20	10	8	24	21	18	10	15
Number of pages	4	3	8	12	15	2	5	7	13	10

What is the correlation coefficient?

6.3 At the start of the 1969 Football season the New York Jets had 39 players. The number of each player and his weight were announced as follows:

Number	*Weight*	*Number*	*Weight*	*Number*	*Weight*
62	230	80	244	28	160
66	255	71	270	46	180
77	265	31	220	26	180
34	235	51	235	48	178
13	179	83	195	18	205
40	180	60	210	12	200
52	250	42	195	67	255
63	230	41	219	86	265
75	285	20	185	89	213
32	207	22	210	15	190
85	245	56	212	65	215
64	248	29	179	43	186
87	230	81	245	11	215

Plot the number and weight.

For the above data

Σ (numbers) = 1971 Σ (numbers^2) = 122,199
Σ (weights) = 8500 Σ (weights^2) = 1,889,100
Σ (numbers · weights) = 447,979.

Now calculate the correlation coefficient.

6.4 The income tax rate schedule for a single person after deductions (for the income year of 1970) was

Income	Tax	Income	Tax
1000	145	10,000	2090
1500	225	14,000	3210
2000	310	20,000	5230
4000	690	26,000	7590
6000	1110	50,000	20,190
8000	1590	100,000	53,090

Plot these values. Calculate the correlation coefficient (this will be time consuming without some type of computation equipment).

6.5 Verify the correlation coefficients of Figure 6.6. For the last three diagrams, Figure 6.6d, e, and f, you will find it easier to use the following columns:

X	Y	f	f · X	f · Y	f · X^2	f · Y^2	f · X · Y
		sum	sum	sum	sum	sum	sum

The frequencies, f, are the number of tally marks in the respective boxes. The correlation coefficient would then be calculated using the following equation:

$$r = \frac{\Sigma f \cdot X \cdot Y - \frac{(\Sigma f \cdot X)(\Sigma f \cdot Y)}{(\Sigma f)}}{\sqrt{\Sigma f \cdot X^2 - \frac{(\Sigma f \cdot X)^2}{(\Sigma f)}} \sqrt{\Sigma f \cdot Y^2 - \frac{(\Sigma f \cdot Y)^2}{(\Sigma f)}}}. \qquad (6.9)$$

Thus, for Figure 6.6d instead of 62 sets of calculations you only need 37.

Section 6.3

6.6 Is the correlation coefficient of Problem 6.1 significant?
6.7 Is the correlation coefficient of Problem 6.2 significant?
6.8 Is the correlation coefficient of Problem 6.3 significant?
6.9 Is the correlation coefficient of Problem 6.4 significant?

6.10 Figure 6.6 has 6 scatter diagrams with the respective correlation coefficients (as well as n). Which have a significant correlation (0.05)? Which are *very* highly significant? Which are essentially not significant?

6.11 To understand the need for a large enough sample size, compute a correlation coefficient for just the first 5 students listed in Table 6.2 (use Table 6.3 to ease your computations). Is this correlation significant? Why is there such a difference between this value and the one for all of the 20 students?

Section 6.4

For Problems 6.12 to 6.15 identify whether one of the measured quantities might cause an effect in the other quantity. If there is no obvious cause and effect relationship, what third or outside factor could be influencing both quantities?

6.12 Refer to Problem 6.1.
6.13 Refer to Problem 6.2.
6.14 Refer to Problem 6.3.
6.15 Refer to Problem 6.4.

Section 6.5

6.16 The quality of coffee is to be gauged by two coffee tasters. They apply a rating from 0 to 10 (with good coffee rated 10, poor coffee rated 0). The coffee has been perked from 1 minute to 8 minutes and the ratings were

Time	Rating	Time	Rating
1 minute	1,1	5 minutes	4,4
2 minutes	3,0	6 minutes	7,5
3 minutes	2,2	7 minutes	8,7
4 minutes	4,2	8 minutes	8,6

Plot these values. Calculate the regression equation. What taste value do you expect for coffee perked $7\frac{1}{2}$ minutes? What rating do you predict for coffee perked 5 minutes? How close is this to the actual values?

6.17 Refer back to Problem 6.2, the stapled pages. Which value belongs on the X-axis? Calculate the regression equation. How many staples do you predict would properly staple 20 sheets? How many staples do you predict would properly staple 10 sheets? How does this value compare to the actual result.

6.18 Refer to Problem 6.1. What is the regression line for predicting errors from speed? Speed from errors? What is the prediction of errors for a 90 word per minute typist? What speed would a typist with 5 errors be typing at?

6.19 The median years of school completed for people in different income levels was reported as

Income	Years of school	Income	Years of school
1500–1999	8.4	6000– 6999	12.1
2000–2499	8.6	7000– 7999	12.3
2500–2999	8.7	8000– 9999	12.5
3000–3999	9.0	10000–14999	12.8
4000–4999	10.6	15000–24999	16.1
5000–5999	11.7	25000–over	16.3

Using the appropriate values for income calculate the regression equation. What does your equation predict for a $10,000 income? $20,000? How close is the predicted value to actual (in the latter case)?

6.20 What do you predict the weight of a new Jet Football player, No 50, would be? (See Problem 6.3). Calculate the regression equation first.

6.21 Refer to the data of Problem 6.4. What is the linear regression equation relating the amount of tax for different incomes? What do you predict the tax on a $10,000 income (after deductions) would be? On a $50,000 income? On a $2000 income? How much different are the actual and predicted values?

Section 6.10

6.22 A basic college algebra test was administered to a group of 20 students. Along with the test grade the heights of the students were also recorded. The 20 grades (out of a possible 100 points) and heights were

Grade	Height	Grade	Height	Grade	Height
85	5' 9"	90	5'11"	75	5' 6"
15	5' 2"	95	5' 8"	45	5' 1"
60	5' 4"	55	5' 4"	75	5' 3"
5	4' 8"	35	5' 1"	80	5' 8"
30	4' 9"	20	5' 0"	40	5' 1"
30	5' 1"	60	5' 7"	35	5' 3"
20	4'11"	25	4'11"		

Plot a graph. Calculate a correlation coefficient and explain this unusual correlation.

6.23 Go back to Problem 6.16. What would you predict that the average taste value would be for coffee perked for 11 minutes? How does this compare to a maximum possible reading of 10?

Several more tests were run and they were

Time	Rating	Time	Rating
9 minutes	6,7	12 minutes	1,4
10 minutes	5,7	13 minutes	1,3
11 minutes	3,5	14 minutes	0,3

Add these values to your previous graph. Explain what has happened.

6.24 Refer to the data in Problem 6.19. Why is it improper to use this data to predict income if you know the years of schooling completed?

6.25 Examine the graph of Problem 6.4 and the equation of Problem 6.21. Explain why the apparent large differences between the predicted values and the actual values. What is the difficulty with assuming a linear regression? Try plotting the data on other graph paper such as semi-logarithmic or log-log paper.

7

Graphing II: Control Charts

We will now find ways of determining if time-dependency or instability might be present in a single set of data. This will lead to something called a control chart and then to an elementary type of experimental design method. We will also begin to look for ways of making interpretations and decisions based on graphs.

7.1 Time-Dependency

Most problems involving data also involve time. I do not mean that it takes time to perform the experiment or to gather the data, but rather that the data is not accumulated all at once. There is some sequential nature about the data. We will look at what might be considered as three levels of time dependency.

1. A nearly instantaneous—one after another—sequence.
2. A short range sequence.
3. A very long range sequence.

(1) Almost every type of process or machine produces things one at a time. Our loaves of bread are made one after the other in a machine. Automobiles come off the assembly line in a sequence. Plastic garbage cans are injection molded one after the other. Paper is made in a continuous roll. This book is printed page after page in a particular order. A poll is conducted by asking one person his opinion and then another person his opinion.

(2) Another way that things may be regarded as time related has to do with slightly larger time intervals. Our officer in Example 2.1 gave tickets on successive days. Commuting to school is performed on a daily basis. Absences are recorded in the same order as class periods. The temperature is different on successive days. Automobiles may be considered as being produced today and

tomorrow, thus lumping groups of automobiles together. Even something like typing errors or pages typed by a secretary would fall into a time sequence.

(3) The third time base could be regarded as dealing with extended periods. We talk about output of eggs or chickens each year, tons of wheat grown each year, number of births a year, number of deaths from cancer or pollution or automobile collisions each year, and so on, and so on. We are concerned about corporate profits not for today but for a quarter of a year. The president of IBM wants to know how many computers will be sold this year or this month. Those who prepare the school budget and the purchasing agent must have estimates of how much soap and toilet paper is needed during the school year.

Several questions might come to mind:

1. Is it possible we might be interested in all three kinds of intervals for the same type of process (or system), and if so, in what ways?
2. What *kind of effect* might time have on our process?
3. How do we *detect* a time dependency?

I will answer the first two questions together with several illustrations, and then we will consider the third question. In each example we will look at how rapid-instantaneous changes, the short range changes, and long range changes might be of concern and might affect a process.

EXAMPLE 7.1

Weather! Everyone talks about the weather and some people are extremely concerned about it.

Instantaneous time sequence: If you are flying a kite, a change in the wind velocity from one moment to another can reduce a successful flight to a grounded one. A pilot might be even more concerned if he has an abrupt change in weather conditions. A surveyor needs to know the temperature because his tape measure expands when it gets hot. A rapid change in the temperature may affect his measurements.

Short range: If we consider day-to-day changes, then whether you should take an umbrella or wear a coat or plan a picnic depends on how much variation there is between today and yesterday or today and tomorrow (assuming that the day to day variation is as good a weather predictor as the weather bureau—as some people will contend).

Long range: If year by year or season by season periods are considered as long range, then a valid question might be how many snow days should we plan for in the next school year calendar? How many degree days might we anticipate is important to the oil producer, the delivery man, or the consumer (in predicting heating cost for his house). How hot the summer will be is important to the electricity company which must provide power for thousands of air conditioners.

EXAMPLE 7.2

Stock market: The stock market is a vitally important source of income and loss for many millions of people. It fluctuates in minute-to-minute, day-to-day, and year-to-year spans (among others). Figure 7.1 shows the change in a stock over 10 of each of these types of intervals.

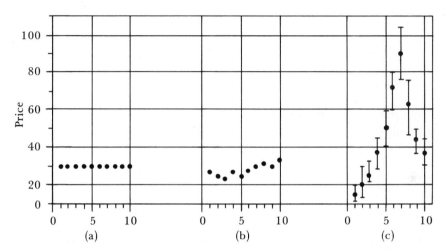

Figure 7.1 Changes in the cost of a common stock over several periods of time. (a) Ten consecutive sales on *one day*. (b) Ten consecutive days—the value plotted is the "average". (c) Ten consecutive years—center point is the "average", the long bar with the two cross pieces is the range during the year.

Instantaneous: The ticker-tape watcher and the broker follow the individual transactions of the stock market and may make buy-or-sell decisions at any moment, sometimes dependent upon a single previous transaction.

Short range: Most investors follow the newspaper summaries of the various stocks traded and are not concerned with very small changes (say $\frac{1}{2}$ or 1 point changes reflecting 50 cent or 1 dollar changes in value).

Long range: There are generally long range patterns that stock prices follow. People willing to put money away in stocks for a long period are interested in these long range changes and projections.

EXAMPLE 2.3 (Cont.)

This example consisted of a set of bread weights. I emphasize this example as an illustration of how time is related to a controllable process.

Instantaneous: Breads baked in a sequence will vary in some way. However, the most similar breads baked will probably be those that come out next to each other.

Short range: The temperature or humidity in the oven may vary from hour to hour or from morning to evening. These variations may affect the baking of the bread, the size of the bread, or maybe just the crust. Nevertheless, such variations will contribute to greater variation amongst breads that are compared over day-to-day or batch-to-batch times rather than next-to-another in the same batch on the same day.

Long range: Some bread is now formed as a continuous ribbon of dough and then cut off every so many seconds. If a slow change takes place in the consistency of the dough, the bread may be smaller or larger at successive times. Tubes through which the dough flows may become clogged as time passes. Certain ingredients may age with time which may affect the bread.

The third question that I anticipated that you might ask was, "How do we detect the time-dependency?" It depends. The primary methods that we will use in this chapter will be *control charts*. They will be useful with any type of process, but generally are to be used with short range time spans. It is a very powerful and widely applicable technique. A second method, often called *time series*, which for the most part is actually regression analysis, is usually applied to long range time spans, although it might be applied to any time length (and in fact may be the only practical method for working with instantaneous time sequences). We have already introduced the concept of regression and have even applied it to some problems that would fall into the category of time series. However, at this time we will disregard the use of regression techniques.

7.2 Variability in Different Time Spans

In looking at the three time spans we should recognize one very important relationship. In each of the examples there is less variation between instantaneous values than with any other sets of values.

EXAMPLE 7.1 (Cont.)
The temperature usually changes very little from minute to minute, so that if I were to take temperature readings every minute for 10 minutes I would obtain very little variation. If however, I then took temperature readings on 10 successive days (all at the same time of day) I would expect much greater variation. Try it!

EXAMPLE 7.2 (Cont.)
The change in price of a stock is usually very little from transaction to transaction (instantaneous) but much more from day to day. So again instantaneous variation is the smallest (refer back to Figure 7.1).

EXAMPLE 2.3 (Cont.)
As I stated during the previous discussion of this example, side-by-side or one-after-another breads will generally be very similar. The one distinction between this example and the previous two is that there is *no* reason for you to

expect that breads baked tomorrow or next week or next year are necessarily more variable than breads baked next to each other. In fact, the mark of a stable, consistent process that is *in control* is that in the long run there is *no* difference between individual-to-individual variation and day-to-day variation or week-to-week variation.

So the smallest variation I could normally expect to have with any type of process will be the individual-to-individual variation—the one-after-another variation. Then let us make use of this small variation to gauge if there is in fact a greater variation over larger time spans. If there is in effect *no* difference between instantaneous variation and perhaps short range (or even long range) variation, then we will say that the process is *in control*.

Now what do I mean by all of this? Look at Figure 7.2. Here I state that we have a process that is in control. The temperature of an oven will vary from

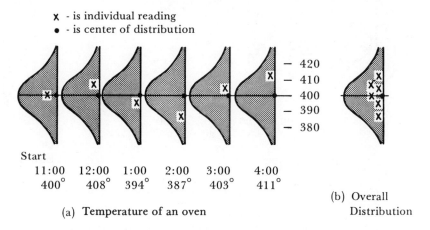

(a) **Temperature of an oven**

(b) Overall Distribution

Figure 7.2 A process that is in control which has a spread around a mean (in this case 400°). The process is the temperature of an oven. (a) We pick a time to start observing and there is a distribution. During each successive time period there is still a spread but the center has not changed even though the individual values may differ. (b) An overall distribution for all of the times. Observe that there is no difference between this distribution and *any* of the distributions at different times.

minute to minute (even when operating perfectly) or it will vary from location to location on the inside. The distribution at any one time is shown in the *start* position. Now we let time pass. At 12 o'clock the process has not changed, although we obtain a temperature reading of 408 degrees. At each subsequent time the distribution has *not* changed, only the particular measurement—and that particular measurement might have been just as different if we had measured it only a minute or two later, instead of an hour or two later.

In addition, if we were to look at the distribution for a long span of time, we would find that there would be *no* difference between the overall distribution and the instantaneous distributions (Figure 7.2b). This is always true with a process that is in control.

As a comparison let us look at an example of a process that is *not* in control, nor can it be controlled—the weather. If I measure the outside temperature during any instantaneous period I would expect to obtain values from a distribution like that at 11 o'clock (my start position) in Figure 7.3a. As time goes on though, that distribution continually shifts its position, so that by 4 o'clock it has

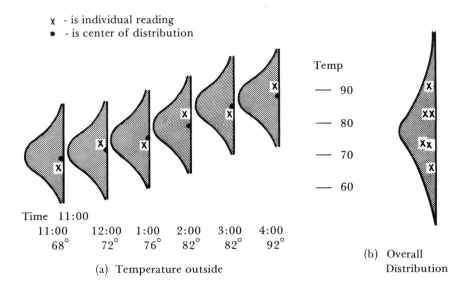

Figure 7.3 A process that is *not* in control—the weather. (a) The distribution changes from hour to hour. Since the distribution shifts, even if the values are close to the center of the distribution they vary considerably from hour to hour. (b) An overall distribution. Observe how broad this distribution is when compared to the other instantaneous distributions.

gone to 88° (from the 68° at 11 o'clock). Thus, the process itself does not remain constant, stable, or in control. I may find that subsequent readings may be the same even though the distribution has shifted, which only points out further that just because two readings are the same, the population or source are *not* necessarily the same. Now look at the overall distribution Figure 7.3b. Do you see how much greater the spread of this distribution is than any of the instantaneous distributions? This is something to beware. If the process is not in control then an overall population will have an exaggeratedly large spread.

7.3 Control

Let us look at the question of control a little closer, and ask the question, "Does all the data seem to come from a *stable* source?" The source *will always have variation* and will always have a distribution of some sort, but by stable I mean that the *distribution* does not change with time. Even when given a process that may *actually be stable*, before you can realistically say it is or it is not stable you must observe and examine it for some time. In addition, just because you see variation does not mean that you will be able to "adjust it" to make it less variable. Let us observe a situation where a stable process can actually be made unstable by altering it.

The temperature of the water in your shower will vary even if the system is perfect. You set the hot and cold faucets and a mixture comes out somewhere between hot and cold *but* with variations from instant to instant. Let us assume that under the best conditions the water has been set at 95° and has a *standard deviation* of two degrees. So I am recognizing that there is variation and I have provided an estimate of its magnitude. By the empirical rule, about 95% of the time the temperature will fall between $\pm 2\sigma$ or ± 4 degrees. So most of the time it will be between 91 and 99 degrees. Let us also say that below 91 degrees it feels very cold and above 99 degrees it feels very hot (see Figure 7.4a).

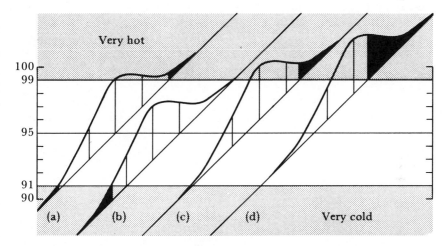

Figure 7.4 The effect of altering the temperature level of the shower. The dark areas show the chance of falling into the very hot or the very cold regions. Note how these areas change as we tamper with the average of the system.. (a) Distribution is stable at 95 degrees. (b) We changed the system by dropping the temperature 2 degrees. (c) We increased the temperature 3 degrees. (d) We increased it 2 more degrees.

On extreme occasions the temperature will be greater than 99 degrees or less than 91 degrees, but no amount of adjusting will completely eliminate these possibilities. The variation and hence the chance of getting cold water or hot water is a property of the system itself.

But you don't realize this or you don't believe it. So, you stick your hand in and you find it registers 97 degrees. That is higher than you want so naturally you lower the temperature to what you call the "proper level." At the same time you have lowered the distribution 2 degrees and that places it in the position of Figure 7.4b, since you have really lowered the *entire* distribution when you adjusted the system. The new average now is 93 degrees, not 95 degrees as it was before.

Notice the amount of area of the distribution that is in the very cold region. You have just increased your chance of being shocked by a very cold blast (remember that the area is equivalent to the chance of ending up in the particular region).

So you stick your hand in again and this time it registers 92 degrees. This is very far from the center so you increase the average by 3 degrees, placing the distribution in position c of Figure 7.4. Your chance of being scalded has now been increased.

You try again with a third check and this time it registers 93 degrees. It is still too cold, so you increase the temperature another 2 degrees. Since you have made so many adjustments you assume it is now fine.

Well, step in.

You don't know it, but by all the juggling and adjusting you have just managed to increase your chance of being scalded from about $2\frac{1}{2}\%$ to about 31% (Figure 7.4d). (Later on, in Chapter 10, we will discuss how the numbers are arrived at, but for now just accept them.)

The moral is simply to learn the capability of the system first. Only a major change in the nature of the system will cause a change in the *random pattern of variation present*. The system variability rarely can be adjusted unless a major change takes place. To change the system might require the installation of a special new mixing valve, thus altering the entire structure of the system, or buying a new boiler, or making sure nobody in the building uses water while you are taking a shower.

7.4 Control Charts

In order to judge whether or not our process is in control we must first determine what constitutes the best measure of inherent variability. In other words, how can we best describe the instant-to-instant variation? Our simplest technique will be to look at a sample of values taken at some "instant." You must realize that it is possible that there may have been some change in the process in that brief time span, but we will have to neglect any such small changes, and attribute them just to random, arbitrary variation. So we will be able to obtain small distributions, and for each of these small distributions we will be able to arrive at a central value and a spread. We must also remember that the measure of spread that we will obtain for each of the various small distributions should be the smallest variations that we will encounter.

We now create a time axis and we plot the values we obtain for *each sample* above the time that we made our observations. We will regard all of the values in a sample as having been accumulated at the same time. Essentially the

Table 7.1 Number of tickets issued. Samples of four with the means and ranges for each day of February

Day	1	2	3	4	5	6	7	8	9	10	11	12	13	14
Values	14	17	18	14	12	15	12	10	11	2	4	9	14	13
	11	9	12	8	10	13	11	8	6	3	3	7	12	11
	13	15	11	11	14	13	19	10	11	4	2	12	13	11
	9	10	11	9	9	11	9	6	9	1	2	8	9	10
Mean of four	11.75	12.75	13.00	10.50	11.25	13.00	12.75	8.50	9.25	2.50	2.75	9.00	12.00	11.25
Range of four	5	8	7	6	5	4	10	4	5	3	2	5	5	3

Day	15	16	17	18	19	20	21	22	23	24	25	26	27	28
Values	15	15	9	10	10	12	14	10	12	10	8	13	14	12
	11	11	5	8	10	13	14	7	9	9	9	12	10	10
	11	8	12	10	9	15	13	9	12	12	10	12	11	17
	8	10	8	10	8	9	6	3	6	7	8	9	9	7
Mean of four	11.25	11.00	8.50	9.50	9.25	12.25	11.75	7.25	9.75	9.50	8.75	11.50	11.00	11.50
Range of four	7	8	7	2	2	6	8	7	6	5	2	4	5	10

points in the pictures I have drawn in Figures 7.2 and 7.3 constitute this type of graph except we only made *one* observation at each time.

I will now use Example 2.1 as my main introduction to control charts.★

Table 7.1 shows the original data for Example 2.1 only slightly rearranged. It now reveals that actually 4 different values of the number of tickets were recorded on each of the 28 days of February. These 4 values constitute a sample for *each* day. The table also includes some calculations for each sample.

As I have already stated, the samples should be considered as small groups of individuals selected in a way that they would be most alike. We talk about a "rational subgroup" or a set of replications or repeated measurements. We might consider that the 4 values or sample obtained on Day 1 (the 14, 11, 13, and 9) represented 4 different periods of time on a road, 4 lanes of a highway, 4 officers, or 4 radar traps. When we discuss other types of problems we could perhaps consider the following as a sample:

1. Five loaves of bread baked during a particular 5 minute interval (thus considering all conditions as fairly constant during that time span).
2. Four different pages typed during a short period (to be checked for errors).
3. Three different temperature readings of our shower (or body temperature) taken in a two minute interval.

We talk of a sample of four automobiles today, four more tomorrow, and so forth; three garbage cans checked this morning, three more this afternoon; two samples of one square foot of paper every 100 feet of production. Also be aware that when we discuss sample we can be considering different degrees of similarity. Our first listing of tickets issued (Table 2.1) was a sample. However, I am now saying that the large sample actually consisted of many smaller, also distinguishable, samples.

Figure 7.5 shows a graph with all the individual values plotted at the various times they were recorded. Here you can see the kind of randomness that takes place. There are some observations that might be made about the data, but in general there is too much overall variation. This masks any characteristic changes. In addition, limits of any type are not too useful on a chart like this.

Two types of new charts or graphs can be constructed. One contains the means of each of the small samples and the other contains the ranges of these same samples. So we will be plotting a graph of central value (mean) and a graph of spread (we will use the ranges since they are so easy to calculate). The samples generally consist of from three to six individuals with four or five being the preferred. In later chapters we will investigate why, but let it suffice to say that for the type of evaluations that the control chart is most effective for, these size samples are the most useful.

★ This is not necessarily the kind of problem that most texts use for discussing control charts. The typical problem deals with a machine or manufacturing process. I think that the control chart is a much more general tool and you should come to understand and consider a process as a very broad concept.

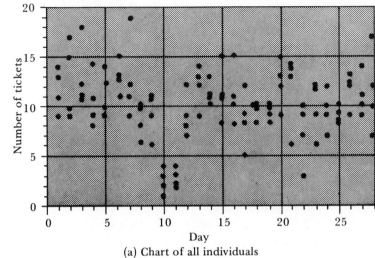
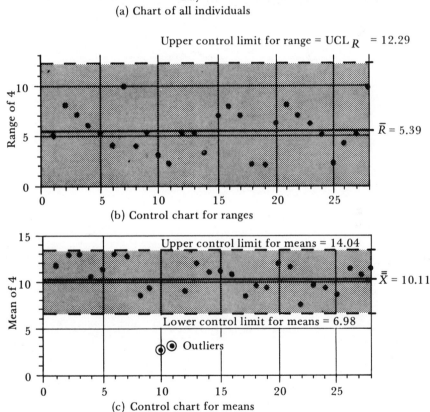

Figure 7.5 Time charts (including control charts) for the number of tickets issued (Example 2.1). Day is day of the month of February. (a) Chart showing all of the individual values. (b) Control chart for ranges of values (sample) for each day. (c) Control chart for means of values (sample) for each day.

Our first *assumption* is that the population or overall distribution from which our samples are taken is reasonably close to a normal or bell-shaped form previously discussed. Later, in Chapter 12, we will discuss just how important this requirement is. If this assumption is valid we can construct limits for the mean and range charts—limits that say that we should rarely expect samples to fall outside them unless something has occurred to the process. The first limits we will consider will be for the range chart.

Range Chart

The range is a measure of variability. What we are looking for is an indication that the variability *for each sample* is consistent. The limits on the range chart are a way of indicating whether something may have caused the variability to have changed. If we obtain individual values outside the limits we construct, then we will become very suspicious that the variability may not be the same from sample to sample. As an extra bonus we can also get a good estimate of the overall population standard deviation.

STEP 1. We first calculate the average or mean of the ranges.

$$\text{Average range} = \bar{R} = \frac{\text{Sum of ranges}}{\text{Number of samples}} = \frac{\sum R}{k} \quad (7.1)$$

where k = number of the samples.

(Refer to Table 7.1.)

$$\bar{R} = \frac{5 + 8 + 7 + \cdots + 5 + 10}{28} = \frac{151}{28} = 5.39.$$

STEP 2. We calculate control limits.

There are two possible control limits, an upper and a lower limit. The limits that we use are found by multiplying the average range by a constant. The constants we will use are called D_4 and D_3 and are found in Appendix Table A.10. The constants depend *only* upon the size of the samples. When the sample size is six or less, the lower limit of the range is 0. Hence, if the values in one sample of six or fewer individuals are all identical (thus resulting in a range of 0) this is still not unusual enough to consider that the variability is out of control.

Hence

$$\text{Upper control limit for range} = \text{UCL}_R = D_4 \bar{R} \quad (7.2)$$

$$\text{Lower control limit for range} = \text{LCL}_R = D_3 \bar{R} \quad (7.3)$$

(the subscript R standing for range; the other letters should be obvious). For samples of four individuals Table A.10 gives:

$$D_4 = 2.28 \quad \text{and} \quad D_3 = 0$$

then, for Example 2.1

Upper control limit for range $= D_4 \bar{R} = (2.28)(5.39) = 12.29$

and

Lower control limit for range $= D_3 \bar{R} = (0)(5.39) = 0$.

Now, what does the control limit mean?

The limit tells us that if the samples are coming from a population in which the *variability is stable*, then there is less than about $\frac{1}{4}$ of 1% chance that a range should go beyond this limit.* If a range goes beyond a limit, the odds would be about 370 to 1 that something suspicious has happened that has changed the system. It is a very strong warning that the samples *have not* come from a population with a constant variation (standard deviation).

STEP 3. Construct the control chart for ranges.

Figure 7.5b shows the ranges plotted along the same time axis we previously used. In addition, I have drawn in the upper control limit. Observe that no values actually exceed it. So we shall consider the ranges, or the variability within each sample, as being reasonably consistent. We can now proceed.

STEP 4. Our bonus is an estimate of the standard deviation of the population and involves a quantity called d_2, where

$$\sigma = \bar{R}/d_2. \tag{7.4}$$

Similarly,

$$\bar{R} = d_2 \sigma. \tag{7.5}$$

The values of d_2 for different sample sizes are also in Appendix Table A.10. If you have at least about 10 or 15 samples, then these constants do *not* depend on the number of samples taken and you can freely refer to the tables. If you are dealing with less than 10 samples you need to adjust d_2. However, we will not worry about this situation in this text.

If we consider n as the sample size then

$$n = 4,$$

and from Table A.10

$$d_2 = 2.059.$$

The standard deviation would now be estimated as

$$\sigma = \frac{\bar{R}}{d_2} = \frac{5.39}{2.059} = 2.61.$$

Observation!

---

* The percentages and odds you will again have to accept more or less on faith at this point, but we will explore this more fully in Chapters 12 and 13.

This is much smaller than the 3.41 we calculated in Chapter 4. Do you have any idea why?

Recall how I showed you that samples from two different populations can create a combined population that is spread out more than the individuals. Look back to Figure 7.3 to see what can be happening. In that figure the population was shifting around to many different locations. The range or standard deviation of the samples at any particular time from the population was relatively small. But combine all the different sample populations and you now have one that is spread out much more. Thus, often the *sample ranges* (or sample standard deviations) act as a better indication of the variability of the system than the overall standard deviation. The next question should be whether this difference in variability is excessive. We will leave the discussion of the methods, and several do exist, for a later time (see Chapter 17, dealing with the F test). However, we still have to examine the means, since they might give us a clue as to whether the distribution seems to be jumping around.

Means Chart

STEP 1. Calculate the average of all the sample means.

$$\text{Mean of sample means} = \bar{\bar{X}} = (\sum \bar{X})/k \tag{7.6}$$

(where $\bar{\bar{X}}$ is read as "X double bar").

Thus, for Example 2.1

$$\text{Grand mean} = \bar{\bar{X}} = \frac{11.75 + 12.75 + \cdots + 11.00 + 11.50}{28} = \frac{283.00}{28} = 10.11.$$

Here you should observe that *the grand mean is exactly the same as the mean of the individuals* which we calculated in Chapter 4. This is *always true* and serves as a convenient check.

STEP 2. Calculate the control limits using the following equations:

$$\text{Upper control limit for means} = \text{UCL}_{\bar{X}} = \bar{\bar{X}} + A_2 \bar{R} \tag{7.7}$$

$$\text{Lower control limit for means} = \text{LCL}_{\bar{X}} = \bar{\bar{X}} - A_2 \bar{R} \tag{7.8}$$

where A_2 is given in Appendix Table A.10 and $\bar{R}$ was already calculated. As with the other control chart constants, A_2 is not dependent on the number of samples, *only* on the *size* or number of individuals in each of the samples.

For Example 2.1

$$\text{Sample size} = n = 4,$$

therefore

$$A_2 = 0.73 \qquad (\bar{R} = 5.39).$$

Then

$$\text{UCL}_{\bar{X}} = 10.11 + (0.73)(5.39) = 10.11 + 3.93 = 14.04$$

and
$$\text{LCL}_{\bar{x}} = 10.11 - (0.73)(5.39) = 10.11 - 3.93 = 6.18.$$

STEP 3. These limits are then drawn on a chart of means.

We now observe in Figure 7.5c that two points are *outside* these limits. As with the range chart we say that if the system is stable the odds are about 370 to 1 against an individual crossing the control line unless something has happened to the process.* Regardless of the exact odds you use, the ones that lie beyond the control line give you reason to be suspicious about the process, and to conclude something may have occurred to change the system. So we go back and question if anything unusual took place on February 10 and 11, the times the process (the number of tickets) went out of control. The result of the investigation revealed that it snowed on the 10th and the area was still snowbound on the 11th. By the 12th however, traffic was back to normal, and so were the traffic tickets.

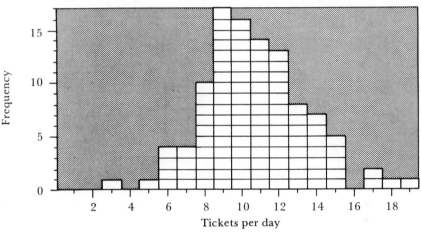

Figure 7.6 Tickets issued (Example 2.1). Histogram which is the same as Figure 3.6 except that the 8 points considered unusual have been eliminated. The result is that the lower mode is also eliminated creating a more "bell-shaped" distribution.

We should now be able to say that there is a significant relationship between tickets issued and weather conditions (particularly if it has snowed). Since we are looking for the stable underlying distribution, we should discard these two samples.

You should note what effect eliminating these eight individuals has on the shape of the histogram we had previously drawn (Figure 3.7). The little hump

---

★ Actually we do not have so great a set of odds. These odds should be interpreted to mean that for some single randomly picked sample the probability is about 0.9973 that it will not cross the limit lines (or the odds are 370 to 1 against *it* crossing a limit). The probability that *all* 28 will *not* cross the limit is smaller. It is $(0.9973)^{28}$ or about 0.93. This represents 13 to 1 odds that *none* will cross the limit—a smaller but still formidable amount.

made the distribution look almost bimodal. Removing the unusual values eliminates that little mode on the left and makes the distribution look even more bell-shaped than before (see Figure 7.6).

EXAMPLE 2.3 (Cont.)

It is now revealed to you that the data originally presented in Table 2.3 was not all accumulated at one time. Rather, it was collected by groups of five loaves a day over a period of twenty consecutive working days. In Table 7.2 I have presented the data by day and included the means and ranges of bread weights for the samples of five loaves for each day. We are interested in making a control chart to determine if the data is coming from a stable process.

Range Chart

STEP 1. Calculate $\bar{R}$:

$$\bar{R} = \frac{0.061 + 0.033 + \cdots + 0.068 + 0.052}{20} = \frac{0.883}{20} = 0.0441.$$

STEP 2. Calculate limits for the $\bar{R}$ chart:

$$\text{Sample size} = n = 5$$

therefore

$$D_4 = 2.11, \quad D_3 = 0,$$

and

$$\text{UCL}_R = D_4 \bar{R} = (2.11)(0.0441) = 0.093$$
$$\text{LCL}_R = D_3 \bar{R} = (0)(0.0441) = 0.$$

Table 7.2 Samples of loaves of bread for twenty work days (five loaves a day)

Day		1	2	3	4	5
Sample values		20.055	20.040	20.019	20.025	20.018
		20.052	20.024	19.993	19.984	19.982
		20.046	20.036	19.968	20.013	19.982
		19.994	20.008	19.987	19.984	20.006
		20.033	20.041	19.997	20.003	20.037
Sample $\bar{X}$		20.036	20.030	19.993	20.002	20.005
Sample R		0.061	0.033	0.051	0.041	0.055

Table 7.2 (cont.)

Day	6	7	8	9	10
Sample values	19.999	19.975	19.985	19.968	20.022
	20.022	20.011	20.002	19.975	19.988
	20.002	20.026	20.009	19.960	19.997
	19.976	19.976	20.000	19.988	20.006
	19.994	19.989	19.970	19.958	20.019
Sample $\bar{X}$	19.999	19.995	19.993	19.970	20.006
Sample R	0.046	0.051	0.039	0.030	0.034

Day	11	12	13	14	15
Sample values	19.981	19.987	20.013	20.018	19.994
	19.990	19.968	20.013	20.009	19.986
	19.965	19.970	20.053	20.026	20.018
	19.977	19.975	20.005	20.024	20.002
	19.986	19.969	20.004	20.003	20.024
Sample $\bar{X}$	19.980	19.974	20.018	20.016	20.005
Sample R	0.025	0.019	0.049	0.023	0.038

Day	16	17	18	19	20
Sample values	20.052	20.017	19.986	19.972	20.006
	19.982	20.017	20.004	20.030	19.983
	20.032	19.955	20.019	20.013	20.001
	20.007	20.011	19.983	19.962	20.035
	20.025	19.999	20.010	20.000	19.993
Sample $\bar{X}$	20.020	20.000	20.000	19.995	20.004
Sample R	0.070	0.062	0.036	0.008	0.052

STEP 3. Construct the $\bar{R}$ chart and place the limits on the chart. Observe that no ranges are outside the limits (Figure 7.7a). Therefore, we can proceed.

STEP 4. Calculate the estimate of the population standard deviation from Equation (7.4):

$$\sigma = \bar{R}/d_2 = 0.0441/2.326 = 0.019 \quad \text{(for } n = 5, d_2 = 2.326\text{).}$$

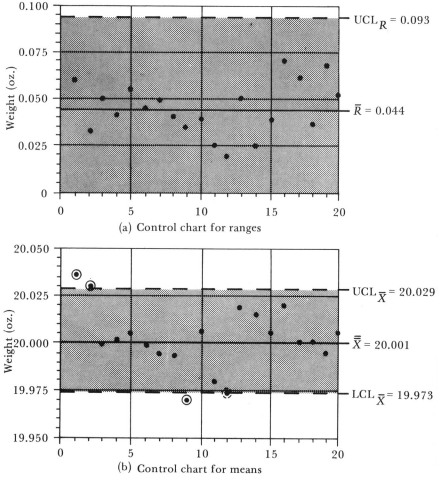

Figure 7.7 Control charts for the weights of Loaves of Bread (Example 2.3). Averages and ranges of 5 loaves for each of 20 work days.

Note that this is much less than the s obtained from the overall data, which may indicate that there is a problem. But it is also a lot easier to calculate if you have to work by hand.

Means Chart

STEP 1. Calculate $\bar{\bar{X}}$ or the overall grand mean of the means of the samples. This is

$$\bar{\bar{X}} = \frac{20.036 + 20.030 + \cdots + 19.995 + 20.004}{20} = \frac{400.021}{20} = 20.001.$$

STEP 2. Calculate limits for the $\bar{X}$ chart.

$$A_2 = 0.58$$

$\text{UCL}_{\bar{X}} = \bar{\bar{X}} + A_2 \bar{R} = 20.001 + (0.58)(0.0441) = 20.001 + 0.028 = 20.029.$
$\text{LCL}_{\bar{X}} = \bar{\bar{X}} - A_2 \bar{R} = 20.001 - (0.58)(0.0441) = 20.001 - 0.028 = 19.973.$

STEP 3. Construct the $\bar{X}$ chart and place the limits on it (Figure 7.7b). Observe that several values are outside the limits. We should be suspicious that something very unusual has been occurring. An investigation is definitely warranted.

7.5 Nonrandom Variation

As you should now be aware, the control chart is a very useful and simple technique to apply to many types of data. With it we are frequently able to identify many of the different kinds of *nonrandom* fluctuations present in a process.

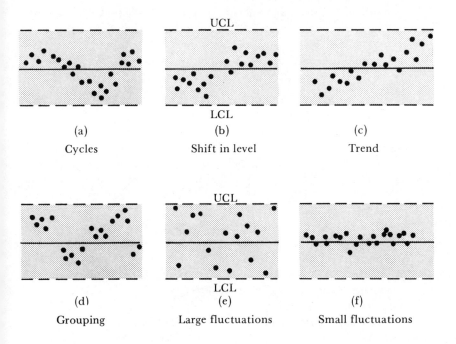

Figure 7.8 Examples of Non-random patterns of variation that may be found on control charts.

I have been stressing the fact that fluctuations or variations are always present and are a part of the process itself. They cannot be eliminated. However, the variation that we can never rid the process of is the random, *unpredictable* individual. What I am looking for *now* are other variations that might be superimposed or added onto the completely random variations. These are causes that we might be capable of identifying and perhaps reducing or removing from the process.

In addition to just plain abnormal individuals, some of the possible kinds of nonrandom variation might be as follows:

1. Cycles (Figure 7.8a): These are repetitive patterns. For example, typing errors that first increase, then decrease, then increase, and so forth.
2. Shift in level (Figure 7.8b): An abrupt change in the population average or variance. Someone changed the amount of time that the red light remains red, and now it takes another minute to get to school.
3. Trends (Figure 7.8c): A tool wears and the size of the pieces produced change. The gas mileage slowly decreases as the spark plugs or the engine wears. Regression techniques may be useful here.
4. Grouping (Figure 7.8d): Frequent changes in level. If the gas station attendant doesn't fill the tank uniformly each time, I will get all kinds of values for miles per gallon.
5. Abnormally large fluctuation (Figure 7.8e): This could mean the system is very unstable or you are working with two distributions.
6. Abnormally small fluctuations (Figure 7.8f): Could indicate the variation within the samples is much too large. Again two distributions may be working together only in a way different than in (5).

You may wonder why the last two charts are indicative of some kind of problem. If the distribution that we are sampling from is pretty close to a normal distribution (if we are using moderate size samples, even of size 5, this may not be too crucial) the control chart that is *in control* will have individual points (samples) that will form a bell-shaped distribution. In Figure 7.9a I have given you a typical control chart and the histogram for the points in that chart. Note how closely the histogram resembles a normal distribution—most in the middle, a few in the outer sections.

Figure 7.9b shows a distribution that I claim has large fluctuations (see (5) above). Look how spread out the histogram is. Too many of the values are in the extreme regions for this to be considered a normal distribution. Similarly, in Figure 7.9c the histogram for a "small fluctuation" control chart shows no values in the outer regions and all very closely packed in the center—another form of non-normality.

Now look back at the control chart of means for Example 2.3 (Figure 7.7b). Several of the types of nonrandom patterns might be associated with this control chart. You might see grouping where several abrupt shifts in the level have taken place. You might observe a trend going downward from day 1 to day 12 with a return to an upper level on day 13. The latter was actually

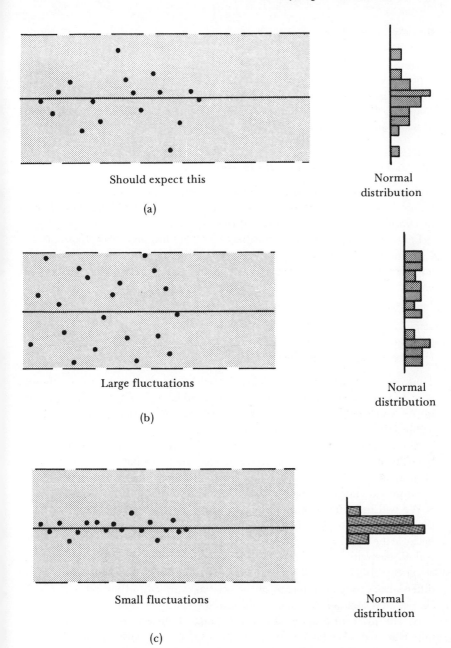

Figure 7.9 Control charts drawn with their histograms. The histogram of all values should be essentially a normal distribution. (a) The expected pattern with a distribution in control and operating typically. (b) Large fluctuations of the control chart. The histogram is highly spread out with very few in the middle and many in the extremes. (c) Small fluctuations of the control chart. The histogram is very highly peaked with none in the extremes or even near the extremes.

the case. It was found that the equipment was completely cleaned every two work weeks (these are 6 day work weeks). The slow clogging of the system altered the flow and gradually caused the weight of the bread to be reduced. Upon cleaning the system after the 12th day, the bread weight jumped back to the higher level. So again we can see how new information and a very simple method permits us to detect some effects that certain conditions cause or lead to in our system.

7.6 A Type of Experimental Design

So far we have been using control charts to look only at samples taken along a time axis. Now I want to enlarge our use of the concept of the control chart by considering other possibilities for that X-axis.

Let us go back to our old friend the officer and his radar trap, Example 2.1. I will now reveal some more information. The samples of 4 were not really haphazard. The numbers recorded on each day were actually in the following order:

Tickets in eastbound lane, morning.
Tickets in westbound lane, morning.
Tickets in westbound lane, afternoon.
Tickets in eastbound lane, afternoon.

Thus on day 1 there were 14 tickets issued in the morning in the eastbound lane, 11 tickets in the morning in the westbound lane, 13 tickets in the afternoon in the westbound lane, and 9 tickets in the afternoon to cars going east (see Table 7.1).

Eliminating the two snow days (the 10th and the 11th of the month) we have 26 values for each of the above 4 categories. We have 4 samples each of 26 individuals. The question now is whether the differences between these 4 samples are large enough to suspect that they are *not* due to a single source.

We will use what is essentially a control chart, only now we will modify the control limits. This is to be considered as a type of experimental design because the data has been accumulated in a previously organized fashion. The technique of analyzing the data to be introduced here is called the *analysis of means*.★ With it we will attempt to determine first, if there is a significant difference between the means of the several samples and second, what is the chance that such a difference could just accidently or randomly occur.

In our prior discussion of control charts I mentioned briefly one very important fact. We have been using the control limits with the idea that there is a very slim chance of a value exceeding such a limit, namely less than 1

★ The more common technique applied to data analysis is called the *analysis of variance*. The analysis of variance (abbreviated ANOVA) is a more complex method and will be discussed in Chapter 19. The analysis of means is a more recent and less well-known method developed by Dr. Ellis Ott of Rutgers University. It is simpler and easier to understand than ANOVA.

chance in about 360. What you still may not have grasped, however, is that this probability level is for a *single* sample. Perhaps it is the last one or the next one, but regardless, it is for *only one sample*. If we were to look back to all twenty or all thirty samples over a certain time, we really cannot say that there is 1 chance in 360 that *none of them* should fall outside the limits. After all, if we looked at 360 samples we would expect that one of them would be outside the limits.

However, *when we are actually attempting to decide if there are significant differences amongst a collection of samples we must adjust the limits*. This is not the same as asking if there is time-dependency or stability. We are actually selecting the samples so that they belong to different categories. Then we try to see if there is a difference in the measured quantities, or the samples, when they are in different categories. If there is *no difference* between the means or averages of the categories then the means of the samples should fall within a pair of limits. We will now discuss one of the types of problems that may be analyzed in this way.

We will *assume* that there is a *known standard deviation* for the things you are measuring (the population). We need to know σ (if it is only derived or calculated from the samples then there should be some modification, which will be discussed briefly). For our example we will use the overall σ that we obtained in Chapter 4, $\sigma = 3.41$.★ The limits are then found as

$$\text{Upper limit} = \bar{\bar{X}} + Z_\alpha \frac{\sigma}{\sqrt{n}}$$
$$\text{Lower limit} = \bar{\bar{X}} - Z_\alpha \frac{\sigma}{\sqrt{n}}$$
(7.9)

where $\bar{\bar{X}}$ = the grand overall average. Z_α is a special factor that we look up in Table A.11. It depends only on the number of *samples* being compared (the number of samples will be designated by the symbol k) *and* the *level of significance* or chance of error we are willing to take (the error will be designated as α). The error is either 0.05 or 5% or it will be 0.01 or 1%. These resemble the level of significance ideas that were used with correlation coefficient.

n = the size of the samples. We are assuming that they are all the same size.

For our example we have

$$\bar{\bar{X}} = 10.68$$
$$\sigma = 3.41$$
$$n = 26.$$

★ Actually we should be calculating the variances of the four samples separately. We would then average them together and find the standard deviation of the new average.

The remaining term is Z_α. Looking up in Table A.11a we find that for our 4 samples (number of samples = $k = 4$)

$$Z_{0.05} = 2.49 \quad \text{and} \quad Z_{0.01} = 3.02.$$

We then calculate *two pairs* of limits: one to be the limits using the $Z_{0.05}$ or the 0.05 limits; and the other limits using the $Z_{0.01}$ or the 0.01 limits. They are

	0.05 limits	0.01 limits
Upper limit	$10.68 + 2.49 \dfrac{(3.41)}{\sqrt{26}} = 10.68 + 1.66$ $= 12.34$	$10.68 + 3.02 \dfrac{(3.41)}{\sqrt{26}} = 10.68 + 2.01$ $= 12.69$
Lower limit	$10.68 - 2.49 \dfrac{(3.41)}{\sqrt{26}} = 10.68 - 1.66$ $= 9.02$	$10.68 - 3.02 \dfrac{(3.41)}{\sqrt{26}} = 10.68 - 2.01$ $= 8.67$

Referring back to Table 7.1 and ignoring the two unusual days (the 10th and 11th) we can *compute the averages for each sample* (averages of 26). They are:

Row 1—eastbound, morning   = 12.42.
Row 2—westbound, morning   = 9.85.
Row 3—westbound, afternoon = 12.08.
Row 4—eastbound, afternoon = 8.38.

Two of the 4 values exceed some limit. Eastbound, morning is outside the 0.05 limit but within the 0.01 limit. It is *significant*, but not extremely or highly significant. Eastbound, afternoon is below the 0.01 limit and is therefore *highly significant*.

A graph of this problem will help illustrate how these limits and values are related. In Figure 7.10 I have drawn a graph and have circled the two points that are outside some limit. From the looks of the graph it appears that two of the averages are high and two are quite low. No averages are in the inner region, where we would expect a few. The net conclusion should be that there is some attributable cause to the shifts in level. Perhaps the traffic pattern is such that more cars drive east in the morning and west in the afternoon. With more cars traveling, there may be a better chance that there will be a larger number of people speeding. Whatever the cause, we have found some significant differences that give us reasons to look further into our process.

In order to make this technique even more useful, and in order to be able to apply it quickly to practical problems, it should be emphasized that you can use the average range of the samples, $\bar{R}$, when you are comparing a collection of small samples. This can ease the calculations considerably by eliminating

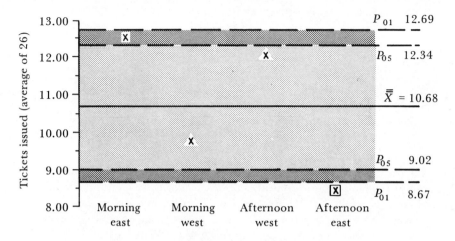

Figure 7.10 Graphical representation of the analysis of means showing the two unusual averages. Number of tickets issued (average per day) for different directions and times of day.

the necessity of computing standard deviations. You would then compute the limits with the following formulas:

$$\text{Upper limit} = \bar{\bar{X}} + Z_\alpha \frac{\bar{R}}{d_2 \sqrt{n}}$$
$$\text{Lower limit} = \bar{\bar{X}} - Z_\alpha \frac{\bar{R}}{d_2 \sqrt{n}}.$$

(7.10)

Actually we should use a constant other than Z_α, but for many problems there would be only relatively small corrections or changes from this Z_α value. The correct multiplier is called H_α and is given in Appendix Table A.11. The instructions for finding particular values of H_α are also given in the appendix.

At this point you now have learned some very useful and easy techniques to use to analyze data. You can determine if there is stability or time-dependency inherent in your data, and now you can plan and perform a statistical analysis of some data. The analysis of means gives you a valuable tool to make comparisons between averages, and the averages are unrestricted as far as the nature of the source of the data or the type of problem.

REVIEW

Time-dependency · Instantaneous · Short range · Long range · Control · Pattern of variation · Control chart · Sample · Sample mean · Sample range · Nonrandom variation · Cycle · Shift · Trend · Grouping · Fluctuation ·

Experimental design · Analysis of means

$$\sigma = \bar{R}/d_2$$

Upper control limit for range = $D_4\bar{R}$
Lower control limit for range = $D_3\bar{R}$
Upper control limit for means = $\bar{\bar{X}} + A_2\bar{R}$
Lower control limit for means = $\bar{\bar{X}} - A_2\bar{R}$.

PROJECTS

1. Using an oral thermometer take your temperature. About 5 minutes later take it again. Do the same about every two or three hours for several days. Plot a control chart. How would you describe the variation? If you eat or drink something between two "immediate" readings or if you engage in some physical activity, what happens to the variability?
2. Measure the distance from the floor to the ceiling along a wall that is at least 10 feet long. Take about 20 readings (about every half foot). Record the values to the nearest quarter of an inch. Note where you measured the height and in what sequence you took the readings. Plot the data. Does there seem to be any sort of relationship?
   Take two more readings at each location. Now plot a control chart. Are the values taken at the same location closer together than those taken at different locations? Are the points in control? Is the wall uniform?
3. If you have been taking readings of the temperature of your refrigerator, then take several readings that are close together in time and consider them as a sample. Plot a control chart.
4. If you have been recording gas mileage, use every three values as a sample and plot a control chart. Is there any type of trend?
5. If you can obtain a micrometer (preferably one that reads to the nearest 0.0001 inches) measure a roll of aluminium foil or wax paper. Take several readings next to each other, some readings at different locations across the sheet, and sets of readings at the beginning, middle, and end of the roll. How do the variabilities of the readings compare?

The following projects require you to have covered the section on Experimental Design (Section 7.6).

6. Using the setup of Project 1, Chapter 6, compare 4 different elastic bands, all of the same size. With *each* elastic band take 5 readings with the same number of coins in the envelope. Analyze to see if there is a statistically significant difference in the stretch of the different elastic bands. What other types of comparisons might be made? Try some.
7. Put the thermometer in different locations of your refrigerator (top shelf-front, top-rear, bottom-front, bottom-rear). Move the thermometer in a random fashion to the different locations as you gather 6 readings of the temperature in each location. Determine if there are significant differences in locations.
8. Compare the number of songs per hour on three different radio stations.

9. Compare the number of commercials per hour on different television channels.
10. Compare the weights of sugar packets for two different manufacturers.

PROBLEMS

Section 7.3

7.1 For the following discuss how and why there might be differences in the variability among different time spans. What would constitute those different time spans? In which cases is the term "control" applicable, and in what way? What would constitute being in control?
   a. The actual amount of aspirin in aspirin tablets.
   b. The test grade on an aptitude test.
   c. Milk production.
   d. Automobile speed while driving along a highway.
   e. The size or quality of your mother's chocolate chip cookies (this is an example of what is called a batch process, where items are made in batches—cookies are made in groups on a cookie tray).
   f. Recall of nonsense syllables (a nonsense syllable is a collection of letters that bear no relation to words, such as, BEW, KOD, REM, PID, ...).
   g. Number of people unemployed.
   h. Rating of beer (taste or flavor).
   i. A Presidential or an attitude poll.

Section 7.4

7.2 Although most of the discussion has centered about three "levels" of time dependency, these levels may be applied to other characteristics. Discuss in what way the following subjects would be expected to reveal different degrees of variation.
   a. Thickness of top soil.
   b. Musical aptitude of twelve-year-olds. (*Hint:* consider identical twins.)
   c. Amount of tire tread wear on a particular brand of tire after 20,000 miles of driving.
   d. Taste of beer (or soda or milk).

7.3 Four diapers were tested for their absorbency. They were dipped into containers of water and the amount of water absorbed by the diapers was recorded. They were then washed and dried (in a clothes dryer). The results were as follows:

	Test						
	1	2	3	4	5	6	7
Water absorbency (ounces)	8.6	8.4	8.5	8.3	8.2	8.1	8.2
	9.1	8.8	8.7	8.7	8.6	8.6	8.5
	8.7	8.8	8.5	8.6	8.3	8.4	8.2
	8.8	8.7	8.8	8.5	8.6	8.6	8.4

212 *Descriptive Methods*

	Test						
	8	9	10	11	12	13	14
Water absorbency (ounces)	7.9	8.0	7.9	7.7	7.6	7.4	7.5
	8.5	8.3	8.4	8.2	8.1	8.1	8.0
	8.3	8.0	8.1	8.0	7.7	7.8	7.7
	7.9	8.2	8.1	8.0	7.9	8.0	7.8

Calculate the average absorbency and the ranges for each test. Plot a control chart.

7.4 4 aspirin tablets are taken from 2 bottles of aspirins on each day. They are carefully analyzed to determine how much aspirin they contain (they are supposed to contain 5 mg.). The results of three weeks of analysis are as follows ($\bar{X}$ = average of 4; R = range of 4):

	Bottle number									
	1	2	3	4	5	6	7	8	9	10
$\bar{X}$	5.00	4.96	5.01	4.99	5.02	4.92	4.99	5.00	4.98	4.97
R	0.08	0.06	0.10	0.05	0.05	0.07	0.03	0.08	0.06	0.04

	Bottle number									
	11	12	13	14	15	16	17	18	19	20
$\bar{X}$	5.02	5.00	5.04	4.97	5.01	4.96	5.03	4.98	5.00	4.96
R	0.06	0.05	0.07	0.09	0.03	0.02	0.04	0.05	0.04	0.07

	Bottle number									
	21	22	23	24	25	26	27	28	29	30
$\bar{X}$	5.00	5.01	5.01	4.96	5.02	5.03	4.98	4.99	5.00	5.00
R	0.03	0.01	0.07	0.06	0.05	0.03	0.07	0.05	0.06	0.04

Plot a control chart. Are all the values in control?

7.5 Soil thickness readings were taken to determine the amount of topsoil on a large plot of land. Three small cores were taken at 20 different locations (each set of cores were inside of a 6 square inch area). The locations of the cores are shown on the map of Figure 7.11. The thickness of the topsoil at the locations is listed below.

	Location									
	1	2	3	4	5	6	7	8	9	10
Soil (inches)	6.5	8.5	9.1	9.5	10.4	6.7	7.0	7.8	9.0	10.4
	6.8	8.6	9.2	9.4	9.9	6.7	7.1	8.1	8.8	10.6
	6.7	8.5	9.2	9.3	10.1	7.1	7.1	8.0	8.6	10.8

	Location									
	11	12	13	14	15	16	17	18	19	20
Soil (inches)	6.9	8.9	9.5	10.5	11.0	6.9	9.1	9.5	10.5	12.1
	6.7	8.7	9.5	10.0	10.8	7.1	9.0	9.4	10.1	11.8
	7.1	8.8	9.5	10.1	10.8	6.8	9.2	9.6	10.0	11.9

Calculate the averages and ranges. Plot and analyze the control charts.

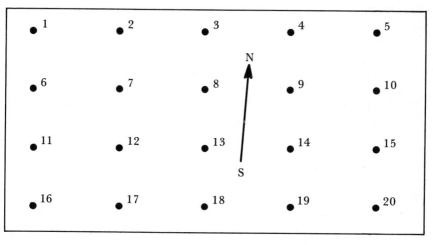

Figure 7.11 Location of the samples of topsoil.

Section 7.5

7.6 Refer to Problem 7.3. Do you detect any nonrandom variation?
7.7 Refer to Problem 7.4. Do you detect any nonrandom variation?
7.8 Refer to Problem 7.5. Do you detect any nonrandom variation?
7.9 For the following items explain why you might expect nonrandom variation to exist in the data.
   a. Sales of bathing suits per week listed over several years.
   b. Egg prices per week over several years.
   c. Pages of advertising per day over several weeks.
   d. Pages of advertising per week for several years.
   e. Milk production per month for several years.
   f. Cars per minute every hour for several days.

Section 7.6

7.10 Three different brands of aspirin were to be tested to determine how fast they dissolve in the stomach. Five tablets of each brand were dropped in mild acid (simulating the stomach) and the lengths of time for them to fully dissolve were recorded.

	Brand		
	Brag	And None As Fine	X
Time to dissolve (minutes)	4.3	5.1	5.1
	4.7	4.7	5.3
	5.2	5.1	4.8
	4.2	4.4	5.7

Does there appear to be a significant difference between brands? (Use the average range as an approximation of the standard deviation.)

7.11 Refer to the diapers in Problem 7.3. Each of the four rows contain the measurements on one diaper (there are 4 diapers). Using the average range (from Problem 7.3) determine the estimate of standard deviation. Is there a difference between the diapers?

7.12 Refer to the data in Problem 7.4. If the first 10 readings were for the first week, the second 10 for the next week, and so forth, is there a difference between weeks? The odd numbered bottles were taken from one machine, the even numbered bottles from another machine. Is there a difference between the amount of aspirin in tablets in the two different machines?

7.13 Refer to Problem 7.5. Examine the location of the samples. Compare the topsoil depth in the two different directions. First look at all the values from 1 to 5, 6 to 10, and so forth, to see if depth varies in a north to south direction. Then look at all the soil samples in each of the five columns (1–6–11–16, 2–7–12–17, and so forth). Is there a difference in the east to west direction?

PART III

Distributions

8

Probability and Distributions

Probability and statistics is often the title of what may presume to be primarily a course in statistics. It is frequently the title of the textbook used in that course. Thus far we have avoided using the terminology of the subject of probability. Of course, I have referred to the chance or probability of certain things happening or not happening, and I have even stressed in Chapter 1 and again in Chapter 6 that this coupling of a probability statement to a decision based on data is a primary advantage and purpose of understanding and using statistical methods. But now we must begin to explore one of the more general relationships between probability and statistics, and begin the study of distribution concepts in preparation for understanding statistical analyses.

8.1 Probability and Statistics

We have come to the point where you should be aware of the necessity of understanding the concept of a distribution and the nature of sampling from that distribution. These two concepts, distribution and sampling, are throughout the study of both probability and statistics. But the kind of initial information that is available and the ultimate purposes of the subjects are quite different.

The study of probability assumes that you know the entire universe or population from which you are going to sample. You are then able to completely determine all of the possible samples that can be taken from that population. As a result you will also be able to state the *chances* that any particular samples will be chosen.

Statistics assumes that you have only a sample and you are then trying to determine the most likely population or source containing the sample.

Figure 8.1 partly illustrates this difference. In Figure 8.1a we have a known population. We know all the individuals. We can then determine all of the

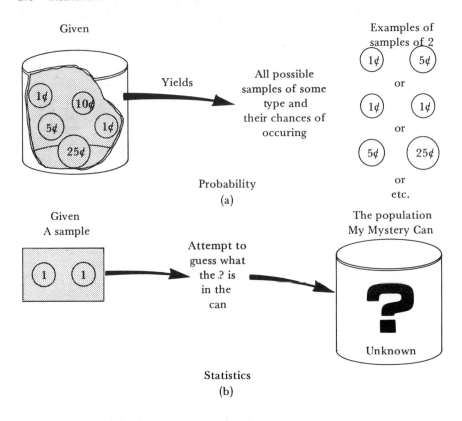

Figure 8.1 One distinction between probability and statistics. How the sample relates to the distribution. (a) Probability starts with a certain known population and samples are described from that population. You are certain of the population—you know *all* the possible samples. *But*, you are uncertain about the *particular* sample. (b) Statistics starts with a sample and tries to guess the population. You are certain about the *sample—uncertain* of the population.

possible samples that may be taken from that population. In this way it becomes possible to determine what probability or chance we might have of obtaining the different values from this population.

In Figure 8.1b we have an unknown population. It *may* be the same one as above, but we do not know that. We look at a sample and then try to infer what population that sample might have come from, *or* we determine what chance there is that the sample might have come from, for instance, the upper population.

I will use an example that you can easily duplicate to more fully explain these concepts. First we will deal with probability.

EXAMPLE 8.1

In the morning, before taking your loose change (the pennies, nickels, dimes, and so forth) off the dresser and placing them in your pocket or wallet, you observe exactly how many of each type of coin you have. This establishes the *complete distribution of the population* that we will now be working with. You find that this morning it consists of 2 pennies, 1 nickel, 1 dime, and 1 quarter (a total of 42 cents). In Figure 8.2 I have drawn a histogram showing this distribution of coins, and I have also included the relative frequencies.

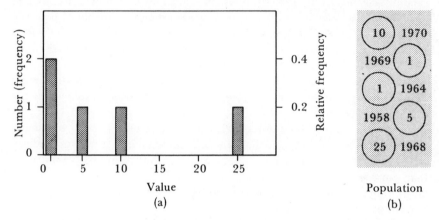

Figure 8.2 (a) Histogram for the population of coins consisting of two pennies, one nickel, one dime, and one quarter. (b) The population.

Because this distribution describes the complete collection of possible individuals and because it also allows us to determine the *chance* that each of these various values might occur or be picked, it is often called a *probability distribution*. Now for *a* definition of probability (other definitions do exist, but we will not discuss them here), we will use the definition:

DEFINITION *Probability is the relative frequency of a particular occurrence.*

Thus, if we assume that all of the coins have an equal chance of being picked out (an assumption that will be discussed at length) then the relative frequency of nickels is 0.20, and therefore the probability of selecting a nickel would also be 0.20, since 20% of the coins are nickels. There is a 0.40 probability of selecting a penny since two of the five coins, or 40% of the coins, are pennies. Similarly the probability of selecting a dime and the probability of selecting a quarter are both 0.20. So we have found probabilities or relative chances for selecting each of the items in the distribution.

Our description of a population and determination of probabilities of the number of individuals in a population has a purpose. This purpose will be to determine what kinds of samples we may obtain from this population and how frequently these samples might occur.

First, we must discuss selection of samples. How many times have you blindly reached into your pocket or wallet and grabbed two coins when trying to come up with a 15 cent tip? This manner of selecting coins is what you are going to keep in your mind as the kind of sample we will be taking. But before we go further we should consider the assumption previously mentioned, *that each coin has an equal chance of being selected*. This is an important consideration since it was used to establish our set of probability values for the chance of picking a particular type of coin. We have already discussed the concept of random selection in Chapter 2 (page 37), and observed the effect of non-random selection in the 1969 draft lottery. The idea of random selection and "equally likely" are very similar and often used interchangeably. I have observed often enough that I may dig into my pocket searching for a nickel for a parking meter only to come out with a quarter, or reach for a quarter for a toll booth and come out with a dime. So I do not trust my sense of feel in selecting coins. If, however, you have a keen sense of touch and are able to distinguish between the various types of coins, we would no longer have a problem involving *random* selection.

If we were talking about dollar bills (such as 2 singles, 1 five, 1 ten, and a counterfeit twenty-five—since there are twenties but not twenty-fives) and you were reaching into your wallet to pay for a fifteen dollar meal, you would agree that you cannot select the bills by feel. So here perhaps you have a better illustration of a random selection (providing, of course, that the bills are not in any order, and you do not look).

Getting back to the question of the samples of coins, let us examine all of the possible pairs of coins that I might pick from the population of the five coins in

Table 8.1 All possible samples of two from the population of Figure 8.2

First pick		Second pick	Sum of the 2 coins or value of the sample (cents)
1964 penny	and	1969 penny	2
1964 penny	and	nickel	6
1964 penny	and	dime	11
1964 penny	and	quarter	26
1969 penny	and	1964 penny	2
1969 penny	and	nickel	6
1969 penny	and	dime	11
1969 penny	and	quarter	26
nickel	and	1964 penny	6
nickel	and	1969 penny	6
nickel	and	dime	15
nickel	and	quarter	30
dime	and	1964 penny	11
dime	and	1969 penny	11
dime	and	nickel	15
dime	and	quarter	35
quarter	and	1964 penny	26
quarter	and	1969 penny	26
quarter	and	nickel	30
quarter	and	dime	35

my pocket. One sample could be the 1969 penny and the dime, another could be the nickel and the dime (our 15 cent tip), a third might be the dime and the quarter. A complete list of all of the samples is given in Table 8.1. The values of the samples are also listed, so the penny and dime make 11 cents, and so forth, and the possible samples range from 2 cents to 35 cents. (As you study the distribution you may question some of the reasons for the way in which the samples were chosen. Some of these reasons may become more obvious as we continue. You might also consult Appendix C for further work on probability.)

We have a total of 20 different samples and each has a value. We can now draw a histogram or probability distribution for these 20 values (Figure 8.3) and we will then be better able to visualize the chance of obtaining different sample values. Thus, for example, we find the relative frequency of 15 cents is only 0.20, so that the probability of randomly picking out a 15 cent tip is only 0.20 (or 1 chance in 5).

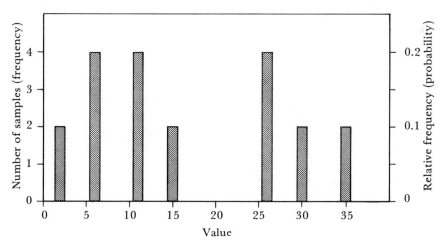

Figure 8.3 Histogram for all possible samples of two from the population of Figure 8.2. (Samples of Table 8.1).

The net result of our work with a *probability distribution* is that we have been able to

1. Identify *all of the possible samples* from the original distribution, and
2. Establish the distribution of samples *with the appropriate probabilities* for the various samples.

As a consequence we can say what is likely or unlikely to occur when we take a sample. We cannot tell which sample *will* be chosen but we can give you *odds* that will tell you what you probably will or will not select.

Now what about *statistics?* Here we do not know what the population is, we only have a sample. We put a hand into our pocket and pick out two coins—perhaps 2 pennies (2 cents). Our problem now is to try to guess what the pocket contains as an overall population. This is where the two subjects—probability and statistics—merge. Our usual practice consists of the following steps.

(1) We make an assumption about what we *believe* the population looks like. One population we might say we think this sample belongs to could consist of the 2 pennies, 1 nickel, 1 dime, and 1 quarter we have been working with. Another population we could consider might be one with twenty-four coins— 2 pennies, 10 nickels, 10 dimes, and 2 quarters. There are innumerable other possibilities that could be the source of the sample, but we will only look at these two for the moment.

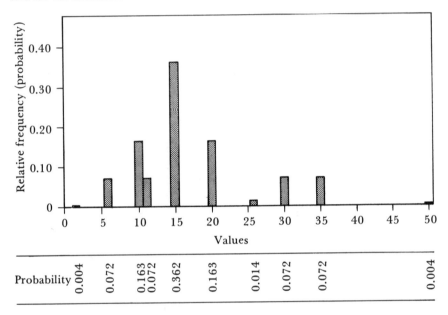

Figure 8.4 Distribution of samples of two from a population consisting of two pennies, ten nickels, ten dimes, and two quarters. The exact probabilities are also shown below each possible value.

(2) We look at the probability distribution of the samples from the theoretical population we believe is the source of our sample. So we look back at Figure 8.3 for the probability distribution of samples from our first possible population. Figure 8.4 shows the distribution of samples from the second population we are also considering.

(3) We compare our *sample* to the probability distribution of the population. We do this by observing what the probability is that we might obtain this particular sample, or a sample this large, or this small, from the distribution. Thus in Figure 8.3 we would find that the probability of obtaining 2 pennies (2 cents) is only 0.1 or a 1 in 10 chance (actually 2 in 20). This is a moderately small chance, although 3 of the 7 possible samples have this probability associated with them. The second distribution, Figure 8.4, is a better illustration.

Here the probability of picking out 2 pennies is extremely small. The probability of obtaining such a sample is only 0.004, or we would only expect 4 times out of a *thousand* to have such a sample. Because of the small chance of a sample of this kind coming from the second distribution we would conclude that *it is highly unlikely that our sample came from a distribution of this kind*. We would extend our statement to say that we believe that the population out of which we picked

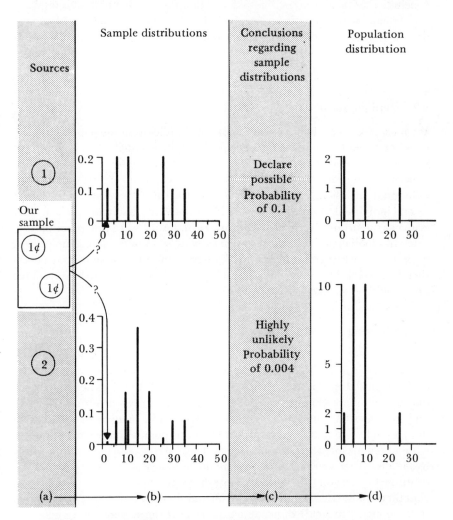

Figure 8.5 Comparing a sample to possible sources. (a) Look at our sample. (b) Look at some possible distributions of samples (1) and (2). (c) Compare the sample to the distribution and conclude that with (1) there is a good possibility that the sample came from it and that (2) is a highly unlikely source of the sample. (d) Extend our conclusion about the sample and the sample population to the population that generated the sample distribution. Therefore, (1) concludes the population might be the source of the sample and (2) is a highly unlikely source of the sample.

our coins, the population of coins in our pocket, does *not* consist of 2 pennies, 10 nickels, 10 dimes, and 2 quarters (the population of Figure 8.4).

In other words we look at the sample and compare it to a theoretical distribution of samples. Depending on how likely it is that our sample may have come from this theoretical distribution of samples, we pass judgement about the population. If we find the sample is not unusual then we would conclude that there is no reason to doubt that the true population is the same as the theoretical population from which the theoretical distribution of samples came. If, however, the sample is an unusual sample we will say it is highly unlikely that we obtained the sample from the theoretical population. Figure 8.5 summarizes this idea.★

8.2 Replacement

We must now go back to the way I took the samples from the population, and consider a couple of important concepts that I disregarded when I listed those samples. For ease of visualizing we will work with the five coins (2 pennies, 1 nickel, 1 dime, and 1 quarter). Each sample was taken by first selecting one individual and then, *while keeping that individual out of the population*, picking a second individual. In other words, I did *not replace* the first item I picked. This is a very important distinction to be aware of, for if I were to replace the first coin after looking at it, *I would have a different collection* of possible samples. For instance, if the first coin I picked was a quarter and I put it back in my pocket then I *could* pick it again, so one of the samples that would have to be considered now would be *two quarters*.

The standard terms used here are

1. Sampling *without* replacement, and
2. Sampling *with* replacement.

Whenever we deal with the first case we *alter our population* as we sample. After we have picked the quarter, the population that remains does not contain the quarter any more (Figure 8.6a). So the next pick *cannot* be a quarter. This limits the possible collection of samples of two. How many actually different samples of five could you have picked from this population when you sample without replacement? If you said one, you are correct. After all, there are only five items. After you pick one, you are down to a population of four. If you continue to pick you will finally exhaust the population and so the sample will contain all of the items of the population and the only possible net sample will be 2 pennies, 1 nickel, 1 dime, and 1 quarter, or the 42 cents.

What about sampling with replacement? Figure 8.6b shows in effect what occurs. In this case we toss the item back into the population and we can pick

---

★ A second aspect of statistics deals with estimation and confidence intervals. You might read Chapter 12.1 at this time to get some initial familiarization with the ideas of estimation, since that is when we pick it up again.

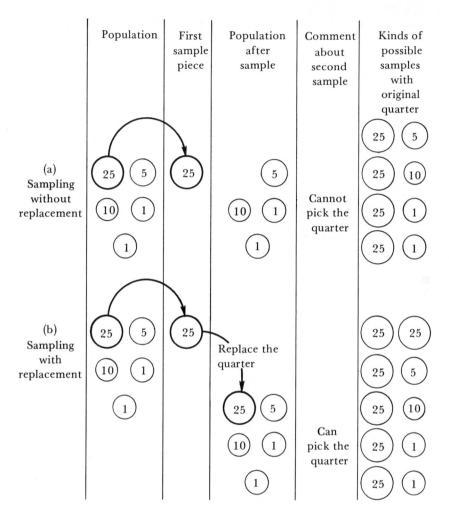

Figure 8.6 Sampling with and without replacement. Taking the population of five coins and selecting the quarter on the first pick, we have two possible sets of samples if we (a) do not replace the quarter or (b) replace the quarter.

the same item again. Table 8.2 lists the total collection of possible samples of two from this population when we sample *with* replacement. Observe that all of the previous samples are included *plus* a few additional ones, namely the five cases of sampling the same item twice. Figure 8.7 shows the distribution of these samples and you may compare this distribution to the previous one in Figure 8.3 to see the effect of replacing the first item before continuing to sample.

We can make a couple of observations. Only if the population is of a definite, finite size, a size we can really count or enumerate, can we be sampling *without* replacement. If the population is effectively (or actually) of infinite size, then removing some sample from the population will not affect the distribution of

Table 8.2 All possible samples of two from the population of Figure 8.2 (2 pennies, 1 nickel, 1 dime and 1 quarter). This time sampling *with* replacement

First pick		Second pick	Sum of the 2 coins or value of the sample (cents)
*1964 penny	and	1964 penny	2
1964 penny	and	1969 penny	2
1964 penny	and	nickel	6
1964 penny	and	dime	11
1964 penny	and	quarter	26
1969 penny	and	1964 penny	2
*1969 penny	and	1969 penny	2
1969 penny	and	nickel	6
1969 penny	and	dime	11
1969 penny	and	quarter	26
nickel	and	1964 penny	6
nickel	and	1969 penny	6
*nickel	and	nickel	10
nickel	and	dime	15
nickel	and	quarter	30
dime	and	1964 penny	11
dime	and	1969 penny	11
dime	and	nickel	15
*dime	and	dime	20
dime	and	quarter	35
quarter	and	1964 penny	26
quarter	and	1969 penny	26
quarter	and	nickel	30
quarter	and	dime	35
*quarter	and	quarter	50

* Additional samples because of sampling *with* replacement.

the population. For instance, it would be difficult to argue that the samples of soil taken from the moon on the Apollo journeys would alter the overall distribution of material (or some element) on the moon. However, plunging oil wells into the ground in Texas and tapping oil for many years by the millions of barrels will alter the distribution of oil in that area (we are hardly replacing the oil). Taking a drop of blood from your finger to do a blood count or taking a few cells of your brain to do a biopsy to determine if cancer is present, will not alter the distribution of blood cells or functioning of brain cells in your body. The loss of a pint or even a quart of blood will, however, significantly change the distribution of blood cells as a frontal lobotomy would affect the processes of the brain.

For the two extremes we have little difficulty in deciding which we are dealing with (with replacement or without replacement); it is the in-between areas that pose problems. As a rule of thumb we consider that we have sampling *with* replacement as long as the ratio of *sample size* to *population size* is less than about 1 to 10 or

$$\frac{\text{Sample size}}{\text{Total population}} \leq \frac{1}{10} = 0.1.$$

Thus when we look at two out of the original five coins we have a ratio of $2/5 = 0.4$ and this is *not less* than 0.1, so we *cannot* consider that sample as one *with* replacement. However, when we look at the sample of 2 coins out of the second population of 24 coins we have a ratio of $2/24 = 1/12 = 0.083$ which is less than $1/10$ or 0.1, so we *could* treat that sample as one taken essentially *with* replacement.

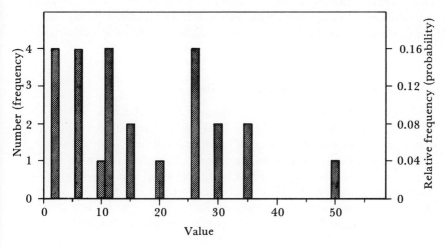

Figure 8.7 Samples of two *with* replacement. Compare this histogram to Figure 8.3.

EXAMPLE 2.3 (Cont.)

Hypothetically, a bread-making machine represents an infinite source of breads. There is a distribution of possible breads that could come out of the machine. The one bread that just came off the machine should not influence in any way the distribution of breads that the next bread will belong to. It does not alter the distribution. The bread is a sample of all possible breads from the machine and can be considered a sample *with* replacement or from an infinite population.

Now let us say we have 50 breads that were produced and are being delivered to a particular store. This too constitutes a population with a distribution (although it is a sample of breads from the machine). If I were to remove one of the breads from this new population, I would be changing the distribution. If I took away 10 breads I would be *extensively* changing this small population, so that we would now be talking about sampling *without* replacement.

Another important type of data that can be regarded as a sample *with* replacement is that of most measurement data. In Chapter 1 I described how taking one measurement is actually only a sample of a whole distribution of possible measurements. But taking that measurement should not alter the distribution (if it does we probably do not have independent measurements and the problem is even more complicated). Hence, we can consider measurements as from an infinite population and therefore as a sample *with* replacement.

From the tone of the past few pages you should have gotten the feeling that it is desirable to be able to work with sampling problems that may be considered to have come from essentially infinite populations. The question of whether we are sampling with or without replacement is sometimes critical in working backwards to determine what type of actual distribution we have, or for *estimating* the population values. However, it is generally only when you get into sophisticated analysis that the distinction will become important, and for the most part we will be dealing only with problems that would be described as sampling *with* replacement.

8.3 Discrete and Continuous Distributions

One type of classification that distributions can fall into has to do with the kinds of possible values that may be obtained. When we sampled from the five coins in my pocket you only had a small set of definite, distinct, *discrete* possibilities. The number of tickets issued, or the absences from class, or broken eggs in a carton, or deaths from slipping in a bathtub, are all illustrations of discrete distributions, because there is a set of exact specific categories into which our individual measurements or samples can fall.

When we discussed the concept of measurement back in Chapter 1 and again in Chapter 3 we observed that most measurements fall into categories only because of the limitation of the instrument we use to make the measurement. We cannot really determine if you had exactly 12 fluid ounces of beer at a temperature of 48 degress which you drank in 3 seconds, because in reality there might have been 12.00683142798143 . . . fluid ounces of beer at 47.8314601273 . . . degrees which you drank in 3.47129814306 . . . seconds. You used discrete measurements that were arrived at by rounding off at some point. The actual measurements or the *possible* values of the distribution may cover an infinity of different values spread out across the number line. We consider this to be in effect a continuous set of possible points, especially if you recognize that the next measurement of fluid ounces perhaps might be 12.00683142798142 . . . or 12.10683142798143 . . . . Then the difference from one true value to another might be only 0.00000000000001 or it could be 0.1. Then we might have over a trillion points to plot if we were to regard *these* values as discrete points, and hypothetically we could continue adding more decimal places and thus create the need for more points (every decimal place multiplies the number of points by a factor of 10). When you plot all of these points along the X-axis you have what is about as close as possible to a continuous line. If at the same time we regard the distribution as a "probability distribution," then we can establish or assign a corresponding relative frequency or probability for each of these points and obtain a continuous line for the distribution itself.

Practical cases, however, do not work with the exact points, but rather with ranges of points and areas of the distribution. This concept will become more apparent when we work with the normal distribution in Chapter 10. So in real situations we almost always work with discrete distributions, although in theory we may be dealing with continuous data.

8.4 Random Numbers

In the beginning of this chapter I asked you to take five coins and start picking samples. Even though the coins were of different shapes I said I didn't have a sufficiently good sense of feel to be able to distinguish them apart. But you might be able to tell them apart, so you should have a way of sampling so that the sample may be regarded as truly random. I offer a solution that involves what is called a *random number table*.

Appendix Table A.16 is a table of many digits grouped into sets of two. They are quite disorganized and should show absolutely *no* pattern, no matter what direction or sequence you choose to follow. You may start in any location and go in any direction, and there is *an equal chance for every digit*, 0 to 9, to appear next. The numbers are completely random. Similarly, you can pick 2 digit or 3 digit numbers. So, after picking a point in the table to start at, the next pair of 2 digit numbers could just as likely be a 27 as a 00. The 3 digit numbers could be 000 or 517 or 638 or . . . .

With a random number table we can construct any distribution we desire, and then sample from the table. We can even alter the distribution as we go along and see what effect the alteration has on the samples. We can simulate all kinds of situations. All we need to do is assign the various digits to the possible circumstances, and allot them in the same proportions as our relative frequencies or probabilities are allotted to the possibilities. So for the coin problem we had 5 coins: 2 pennies, 1 nickel, 1 dime, and 1 quarter. We established the probability of picking a penny as 0.4, a nickel as 0.2, a dime as 0.2, and a quarter as 0.2.

To split up the ten digits we would simply make the first four digits correspond to pennies, the next two to nickels, the seventh and eighth to dimes, and the ninth and tenth to quarters. We will say that if the digit that appears is a 0, 1, 2, or 3 it means that a penny was picked. If a 4 or 5 appears it means a nickel, a 6 or 7 means a dime, and an 8 or 9 indicates a quarter was picked.

We now go to the table. I will start in the upper left corner of the first page. The first digit is 1. Therefore, a penny. Now comes a question: Are we sampling *with* or *without* replacement? If the answer is *with*, then we put the penny back. The next pick is a 1 and that means penny again. So our sample is 2 pennies. Our next sample (with replacement) consists of a 1 (a penny) and a 6 (a dime). The third is a 4 (a nickel) and a 3 (a penny). Observe the ease of sampling this way; and it is *random*.

Now what if we wanted to sample *without* replacement? All we need to do is remove the digits corresponding to the item we first sampled (I don't really mean remove by erasing but rather by ignoring). So the first digit was a 1. Then we need to eliminate *a* penny from the distribution. To do this we can ignore two of the four digits corresponding to the penny. We might choose to ignore the 0 and 1 if the number appearing was a 0 or 1, and the 2 and 3 if the sample was a 2 or 3. So we now eliminate 0 and 1. The next two digits listed are both 1, so we just *ignore* them and go on in the table until we find a digit we *can* use, a 6 (dime) in this case. Then the first sample would be a penny and a dime. The second sample is a 4 (a nickel) and a 3 (a penny) as before (with replacement). The third sample is a 6 (dime) and a 3 (penny). The fourth sample is a 1 (a penny) and an 8 (a quarter).

Let us look at another problem.

EXAMPLE 8.2

Cars pull up to a very small ferry with 1, 2, 3, 4, 5, or 6 occupants. Based on various surveys of automobile passengers it is believed that the probabilities associated with these number of occupants is 0.2, 0.3, 0.2, 0.15, 0.10, and 0.05 respectively. The ferry boat owner would like to know the number of people he could be expecting to have on his ferry if the ferry can only carry five cars at a time.

Now, if the probability of 1 passenger is 0.2 then 20% of the digits will need to be assigned to 1-passenger cars, 0.3 or 30% of the digits to 2-passenger, and so forth. A total of 100 digits will suffice (since we have two digit probabilities or percentages) so that 20 are to go to 1-passenger, 30 to 2-passenger, etc.

We can prepare the following table showing the number of digits and the actual digits to use for each possible number of passengers:

Number of passengers	There are these number of points	Then use these digits
1	20	00–19
2	30	20–49
3	20	50–69
4	15	70–84
5	10	85–94
6	5	95–99
	Total 100	

Note that none of the cases has more than 2 digits, and no pair of digits falls into more than one category.

Now let us take a sample to see how the random number table works. We will start in the upper left corner of the second page of digits. Here we find (by going from left to right)

$$40, 60, 31, 61, 52.$$

They would represent in passengers

$$2, 3, 2, 3, 3.$$

This is because the 40 falls in the 20–49 range and stands for a 2, 60 falls into the 50–69 range and stands for a 3, the 31 falls in 20–49 and stands for a 2, and the last two fall in the 50–69 range or are 3. Thus the total number of passengers would be 13.

For the next sample let us use the next row:

$$40, 94, 15, 35, 85,$$

which corresponds to

$$2, 5, 1, 2, 5,$$

passengers in each car, or a total of 15 passengers. If we were to repeat this selection process we could obtain a whole collection of sets of 5 cars, and we could determine a distribution of number of passengers for trips, providing, of course, that the assumptions of the probability values are valid.

You should be able to see by now how we could elaborate even more. A few exercises will provide you with some opportunity to try some fancier simulations and see how real processes can be duplicated and observed through this random operation. The one problem that exists in real life is that the number of trials desired to get a large enough set of samples may be very time consuming. As the problems get very sophisticated we tend to use computers to generate the random numbers and even make the assignments and do the computations of statistics for us. So with the right kind of programming we might just obtain a set of number of passengers, the average number of passengers, and maybe the standard deviation of number of passengers. Some of this can also be done with library programs on time-shared terminals. It is a very useful and easily understood technique for duplicating certain processes and perhaps even avoiding great expenditures of money on projects that have not been developed.

REVIEW

Probability · Probability distribution ·

Sampling with replacement · Sampling without replacement ·

Discrete · Continuous · Random number ·

Random number tables

PROBLEMS

Section 8.1

8.1 A pond contains 6 fish. 2 weigh 10 ounces, 2 weigh 15 ounces, 1 is 6 ounces, and another is 25 ounces.
   a. What is the distribution of all possible samples of total weight for two fish?
   b. For all samples of three fish?
   c. If I have to toss back all fish under 12 ounces, what is the distribution of weight for two fish?

8.2 The costs for the textbooks for six of the courses at a college was:

Mathematics	$8	History	$6
Chemistry	$9	German	$7
English	$10	Art	$8

   a. A part-time evening student is taking two of these courses. What is the distribution of possible total costs for his books?
   b. If a student claimed his books cost him a total of $16, is this unusual?
   c. What do you say if the student claimed his books cost him $20?

8.3 If the courses listed in Problem 8.2 were given on the following evenings—mathematics, chemistry, and English on Monday; history, German, and art on Wednesday—how much does this affect the possible distribution of costs for books?

8.4 A book salesman claims book prices are roughly: $9 40%, $8 and $10 20%, $7 and $11 10%. Use this distribution as an alternate population for comparison. Now if a student claims his two books cost $16, which population do you believe in? How about a cost of $20?

Section 8.2

8.5 Refer to Problem 8.1. When is this sampling with replacement and when is it sampling without replacement?

8.6 Refer to Problem 8.2. Would you consider this as sampling with or without replacement? How about Problem 8.4? If they are different, explain why.

8.7 For the following examples explain how we would sample with replacement, and why sampling with replacement might not be advisable.
   a. A poll is taken on preference for Presidential candidate.
   b. The number of pages assigned by English teachers is sampled.
   c. A milk dealer is interested in determining the portion of families willing to have his milk delivered.
   d. Beer is to be tasted for quality and rated by baseball players (consider two ways of replacement and the different effects).

Section 8.3

8.8 Which of the following types of measurements (or data) would be considered as discrete, and which as continuous?

a. number of scratches on a car
b. length of a scratch on a car
c. spelling errors in a book
d. words in a book
e. sourness of a lemon
f. number of apples on a tree
g. weight of the apples on a tree
h. number of hours of television watching.

Section 8.4

8.9 Assume the probability of the number of cars coming to the ferry at the time of departure is as follows:

	Probability		*Probability*
0 cars	0.05	4 cars	0.20
1 car	0.10	5 cars	0.20
2 cars	0.15	6 cars	0.05
3 cars	0.20	7 cars	0.05

Now use the random number table to determine how many cars will arrive. Then find the distribution of passengers. You can go through this whole process several times to obtain a distribution of passengers on the ferry. How does it differ (or how would you expect it might differ) from the distribution in Example 8.1?

8.10 The owner of a one-minute car wash has observed that the number of cars arriving each minute is about:

0 cars	20% of the time
1 car	30% of the time
2 cars	20% of the time
3 cars	20% of the time
4 cars	10% of the time

Using the random number table determine how many cars will arrive each minute for a two-hour period. When the first car arrives note that it will take one minute to leave. Then the second car can be washed. Record the time *each car* arrived, the time it was washed (or left) and how much delay there was (this is the difference between the arrival and the departure time). What is the average delay for cars? How big a lineup was the longest one you had?

9

The Binomial Distribution

In the last chapter I described a difference between the areas of probability and statistics, that is, how they rely upon different relationships between samples and populations. Our main purpose in this text is to study the topics and concepts of statistics, and that implies we will be mainly concerned with making *inferences or decisions* about the source of the data that we may have gathered or are about to gather. But in order to infer what the population *might* be, we must be able to describe what samples can come from various hypothetical populations, hypothetical populations that might serve as models for our problems.

9.1 The Need for a Theoretical Distribution

Recall back to the last chapter how *first* we looked at a theoretical distribution of samples from some populations (in that case, of coins), even *before* we took our sample (see page 221). We then compared our sample to the collection of *possible* samples that *might* have arisen from that particular population and decided if it was likely or unlikely that our sample might have come from that population. This type of reasoning process is followed throughout the study of statistics. But we need to develop a hypothetical population to compare against, and preferably it should be something that can be shown to have some reasonable connection to the source of the actual data.

In this chapter we will look at one *type* of distribution that might serve as a theoretical model for a real process. I will present a mathematical equation that is a way of relating a set of X-values (or samples in these cases) to probabilities. In this case we will then be able to express what the chance of selecting all of the various samples would be for the possible source or model. Only after arriving at this theoretical distribution can we begin to assess or compare our sample with the hypothetical population, or for that matter with any theory about the nature of its source.

9.2 Binomial Distribution

The theoretical model that we will now discuss is called the *binomial distribution*. It serves as a way of describing how *categorical* data may be distributed. If you recall back to Chapter 2 you will remember that we discussed the concept of *attribute* data in Example 2.5, and later classified this type of data as the simplest of the different forms of measurement scales. We will now look at the simplest case of categorical scales—one with only *two* possible categories.

Examples 2.7 through 2.10 all had only two classifications.★ If we look back to Example 2.5 we find that for *each day*, each student has a possibility of being *present* or *absent*. So there are two possibilities for each student on each day. Each of the numbers in Table 2.5 served as a sample of 25 students, each of whom may have been present or absent. In effect what we actually have done is to assign values to the two conditions; we have assigned a numerical value of 0 for "present" and 1 for "absent," and we have totalled the number of 0's and 1's for all the individuals who are enrolled in the course and who thereby constitute the members to be samples. This sum is the numerical value of our sample.

Now as we look at many more days we assume again that each individual may or may not come to class. With each student independently making that decision we arrive at a new overall sample value (maybe the first day had 0 absences but the second day had 1 absence). So much for the real data and the real process. If each student has the same chance of deciding to cut class (or to be ill), and there is no collusion or ganging up on the teacher, then there exists a mathematical model, a theoretical distribution of sample values that we can derive and later compare with our set of samples. Furthermore, if the theoretical model serves as a satisfactory substitute for the data, then we may say that it in effect duplicates the process and we will then be able to determine what the *relative chances* of occurrence are for *all* of the different numbers of absences. Some further examples of observations, measurements, or phenomena that may be classified as having only two possible results follow.

EXAMPLE 9.1
 a. A coin may come up *heads* or *tails*.
 b. A poll may permit two responses, *in favor of*, or *against*.
 c. A person *may receive* or *may not receive* a ticket.
 d. An item may be *defective* or *not defective*.

★ Refer back to page 40. Note that Example 2.6 could be included if we ignore withdrawals (perhaps by calling them withdraw pass and withdraw fail and include them with the pass and fail). As a reminder, Example 2.7 is finish or not finish an auto race, Example 2.8 is loaves of bread above and below minimum standard weight, 2.9 is men who are 6 footers or less than 6 foot, and Example 2.10 is temperature above or below the average.

e. A traffic light may be *red* or *green*.
f. A customer may *buy* or may *not buy* something.
g. The radar trap may be *working* or *not working*.

Now that we have the two possible cases, we must consider what constitutes *a sample*. The value of the sample will then serve as the X-values of our theoretical distribution, which will be called a binomial distribution. It will be the variation of these sample values that will most concern us.

EXAMPLE 2.6 (Cont.)

We may have 23 pass and 5 fail; thus the X- or sample value is 23 and the sample size is 28.

EXAMPLE 2.7 (Cont.)

Out of 42 starters of the race, 18 finish. Then $X = 18$ and the sample size $n = 42$.

EXAMPLE 2.8 (Cont.)

10 loaves are weighed and 1 is below the minimum standard weight. Then we may say $X = 9$ while $n = 10$ (or $X = 1$ while $n = 10$, depending upon which you are counting).

EXAMPLE 2.9 (Cont.)

The starters on our basketball team have 4 six-footers. Therefore $X = 4$ and $n = 5$.

EXAMPLE 2.10 (Cont.)

The temperature has been above average every day this week. Then $X = 7$ and $n = 7$.

The next samples should point out what *can* occur.

EXAMPLE 9.1 (Cont.)

a. If we were to toss a coin 5 times we could get, 0, or 1, or 2, or 3, or 4, or 5 heads. Then these are all of the possible X-values, or all possible samples of 5 ($n = 5$).
b. You poll 50 people; then $n = 50$. The possibilities are none in favor, 1 in favor, 2 in favor, . . . all the way up to 50 in favor. *Each* case can occur. Some will be highly unlikely while others will have a good chance of happening.
c. We watch 10 people. Then none, or 1, or 2, or . . ., or all 10 can receive a ticket.
d. If we try 6 batteries we may find that none work, 1 works, 2 work, 3 work, . . . or all 6 work.
e. You arrive at the traffic light 8 times. It can be red none of the times, or 1 time, or 2 times, . . . or 8 times.

f. Of 30 customers who enter a store none, 1, 2, 3, . . . on up to 30 may buy something.

g. The radar trap may be working 0, 1, 2, 3, 4, 5, 6, *or* 7 of the days this week.

So we must realize that a particular sample, as Examples 2.6 through 2.10 illustrated, can have only one value. However, many possible values can occur, although the number of possibilities is limited by the sample size (obviously if you start a race with 42 cars you don't expect 45 to finish). The recognition of the fact that this variety of values is possible is the beginning of the development of the model called the binomial distribution.

I have said several times that we want to determine a theoretical model which, given some assumptions, might possibly indicate to us the chance of having 23 passing students, or 18 cars finishing the race, or 9 good loaves of bread, or 4 six-footers, or 7 days of hot weather. Or we might want to know the entire model: what the entire distribution of heads, or people in favor, or tickets, or defective batteries, might likely be. We might want to know the *chance* of not being stopped by any red lights, or being stopped only once, or twice, or three times, or four times, and so forth.

This leads us to make some assumptions to arrive at the underlying concept that we will use to establish values for the chances or the probabilities. Thus, we must describe the model we will use, remembering that we need probabilities for each X-value in order to establish this hypothetical or theoretical probability distribution or model.

Our first assumption is that each of the individuals in the sample has the *same chance* or the same probability of falling into the two categories. I have already said (in Example 2.5) that each and every one of the students is to be considered as having the same chance of being absent. I am assuming each student in the class now has the *same chance* of passing, each auto in the race has the *same probability* of finishing, each loaf of bread in the sample has the *same chance* of being above the minimum weight. *And*, furthermore, this *same* probability is to be extended to *all* the individuals in *all* the possible samples. So when tossing coins the chance of a head occurring will not change during *all* the tosses for the first sample or the second sample or *any* sample. The decision or chance of coming to class is *assumed* to be the same for *all* students on *all* different days.

In addition, each trial, each person polled, each item looked at must be independent of every other trial, person, or item. Thus, when we toss a coin and get a head we assume that this will not influence our chance of getting a head or a tail on the next trial. When we ask an opinion on a poll we are assuming that the next person polled has not been influenced by the previous person or will not influence a future individual. As you might observe, this last case requires careful control of the way the data is accumulated, otherwise valid interpretations are virtually impossible to make. (Checking day-to-day temperature could very well not involve independent measurements. I would suspect that the chance of it being hot tomorrow would be greater if it were hot today than if it were cold today.)

Nonetheless, if we use the binomial distribution we are assuming that independence is present, and not only within the sample, *but* from sample to sample (day to day, and so forth).

Now to identify the probability values. We actually have to assign a value to *both* of the possible occurrences, but knowing one of them we can quickly compute the other.

In general, we may determine the probability either by guessing, by using past data, or by our experience. Or we may use a sample to approximate the value. If we use a sample (including any previous samples) we will say that

$p =$ the long range probability or chance of *one* of the occurrences

$$= \frac{\text{Number of times one of the occurrences took place}}{\text{Total number of occurrences}}. \tag{9.1}$$

Observe that the total number of occurrences need not be just one sample but could include any number of trials or individuals as we will see in the following examples.

EXAMPLE 2.5 (Cont.)

Of the 750 possible attendances (25 students on each of 30 days) there were a total of 48 absences. So the *total number of occurrences* is 750, and the *number of times one of the occurrences took place* is 48. Then the long range probability of *that* occurrence (absences) is $48/750 = 0.063$ (or about 6.3% are absent). The probability associated with the second possibility would be $702/750 = 0.937$.

EXAMPLE 2.6 (Cont.)

Of the 530 students who have taken the course (over several years and in 21 sections) 69 have failed the course. So the overall, long range probability of failure is $69/530 = 0.130$ (or 13% fail the course). Then $461/530 = 0.87$ is the probability of passing the course.

EXAMPLE 9.1 (Cont.)

a. If we have a fair coin, we would say that there is a 50–50 chance of a head appearing. So out of 100 tosses we guess that 50 should be heads. Therefore we say that $p = 50/100 = 0.50$ that a *head* will occur on any given toss.
b. We may assume for some reason that an individual has a p of 0.60, or a 60% chance of saying "yes, I am *in favor*." We may have taken a preliminary sample of 250 people where 150 (or 60%) said they were "in favor."
c. If one person out of a hundred gets a ticket, then $p = 0.01$ or the probability is 0.01.
d. Out of 1800 batteries tested we found that 720 were defective; then the chance of a *defective* would be 0.4.
e. If the light is red a quarter of the time, then the probability of the light being red is 0.25. Here we have a slightly different source than in the

previous examples. In order to define the p, we should modify the equation to read

$$p = \frac{\text{Span of time the light is red}}{\text{Total span of time considered}}.$$

The distinction here is that we must consider time as a continuous line. We should think of the length of that line as the total number of possible occurrences. One portion of that line is the time that the light is red, and that would be considered the times *red* occurred. The remaining time would be green. (I am ignoring the time that the light is amber, especially since most drivers also ignore it.) Thus, if the total time to be considered were eight minutes, the line below could represent this total time span.

```
0   1   2   3   4   5   6   7   8
              minutes
```

If the light was red for a portion of that time, say two minutes, we could show it as

```
0   1   2
```

The proportion of length of red to the total length would be

$$\frac{\text{Red}}{\text{Total}} = \frac{\rule{1em}{0.4pt}}{\rule{4em}{0.4pt}} = \frac{1}{4}$$

If we know the probability, p, of the occurrence of one of the items, then the probability that the other will happen is 1 minus p, and is called q. So

$$q = 1.00 - p. \qquad (9.2)$$

We have already found the two probabilities for Examples 2.5 and 2.6. If you go back to these two examples you will find that the two probabilities in each example added up to 1 (or if you subtracted one of the probabilities from 1.00 you would have found the other). Now let us look at the remaining examples.

EXAMPLE 9.1 (Cont.)

a. The probability of a tail is

$$q = 1 - p = 1 - 0.50 = 0.50.$$

Observe that if the two items are *equally likely* to happen, *they have the same probability*.

b. The probability of an individual saying "no" is $1 - 0.60 = 0.40$.
c. If $p = 0.01$, then the probability of getting no ticket is $1 - 0.01 = 0.99$.
d. The chance of a good, nondefective battery is 0.6.
e. If the probability of *red* = 0.25 then the probability of *green* is 0.75.

Now, we are really interested not in the chance that *you* will be absent, or *you* will fail the course, but rather in the chance that one, *or* two, *or* . . ., and so forth, people would be absent out of the entire class, or that various numbers of students might fail the course. We must now determine what the probabilities are for each possible kind of sample. The formula to be used depends upon the values of p and n (the sample size). (Remember that $q = 1 - p$.) The probability of one of the sample values (one of the X values) occurring is

$$\text{probability of } X \text{ occurrences out of } n \text{ trials} = \frac{n!}{X!(n-X)!} p^X q^{(n-X)}. \quad (9.3)$$

where $n!$ is called n factorial. (The factorial, !, means to multiply together every integer up to the number n. So $1! = 1$, $2! = 1 \cdot 2$, $3! = 1 \cdot 2 \cdot 3 = 6$, $4! = 1 \cdot 2 \cdot 3 \cdot 4 = 24$, and $n! = 1 \cdot 2 \cdot 3 \cdot 4 \ldots n$).

This is very tiresome to calculate for even a few values, especially if p is not an easy number to handle.

EXAMPLE 9.1 (Cont.)

a. If you wanted to know the chance of 3 heads occurring in 5 tosses you would have to calculate

$$\text{Probability of 3 heads out of 5 tosses} = \frac{5!}{3!(5-3)!} \left(\tfrac{1}{2}\right)^3 \left(\tfrac{1}{2}\right)^2$$

$$= \frac{1 \cdot 2 \cdot 3 \cdot 4 \cdot 5}{(1 \cdot 2 \cdot 3)(1 \cdot 2)} \cdot \left(\tfrac{1}{8}\right)\left(\tfrac{1}{4}\right) = 10\left(\tfrac{1}{32}\right) = 0.3125.$$

This would mean that you would expect that 31.25% of the time three heads will appear.

e. What is the probability that you will get stopped by five traffic lights? Then n is 8 (look back to page 237), X is the 5 we are looking at, p we found to be 0.25 and q is 0.75. Then

$$\text{Probability of 5 out of 8} = \frac{8!}{5!(8-5)!} (0.25)^5 (0.75)^3$$

$$= \frac{1 \cdot 2 \cdot 3 \cdot 4 \cdot 5 \cdot 6 \cdot 7 \cdot 8}{(1 \cdot 2 \cdot 3 \cdot 4 \cdot 5)(1 \cdot 2 \cdot 3)} (0.25)^5 (0.75)^3$$

$$= \frac{40320}{(120)(6)} (0.0009766)(0.2373047)$$

$$= 0.0231.$$

I think you can appreciate the magnitude of work involved in calculating just one simple probability value. In order to make life easier for us when we need a complete binomial distribution, or even a few terms, tables have been constructed. A few of the important tables are provided in Appendix Table A.8. Here you only need to know the n and the p and you have the complete probability distribution for all of the X values. Of course you do not have every possible p or every possible n value. Think how large the table would have to be if you wanted every p from 0.01 to 0.99 for just an n of 100. Observe that your tables do not have values of p for either Examples 2.5 or 2.6. We will have to turn to other means to arrive at these distributions. Either we will have to calculate them by hand (which I would dread) or with the aid of a calculator, or with a computer (perhaps with a library program as will be demonstrated soon), or in some cases by an approximation that will come pretty close to the real distribution. This last technique will be discussed later in this chapter and in the next chapter.

EXAMPLE 9.1 (Cont.)

a. To find the probability of three heads look in Appendix Table A.8 as follows:

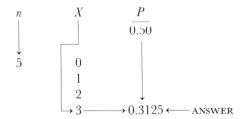

e. Here I would like the entire distribution, so we will look up all of the possibilities under $n = 8$. I want the probability that I will be stopped by red light once, or twice, or up to 8 times.

n	X	p 0.25
8	0	0.1001
	1	0.2670
	2	0.3115
	3	0.2076
	4	0.0865
	5	0.0231
	6	0.0038
	7	0.0004
	8	0.0000

The probability for 5 red lights is right there—0.0231—with no calculations! You can also see that the chance of the light being red 8 times is less than 1 in 10,000. (0.0000 does not mean 0.0000000.... It says that it is closer to 0.0000 than to 0.0001.) There is less than a 3% chance of being stopped 5 times and about a 31% chance of being stopped 2 times, and so on.

Next, with these values we can draw a histogram (Figure 9.1). This histogram is our theoretical distribution or model of the number of occurrences when n is 8 and p is 0.25.

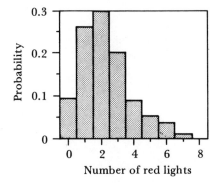

Figure 9.1 Distribution of the number of red lights out of a total of eight traffic lights. The probability of a red light is 0.25.

EXAMPLE 9.1 (Cont.)

b. Here we were taking a poll of 50 people and we felt that 60% would be in favor of the issue (or each person had a probability of 0.6 of answering favorably). Now I ask: What percent of the samples of 50 will show a majority in favor of the issue?*

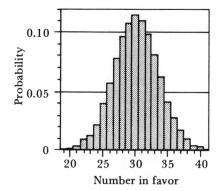

Figure 9.2 Distribution of *number of people in favor* of an issue. Fifty people voting, each with a probability of 0.6 being in favor.

The first step must be to come up with the entire distribution of *number in favor*. But your tables stop at $n = 30$, so they are useless for this problem. Since the calculations are simply enormous, I consulted other tables.† In Figure 9.2 I have drawn the distribution. Look at the distribution of *samples*. Some samples will only have 20 or 21 people in favor out of the 50 that are asked. These samples will *not* indicate that the majority is *in favor*. Only those samples with X-values of 26 through 50 will indicate that the majority is *in favor*.

---

* Note that this also begs the question "What proportion of the samples will have a sample p that is greater than 0.50?"
† H. G. Romig *50–100 Binomial Tables*, (New York: John Wiley & Sons, Inc, 1963).

To find the probability or percent of times that the sample will show a majority in favor, we merely add up all of the heights above all of the values from 26 to 50 and we find that this is 0.90.★ Thus there is a 90% chance that the poll will have the majority in favor of the issue.

If you look further at Figure 9.2 you should see how close the lines on the histogram are. This distribution is beginning to resemble a smooth continuous distribution. With large samples, large n, we will find that this smoothness will become even more pronounced and we will find that we will be able to use some other distributions to approximate the binomial distribution.

Some other observations should be made about the binomial distribution. Figure 9.3 shows a collection of histograms for different values of n and p. Regardless of the n, when $p = 0.5$ the distribution is symmetrical and centered at the middle of the set of possible values. When p is at one of the extremes, either around the 0.1 or the 0.9, the distribution is heavily skewed. However, as the sample size increases, this skewness decreases. When the p is small, the samples consist mostly of small values, and when p is large the samples tend to be mostly large values. These factors are true pretty much regardless of n. As we observed before, as n increases the distribution begins to smooth out, regardless of p. We just noted that as n increases, the skewness decreases. In addition, the distribution of samples covers much less of the total possible sets of samples. For $p = 0.5$ and $n = 5$ there is at least a 3% chance for the most unlikely possible values to occur (the 0 and the 5), so *all* the values have at least some small chance of being observed (3% chance is small but not impossible). For $n = 20$, however, the span covering those values with at least a 3% chance of being observed goes from 6 to 14, only *half* of the possible values. There is some chance for the other values to appear, but it is much less, and so basically the distribution covers much less of the whole set of *possible* values.

9.3 Means and Standard Deviations

The next thing we should determine about these distributions are the descriptive statistics—the *mean* and the *standard deviation* (or variance). Since we are talking about a theoretical distribution, we will use *population* symbols. We will call the mean by the letter μ (the Greek letter called mu), and the standard deviation by σ (sigma). Then the variance will be σ^2 (sigma squared).★★ We

---

★ This is most easily found by determining the chance of being defeated. This is found by adding the probability of 19, 20, 21, 22, 23, 24, and 25 being in favor—or $0.000 + 0.002 + 0.003 + 0.005 + 0.010 + 0.015 + 0.025 + 0.040 = 0.100$. Then the probability of a majority is $1.000 - 0.100 = 0.900$.

★★ The fundamental formula used to calculate this variance would be

$$\sigma^2 = \frac{\Sigma (X - \mu)^2}{n}.$$

Note that we use μ not $\bar{X}$ as the mean and n not $n - 1$ for the denominator. These are used because we are working with a population *not* a sample. Refer to page 106 for our original discussion of this idea.

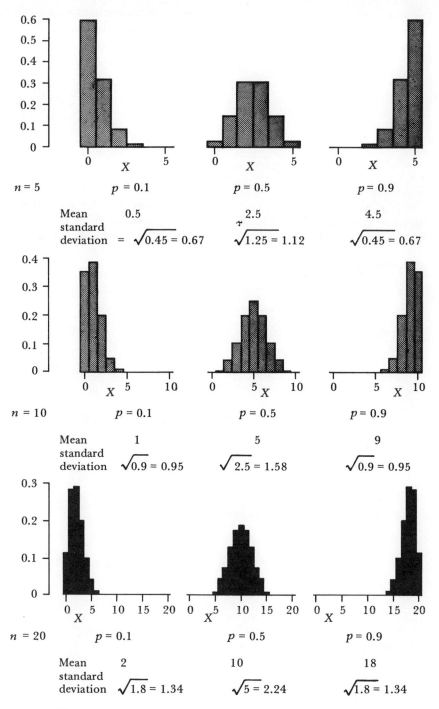

Figure 9.3 Binomial distributions for different values of n and p.

will not bother to derive the following formulas for calculating the mean and standard deviation of a binomial distribution, as they are fairly straightforward.

Mean
$$\mu = n \cdot p. \tag{9.4}$$

Variance
$$\sigma^2 = n \cdot p \cdot q = n \cdot p \cdot (1 - p). \tag{9.5}$$

Standard deviation
$$\sigma = \sqrt{n \cdot p \cdot q} = \sqrt{n \cdot p \cdot (1 - p)}. \tag{9.6}$$

Using Equations (9.4) through (9.6) we can quickly find the two values that describe the distribution, given only the population, p, and the number of items n. The mean, μ, and the variance, σ^2, are considered fixed constants for the particular distribution and are established once we know n and p.

EXAMPLE 2.5 (Cont.)

n was 25 and p was 0.063. Therefore,
$$\mu = n \cdot p = (25)(0.063) = 1.57.$$

Then there is an average of a little over $1\frac{1}{2}$ absences each period.

The variance
$$\sigma^2 = n \cdot p \cdot q = n \cdot p \cdot (1 - p) = (25)(0.063)(0.937) = (1.57)(0.937) = 1.47.$$

Then
$$\sigma = \sqrt{1.47} = 1.21.$$

EXAMPLE 2.6 (Cont.)
$$n = 28, \quad p = 0.130.$$

We expect the number of failures to be equal to
$$n \cdot p = (28)(0.130) = 3.64$$

(if the 3 failures we experienced is pretty close to the average).
$$\sigma^2 = (28)(0.130)(0.870) = (3.64)(0.870) = 3.16.$$

Then
$$\sigma = \sqrt{3.16} = 1.78.$$

Note that once you find the mean $(n \cdot p)$, to find the variance you only have to multiply the mean by $1 - p$ (or by q).

EXAMPLE 9.1 (Cont.)

a. The mean number of heads in five tosses is
$$\mu = n \cdot p = 5 \cdot (0.5) = 2.5.$$

The variance
$$\sigma^2 = n \cdot p \cdot q = 5(0.5)(0.5) = 1.25$$
and the standard deviation
$$\sigma = \sqrt{1.25} = 1.12.$$

b. The mean number of "yes" votes is
$$n \cdot p = (50)(0.60) = 30.$$
So on the average 30 people will vote for the issue.
The variance
$$\sigma^2 = n \cdot p \cdot q = (50)(0.60)(0.40) = 12$$
and the standard deviation
$$\sigma = \sqrt{12} = 3.46.$$

c. The mean number of tickets is
$$n \cdot p = (10)(0.01) = 0.1$$
and
$$\sigma^2 = n \cdot p \cdot q = (10)(0.01)(0.99) = 0.099,$$
or
$$\sigma = 0.31.$$

If you observe that the average is only 0.1 ticket, you should realize that this must indicate that most of the time *no* tickets are issued.

e. The average number of red lights is
$$(10)(0.25) = 2.5$$
and
$$\sigma^2 = (10)(0.25)(0.75) = 1.875,$$
or
$$\sigma = 1.37.$$

9.4 Approximation to the Binomial Distribution

Now that you have an ability to compute the mean and standard deviation relatively easily you should be able to reproduce approximately the distribution using the empirical rule, right? Wrong. Only sometimes. Look at Figure 9.3 again. I have provided the mean and standard deviations for each of the distributions. Go only one standard deviation to the left of the $n = 5$, $p = 0.1$ distribution and where are you? You are out of the distribution because it is so heavily skewed. So for some of the distributions the empirical rule is not useful at all, namely with the skewed ones. However, the center distributions, those with p around 0.5, will follow pretty closely the empirical rule. In the next chapter we will discuss this concept further by considering what is called an *approximation to a binomial distribution*.

When we were building up a binomial distribution we calculated or found from tables the probabilities for the various possible outcomes. This was fine for small numbers of trials or a limited set of individual probabilities. When the probability of occurrence, p, is not an easy number to work with, such as 0.34, or is not to be found in the table (again a p of 0.34), we must go through extensive calculations even for small values of n. When n is very large, even only as large as 25 or 50, the calculations are immense. The net result is that you would hope to be able to avoid the necessity of using the binomial distribution in these cases. It is partly in order to solve these difficulties that the normal distribution will be introduced. Of course, it will have much broader use than just as a substitute for the binomial, but this will actually serve as a valuable way of understanding and of using the normal distribution.

REVIEW

Theoretical model · Binomial distribution · Independence · Equally likely · p · $q = 1 - p$ · $\mu = np$ · $\sigma^2 = n \cdot p \cdot q = n \cdot p(1 - p)$ ·

Approximation to the binomial distribution.

PROJECT (Class)

Determine the total number of possible periods in a week at your school (if there are 6 periods a day and 5 days a week there are 30 periods). Call this C. Assume that all students have an equal chance of having a class in any of those periods (except for this statistics class). For the sample of students in the class (n = sample size = class size) determine the number of students who have classes during each period. Compare the distribution of number of students having each class (the number having class first period Monday, second period Monday, and so forth) to a binomial distribution with $p = 1/C$ and n = sample size.

PROBLEMS

Section 9.1

9.1 In what different ways could you classify the following as binomial distributions?
   a. Rolling a pair of dice.
   b. Making a call with a telephone.
   c. Spraying bugs with an insecticide.
   d. Starting a car.

9.2 In what way might the assumption that p is the same for all individuals be incorrect for the following examples?
   a. Absences in a particular class.
   b. Students' score on a test.
   c. Matches igniting when struck.
   d. Babies crying in the nursery of a hospital.
   e. Insect being killed by a bug killer.
   f. Being stopped by a red light.

9.3 How does the question of replacement or nonreplacement affect the assumptions of a binomial distribution? (For example, consider polling students for their attitudes on some issue.)

9.4 In flipping a coin it generally is assumed that there is a 50-50 chance of a head. What effect does the possibility of the coin standing on edge or rolling under the table have on this? How can you still describe this as a binomial distribution?

9.5 On a roulette wheel there are 18 reds, 18 blacks, and 2 noncolors (0 and 00). In what way can we still regard this as a binomial distribution?

9.6 Using the tables determine the following probabilities.
   a. For $p = 0.5$ and $n = 6$ what is the probability that $X = 3$?
   b. For $p = 0.5$ and $n = 6$ what is the probability that X is less than 2?
   c. For $p = 0.4$ and $n = 8$ what is the probability of $X = 6$?
   d. For $p = 0.4$ and $n = 8$ what is the probability that X is at least 5?
   e. For $p = 0.6$ and $n = 12$ what is the probability that X is between 4 and 8?
   f. For $p = 0.6$ and $n = 9$ what is the probability that X is greater than 7?

9.7 A bag of seeds states that 70% of the seeds will germinate.
   a. What percentage will not germinate?
   b. If I plant 5 seeds, what is the chance that none of the seeds will *not* germinate?
   c. If I plant 10 seeds what is the chance that all of the seeds will germinate?
   d. For 10 seeds, what is the probability of: all germinating, 9 germinating, 8 germinating, ..., none germinating?
   e. If I need at least 5 plants for an experiment, what is the chance that I will not have 5 plants if I plant 10 seeds?
   f. If I need 5 plants and plant 15 seeds, what is my chance now of having enough plants?

9.8 A study at a large university revealed that students in class were really thinking about sex almost 60% of the time. What is the probability that at a given time more than 15 out of the 20 students in my class are thinking about sex? What is the chance that a majority of the students are not thinking about sex (and I hope thinking about the lecture)?

9.9 On a true-false test, if you just guess the answer, we can say that your chance of getting a correct answer is 0.5.

a. If there are only 4 questions what is the probability of getting all correct by guessing?
b. If there are 20 questions what is the probability of getting all correct by guessing?
c. If just passing on a 20 question test is 15 correct, what is the probability of passing by just guessing?
d. What is the probability of failing the test by just guessing?
e. If an A is 19 or 20 correct, what is the probability of getting an A?

9.10 On the true-false test I assume that a person who understands the material has a 0.8 chance of getting a correct answer.
a. What is the probability of getting all answers correct if there are only 4 questions (and you understand the material)?
b. What is the probability of getting all answers correct on a 20 question test?
c. What is the probability of passing a 20 question test (if 15 is passing)?
d. What is the probability of getting an A (if an A is 19 or 20 correct)?

9.11 It has been assumed at State College that 20% of the students will fail to complete the first year. Seventeen students are admitted from XYZ High School.
a. What is the probability that 3 students will fail to complete the first year?
b. What is the probability that 9 or more students will not complete the first year?
c. What is the probability that 16 will complete the first year?
d. What is the probability that a majority will complete the first year?

Section 9.2

9.12 The probability of a particular seed germinating was 0.7 (Problem 9.7).
a. If I plant 5 seeds what is the average number of seeds I would expect to germinate?
b. If I plant 10 seeds how many should I expect to germinate? What is the standard deviation of the number of seeds germinating?
c. If I were to plant a bag of 10,000 seeds, how many should I expect to germinate?
d. Assume I plant 10,000 seeds. Using the empirical rule, 95% of the time I should have what possible numbers of seeds germinating?

9.13 If 60% of the time students are thinking about sex (see Problem 9.8),
a. What would be the average number of students in my class of 25 that are thinking about sex?
b. What would be the average number thinking about sex in a large lecture hall with 200 students? What are the 2 sigma limits for the number of students thinking about sex (in the lecture hall)?

9.14 What is the average number of correct answers for a student who guesses on a 20 question true-false test (Problem 9.9)? How would he do on a 100 question test? What is the standard deviation of correct answers on a 100 question test?

9.15 If a person taking a 100 question test understands the material (and therefore the probability of passing is 0.8) how many correct answers should he expect to get? What will the standard deviation of the number of correct answers be?

9.16 If 200 students from XYZ High School go to State College, how many would be expected to pass (see Problem 9.11)? Over the course of 100 years the guidance department would expect that the number who would pass would be within what limits for about 99 of those years?

10

The Normal or Gaussian Distribution

In the previous chapter we worked with distributions that served as models for categorical data. We dealt with samples that had integer values and limited sets of possible values. Now we will turn our attention to measurement data with numerical scales. No longer will a sample merely consist of a single number (remember that a sample from a binomial distribution could be 2 heads out of 5 tosses of a coin or 1 absence in a class of 25 students). Rather, each of the individuals in the sample may have a *different* value. Thus, we will obtain a *distribution of values* within the sample itself. One common model for this type of distribution is our bell-shaped, normal distribution, the one that served as the basis for the empirical rule. Later, in Chapter 12, we will also find that means (averages) of *samples* will fall into a normal distribution and this will be an even more important reason to study this theoretical model. For now, however, we will consider the normal distribution as a model for a collection of individual values—and many collections of measurement data yield distributions that very closely resemble the normal distribution.

10.1 The Nature of the Normal Distribution

This distribution, like all others we deal with, has a mean and a standard deviation.★

★ Unlike the binomial distribution, in order to fully describe a normal distribution we need to specify both the mean and the standard deviation. Special circumstances involving the formula that defined the binomial distribution required only one parameter or constant be present in order to establish a complete probability distribution or curve for the binomial distribution. Once one of the probability values was identified (and the *n* of the sample was known), the distribution could be constructed, since the mean and the variance were related to each other and to the probability, *p*.

With the normal distribution, if we specify the mean, μ, and nothing else, we are going to be left without a scale. Figure 10.1 shows a distribution with mean, μ, equal to 50, but neither the variance nor the standard deviation has

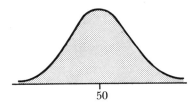

Normal distribution

Figure 10.1 Normal distribution.
Mean = μ = 50. Variance unknown.

been specified. This is equivalent to saying that we have a distribution of problems assigned for each week of class, or the lengths of women's skirts, or the ounces lost while on a special diet—all with an average of 50. However, without a variance or standard deviation we *cannot* fill in any other values since we know nothing about the spread of this distribution or the way in which the individuals are dispersed across the scale. Thus we cannot tell if almost all the weeks have 50 problems assigned, if the lengths of skirts are all close to 50, or if the weight loss varied all the way from nothing to 100 ounces. Remember that this distribution is the basis of what we called the empirical rule, which gave us an indication of what percent of the data fell between certain values of standard deviation. But if we do not know the standard deviation, we are at a loss to know where the limits are to be. In order to work with the complete theoretical distribution it becomes even more vital to know the standard deviation (or variance).

Throughout this chapter I will use a problem involving the distribution of the amount of weight lost by a collection of people using various diets.

As we look to Figure 10.2 the nature of the distribution and its reliance on a value of standard deviation becomes apparent. Here we can see what is the relative chance various values have of appearing, since the height above each value on the scale indicates, as always, the probability of occurrence of that value. Here we see that most people will lose from 45 to 55 ounces, with a few only losing around 40 ounces and some lucky people losing up to 60 ounces.

As we change the variance (and hence the standard deviation) we alter the scale (it may be we changed the diet). Thus, in Figure 10.3, as the variance is increased from 25 to 100, the scale is likewise changed. A point that appears to be approximately the same distance away from the center actually is numerically further away in Figure 10.3. Compare the X-value of 40 in Figure 10.2 with the X-value of 30 in Figure 10.3. They both appear to be at the same

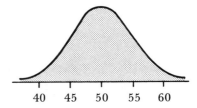

Figure 10.2 Normal distribution. Mean = μ = 50. Variance = 25. Standard deviation = 5.

Normal distribution

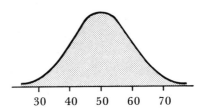

Figure 10.3 Normal distribution. Mean = μ = 50. Variance = 100. Standard deviation = 10.

Normal distribution

relative locations on the curve. However, in the first case it is only 10 units from the center, while in the second case this point is 20 units from the center. When the variance is greater we find pinpointing the individual loss is more difficult, so that a person with the new diet has the same chance of losing 30 ounces as the old dieter had of losing 40 ounces. The same is true on the high end of the scale.

We can generalize this observation by saying that the greater the variance, the larger the values along the scale. Similarly, if we decrease the variance, we will obtain a smaller scale along the X-axis—the base of the graph. In Figure 10.4 I have drawn a normal distribution that now has a smaller variance (a third diet). Our generalization mentioned above about the variance still holds.

Let us superimpose all three curves on a single set of scales. Since each of these distributions is a probability distribution, the area under each curve must be the same (and must be equal to 1 unit). If we use the scale of Figure 10.2 as a standard (the one with a mean of 50 and a standard deviation of 5), then when I superimpose the next graph I will have to stretch the curve. This in turn will cause it to shrink in height (see Figure 10.5). The third graph we discussed (the one with the smaller standard deviation—Figure 10.4), because it is on a smaller scale, would be compressed inward in order to squeeze it on this new scale. This would cause the height to increase. Remember, this is because *all three graphs must have the same area*. Let me remind you again that

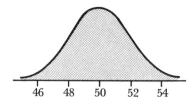

Normal distribution

Figure 10.4 Normal distribution. Mean = μ = 50. Variance = 4. Standard deviation = 2.

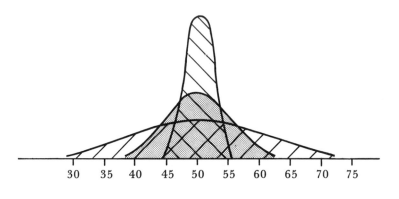

Normal distributions

Figure 10.5 Normal distributions. Mean = 50. Figures 10.2, 10.3, and 10.4 are superimposed on the same scale. All areas are equal.

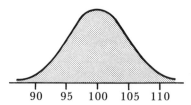

Normal distribution

Figure 10.6 Normal distribution. Mean = μ = 100. Variance = 25. Standard deviation = 5.

these curves are to be interpreted as though they are histograms. The relative heights represent relative frequencies with which the particular X-values occur. The more compressed curves would have most of their values close to the mean, while the others, which are much more spread out, would have less of a chance of having values that are actually near the mean.

Thus far we have looked at several different normal distributions, but all had the same mean or central value. It is not necessary that the mean be equal to 50. Figure 10.6 shows a normal distribution with mean equal to 100 and variance equal to 25. This does not look much different than Figure 10.2 except for the scale. Figure 10.7 shows a distribution with mean = 0 and variance = 100, and resembles Figure 10.3 except for the scale. Each of these

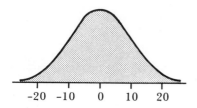

Normal distribution

Figure 10.7 Normal distribution. Mean = μ = 0. Variance = 100. Standard Deviation = 10.

could be a consequence of additional different diets that yield different results. The last diet appears to cause no change, and may just reflect random change in a control or a maintenance diet. (Note that a minus value would indicate that weight was gained not lost.)

I have taken the pairs of distributions that have identical variances and placed them on the same axis. Figure 10.8 contains both distributions with σ^2 (variance) = 25 (from Figures 10.2 and 10.6). Figure 10.9 contains the two distributions with variance of 100 (from Figure 10.3 and Figure 10.7). I then drew all four of these distributions on a single axis (Figure 10.10). You

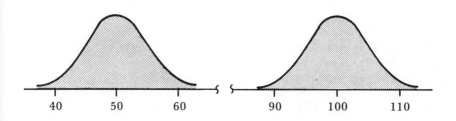

Normal distributions

Figure 10.8 Two normal distributions having same variance (25) but different means. Figure 10.2 and 10.6 are shown on the same axis.

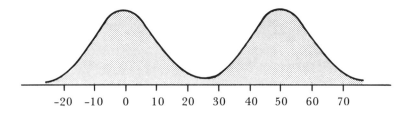

Normal distributions

Figure 10.9 Two normal distributions with variance of 100 and different means. Figures 10.3 and 10.7 are shown on the same axis.

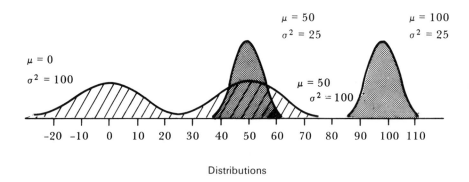

Distributions

Figure 10.10 Distributions of Figures 10.8 and 10.9 on one scale. The scale has been adjusted to allow all four distributions to fit on a single scale. All areas are equal and equal to previous curves.

should now be aware that many different kinds of normal distributions can be drawn—as many different distributions as can arise in real life. They resemble each other in certain ways, but with each new value of μ (mean) or σ^2 (variance), we will effectively be drawing a different picture, or at least using a different scale.

Now if you have been observant, you may have realized that something has been avoided. None of the curves thus far presented have had scales for the probability or Y-axis. There is good reason for this. In order to calculate the probability values (or the heights) at each particular X point, we must know what relationship exists between the X and the probability values. The formula for the normal distribution is

$$\text{Probability of } X = \frac{1}{\sqrt{2\pi}\,\sigma} e^{-[(X-\mu)/\sigma]^2/2} \tag{10.1}$$

where $\pi = 3.14159\ldots$
$e = 2.71828\ldots$
$\sigma = $ standard deviation of the distribution
$\mu = $ mean of the distribution.

To calculate a large enough set of points to draw the curve accurately is extremely tedious. Since we will not usually be concerned with the actual heights we will not pursue the matter, and we will not be using the formula directly.★

It should now be mentioned that with a continuous distribution the probability associated with an exact value is zero. We can compare this to the problem of measurements and the concept of a mathematical point, where we can always question the exactness of a value by going one decimal place further. However, when we work with practical problems we work with a scale, and each value has a pair of bounds. These bounds effectively are the class limits we encountered in our work with histograms and measurement data in Chapter 3.

To determine the probability of a particular value or set of values, we are really determining the area under the curve between two limits. This again is because the *area under the curve is equal to the probability* associated with the collection of values between those limits. So we really don't care about determining your weight loss in thousanths of an ounce—the nearest half an ounce is more than satisfactory.

In Figure 10.12a I have drawn a normal distribution that averages 0 (no change) with a variance of 1 (so that most people seem to vary up or down within about 2 ounces). On that distribution I have shown the probability line for an *X*-value (or weight loss) equal to 0.5 ounces. Looking at this line under a magnifying glass, Figure 10.12b, you can see more graphically the limits being discussed. There is a definite area that you can observe and calculate. Hence, the probability would be the product of the width along the *X*-axis times the approximate average height along the probability axis (the *Y*-axis).

★ If the need should arise to identify the heights, there is actually a relatively simple way around the problem. We shall be discussing the special normal distribution with mean = 0 and variance = 1 (standard deviation = 1). For this distribution, tables of probability values do exist. Appendix Table A.13, called Ordinates of Normal Distribution, lists such heights. If you look carefully at the formula for the probability Equation (10.1), you may realize that the $(X - \mu)/\sigma$ portion reduces any distribution to a mean of 0 and a variance of 1. We will be discussing this at length in this chapter. The only part that will vary in the equation will be the σ in the denominator of the multiplier; the rest is the same for *all* normal distributions. Hence for different values of mean and/or variance we first calculate $(X - \mu)/\sigma$, look that quantity up in the table to determine *Y*, and multiply the *Y* by $1/\sigma$. Thus for a distribution of mean 0 and variance 1 we obtain Figure 10.11a. For a mean of 0 but variance 100, the height corresponding to the middle point (the value of 0.4 in the previous distribution) is now divided by $\sqrt{100}$ or 10, the standard deviation, (see Figure 10.11b). As a third example we consider a mean of 0 and a variance of 4 (see Figure 10.11c). Figure 10.11d shows all three of these distributions on the same graph with a single scale.

260 Distributions

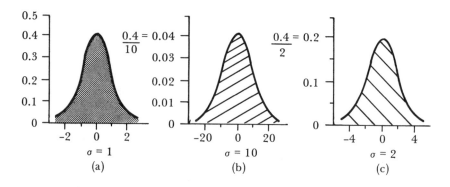

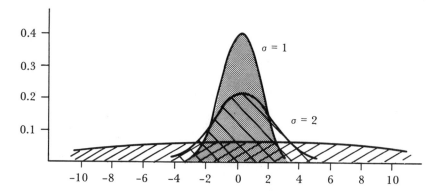

Figure 10.11 Relationship of the Y-axis to the variability of a normal distribution.

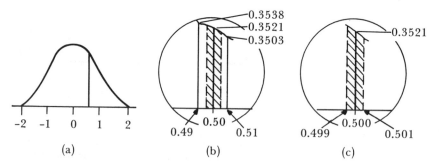

Figure 10.12 Finding the probability of a single value for a normal distribution (mean = σ, σ = 1). (a) The probability of 0.5 is the area above the point (area of the line). (b) A blowup of the line showing the possible area associated with the line. (c) A further blowup showing smaller the width of the line.

In this example, the area would be approximately

$$\text{Area} = \text{Height} \cdot \text{Width}$$

since we have

$$\text{width} = 0.505 - 0.495 = 0.01$$

and since the average height is about 0.3521 (from Table A.13), we have as the area

$$\text{Area} = 0.3521 \cdot 0.01 = 0.00352.$$

So the probability of 0.50 ounce loss would be 0.00352, very small indeed.

If the nearest subdivision was now changed to 0.001, as in Figure 10.12c, the probability would still be the area. But now

$$\text{width} = 0.5005 - 0.4995 = 0.001$$

and

$$\text{Area} = \text{Height} \cdot \text{Width} = 0.3521 \cdot 0.001 = 0.000352.$$

The probability is now $\frac{1}{10}$ of the previously obtained value, even though both represent in some way the probability that the value 0.5 ounces will occur. It should be apparent that we could carry this idea indefinitely, getting smaller and smaller areas. Hence, we must always question what is meant by an *individual value* from a normal distribution (or for that matter, any continuous distribution).

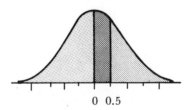

Figure 10.13 The probability of a value from 0 to 0.5 is the shaded area.

We will now generalize the problem and make it more practical. Whenever we discuss probabilities of certain values from normal distributions, we will use the word *between*. In all cases we will be required to specify two points. We will therefore refer to the *probability of a value* (call it X) as that area under the normal curve *between two points*, two points that somehow include the X-value.

In Figure 10.13 we would say that the "probability that the variable X (weight loss) will fall in the *range of possible values* from 0.0 to 0.5 is equal to the area (shaded) under the curve for that range."

262 *Distributions*

To find that area we will use *tables*. The tables are not perfect, but they generally are more than sufficient for any and all problems you will ever encounter. The table may be constructed in several different ways, but we will mainly discuss the tables used in this text.★

By now you should realize that a multitude of different possible normal curves can be drawn. Yet, merely one equation will accommodate all the different μ- and σ-values. Even more surprising, only *one table* is needed to describe all of these curves. How come? Look back to Equation (10.1) and carefully examine the part in the parentheses—the $(X - \mu)/\sigma$. This is the only place that the individual X-values appear in the equation, and we are doing something to them. We are adjusting them. The effect is

1. To reduce the distribution center to a *center of zero*, and
2. To change the standard deviation to a value of 1 (by dividing by σ).

This is called *standardizing*. We have essentially changed the scale, or superimposed another scale on the original one. This new scale reads with a zero at the mean and has one unit for each standard deviation.

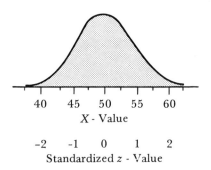

Figure 10.14 Relation of X-value to z-value. The difference is merely the scale.

In Figure 10.14 I have taken the curve in Figure 10.4 (mean 50 and variance 25) and superimposed another scale on the same axis. The table will now give

---

★ The theoretical technique for finding the exact area under a curve uses a process called integration. This is a part of calculus. Since you may not have encountered the calculus we will not pursue the matter. Let me merely say that you would perform an operation called integration, which is much like summation, and is represented by the large symbol $\int$ in front of the equation we are integrating (or determining the area under). Since we can usually determine particular areas between limits it is a very powerful and useful technique. Even if you do know some calculus, and feel energetic and are in an inquisitive mood, let me warn you. There are no elementary methods to calculate the integral of Equation (10.1). By indirect and generally advanced procedures we can find ways of assuring that the area equals 1.00. The mysterious $1/\sqrt{2\pi}$ is a consequence of this procedure. However, it is only by what are called numerical approximation methods that we can calculate individual areas.

us all the probabilities for the *standardized normal distribution* (mean $= 0, \sigma^2 = 1$) and since this also corresponds to *our* normal distribution (mean $= 50$, $\sigma^2 = 25$), we automatically have probabilities for our distribution.

What I have been saying is that we can adjust the scale on any normal distribution to correspond to our standardized normal distribution. If we then work with the new scale, we can work with a single universal table. This universal scale, as I have said, will possess a central value of 0 and variance or standard deviation of 1. We will label the values on this new scale *z-values*. Thus we will have two scales side by side, and we transfer or change each X-value to a corresponding z-value. This is a very elementary kind of transformation, but a very important one for statistics.★

We will make our interpretations using the z-values, and we do this by referring to probabilities associated with the particular z-values. However, since the normal distribution is a continuous distribution, probabilities will always refer to areas. An area must have boundaries. Since we are discussing area under a curve, we need only to specify the end points. One of these must be the z-value. The other end point traditionally is one of two values, either 0 or $-\infty$ ($-\infty$, or negative infinity, means going to the left endlessly). When we use negative infinity, we are considering the case in Figure 10.15a. This is a true *cumulative probability* value where the value associated with z is the area *up to* z.

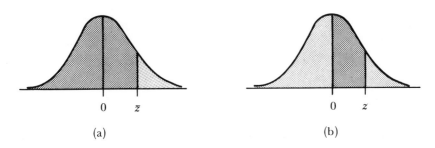

Figure 10.15 Cumulative areas for normal distribution. (a) True cumulative—lower limit is $-\infty$. (b) Area used in many tables—lower limit is 0.

More frequently we use tables that provide the area or probability of a value between 0 and z (Figure 10.15b). Since the normal distribution is symmetric, we can consider the area from 0 to $+z$ as the same as the area from $-z$ to 0 (see Figure 10.16). Appendix Table A.1a contains a table of values for

---

★ You use a kind of transformation every time you work a slide rule. Here you change each number into its logarithm in order to perform a complicated calculation by a simpler process. Many other kinds of transformations can be applied to practical problems in order to simplify them or permit us to make more sense out of them.

areas from the mean to z (or from 0 to z). These are listed in Column B. In Column C we also have the areas above $+z$ (the area in the tail of the distribution). Note that this is also the area up to $-z$. (We will gradually develop these concepts so don't panic.) The table goes up to $z = 3.25$ in intervals of 0.01 and then up to $z = 4.00$ in larger intervals.

Figure 10.16 The area from 0 to $-z$ is the same as from 0 to $+z$.

10.2 Using the Tables

For each of the next nine problems we will explore how you use the tables to find probabilities, or areas, of a normal distribution. We will work these problems for a normal distribution with mean 0 and standard deviation 1, and you can think of the values (the z-values) as weight loss, or just as z-values. You can think of the problems as discussing chances of having particular weight loss, or you can try to generalize and let the distribution be any type of problem you might prefer.

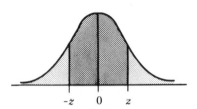

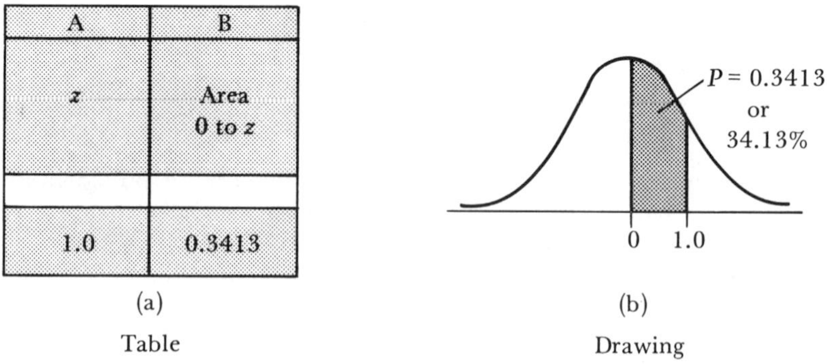

A	B
z	Area 0 to z
1.0	0.3413

(a)
Table

(b)
Drawing

Figure 10.17 Finding the probability (or area) for a z-value of 1.0 (to 0.0). Example 10.1.

EXAMPLE 10.1

Find the probability of a value being in the area from $z = 0$ to $z = 1.00$.

I have drawn part of the table in Figure 10.17a. The left column is for the decimal value of z. In Column B we find that the value for the area is 0.3413. This means that 34.13% of the area is in the range from $z = 0$ to $z = 1.00$ and thus there is a 34.13% chance our value is in this area. In Figure 10.17b we have the distribution showing the area that we are considering.

EXAMPLE 10.2

Find the area between $z = 0$ and $z = 2.34$.

We look down Column A until we reach 2.34. Alongside the 2.34 we find in Column B that the area between $z = 0$ and $z = 2.34$ is 0.4904 (Figure 10.18). Note that the area between $z = 0$ and $z = -2.34$ would be the same (0.4904) as seen in Figure 10.18c.

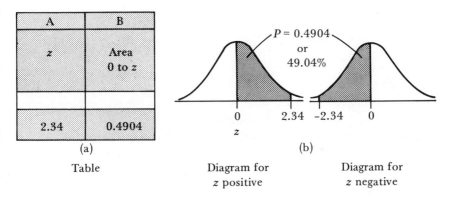

Figure 10.18 Finding a two-decimal place with a positive z-value probability and a negative z-value probability.

Now remember that the normal distribution is symmetric and therefore half of the area is above the mean while half is below the mean. Then the area above or below the zero point is 0.5000. We will use this observation to come up with *complete cumulative distributions*. If z is *positive*, say $z = 1.22$, then the total cumulative area is the sum of the amount *below* $z = 0$ and the *amount between* $z = 0$ and $z = 1.22$.

EXAMPLE 10.3

Find the area below $z = 1.22$ (the probability of z being less than 1.22).

The area from $z = 0$ to $z = 1.22$ is 0.3888. The total cumulative area = $0.5000 + 0.3888 = 0.8888$ or we can say 88.88% of the values are below $z = 1.22$ (see Figure 10.19).

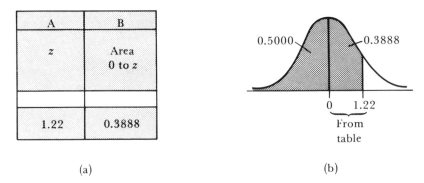

Figure 10.19 Finding the cumulative area below a given positive z-value. We must add the two areas.

If we have *z negative* then for the cumulative area we really want the area *below* the particular *z*-value. This time we must realize that we are to *subtract* the area we read in the table from 0.5000. This will leave us the amount in the lower tail, and give us the cumulative area we want to know. Our next example will illustrate.

EXAMPLE 10.4

Find the cumulative area for $z = -0.88$ (see Figure 10.20).

The first way is to look up the area for $z = +0.88$ and obtain in Column B a value of 0.3106. If we subtract this from 0.5000 we get 0.1894. This latter value is the cumulative area for $z = -0.88$. Notice that it is the left tail of

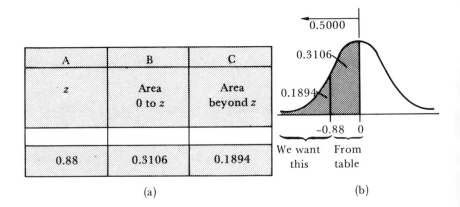

Figure 10.20 Finding the cumulative area below a given negative z-value. We must subtract the value in the table from 0.5000.

the distribution that we are actually looking at. But *we can also read this value directly in Column C,* since Column C is the area *beyond* the z-value.

You should begin to become aware of the usefulness of the little picture. By identifying the approximate locations on the graph you should be able to rapidly determine the proper procedure needed to calculate the desired area. We will now cover additional major types of problems. They all follow from the previous examples.

EXAMPLE 10.5

Find the area above $z = 1.65$ (see Figure 10.21).
 We either

1. Determine the area between $z = 0$ and $z = 1.65$; this is 0.4505; if we subtract this from 0.5000 we have our desired value—0.0495; or
2. We look in Column C to read the value directly.

A	B	C
z	Area 0 to z	Area beyond z
1.65	0.4505	0.0495

(a)

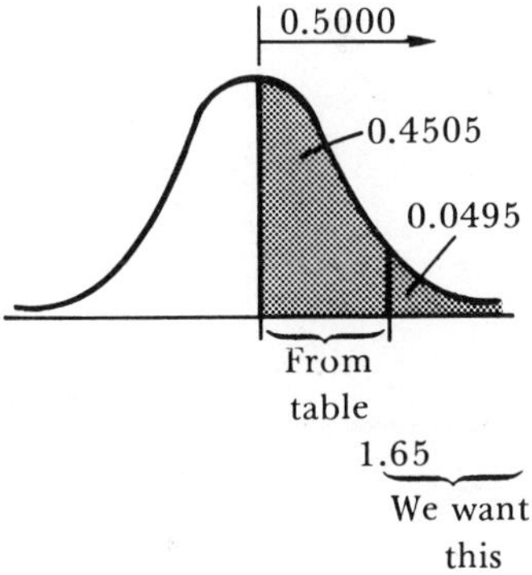

(b)

Figure 10.21 Finding area of a tail $z = 1.65$.

So just about 5% of the values are above a z of 1.65. This is a very common problem and is referred to as "finding the area in a tail." You should also note that there is no real difference in the technique for finding the area in either of the tails.

The reason I am emphasizing the subtraction method of finding areas in the tails is that many tables you may encounter are not as convenient as Table A.1a—they may not have a Column C to read the areas in the tail directly.

EXAMPLE 10.6

What is the probability of a value falling between $z = -0.75$ and $z = +1.25$? (See Figure 10.22.)

This is a problem with *two limits*. We must find the areas for both of these values. First, look at the figure; then decide what we do with the two quantities when we find them. We will add them together to get the total area between

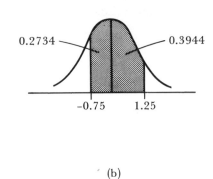

A	B
z	Area 0 to z
0.75	0.2734
1.25	0.3944

(a)   (b)

Figure 10.22 Finding the area between two z-values. Here we add 0.2734 to 0.3944 to get 0.6678.

the two limits. We obtain $0.2734 + 0.3944 = 0.6678$. So about $\frac{2}{3}$ of the values will fall in this range.

EXAMPLE 10.7
Now, what is the probability of a value being between $z = 1.30$ and $z = 1.89$ (see Figure 10.23).

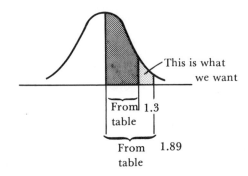

A	B
z	Area 0 to z
1.30	0.4032
1.89	0.4706

Figure 10.23 Finding the area between two z-values. The area is 0.4706 minus 0.4032 or 0.0674.

Look at the figure and observe how the area associated with the larger z-value, the 1.89, extends all the way back to zero *and* also includes *all* of the area up to $z = 1.30$. But we do not care about the area up to $z = 1.30$, so we must subtract this area from the larger one. This is true *whenever the two z-values have the same sign* (both are plus or both are minus z-values). Since for a z-value of 1.89, $p = 0.4706$ and for a z-value of 1.30, $p = 0.4032$, the area between is the difference, or $0.4706 - 0.4032 = 0.0674$.

Let us now reverse the type of problem. We will now try to find z-values between which a given amount of area or percent of individuals lie.

EXAMPLE 10.8

We are now to answer the question, "17% of the area is enclosed between what value of z and 0?" (Figure 10.24.)

A	B
z	Area 0 to z
0.44	0.1700

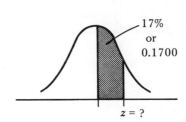

Figure 10.24 Given an area of 17%, find the corresponding z-value. We work from inside the body of the table to the margins to get a z of 0.44.

Now we work from the *inside* of the table, in Column B, to find the probability designated. We then look across to Column A for the corresponding z-value. Thus, we look into the table to find 0.1700, since this is equivalent to 17%. In Column B we are fortunate to find the exact number 0.1700 (this is not too common an occurrence). In Column A we find that it corresponds to a z of 0.44. So 17% of the area lies between $z = 0$ and $z = 0.44$.

EXAMPLE 10.9

What is the z for which 15% of the values are in the upper tail? (Figure 10.25).

This is a more common problem, involving the determination of the z-value for a percent or probability of values falling in the tail (or tails). Well, if 15% are to be in the tail, then 50%–15% (0.5000–0.1500) or 35% (0.3500) are included between 0 and z. It is the 0.3500 that we would look up in Column B (or the 0.1500 could be looked up in Column C). Search as hard as you can, you won't find 0.3500. So we need to use *interpolation* to arrive at a precise value of z.

We interpolate simply by observing first that there are 0.0023 units between the two quantities that *are* in the table and are nearest to the 0.3500 that we are looking for (the values are 0.3508 and 0.3485, the difference is 0.0023). Second, we observe that there are 0.0015 units between our *desired value* (the 0.3500) and the smaller one in the table (the 0.3485) (0.3500 minus 0.3485 = 0.0015). Our value is then 0.0015/0.0023 or 15/23 of the way between the two

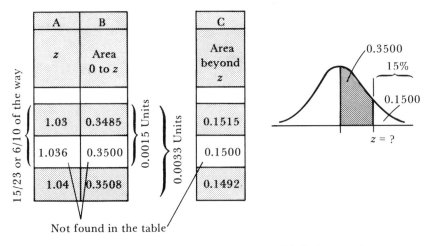

Figure 10.25   Given a desired area in the tail, 0.1500, find the z-value. We must find the z-value for 0.5000 minus 0.1500 = 0.3500. We find either 1·036 or 1·04.

values that we found in the table. This is 0.65 or about $\frac{6}{10}$ of the way between the values. *It is also $\frac{6}{10}$ of the way between the two corresponding z-values.* Since the z-values are 1.03 and 1.04 then our value is $\frac{6}{10}$ of the way beyond the lower value, or it is *1.036*. Generally we are not interested in that third decimal place, so we will not need to go to all the trouble of interpolation. Instead we look to see which one of the two possible values ours is closer to, and we then round off to the closer one. We would observe that the 0.3500 is closer to the 0.3585 than to the 0.3485 (it is only 0.0008 units from the first while it is 0.0015 units from the second). Therefore we will use the z corresponding to the 0.3508 and we will say that $z = 1.04$.★

Finally we need to discuss the cases where we desire to find the two z-values between which a particular percentage of values is included. Generally, we will split the amount evenly between the upper and lower halves. Then, because the split is in equal amounts, the distance from the center to the limits is the same for both directions, and therefore, except for the sign, the z-value will *be the same* in both directions. So we only need to find *one* z-value (in the same manner as in Example 10.5). In Figure 10.26a you can see that when we ask for the z that encloses 70% of the values, we actually pose the problem of finding the z for which 35% is between 0 and one of the limits. Thus we end up with two limits that are equal in magnitude. Since the z for which 35% of the values lie between it and 0 is 1.04 (from Example 10.9) we would say that the two limits which enclose 70% of the area are $z = +1.04$ and $z = -1.04$.

★ We could also work directly with Column C in the table. Here we interpolate between 0.1515 (for z = 1.03) and 0.1492 (z = 1.04). Thus, 0.1500 is 0.0015 less than z of 1.03 and this is 0.0015/0.0023 of the way between them (as before). You have to be careful to remember that the area in the tail *decreases* as the z increases.

In a similar manner, to calculate limits for which an equal amount of area is in each tail, say 5% in each tail or a total of 10% outside the limits, we again look at *one* of the limits or one side of the distribution. We would then find the z-value for which 5% were above (or 45% from 0 to z). This limit would be 1.645 and we would say that the two z-values were $z = -1.645$ and $z = +1.645$.

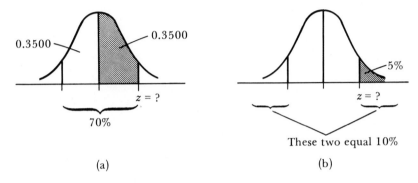

Figure 10.26 Finding limits for two-tailed problems. (a) Given a percent between the limits, divide the area in two. (b) Given a percent in the two tails, divide the area and find the limit for one as before.

Okay, so much for z-values, but what about a typical normal distribution? We already talked about converting individual values from any distribution to z-values or vice-versa, from a z table to another normal distribution. We merely apply the formula

$$z = \frac{X - \mu}{\sigma} \qquad (10.2)$$

or its counterpart

$$X = z \cdot \sigma + \mu \qquad (10.3)$$

to convert in either direction. We can then employ the z table for computing *all* areas or *all* probabilities for *all* normal distributions.

EXAMPLE 10.10

The diet used by the Hillsboro Chapter of the Weightcontrollers Society leads to an average (mean) weight loss of 50 ounces every two weeks. The variance among dieters is 100 (then $\sigma = \sqrt{100} = 10$).

What percent of the dieters will loose between 44 and 64 ounces?
First we convert these two limits to z-values using Equation 10.2.
For $X = 44$

$$z = \frac{44 - 50}{10} = \frac{-6}{10} = -0.6,$$

and for $X = 64$

$$z = \frac{64 - 50}{10} = \frac{-14}{10} = 1.4.$$

Next we consult the z table to determine the percent that lies between a z of -0.6 and $+1.4$. We find 0.2257 for the area from 0 to -0.6 and 0.4192 from 0 to $+1.4$. The area between -0.6 and $+1.4$ is the sum of these two values (see Figure 10.27) and

$$\text{Probability} = 0.2257 + 0.4192 = 0.6449.$$

Thus about 64.49% of the dieters should lose between 44 and 64 ounces.

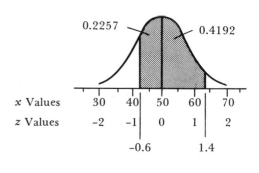

Figure 10.27 Given a normal distribution with mean = 50, variance = 100, what is the probability of a value between 47 and 57. (First find the z-values, then consult the table as before to determine the areas.)

EXAMPLE 10.11

This time let us perform the reverse problem by finding an X-value.

We decide we want to know what minimum weight loss we should guarantee that 90% of the dieters will lose. This question amounts to asking "above what value are 90% of the individuals?" (Note that it also is the same as asking below what value are 10% of the individuals.)

We start by determining the z-value for which 90% of the values are above. This is equivalent to asking for a z for -0.4000, since we are including the entire upper 50% in our area, along with an amount on the lower end, 40%, which makes up the difference (see Figure 10.28). Consulting Appendix Table A.1a we find that the z for 0.4000 corresponds to an approximate value of 1.28, which is to be -1.28 for our problem (see the figure again).

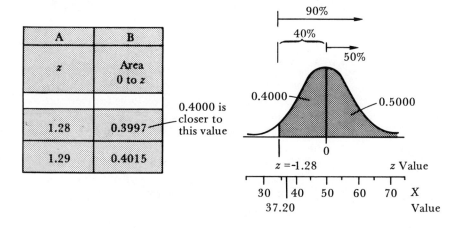

Figure 10.28 Above what value is 90% of the distribution? (Where mean = 50, σ = 10.)

Substituting this z-value into Equation (10.3) we will be able to find the X-value, or

$$X = z \cdot \sigma + \mu$$
$$= (-1.28)(10) + 50$$
$$= -12.80 + 50$$
$$= 37.20.$$

Thus 90% of the people should lose more than 37.20 ounces.

The thing to remember is that there are two steps—one is to *find the z for the desired area;* the second is to *find the X-value that corresponds to the computed z-value.*

Two additional observations should be made. Generally we only use a few important z-values. Eventually you may even learn some of them by heart. However, Appendix Table A.1b provides a simplified table of *selected* values that are most frequently used. You will probably find this table to be more valuable than the more complete tables (such as Table A.1a).

Finally, I hope you have seen the extreme value of drawing the little pictures of the distribution when working problems. They often assist you in quickly and correctly arriving at answers.

10.3 The Normal Approximation to the Binomial Distribution

In the last chapter I said that many times we can replace a binomial distribution by another distribution, one that is simpler to work with. We observed that when a binomial distribution was centered around a p of 0.5 it was fairly symmetric, and the empirical rule worked reasonably well. We can be more specific and state that if you are given the *n and p* of a binomial distribution then whenever

1. The mean or $n \cdot p$ is greater than 10, and
2. The variance or $n \cdot p \cdot q = n \cdot p \cdot (1 - p)$ is also greater than 10.

We can replace the binomial by a normal distribution with the *same mean and variance*.★ Thus we can refer to the normal distribution table to determine the nature of a binomial distribution.

The first thing you should realize is that the binomial was a discrete distribution, with exact probabilities for each circumstance or sample. This is not the case with the normal distribution. The normal distribution is a continuous distribution, where at best we can provide probabilities associated with intervals. So we have to ask, just what is the purpose of having the distribution in the first place? Usually you will find that we will not be too concerned with knowing what is the chance that a particular value will occur. We will generally be more interested in ranges or sets of values that would include various percentages of the distribution. Thus we will desire to know where are the upper 5%, or the lower 10%, or the middle 95% of the individuals. We found these ranges separately for both the binomial and the normal distributions. We will now go on the assumption that both are essentially the same. Look back to Figure 9.2 and you will see that if we were to draw a smooth curve through all of the peaks of that distribution we would have one that looks pretty much like a normal distribution.

EXAMPLE 9.1b (Cont.)

A poll of 50 persons ($n = 50$) consists of answers *in favor* or *against*. The probability of being in favor is believed to be 0.60. Therefore,

$$n \cdot p = (50)(0.6) = 30.$$

This is larger than 10, so the normal distribution should be a good approximation (in addition the variance $= n \cdot p \cdot (1 - p) = (50)(0.6)(0.4) = 12$, which is also larger than 10).

We will now use a normal distribution with mean, $\mu = 30$ and variance, $\sigma^2 = 12$. The standard deviation will be $\sqrt{12} = 3.64$.

The one question we had previously asked was what is the chance of a majority being in favor of the issue, or what is the probability of a value being *greater than 25*. We can ask the question again, only this time using the probabilities related to a normal distribution, namely the area above or beyond the value corresponding to 25. However, we must recognize that the binomial distribution is a discrete distribution, and when we use the normal distribution we must properly label the limits we are to use. The true limit is not 25, but is

---

★ When the mean is less than 10 we generally can use another distribution, called the Poisson, to approximate the binomial. We will not consider the Poisson distribution here. You may also encounter other rules which may be equally acceptable.

actually 25.5. Thus we use a z that corresponds to an X of 25.5. Using Equation 10.2 we find that for $X = 25.5$, $\mu = 30$, and $\sigma = 3.64$

$$z = \frac{X - \mu}{\sigma} = \frac{25.5 - 30}{3.64} = \frac{-4.5}{3.64} = -1.24.$$

Looking in Table A.1a for the area above $z = -1.24$, which also corresponds to that area below $+1.24$, we find the area between $z = 0$ and $z = -1.24$ is 0.3925, so that the total area above $z = -1.24$ would be $0.3925 + 0.5000 = 0.8925$ (see Figure 10.29).

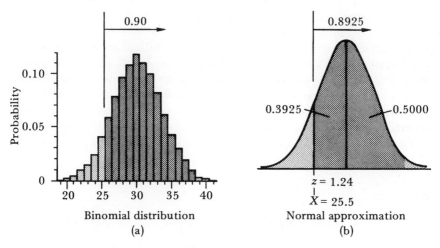

Figure 10.29 Comparison of the binomial distribution and the normal approximation. (a) The actual binomial distribution of the number in favor (Example 9.1b). The shaded area is the portion of samples that would indicate a majority in favor. (b) The normal approximation. Observe how close the proportions (or probabilities) are.

In Chapter 9 when we worked with the exact binomial distribution, we obtained a probability of 0.90, so we are off by a mere $\frac{3}{4}\%$, and notice how much easier the calculations are.

So whenever we want to use a normal distribution in place of a binomial, and that will probably be most of the time we run into a binomial, we will work with the normal table the normal distribution by substituting

$$n \cdot p \quad \text{for } \mu \text{ (the mean)}$$

and

$$\sqrt{n \cdot p \cdot (1 - p)} \quad \text{for } \sigma \text{ (the standard deviation)}.$$

10.4 An Interpretation of the Normal Distribution

An interesting description of why certain processes or types of data appear to behave as though they were normally distributed can be developed using the concept that the normal distribution is an approximation of a binomial distribution. If we measured the heights of all of the corn stalks in a corn field we would get a certain distribution. Moreover, it would quite likely resemble a normal distribution. What could cause such an occurrence? We could start to enumerate reasons; perhaps starting with the idea that of the thousands of genes that might influence height we could categorize each gene as causing either a large or a small effect. Therefore, we could regard the effect of genes on height of corn as a binomial distribution with a large n. Another influence might be water. The amount of water delivered to each plant would vary from plant to plant and from day to day (thus perhaps too much, or too little each day). This results in much more possible variation to be considered. Similarly, there is the effect of sun, which may vary from location to location and day to day. Crowding, fertilizer, and other ingredients of the soil and air, all contribute to the overall growth of the plant and all vary from plant to plant. Since all of these possibilities may or may not help increase the growth of the plant we can get an exceedingly large number of individual influences, with an overall effect being a binomial distribution with an n so large we effectively have a continuous distribution that resembles a normal distribution. You might be able to think of some other similar types of problems that could be explained in this way (I.Q.'s for instance).

REVIEW

Standardized normal distribution · z-value · Normal table

$$z = \frac{X - \mu}{\sigma}$$

Normal approximation to the binomial distribution

PROBLEMS

Section 10.2

10.1 Find the probability of a value falling in the following portions of a normal distribution:
 a. Between $z = 0.20$ and the mean.
 b. Between $z = 2.41$ and the mean.
 c. Between $z = 1.47$ and the mean.
 d. Between $z = -0.42$ and $z = 0.42$.
 e. Between $z = -0.61$ and $z = 1.35$.
 f. Between $z = 0.87$ and $z = 0.65$.
 g. Beyond $z = 1.42$.
 h. Beyond $z = 0.15$.
 i. Smaller than $z = 1.50$.
 j. Smaller than $z = -0.50$.

10.2 Find the z-value or values in which the following areas are contained:
   a. The 10% between z and the mean.
   b. The 15% beyond z.
   c. The middle 25%.
   d. The 40% below z.
10.3 Find the specified proportion of the following normal distributions:
   a. $\bar{X} = 25$, $\sigma = 10$; area between $\bar{X}$ and 30.
   b. $\bar{X} = 25$, $\sigma = 10$; area between 20 and 30.
   c. $\bar{X} = 25$, $\sigma = 25$; area between 20 and 30.
   d. $\bar{X} = 50$, $\sigma = 5$; area between 43 and 61.
10.4 A test has been shown to result in a distribution of grades with an average (mean) of 65 and a standard deviation of 12. I decide that about 15% of the class should fail on this test. What should be my bottom cutoff for passing? If I decide that only 5% will get A's, what is the cutoff for an A?
10.5 My electric bills have averaged $12 a month, with a variance of 9.00. What is the chance that my bill might be more than $16? Ninety-five percent of the time my bill will be at least how much?
10.6 Fire drills have been conducted at a school for several years. The time to evacuate has been approximately normally distributed with a mean of 4 minutes 15 seconds, and standard deviation of 40 seconds. A phone call is received that a bomb is to go off in $3\frac{1}{2}$ minutes. What is the probability that the building will be fully evacuated? (*Hint:* Convert everything to seconds.) If the caller said 5 minutes, what is the chance of full evacuation?
10.7 The life of a light bulb is normally distributed with an average (mean) life of 100 hours and variance of 36 hours.
   a. What percent will last more than 110 hours?
   b. What percent will last between 85 and 95 hours?
   c. 15% will burn out before what length of time?
10.8 One pound bars of butter have actually been weighing an average (mean) of 0.98 pounds with standard deviation of 0.04 pounds. What is the chance of you buying a bar of butter that is at least 1 pound?
10.9 A tree I just planted is supposed to grow to 25 feet tall in 5 years (with $\sigma = 3$ feet). What is the chance it will be taller than my house, which is 30 feet high? What is the chance it will not reach the bottom of a window that is $24\frac{1}{2}$ feet high? If the window is 3 feet high what is the chance that the top of the tree will be in front of the window?
10.10 The family income in Executive Township is $15,000, with a standard deviation $4000. A sociology researcher randomly picks a person to interview. What is the chance that this person will come from a family with an income of less than $10,000? With an income of over $25,000?
10.11 The time between buses (or subway trains) is 19 minutes, with a standard deviation of 4 minutes. If I just missed the last bus:
   a. What percent of the time will I wait between 15 and 20 minutes?

b. 90% of the time I should not expect to wait longer than how many minutes?

c. Less than 5% of the time I will catch a bus as soon as how many minutes?

Section 10.3

10.12 If I plant 10,000 seeds (Problem 9.12) what is the probability that more than 7500 will sprout? Is the correction for the discreteness of the distribution really necessary in this problem (would you really care about 1 seed)?

10.13 What is the chance that the majority of the students in a large lecture are thinking about sex (see Problem 9.13)?

10.14 If there are 100 questions on a true-false test, what is the chance that someone who guesses would get at least 65 correct (see Problem 9.14)?

10.15 If a person knows the material on a true-false test, what is the chance he will fail if failing is less than 65 correct (see Problem 9.15)?

Section 10.4

10.16 Which of the following distributions would most likely be normally distributed? How else might the distributions be expected to look?
a. Postage on mail delivered to your home.
b. Cost of telephone calls out of your calling range.
c. Age in months of a baby when speaking his first words.
d. Averages of samples from a normal distribution.
e. The answer on an arithmetic problem involving many additions, multiplications, subtractions, and divisions.
f. The number appearing on a single roll of a die.
g. The age of a person at the time of death.

PART IV

Inference and Hypothesis

A Nonmathematical Approach to Hypotheses

One of the main purposes of the gathering of data is to come to some kind of decision. The data is nice by itself, but it is useless unless we are to make some evaluation of it. One of the principle kinds of decisions we make, or questions we answer, involves the action we take with a hypothesis.

11.1 What is a Hypothesis?

Basically, a hypothesis is a speculation. It is a statement that you are interested in proving or disproving, declaring correct or incorrect. The hypothesis may deal with any kind of endeavor or question from "Blonds have more fun" to "Educable mentally retarded children progress faster in reading level when integrated in regular classrooms than when kept in special classes" or "The endurance limit of prestressed reinforced concrete is greater than nonprestressed reinforced concrete" or "The galaxy was formed by a 'big bang'."

Formulating a hypothesis is actually a first step. The gathering of data and the decision about it is the next important step. The final decision-making process should depend on the kind and amount of data that we have, and ideally the reasoning will involve some statistical analysis of the data that has been obtained.

In some cases we can have a hypothesis with no data. If this is so because there is no data or we cannot obtain data, then the hypothesis is not of concern to statistics. However, in this chapter we *will* look at some problems without data, but *only because* certain concepts may be developed without data and at this point you may actually be confused by the use of data.

Our goal is to take some data and a hypothesis and make a decision using an appropriate technique. The ultimate decision will be whether we agree or disagree, accept or reject, a particular hypothesis. Let us see how we go about this business of experimentation with or evaluation of a hypothesis (but without using data).

As a researcher or investigator we will propose a hypothesis, which is a formal question we intend to resolve. (I assert that you need not be engaged in actual research to be concerned with some speculation. I hope you soon will begin to turn many different kinds of questions into hypotheses, your projects for one.) Next we would gather some data. Then we would use the data to decide if the hypothesis is true or false—but remember, we can only make *one* decision—true or false. Now, *how* did you arrive at your decision? Was it by "common sense" or because "it looked good" or "it was obvious."?

If so, perhaps your decision is correct. Fine. But what if it isn't correct? You really won't be able to tell if it is not correct. But can you state anything about the chance that you might be wrong? Can you tell me *what is the probability that you might be wrong?*

Here lies the important difference between answering a hypothesis with "common sense" as opposed to the use of statistical methods. In either case *we may be correct in our decision* (the key word is *may*). However, using the principles of probability and statistics we will have an opportunity to disclose, and even limit, *our chance of making certain kinds of errors*. This is an extremely vital factor to be aware of.

11.2 Kinds of Errors: Decision Problems

Let us look at what these errors I am alluding to are (there will be no data yet).

EXAMPLE 11.1

Consider the following situation you may face in traveling a particular road.

There is a small hill on a one lane road. You cannot see over the hill, nor could you stop quickly enough if there were an obstacle just on the other side (a stuck car, the end of the world). You therefore may formulate a hypothesis as you approach the top of the hill. (It may be so rapid you don't even realize you are doing it yourself.)

Hypothesis: There is no car stuck on the other side of the hill.

If I *accept* the hypothesis, then I will drive right on through at the legal 60 miles per hour. If I *reject* this hypothesis then I am saying that I believe that there is an obstacle ahead, and I will slow down considerably. But there may be a consequence of slowing down. If I slow down I may miss a meeting that could have led to a new important job. So there are alternatives to choose and consequences to each alternative.

First, we should realize that *there may actually be a car on the other side of the hill*, or, there may *not* be a car stuck ahead. These two cases exist for the problem, but *we are not able to say what is correct* as we approach the hill. If we knew which was the correct case, there would be no difficulty and the problem would be trivial. A person flying in a helicopter watching us and the hill would be able to say which is correct. But we are not in the helicopter, we are in the car.

Since there are two possibilities for the actual case, let us treat them separately.

1. *If the hypothesis actually is true:* Then there would be no car stopped on the hill. The problem for us is *still* to decide if there is a car on the hill. We must make a decision. What two alternatives are available to us?
   We could:

   a. accept the hypothesis—say there is no car ahead and ride on through;
   b. reject the hypothesis—say there is a car ahead and slow down.

   In Figure 11.1 we examine the *consequences* of the decisions. When the hypothesis is true and there actually is no car ahead, then the two faces of the display show the correctness of our decision. The expression only describes our feeling with regard to the decision or the hypothesis.★

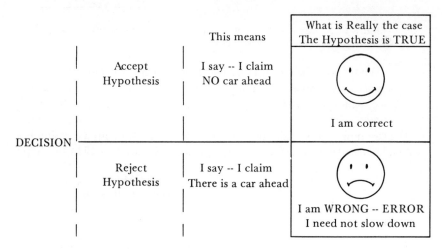

Figure 11.1 The two decisions to make with a hypothesis that is TRUE. The hypothesis is—There is NO car ahead.

2. *If the hypothesis is actually false:* Then there is a car ahead, and we have the second case. But *our decision problem has not changed.* There still are only two alternatives—accept or reject the hypothesis. Figure 11.2 shows the effect of our choices—the faces are now reversed. This time the sad face indicates accident ahead. As before, when we accept the hypothesis, we do not slow down. But, since this time the hypothesis is wrong, namely, there is a car ahead, we misjudge with great consequences—an accident. (I emphasize though, that the sad face is *not* because of the accident, it is because we decided that the hypothesis was true when it was not true.)

★ The consequences of accepting the hypothesis even though they result in ☺, may be such that we would not be pleased. However, with respect to the decision, we would be pleased. It is this satisfaction with the decision that the face is showing.

284 *Inference and Hypothesis*

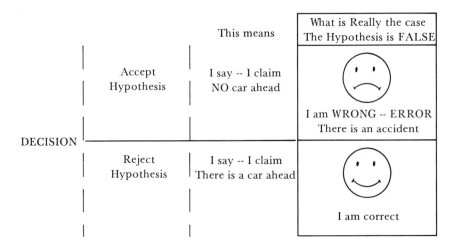

Figure 11.2 The two decisions to make with a hypothesis that is FALSE. The hypothesis is—There is NO car ahead.

What about the happy face? He made the correct decision as far as the *hypothesis* was concerned. But he missed the meeting. As for the indicated face, again *it only concerns the hypothesis, not any indirect* (or direct) *consequences.*

We can now combine the two previous cases and draw up a decision box, which I will call our "four faces of consequences," Figure 11.3. The two happy

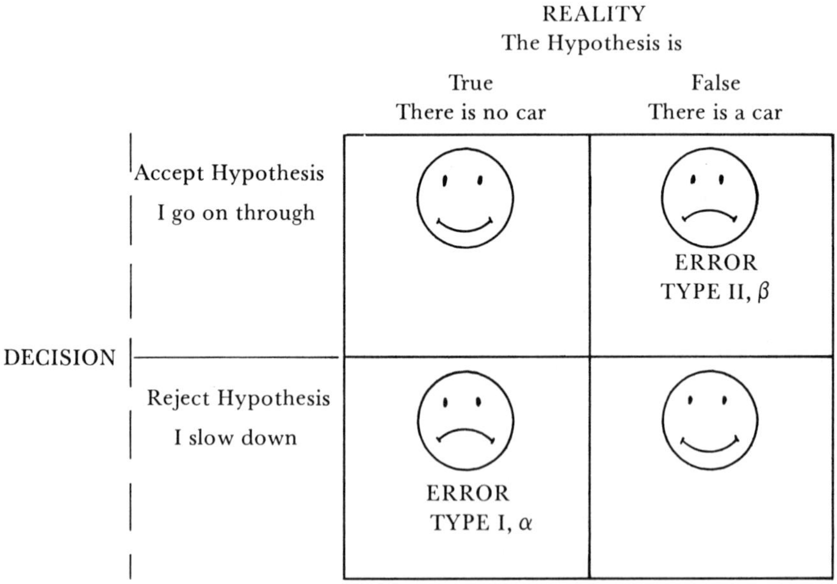

Figure 11.3 A combination of the two possibilities of reality. The "four faces of error" for our hypothesis, "There is NO car ahead."

faces indicate correct judgements—correct decisions. The two sad faces we call error. Since the two types of errors will become very important we have names for them:

Mistake Type I: The times *we reject* when we should not—we designate this occurrence by *Type I, alpha* (α) error, or *Error of the First Kind*.

Mistake Type II: The times *we accept* when we should not—we designate this occurrence by *Type II, beta* (β) error, or *Error of the Second Kind*.

(I suggest you go back over the notion of these two kinds of error several times. The idea is not a simple one, and I would recommend rereading this and the subsequent presentations as well.)

EXAMPLE 11.2

As another example, let us modify Example 11.1 as follows.

Hypothesis: There is no radar trap ahead.

If I accept the hypothesis, I will continue traveling at a rate exceeding the speed limit. If I reject the hypothesis, then I slow down and I will miss my meeting, which may cost me the new job.

Generally another term is introduced at this point. I have been saying that we may accept or reject the hypothesis. In the cases when we *reject* the hypothesis I have been just writing down what appears to be a logical statement. We usually call this statement an *alternate hypothesis*, and it is always given along with the hypothesis we are questioning. So in this case we could say that the alternate hypothesis is: There is a radar trap ahead. (Although this alternate is obvious there will be future cases where there may be a number of alternatives with different consequences or chances of occurring.)

Now let us look back at the consequences of accepting or rejecting the hypothesis. Figure 11.4 shows them graphically. Again, we can have two ways of making an error. This time, however, the relative weights or degrees of importance we may choose to give the errors will perhaps be reversed. Here the two possibilities of error are as before, namely:

Type I (alpha error): *Reject* when we should not reject; that is, I *say that* I expect a speed trap and I *slow down when there is not a trap*.

Type II (beta error): *Accept* the hypothesis when we should not accept; that is, I say that *there is no radar trap* when there actually is one, with the result that *I get caught*.

Now, how much weight or emphasis might we place on the two alternative kinds of errors? This leads to the question of which mistake would cost more. Let us look at one possibility.

> A ticket is worth $10; missing the meeting will result in a loss of a $15,000-a-year job. (You might reason that getting a ticket would also entail missing the meeting, but I haven't told you of our progressive system here. The ticket is mailed to your home and you can pay by mail.)

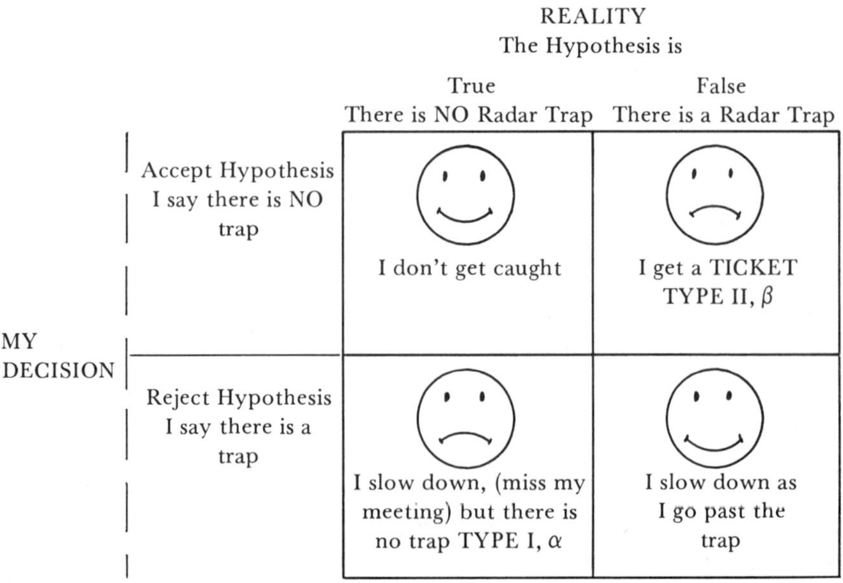

286 Inference and Hypothesis

		REALITY The Hypothesis is	
		True There is NO Radar Trap	False There is a Radar Trap
MY DECISION	Accept Hypothesis I say there is NO trap	😊 I don't get caught	☹ I get a TICKET TYPE II, β
	Reject Hypothesis I say there is a trap	☹ I slow down, (miss my meeting) but there is no trap TYPE I, α	😊 I slow down as I go past the trap

Figure 11.4 The "faces of error" for the hypothesis, "There is NO radar trap ahead."

Here, the Type I error, which was slowing down when we didn't have to, is a situation we really want to prevent. The Type II error we will be more readily willing to accept under these circumstances. From a purely monetary basis it is worth the risk of getting the $10 loss to make the meeting and get the job.

We must look at the weights we place upon the two possible types of errors that we can make. First, we should realize that *we can never eliminate both errors at the same time*, unless we know something special, like the radar unit is being repaired, or the only patrolman in the area has the flu, in which case we need not bother with this discussion. If we are not privy to any confidential material, then at best we can eliminate only one error. But doing that means we will *always* choose one of the alternatives, and it hardly seems reasonable to call that a decision. (I realize I am forgetting the moral issue of breaking the law, but, from within the academic confines we can analyze the problem as one of economics only. This is a deliberate avoidance.)

Consequences are often extremely vague and may include some which we cannot directly analyze. Those that are simple are the ones to which we can assign a monetary value, or from which we may see direct consequences—good or bad. Thus, getting a ticket may also lead to losing a license, which may mean that we would not be able to take the job anyway. It might mean no dates or extra carfare or any number of things.

There often are nonmeasurable or nonequatible consequences of a decision such as a psychological trauma or perhaps future unpredictable circumstances (maybe the job wasn't what you wanted and you would have been better off missing the meeting after all). There might also be many other possible occurrences, such as a tire blowing out because of excessive speed, or a car suddenly stopping in front of you, all of which cannot be reasonably considered simultaneously. Most frequently, however, whenever we make a decision, we have already done some simplification that we may not even be aware of, eliminating many factors and making numerous assumptions.

After all of this we still have not made the decision; the problem of assigning actual weights to the possible decisions is what is still remaining. As you have already seen, in this example we wanted to limit the Type I error, since we mainly were interested in making the meeting. In the previous example we wanted to eliminate the Type II error, since we did not relish the thought of a possible fatal accident. Often a problem may exist where the best decision must be a balance between the two errors. Under these circumstances, you will have to assign the weights yourself.

So far we have not used any statistics or statistical principles, since we have no data or numerical relationships to aid in our decision. We will soon attempt to find ways of describing the errors by numerical methods. We will be discussing the probability or the odds of making those errors. In the present problems we have not been able to do this, but our goal is to assign probabilities to the chance of making the two wrong decisions.

EXAMPLE 11.3

I claim to be a marksman, an excellent shot with a rifle. You do not agree and I feel I must prove myself, so we come up with a hypothesis and a way of deciding how to make our decision of whether I am a marksman or just an average shooter (for sake of argument we will assume that an average shooter is anything from mediocre to reasonably good—but not a marksman).

1. *Hypothesis:* There is no difference between me and what is to be considered an average shooter—or *I am average.*
2. *Alternate hypothesis:* There is a difference, *I am a marksman.* (Note that this alternate hypothesis will be the statement I will say is true and will be the one I will use *if* I declare that the hypothesis above is *not* true.)
3. I will take five shots. We will have to agree on some criterion for instance:
    If I hit the target only 0, 1, 2, 3, or 4 times we *accept* the hypothesis—we say, "I am average."
    If I hit the target all 5 times we will *reject* the hypothesis and *accept* the alternate hypothesis—we will say, "I am *not* average, I am a marksman."
4. There is still the possibility of error.

    Type I error: I may be an average shot *but* I happened to be lucky for this one set of 5 trials. I might hit the bullseye 5 times even though I am not a true marksman. When I score 5 hits you will call me a marksman *but* you will be wrong. This is the α or Type I error. We would like to give it a numerical value. However, unless we can do more and describe just what is the chance that an average person like me can get 5 hits, we cannot assign a value to α. The kind of thing I am referring to might be as follows:

Assume that as an average shooter you have a 50–50 chance of hitting the target. In other words, my chance of hitting on any shot is equivalent to flipping a coin. Now what is the chance of getting 5 hits? Referring back to the binomial distribution with a p of 0.50 and an n of 5 we find that the probability of getting 5 hits is in the order of 0.03 (or about 3% of the time). This 0.03 would be the α error or the chance of making the *wrong* decision, since the 3% of the time that I would hit on all 5 shots would deceive you into believing that I am a marksman. Remember also that there is a very important assumption that has been made—the chance of an average person of making a hit has been stated as $\frac{1}{2}$ *and if that assumption is wrong so will the α error be wrong.*

Type II error: I may be a marksman but I have a bad round of shots and get only 4 hits. So you say I am average. You are wrong and this error is the β or Type II error (and maybe I will slug you for not believing me in the first place). Again in order to assign an actual value to the chance of a Type II error we must make some assumptions regarding the nature of a marksman's shooting—the chance of a marksman of getting 4 or less hits out of 5 trial.

Perhaps we would make the *assumption* that a marksman hits 95% of the time (with each shot independently having the same chance of a hit). Again we have what amounts to a binomial distribution with p equal to 0.95 and n equal to 5. The probability that I have less than 5 hits would be 0.23, or a 23% chance that the marksman will *not* get 5 good shots and

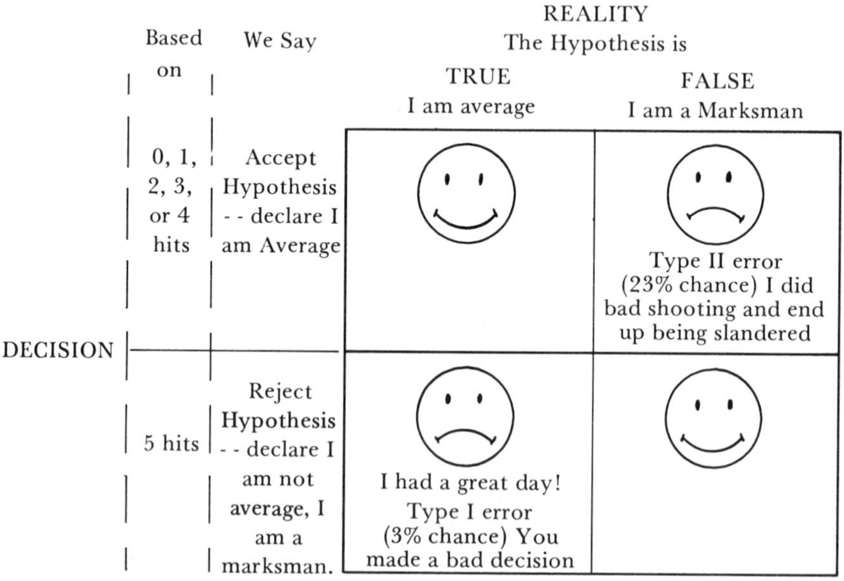

		REALITY The Hypothesis is	
Based on	We Say	TRUE I am average	FALSE I am a Marksman
0, 1, 2, 3, or 4 hits	Accept Hypothesis -- declare I am Average	☺	☹ Type II error (23% chance) I did bad shooting and end up being slandered
5 hits	Reject Hypothesis -- declare I am not average, I am a marksman.	☹ I had a great day! Type I error (3% chance) You made a bad decision	☺

Figure 11.5 Decision box for Example 11.3, determining if I am a marksman.

be labeled as an average shot. Thus the chance of a β error occurring would be stated as 0.23.
5. Now I do my shooting. Let us look at two possibilities.
   a. I get 4 hits.
   b. I get 5 hits.
6. We make a *decision*. (See Figure 11.5 for the decision box.)
   a. We could *accept* the hypothesis and say that there is *no* difference between me and an "average" shot. There is about a 23% chance that this decision (saying that I am average) is wrong (provided that the assumptions were correct). So you cannot rely too heavily on your decision since there is a good chance that it is wrong.
   b. We *reject* the hypothesis (that I am average) and *accept* the alternate, thus saying that "I am a marksman." This statement (given the various assumptions) has only a 3% chance of being wrong. So I am much more confident in this decision, if it is the one I am to make, but I still might be wrong.

If we had used the same general assumptions but had altered the decision criterion to something else, we would also have changed our α and β errors. For example, if we made the rejection criterion, item (3), as *4 or 5* instead of 5, then the chance of an α error would have changed to about 0.19 and the chance of a β error to about 0.02, thus approximately reversing the levels or magnitudes of the two errors. In this case you would be calling a lot of people marksmen when they do not deserve such a label (they are just average).

You must also be aware of the fact that we have selected only one alternative. We have not even considered anything in-between as a possible case. When we begin to apply actual distributions to our assumptions we will see how to gauge better the change in error resulting from the change in the various alternate hypotheses.

EXAMPLE 2.1 (Cont.)

Try to place yourself in the patrol car and consider the question that the patrolman must ask each time he closes up the radar trap and goes home at night.

Tickets issued at a particular location on Route 10 have been falling within what appears to be a stable distribution (our control charts have shown that to be true). The decision today must be whether or not something may have occurred to change the distribution.★

★ You may question what could change the distribution. Several years ago a news article appeared recounting a situation where a recent recipient of a ticket turned around and parked a short distance before the radar trap. He displayed a large sign that read "Radar Trap Ahead." Much to the dismay of the waiting patrolmen, there was a sudden, marked drop in the number of speeders passing their location. Investigation soon turned up the cause of the major change in the rate of violations and the "informer" was promptly removed.

A more typical situation is one that involves a gradual day by day change or a "learning process." As drivers realize there is a good chance of getting caught by a speed trap, they begin to learn to avoid speeding in certain areas, and gradually a smaller number of tickets are issued in those locations. If one applies a control chart to this case a trend may appear.

Now our patrolman is to formulate a hypothesis.

1. *Hypothesis:* There is no change in the average number of tickets issued, or, another way of putting it: The distribution of tickets issued has not changed.

    We next need to decide the alternate hypothesis we are concerned with. Three possibilities may exist:

    a. I care only if there is a decrease in tickets (I will then change locations).
    b. I care only if there is an increase (I will get a bonus).
    c. I am concerned if there is a change either way (combination of the two possibilities).

From these three possibilities one must be selected, and this alternate hypothesis must be written down. It makes up what we call the rejection statement and is the one we will agree upon if we reject the hypothesis. Later we will find that the selection of the alternate will influence our entire problem, especially the determination of limits (as it did in Example 11.3). We will choose the first possibility, a, in this case, so:

2. *Alternate hypothesis:* There *is* a change—a decrease—in the average number of tickets issued. (Later we will encounter what will be called one- and two-tailed tests which will be used depending on the particular alternate chosen.)
3. On the basis of today's (or past) results I must arrive at a set of criterion to *accept* or *reject* the first hypothesis. This is our *test* method. It will depend somewhat on which possibility, a, b, or c, we consider in forming the alternate hypothesis. Perhaps we will use as our test criterion the control chart of Chapter 7 (p. 195). If a point is above the lower limit we *accept the hypothesis* and don't change the status quo. If a point is *below* that *lower limit* then we *reject the hypothesis, accept the alternate hypothesis,* and move locations.
4. Of course, there is a possibility of *error*. If we *accept* the hypothesis we may be wrong (this is β error). If we *reject* the hypothesis and *accept* the alternate hypothesis, we may also be wrong (this is α error and there is about 1 chance in about 370 of it occurring).
5. We look at the observations we made today and compare them to the criterion for accepting or rejecting the hypothesis. We see if today's average number of tickets is above or below the lower control limits.
6. We make our decision.

    a. If the sample from today is *above* the lower limit, we *accept* the hypothesis; we say "there is *no* change." But we have a chance of being wrong (a β error). This means that there might actually have been a decrease in the tickets issued, but we will not recognize or detect the change and I will not go to a different location even though I should.

b. If the sample from today is *below* the lower limit, we *reject* the hypothesis, accept the alternate hypothesis, thus saying "there *is* a decrease in the average number of tickets." Again there is a chance I am wrong (an α error), and there may actually be *no* change in the average number of tickets being issued. My error will result in my going someplace else to set up my radar trap, even though I should stay at my present location.

We have now begun to formalize our approach to analyzing a simple decision problem. In future chapters we will discuss hypothesis testing and errors in much greater detail.

11.3 The Null Hypothesis

Throughout this chapter the hypothesis has been written in a particular way. It has always been stated as "there is *no* difference," "no change," or "no radar trap ahead." The use of the word *no* has been intentional. Whenever we write a hypothesis with the idea of performing a statistical test we write it in this form of a *no* statement, which is called a *null hypothesis*.

In general we have two possible ways of writing a hypothesis. We can state either:

1. There is a difference, or
2. There is *no* difference.

Let us examine the first case to see why we do not use this form.

Most real problems, including those that we will discuss from now on, consist of decisions involving a value that is measured on a scale, and a large portion of these are measured on continuous scales. The officer asks whether there has been a change from the usual average of 10 tickets. The marksman is defined as one who has a 95% chance of hitting the target. We might ask if the chance of a student being absent has changed, if the aptitude of a group of students is increased, if the size of breads has decreased, and so forth. All of these examples involve population characteristics that can vary slightly or to large degrees. More important, however, you should observe that it is virtually impossible to reasonably declare that there is *no* difference, because in fact there almost certainly is *some* difference or change. Today's population of possible tickets was undoubtedly not exactly 10. No marksman hits a perfect 95% all of the time (nor are all marksmen hitting 95%). I expect the chance of a student being absent to vary somewhat or the aptitude to change to some degree or the population of breads to vary at least slightly. Thus in this form we are almost of necessity required to accept the hypothesis that there is a difference in all cases.

This is why we instead state the hypothesis in the form of a "no difference" question and then attempt to determine if it is possible to reject this statement. If we can reject the null hypothesis we can then state the alpha (α) error and so can truly measure our risk in accepting the alternate hypothesis. Furthermore, the alternate hypothesis is the statement we are actually interested in accepting or proving.

When we are not able to reject the null hypothesis, I have been stating we "accept the hypothesis." You should be aware that although this seems to be a logical and convenient statement, it is not entirely correct. You must always bear in mind that the population value is almost never exactly what we declare as the hypothesized value, there almost always is some small difference between the true population value and the hypothesized value. Thus, when someone states that he accepts the hypothesis (the null hypothesis) he is really saying that he does not have statistical evidence to *reject* the hypothesis. But, he has *not proven* statistically or otherwise that the hypothesis is correct!

REVIEW

Hypothesis · Alternate hypothesis · Null hypothesis · Error

Type I (alpha) · Type II (beta) · Accept · Reject ·

Decision · Criterion

PROBLEMS

11.1 When a person is on trial the jury formulates a hypothesis. What is the hypothesis that our judicial system operates under? What is the alternate hypothesis? What are the two errors that the jury can make? What are the consequences of these errors? How does society consider or compare the importance of these errors?

11.2 The President of the United States may sometimes be confronted with a statement that there is an impending nuclear attack. State the null hypothesis. What is the alternate hypothesis? What types of errors are possible? What are the possible consequences of each?

11.3 A test for cancer is administered. What are the two possible hypotheses? Which is preferred? What errors can be made? Which is the alpha and which is the beta error? What might be the consequences of the errors?

11.4 When buying a stock what are the possible hypotheses? Do you have a single alternative or are there many possible alternates? What kinds of errors can you make? Which would be labeled as an alpha and which as a beta error? Can we assign any relative chance to these errors?

11.5 A biologist is studying the effect of a new fertilizer on corn plants. He is hoping that the yield in bushels of corn per acre will increase substantially. What is the hypothesis that he should propose? What is his alternate? Is the alternate to be concerned with a decrease in yield? Could this be thought of as a one-sided or a one-directional test?

11.6 A stream is being monitored for the level of a particular pollutant. A large paper mill has been considered as the prime source of the pollutant. Federal law provides a large fine for polluting the waterway and the paper mill has been installing equipment to reduce the amount of pollution. From the standpoint of the federal inspector, what is the hypothesis? The alternate hypothesis? The alpha and beta error and the consequences of the errors? What might be reasons for such errors?

11.7 When you buy a box of grass seeds, it states a percent germination (for example, 90%). Would you expect that exactly 90% would germinate? How would you state the hypothesis? What alternate hypothesis would you suggest? Is it a one- or two-directional question? What would be a possible limit for a test value? How would you arrive at a statistic? (*Hint:* first think about planting a sample.) What are the possible decisions you can make? What are the errors?

★If you have covered the binomial distribution, then how might you apply that distribution to determining the test limit?

11.8 A broad and important portion of applied statistics is an area called *acceptance sampling*. Based on a sample, a manufacturer or a supplier of some item decides if his overall batch of items is acceptable to ship. Similarly, based on a sample, you as a purchaser decide if you will accept the product.

Ammunition is produced at an arsenal. When the arsenal is ready to ship some ammunition a few of the items are fired. Why does the arsenal test only a sample?

a. Based on the number of acceptable firings the arsenal decides to ship or not to ship. What are the alpha and beta errors, and consequences of them (this ammunition cannot be repaired)?

b. If the ammunition is shipped, the receiving department of the Army tests some of the ammunition. Based on the sample firings it decides to accept or reject them (send them back). What are the possible errors and the possible consequences? Is there really any difference?

11.9 A sportsman who buys rifle shells must make a decision similar to the Army's decision in Problem 11.8. What are the possible errors the sportsman can make? How does the seriousness of his possible errors compare to that of the Army's? Who has a more serious possible consequence as the result of a bad bullet?

11.10 A nail manufacturer states "less than 1% are bad nails." How would the nail manufacturer check to assure this statement is correct? Might he make an error? If he rejected the fact that "less than 1% are bad nails" what would this cause him to do? What might be the consequence of shipping a lot of nails that had more than 1% bad or defective nails?

11.11 A carpenter is about to buy 50 pounds of nails and he examines a handful of 30. He says to himself: "If I find 2 bad ones I won't buy; if I see 1 or no bad ones I'll risk them." Determine the following:
a. His hypothesis.
b. His alternate hypothesis.
c. His limits.
d. His test statistic.
e. His possible decisions.
f. His chances and beta error.
g. The consequences of the possible errors.

12

Sampling

Once again I must make a vital inquiry and ask: "What generally is the most important question about any problem?" It should be: "What is unknown and how can we better know it?" As a way of exploring this peculiar question I offer you my own personal Mystery Can (Figure 12.1) and I pose a problem. Look at that can and tell me what question you should ask about it.

I hope the question is, "What's in the ——— can?"

And I answer, "Take a peek, take a pick, or take a guess, but only take *one*."

Figure 12.1 The mystery can—and the big question, "What is inside?"

12.1 What is Sampling?

We have here a fundamental issue that concerns all fields of endeavor and that provides a basic reason for the existence of the subject of statistics. This is because the can is merely a symbol of much of what you do not know. You desire to know fully what is inside that can, but neither I nor the can nor anyone else will reveal fully the contents of that can.

You do not have to limit the concept or idea of this symbol to things that can fit inside it. I prefer to think of the symbol as capable of engulfing whatever population you are interested in. It might contain a million nails, or a million people. It might have all the coins that have ever been produced or that might ever be produced. We might pretend it holds an infinite population of measurements of the length of the Brooklyn Bridge; dishes broken each week in a busy restaurant; number of scratches on a new car; or the weight loss by dieters on a new diet. Or it might only be a collection of pieces of carboard with numbers on them. Needless to say it contains something, and I will identify that something as our *parent population*. I will claim that it contains the complete population out of which we will draw a sample.

So let's take a sample. Let's also say that the population stands for the number of residents in one-family houses. The population then is the complete distribution of *all* possible numbers of occupants of one-family houses.

I stick my hand into the can and pull out a number—this symbolizes my knocking on a door and asking how many people live in that house. You can do it also. "Sampling Pieces—I," Appendix Table A.18, lists the values and the order of sampling that I am working from. I am starting from the upper right corner and go to the left as I sample the individuals. You should make the pieces by following the instructions in Table A.18 and then place them in a plastic bag or a can. You will use these pieces in future exercises and right now we will also use the values.

So we take the first sample.

It is a 6.

Okay, now what?

Well, you know more than you did before. You know there is a 6 in the population. You know that a house may contain 6 residents. That is useful, valuable information. *But are you willing to guess what the population is?* what the distribution of *all* houses is?

I hope you said *no*. If you didn't, think it over.

Now, remember one thing. Whenever you take a single measurement of any type, you have only picked one out of the Mystery Can. You are not ready to do justice in estimating or guessing what is inside that can—not unless you know other information that you have not disclosed. Unfortunately, too many people arrive at conclusions after looking at that first value (and do not have additional relevant knowledge to aid them).

With only the one value as a sample, we can only make a poor guess at where the center of the population lies. We can say absolutely *nothing about its spread* or shape. Without an understanding of the spread or shape of the distribution, we are at a loss to say anything about the kind of number that should appear next.

So we knock at another door. We pick another value from the can. What do we get this time?

8.

Now we know considerably more. We can talk about both a mean *and* a variance. We can say that the mean is

$$\frac{6+8}{2} = 7$$

and the variance is

$$\frac{(6-7)^2 + (8-7)^2}{(2-1)} = \frac{1+1}{1} = 2.$$

(The standard deviation is therefore 1.4.) Can we guess now what the population in the can is? We can try, but I wouldn't be too *confident* about my statement.

So we go to a third and a fourth house and see what we get. Here we find 3 and 5 people. Now the mean is

$$\frac{22}{4} = 5.5$$

and the variance is

$$\frac{(0.5)^2 + (2.5)^2 + (-2.5)^2 + (-0.5)^2}{3} = \frac{13.0}{3} = 4.33.$$

(The standard deviation is 2.08.) Now will you venture a guess about that can? How confident are you about what is in that can or what the population actually is?

You should be more confident now than after any of the previous observations. The main point, however, is that you still have no way of measuring or estimating what your confidence should be. What should you give in the way of odds that you will get 5 people in the next house or at any other time? What is the mean and the variance of the population? How close to the population values might I reasonably expect this sample mean and variance to be? These are questions we will now begin to try to answer.

12.2 Distributions of Sample Values

Whenever we take samples from a distribution we create another new distribution. We call this new distribution a *sampling* distribution. A control chart is one example of a sampling distribution. In Chapter 8 we looked at some probability distributions and distributions of samples from them. The binomial distribution was a similar type of distribution. An interesting effect is that some of the new distributions may be more important than the population itself, as was the case of control charts where we worked with samples and not individuals.

In order to explore the concept of sampling more fully we need to start with some kind of a distribution from which we will take samples. Let us consider that our Mystery Can contains a population. We can also think of this population's distribution as a histogram where all the individuals have been stacked at their appropriate X-values. Now we remove a set of individuals. We sample at random and look at 4 values (see Figure 12.2)

$$6, 8, 3, \text{ and } 5.$$

We can observe the mean, median, variance, and standard deviation of this sample (mean = 5.5, variance = 4.33). (We can divide by n and get σ^2 or divide by $n - 1$ and get s^2, and each will provide a different number.) This is just one of the many possible samples we could take. We could take a second sample and get

$$5, 3, 5, 7$$

We now have a mean of 5.0 and a variance (s^2) of 2.67, a median of 5.0, and so forth. Both the mean and the variance are different from the mean and

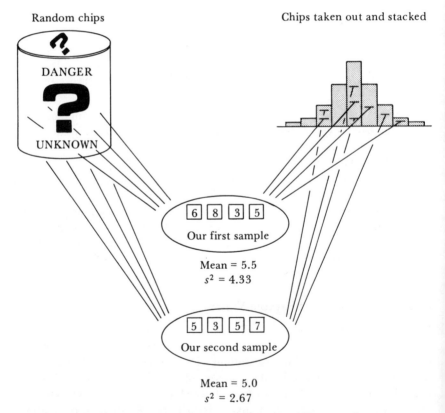

Figure 12.2 Taking samples from a distribution. When you observe a random distribution you can also think of the individuals stacked at their appropriate location on the X-axis.

Table 12.1 Samples from an unknown population. Various statistics are computed for each sample

Individual X-values in samples of four				Means $\bar{X}$	Variance		Standard deviation		Median	Midrange
					s^2 (use $n-1$)	$\hat{\sigma}^2$ (use n)	s	σ		
6	8	3	5	5.50	4.33	3.25	2.08	1.80	5.5	5.5
5	3	5	7	5.00	2.67	2.00	1.63	1.41	5.0	5.0
6	5	4	6	5.25	0.92	0.69	0.96	0.83	5.5	5.0
6	0	3	3	3.00	6.00	4.50	2.45	2.12	3.0	3.0
4	7	3	8	5.50	5.67	4.25	2.38	2.06	6.0	5.5
3	9	8	5	6.25	7.58	5.69	2.75	2.39	6.5	6.0
6	5	6	5	5.50	0.33	0.25	0.57	0.50	5.5	5.5
9	7	5	4	6.25	5.58	4.19	2.36	2.05	6.0	6.5
1	5	4	5	3.75	4.25	3.19	2.06	1.79	4.5	3.0
5	7	2	6	5.00	4.67	3.50	2.16	1.87	5.5	4.5
5	3	7	3	4.50	3.67	2.75	1.91	1.66	4.0	5.0
7	7	3	6	5.75	3.58	2.69	1.89	1.63	6.5	5.0
5	7	6	4	5.50	1.67	1.25	1.29	1.12	5.5	5.5
6	7	5	7	6.25	1.58	1.19	1.26	1.09	6.5	6.0
4	8	3	4	4.75	4.92	3.69	2.22	1.92	4.0	5.5
2	5	5	4	4.00	2.00	1.50	1.41	1.22	4.5	3.5
5	4	6	6	5.25	0.92	0.69	0.96	0.83	5.5	5.0
7	6	5	1	4.75	6.92	5.69	2.63	2.39	5.5	4.0
3	5	8	8	6.00	6.00	4.50	2.45	2.12	6.5	5.5
7	4	3	4	4.50	3.00	2.25	1.73	1.50	4.0	5.0
6	6	5	4	5.25	0.92	0.69	0.96	0.83	5.5	5.0
7	7	4	5	5.75	2.25	1.69	1.50	1.30	6.0	5.5
6	5	9	6	6.50	3.00	2.25	1.73	1.50	6.0	7.0
6	4	6	5	5.25	0.92	0.69	0.96	0.83	5.5	5.0
2	5	4	6	4.25	2.92	2.19	1.71	1.48	4.5	4.0
4	4	8	6	5.50	3.67	2.75	1.92	1.66	5.0	6.0
7	4	6	6	5.75	1.58	1.19	1.26	1.09	6.0	5.5
5	3	6	6	5.00	2.00	1.50	1.41	1.22	5.5	4.5
4	4	3	7	4.50	3.00	2.25	1.73	1.50	4.0	5.0
3	4	2	5	3.50	1.67	1.25	1.29	1.12	3.5	3.5
5	8	4	4	5.25	3.58	2.69	1.89	1.63	4.5	6.0
5	5	3	6	4.75	1.58	1.19	1.26	1.09	5.0	4.5
4	5	5	4	4.50	0.33	0.25	0.57	0.50	4.5	4.5
6	4	5	6	5.25	0.92	0.69	0.96	0.83	5.5	5.0
4	8	4	5	5.25	3.58	2.69	1.89	1.63	4.5	6.0
8	4	7	3	5.50	5.67	4.25	2.38	2.06	5.5	5.5
4	5	7	7	5.75	2.25	1.69	1.50	1.30	6.0	5.5
3	5	7	2	4.25	4.92	3.69	2.22	1.92	4.0	4.5
6	4	3	2	3.75	3.58	2.69	1.89	1.63	3.5	4.0
6	3	5	3	4.25	2.25	1.69	1.50	1.30	4.0	4.5
5	5	7	5	5.50	1.00	0.75	1.00	0.87	5.0	6.0
5	6	5	5	5.25	0.25	0.19	0.50	0.44	5.0	5.5
4	6	10	6	6.50	6.33	4.75	2.52	2.18	6.0	7.0
2	6	5	6	4.75	3.58	2.69	1.89	1.63	5.5	4.0
3	4	4	6	4.25	1.58	1.19	1.26	1.09	4.0	4.5
2	7	4	4	4.25	4.25	3.19	2.06	1.79	4.0	4.5
5	2	4	7	4.50	4.33	3.25	2.08	1.80	4.5	4.5
4	6	5	6	5.25	0.92	0.69	0.96	0.83	5.5	5.0
4	2	6	5	4.25	2.92	2.19	1.71	1.48	4.5	4.0
1	3	5	5	3.50	3.67	2.75	1.91	1.66	4.0	3.0

variance of the previous sample. We should not be surprised that these descriptive statistics are different, since the samples themselves are different.

If we look at a third sample of 4 what do we obtain?

$$6, 5, 4, 6.$$

The mean of this third sample is 5.25 and the variance is 0.92.

We are now building up a little set of means and variance, each different. Any of the three is as good a sample from the population as any other. You can take your pick. If we now continue to sample and look at 50 samples of 4 individuals each, we might get the whole collection of different means and variances in Table 12.1. Look them over to see how many are the same. Remember that they all came from a common population, a single source, yet there are so many different kinds of samples that have been generated.

Now that we have sets of numbers we can draw histograms, calculate some more descriptive statistics (means, and so forth), and in general look at some *new* distributions. The distribution of the means of these samples is drawn in Figure 12.3. For the sake of discussion I have also drawn the distribution of the individuals. For comparison sake I will tell you that the *population mean, variance, and standard deviation* are 5.0, 2.94, and 1.713 respectively.

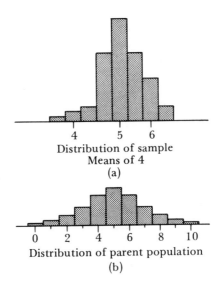

Figure 12.3 Distribution of samples from the given population.

Observe that the histogram for the means of the samples has roughly the same center as the population. This tells us that if we look at the *sample means* (the averages of the samples), in the long run they will be the same as the population mean. Two ideas are at work here. First, we are discussing the concept of estimation. The sample mean is being used to *estimate* the population mean. For many problems it happens to be the best estimate that we can use for guessing what the true population mean is.

The second observation in this case is that the estimate is *not biased* (is unbiased). When I said that on the average the distribution of sample means has the same center as the true population mean, I meant it was an unbiased estimate. This is a very good condition; we usually strive to find and use unbiased estimators. In this way, although we do not expect to get an exact estimate on any one individual sample, on the average, in the long run, we will get the correct value.

As with means, we calculate sample variances in order to estimate what the population variance would be. We already calculated values of the variance for our first three samples. They were all different. In Table 12.1 the variance for each of the 50 samples is given, again with many different values. I have provided two different columns for the variance measures, one labeled s^2 (found by dividing by $n - 1$) and the other labeled $\hat{\sigma}^2$ (sigma–hat–squared

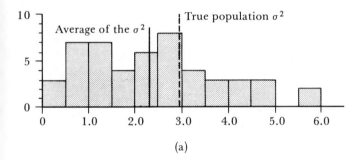

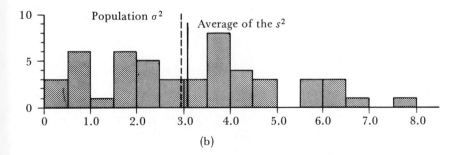

Figure 12.4 Comparison of the distributions of the sample σ^2 and s^2 from a population with true $\sigma^2 = 2.94$. Shown are fifty samples of size 4.
(a) Distribution of sample σ^2 (sample size = 4). (b) Distribution of sample s^2 (sample size = 4).

—it is the estimate of σ^2 found by dividing by n). I am doing this to try and illustrate why I have previously said that we use s^2 for estimating the population variance. In Figure 12.4 we have a histogram for these two distributions. I want you to observe two things here. The first is how much biased $\hat{\sigma}^2$ is. The mean of the $\hat{\sigma}^2$'s is much smaller than the true σ^2 of the population. This is an excellent example of a biased estimator. Most of the time the $\hat{\sigma}^2$ of the sample is considerably smaller than the true population variance. This is not the case with the distribution of s^2. In the long run s^2 will average out to be approximately the same as the true population variance, so we use the sample s^2 to estimate the population variance.

A second thing to observe is how the distributions of s^2 and $\hat{\sigma}^2$ are skewed. Recall back to Chapter 4 where we noted that a skewed distribution tended to have its mean in the tail. The distribution of sample $\hat{\sigma}^2$ is skewed with its tail on the high number side. Not only is it biased, with the mean of the sample $\hat{\sigma}^2$ smaller than the actual population's variance, but the distribution is skewed so that this mean is not centered. It is more to the right of the bulk of the distribution, indicating that there will be many more times that $\hat{\sigma}^2$ will be less than the mean of the σ^2 than greater than this average, so there will be even more values of $\hat{\sigma}^2$ that are below the population variance. All this leads to our decision to use the s^2 as an estimate of the population σ^2.

We should also note something that often seems very peculiar. Even though the s^2 is an unbiased estimate of the population variance (σ^2), s is *not* an unbiased estimate of σ (the population standard deviation). So if we calculate s from s^2 (by taking the square root) we must also multiply this value by a constant in order to get an unbiased estimate of σ. This arises in part because we alter the shape of the distribution when we take square roots. When we change the shape of the distribution we usually change the position of its center. This is particularly true when we change the amount of skewness of a distribution. I want you to look carefully at Figure 12.5. Here I have shown how a scale is changed. In Figure 12.5a I have drawn two scales, one for some X^2-values, and the other for the square roots of them (the X-values). The squared numbers are drawn on a linear scale (with a regular ruler) as would be done with a standard histogram, in this case a histogram of X^2-values, or variances. Then the lines are drawn to the square roots of these numbers on the lower scale. Notice how the square roots of small numbers are spread apart whereas the square roots of successive larger numbers are much closer together.

In Figure 12.5b I have drawn a hypothetical skewed distribution. I have drawn it on a scale corresponding to the upper, linear scale of Figure 12.5a going from 1 to 15). I then determined the area of each rectangle and fitted the area to the associated rectangle on the square root scale (from 1 to 4). This second histogram is shown in Figure 12.5c. Look how the shape has changed; it is almost symmetric. This technique of changing the shape of a distribution by changing a scale is used quite frequently in actual practice and is called a *transformation*. The location of the center of this distribution is not the same as the center of the square root of the above distribution. What I am leading up to

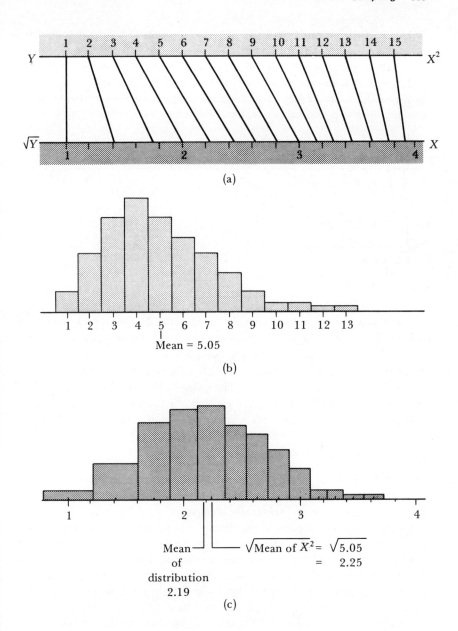

Figure 12.5 Transformations and the effect on a distribution.
(a) Transformation of values from one scale to another. Lower scale shows square roots of upper scale. Note the change from linear or equal separations on upper scale to unequal, nonlinear lower scale. (b) Hypothetical distribution of X^2-values. (c) Distribution of X-values where each value of X was calculated from the X^2-distribution (b) and the areas were approximately the same. Note the change in shape.

is that the distribution of s is not an unbiased estimate of σ, even though s^2 is an unbiased estimate of σ^2.

Another, perhaps simpler, illustration consists of only the five values 1, 2, 3, 4, and 5, and their squares 1, 4, 9, 16, and 25 (we will work in the other direction this time). From a transformation viewpoint we would show the two sets of values as points on two scales as follows:

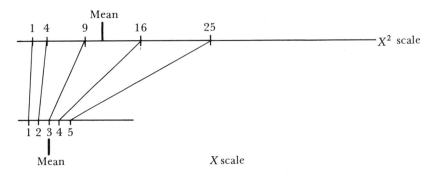

The mean of the X-values is 3. The mean of the X^2-values is 11. You might believe that the square root of the mean of the X^2 should equal the center of the X. Well—$\sqrt{11} = 3.32$, not 3.0. So don't believe it.

Thus we need a correction factor (which we hardly ever use, but nevertheless it exists and it confuses people). The values are given in Appendix Table A.9b(2) and depend only upon sample size. As you can see, if you look at the table, in most cases the correction factor is so small we just disregard it.

12.3 Central Limit Theorem

One of the most important theorems of statistics is called the central limit theorem. We cannot derive it or prove it in this text, but we can try to develop it in an intuitive way and then state what it is.

In Figure 12.3 we looked at the distribution of means of samples of 4. We also have the distribution of the individuals, the parent population. As you can readily see it is spread out much more than the sampled means. In Figure 12.6a I have redrawn the histogram for the entire population. If we think of the parts of the histogram that make up this distribution as the pieces in our Mystery Can, then we can consider that a random sample is simply a set of points or individuals from this population (see Figure 12.2 also). Do you remember what the distribution (histogram) represents when we use relative frequencies as the vertical axis; or if we consider the areas of the histogram? They stand for *probabilities* or chances that we will obtain the particular values when we pick at random from the distribution. Assume we select one chip at random (we pick one house). The chance that it could have an extreme value, say a 10, from our population is very, very small. Here, it has a probability of 0.005, or 1 chance in 200 of being picked, since there is only one 10 in our distribution of 200 items.

Now if by chance we did pick a 10 we would say that it was quite unusual. But let us pick a second time and ask the question, "What is the chance that both cases are 10" or "What is the chance that the average of our selections will be a 10?" If we consider this as sampling with replacement (and we toss the chip back, or we say that the next house we see has the same chance of having 10 people as the one we already looked at) then it can be shown from elementary probability that the chance of both coming out as 10 is the product of the probabilities of the individual samples.* Then the probability of both coming up as 10 is *0.000025* or about *1 chance in 40,000* ($0.005 \cdot 0.005 = 0.000025$). Similarly, the probability of a 9 coming up on a single try is about 0.015 (about 1.5% of the time, since there are only three 9's). Now how about the chance of an *average* of two being 9? It can be shown that this would be about 0.000725.† The chance goes from about 1 in 66 for a single 9 to about 1 in 1380 for an average of 9. So if we look at averages of two selections from a population we should not expect to see as many large values for the averages as we would see for the individuals. If you happen to get one very large value on the first trial you will be likely to get something to balance it on the next trial.

Look at the histogram of samples of 2 from our population of houses (Figure 12.6b). As you can see no averages of two were up around 8, 9, 10, or as small as 2, 1, or 0.

A similar argument can be applied to larger sized samples. Thus, for a sample of 4 to have an average as large as 10 would require four 10's, and the probability of such an occurrence would be in the order of 0.0000000006 or 1 chance in 1,600,000,000. This would be something like the probability of a meteorite striking your roof and putting a hole in it.

Looking at larger and larger samples, the distribution of their means will tend to shrink more and more. We could find many ways to get an average of 7 with say, 8 individuals, but we would find that the probabilities of each way would be so small that the chance of getting an average of 7 would still be very small. Inspect Figure 12.6 and try to see how the distribution gets narrower as the sample size increases. To go to the ultimate extreme, if we were to look at an infinitely large sample (as large as possible) then in effect we would be looking at the entire population, and there would be only *one* average, and obviously no spread.

There is a theoretical relationship between the variance of the population and the variance of the distribution of samples. It is:

$$\sigma_{\bar{X}}^2 = \frac{\sigma^2}{n} \qquad (12.1)$$

---

* If you desire more information on probability and computing probabilities, see Appendix C.
† Here the probability would be equal to
  Probability(9 and 9) + Probability(8 and 10) + Probability(10 and 8)
  = (0.015)(0.015) + (0.05)(0.005) + (0.005)(0.05)
  = 0.000225 + 0.00025 + 0.00025
  = 0.000725

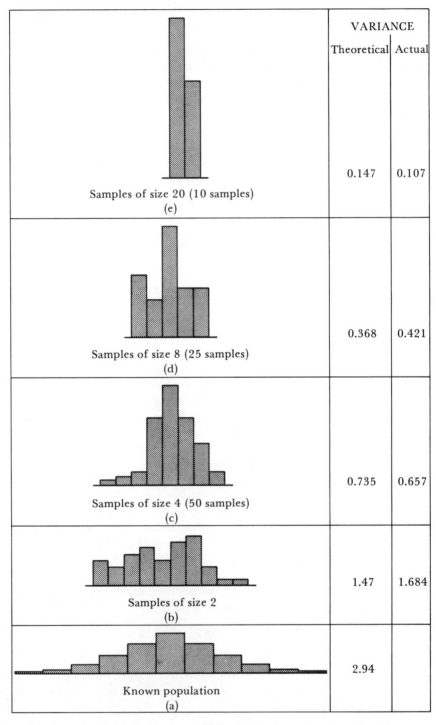

Figure 12.6 Distributions of sample means from a known population. Observe the shrinking of the distributions as sample size increases.

where $\sigma_{\bar{X}}^2$ is the variance of the sample means—the variance of the new distribution. This is read as "sigma-sub-X-bar square". The square root of this value, $\sigma_{\bar{X}}$, is often referred to as the *standard error of the mean.*

σ^2 is the variance of the population itself.

n is the sample size.

Note that if n is 1 then $\sigma_{\bar{X}}^2 = \sigma^2$. So you could say that the population is actually a distribution of samples of size 1.

Since σ^2 equals 2.94, we would expect the variance of the distribution of samples of 2 to be

$$\sigma_{\bar{X}}^2 = \frac{\sigma^2}{n} = \frac{2.94}{2} = 1.47.$$

This compares to an s^2 of 1.684 that I calculated for 50 samples of 2 individuals taken from the population.

Continuing in this fashion, alongside the distributions in Figure 12.6 I have shown the theoretical and the actual variances for some sets of random samples from the population. Observe how close most of the actual and theoretical or expected variances are to each other.

The importance of the formula, as well as the whole idea of these distributions of samples, is that the samples are spread out much less than the population itself. And we can state *how much* they are expected to be spread out. This will permit us to put limits around our estimates and allow us to make statements about how confident we are in our estimations (and say how approximate our approximations are).

A fairly obvious observation is that if our original distribution is a normal distribution then the *sample means* are also normally distributed. So not only can we say what the variance of the sample means will be, but we can also say what is the shape and the specific probabilities associated with the entire distribution.

But there is an even more significant, yet not so obvious fact about the distribution of sample means. In Figure 12.7 we have the histograms of two populations which are quite far from being considered normal distributions. One is rectangular or flat and the other is a highly skewed triangular distribution. A large number of samples of 4 were randomly drawn from these two distributions and the means of each of the samples was calculated. Histograms for the means of the samples were then drawn and I have also superimposed a smooth normal curve on top of the histogram. Observe how extremely close the distribution of sample means approaches a normal distribution. Thus we come to the Central Limit Theorem:

CENTRAL LIMIT THEOREM *For large samples from almost any population, the distribution of means can be approximated by a normal distribution (or approaches a normal distribution) whose mean is the same as the population, but whose variance is σ^2/n.*

Although by large samples we generally mean from about 10 to 20, we have already seen that often samples of only 4 will be sufficient. This is a particularly useful concept for control charts where we are taking numerous samples but are often unsure of the underlying population distribution. As a result, for approximate analysis of most problems we will find it unnecessary to worry about the actual population shape. However, one criterion is that the population must have a finite variance. This may seem strange, but there are some weird distributions. There is even one called the Cauchy distribution that has practical applications and does not have a mean.

Let us look at a few samples to see what this means.

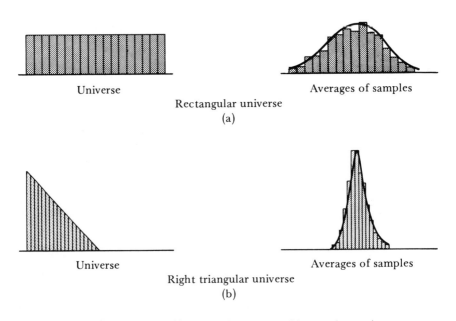

Rectangular universe
(a)

Right triangular universe
(b)

Figure 12.7 Illustration of how sample averages of four tend toward normal distributions even when the universe or population is far from normal.

EXAMPLE 2.1 (Cont.)

Let's return to our traffic ticket problem. We observed in Chapter 7 that we actually had samples of 4 readings taken on 28 different days. When we drew a control chart we computed the averages of each of these samples. If we look at all of the sample averages we should now be able to say

1. The sample means should be approximately normally distributed.

2. The variance of this distribution should be approximately

$$\sigma_{\bar{X}}^2 = \frac{\sigma^2}{n} = \frac{11.62}{4} = 2.90,$$

since the σ^2 of the distribution was found to be 11.62.

Disregarding the two days we found to be unusual, I have drawn the distribution of the $\bar{X}$'s in Figure 12.8. If we eliminate the two unusually low days and recalculate the mean we obtain a new overall mean of 10.68, which should also be the mean of the $\bar{X}$'s. Calculating the actual variance of the *samples* we then obtain

$$\sigma_{\bar{X}}^2 = 2.54.$$

This compares quite favorably with the theoretical value estimated from the overall σ^2 (the 2.90 above).

Now since the distribution of samples is supposed to be approximately a normal distribution, we can also say that

$$\sigma_{\bar{X}} = \sqrt{\sigma_{\bar{X}}^2} = \sqrt{2.90} = 1.70.$$

Then about 95% of the $\bar{X}$'s should be within $\pm 2\sigma_{\bar{X}} = \pm 2(1.70) = \pm 3.40$ of the overall mean. The one and two sigma limits have been drawn on the distribution in Figure 12.8. We find that within the $\pm 2\sigma_{\bar{X}}$ limits (or from 7.28 to 14.08) there are all but 1 of the sample means, and that one is just a borderline case. We would expect 95% of the samples to be within these limits, and we actually have 25/26 or about 96%.

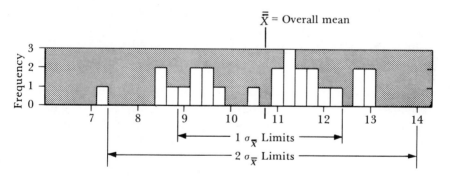

Figure 12.8 Distribution of 26 sample means of four. Assuming a normal distribution, we have one- and two-standard deviation limits, where $\sigma_X = \sigma/\sqrt{n} = 3.41/2 = 1.70$. Within one σ_X limit we expect 68% of averages—we have 0.8/26, or 69%. Within two σ_X limits we expect 95% of averages—we have 25/26, or 96%.

The one sigma limits are 8.98 to 12.38, and here we find 18 of the 26 averages, or 69%. We expect about 68%. So in both cases we have exceptionally good agreement between the actual and theoretical numbers of sample means that lie within some of the limits of a theoretical normal distribution.

The most significant part of this discussion, which you should be aware of, is that we are now finding a way of predicting where certain sets of averages should be located. Because certain important distributions shrink as the sample sizes increase, we will find that our ability to make predictions within narrow limits will increase as our sample sizes increase. This will be explored and used many times over in most of the future chapters. In addition, we have now determined what the distribution of sample means *should* look like, so we have another theoretical probability distribution to compare samples of data against.

Since the standard error, $\sigma_{\bar{x}}$, is used quite frequently, a chart has been prepared for you to rapidly find the standard error when you are given n and σ. This chart is in the appendix and is labelled Appendix Table A.6.

12.4 Selecting Estimators and Efficiency

In previous discussions we found that it was easier to calculate certain descriptive statistics than others. If the data is ordered, it is much simpler to find the median than the mean. Even if it isn't ordered, the midrange is also much easier to determine than the mean. I don't think there is any question that the range is much quicker to calculate than the standard deviation or the variance. Now we are back to the question of when and why each should be used.

We must first reassess what the purposes of our calculations are. Before we thought of them as ways of reducing and describing the data. Since, we have been introduced to the idea of estimation, and we have found that the mean and variance (s^2) of a sample were good estimators of the parent population mean and variance. Now, why not use the median or midrange, or range as estimators? We already have. We used the range to estimate σ way back in Chapter 7. But now I want to generalize and point out difficulties or advantages between various competing measures of central value or spread.

Initially we decide upon the things we are to estimate. Since most of the theoretical distributions have as parameters or constants the mean and/or the variance, it will seem logical that we will most frequently desire to estimate these two quantities (that is, you need to know μ and σ^2 to describe which normal distribution you are working with).

We have already seen that the mean of the sample is an unbiased estimator of the population mean. This meant that in the long run using sample means will average out as the correct value for the population mean. It so happens that for those cases when the population is symmetric, the median and the midrange are also both *unbiased* estimators of μ, the population mean, so we can use either of these simpler methods to obtain an unbiased estimate. The one drawback is that the distribution of midranges or medians will be more spread out than that of the mean. In other words, not only does the mean hit the true center on the average, but *generally when it does not hit the center it is still closer than the others*. This characteristic is called *efficiency* and the mean is the most efficient of these three estimators.

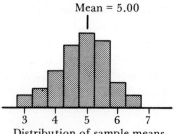

Mean = 5.00

Distribution of sample means

Variance = 0.657
Standard deviation = 0.782
Theoretical variance = 0.735

(a)

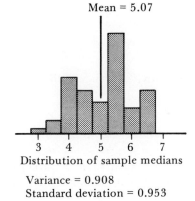

Mean = 5.07

Distribution of sample medians

Variance = 0.908
Standard deviation = 0.953

(b)

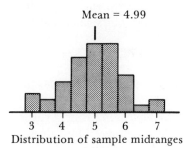

Mean = 4.99

Distribution of sample midranges

Variance = 0.862
Standard deviation = 0.928

(c)

Figure 12.9 Comparison of the distribution of sample means, medians, and midranges. Fifty samples of size 4 from the same population are used.

By most efficient we mean that the distribution of that estimator has the smallest or tightest variance for all possible estimators. Let's look at the distributions of the means, medians, and midranges of the samples of the number of people in houses. We already drew a histogram and calculated the overall grand mean and variance for the *means* of the samples of 4. Now we should do the same for the medians and midranges of those samples. In Figure 12.9 I have drawn the distributions of all 3 types of averages and I have included the means and variances of each of these three distributions. Observe that all three essentially have the same center. This tells us that the three estimators are all unbiased. But look at the spreads. The sample *means* have the least spread, the *smallest variance*. The variance of the means is 0.657 (the theoretical value is 0.735 so the actual is somewhat smaller than expected). The variance of the *medians*, however, is 0.908 and the variance of midranges is 0.862, both appreciably larger than either the actual or the theoretical variance of the means, even though they may not look large.

Mathematically we will work with a term called relative efficiency which is a ratio and is described as

$$\text{Relative efficiency} = \frac{\text{Smallest variance}}{\text{Any other variance}}. \quad (12.2)$$

Thus the relative efficiency of the median for this problem (using the actual values) is

$$\text{Relative efficiency} = \frac{0.657}{0.908} = 0.724$$

(using the theoretical value of variance of means relative efficiency will be $0.735/0.908 = 0.809$). Similarly the relative efficiency of the midrange is

$$\text{Relative efficiency} = \frac{0.657}{0.862} = 0.762$$

(again using the theoretical variance of means the relative efficiency = $0.735/0.853 = 0.862$).

From the Appendix Table A.9 we find that for samples of 4 we should expect an efficiency of 0.823 for both the median and the midrange. Our values are moderately reasonable. Note that perfect or equally good estimators have an efficiency of 1.00, since their variances would be identical.

As you look at the table you should see that efficiency depends on sample size. I have taken the samples of 8 from Figure 12.6 and have calculated the medians and midranges for these samples. Histograms for each of these three distributions of average are provided in Figure 12.10 for comparison. The variances and relative efficiencies are also listed beside the distributions. Note how much more spread out the midranges are than the medians. As the sample size increases this spreading will occur, and the efficiency of the midrange will drop, eventually to zero. The median, however, approaches a constant efficiency.

Other estimates may prove to be more efficient than either the midrange or the median for various size samples. These depend in some measure on how many individuals you want to work with. If you carefully inspect the table you should notice that the efficiency of the median fluctuates. This is mainly because half of the time the median requires the use of 2 values and the rest of the time only 1 value. When more values are used, the estimates are more

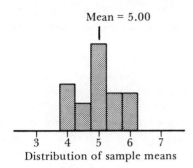

Variance = 0.420
Theoretical variance = 0.368
(a)

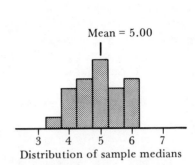

Variance = 0.520
Efficiency (based on actual) = 0.810
Efficiency (based on theo.) = 0.718
Theoretical efficiency = 0.743
(b)

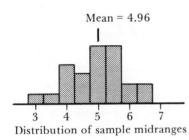

Variance = 0.790
Efficiency (based on actual) = 0.530
Efficiency (based on theo.) = 0.466
Theoretical efficiency = 0.610
(c)

Figure 12.10 Sample means, medians and midranges 25 samples of size 8.

efficient—this is one reason why the mean, which uses all of the data, is the most efficient.

Another interpretation of efficiency is that it gives you an indication of how large a sample you would have to use of an inefficient estimator to get as good an estimate of the population value. For a relative efficiency of 0.724 this says that you only need to use a sample of 724 if you are going to calculate a mean, but you will need 1000 in your sample if you calculate a median. Both of these two samples will give equally reliable estimates of the population mean, but one, the median, requires a much larger sample to get the same reliability in the estimation.

We can turn this idea around a little bit more for another approach that may be easier to understand. Figure 12.11a shows a hypothetical distribution of the *means* of a large collection of samples of size 5. Observe the spread of this distribution. Now let us look at the distribution of medians of the same samples. Figure 12.11b shows medians where the samples were also of size 5. This distribution is much larger than that of the sample means. If we increase the sample size, we will expect eventually to obtain a distribution of medians that will be about the same size as that of the means. Figure 12.11c shows a distribution of sample medians, this time of samples whose size is 7. Now the distributions of means and medians are just about identical. We would determine the relative efficiency based on the sample sizes for the two identical distributions, or

$$\frac{\text{Size of samples of means}}{\text{Size of samples of medians}} = \frac{5}{7} = 0.71.$$

This compares quite favorably with the theoretical efficiency of 0.697 (for medians of 5).

When it comes to estimating the variance, several different kinds of values are again available. For small samples the range is undoubtedly an excellent estimator. The efficiency of the range is as large as 0.890 for samples as large as 8, and for samples of 4 it is 0.975. However, for larger samples or for easier calculations with higher relative efficiencies, other estimators are available.

For large samples, estimates of the mean and variance that use several percentile values can be stated.* With only 10 values a mean can be estimated with 0.97 relative efficiency and a variance with 0.92 relative efficiency. This is particularly nice if you have ranked the data and maybe have 100 or more values to work with.

Although I mention that many estimators can be used most of the time, generally we either use one of the standard substitutes, or else we beg, borrow, or steal some sort of computation equipment. The underlying principles to be considered are first, how much time and money does it take me to get the data, and second, how much is then involved in the computation and the analysis. Where computers or even calculators are available, we will be inclined to stay with the mean and variance. When very limited amounts of data have been obtained, when the data is costly to secure, when we must destroy that which we measure, we will tend to treasure the data and attempt to gain all the information that is possible. On the other hand, when we need rapid approximations or have cheap data, we may want to use other calculations. Sometimes we will desire to delegate the gathering and continual analysis of certain types of data to others. We may give a control chart to a supervisor or a machine operator with instructions for computing values and for recognizing when to make adjustments or to call for help. Here we would want simple straight-

---

* For a table of such estimators see the *Handbook of Tables for Probability and Statistics*, Second Edition ed. by William Beyer (Cleveland, Ohio: Chemical Education Rubber Co., Easton, Pa., 1968).

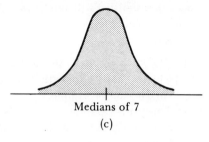
Medians of 7
(c)

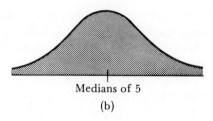
Medians of 5
(b)

Figure 12.11 Comparison of distributions of means and medians. Observe that for samples of 5 the distribution of medians (b) is larger than that of means (a). If the samples are increased to 7 individuals the distribution of medians now shrinks to almost the same size as the distribution of means of 5. (c) Thus we can say that calculating means of 5 can do the same job that requires a sample of 7 if we want to use medians.

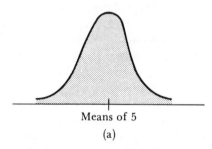
Means of 5
(a)

forward methods both for computation and explanation purposes, and standard deviations are not exactly easy to compute or to explain.

12.5 The Sampling Distributions: t, χ^2, F

Earlier in this chapter we found that a collection of means of samples taken from a population would be *normally distributed*. In addition, we found that these samples had the same overall mean as the population and a standard deviation, called the standard error, equal to the population standard deviation divided by the square root of n.

In the course of our discussions we have observed that other types of statistics can be calculated from samples. We now want to look at a few of these statistics and the nature of their distributions. We will find later that the distributions of these other statistics will also be valuable in making decisions about the source of our samples or the significance of our data. The three distributions we will discuss will be:

1. The *t*-distribution—this will be used for examining the *means of samples* when we do *not* know the population variance and must work with the variance of the sample.
2. The χ^2-distribution—this will be related to the *distribution of variances* from a given population and is used to compare the variance of a sample to a known variance (it will also have other uses that will be discussed).
3. The *F*-distribution—this is a distribution used to decide if *2 samples* have come from a population with the same variance.

We will now discuss these three distributions. Bear in mind as we go through the examples that the distributions are *not* restricted only to the uses or statistics that are discussed here. They have their own mathematical descriptions and may appear as descriptions of additional distributions.

All three distributions depend upon something called *degrees of freedom* (abbreviated *df*). In most of the problems that we will encounter the degrees of freedom will be equal to $n - 1$ (the sample size minus one). However, since this is not true in all cases, the tables are all labeled in some way with degrees of freedom (*df*). Whenever you are to use a value of degrees of freedom that is *not* $n - 1$ it will be clearly emphasized.

The concept of degrees of freedom is somewhat confusing the first time it is encountered. Basically, it is a value that tells how many of the quantities may be independently chosen. For example, if I am allowed to pick any 5 numbers, then there is *no* restriction on *any* of the numbers. Thus, I might choose 1, 3, 5, 2, and 6. Since all of them have had the freedom to vary, we say that the total number of values that may freely vary, the degree of freedom, is 5 (one for each number I was able to freely pick).

If we add a restriction such as the sum of the numbers is to be 18, then I can no longer freely pick *all* the individuals. If I were to pick the first 4 values in the same manner as before (1, 3, 5, 2) then in order to obtain a sum of 18 the 5th value *must* be 7. So we now only have 4 values that may be freely chosen and we state that the degrees of freedom are 4. In a similar manner we tend to "use up" some of the freedom of certain statistics to vary when we use in other statistics and tests means and variances that were calculated from the data.

(1) The *t-distribution*, also called the Student *t*-distribution, is a symmetric distribution that resembles the normal distribution (see Figure 12.12). The *greater the number of degrees of freedom, the closer it approaches or becomes the same* as the normal distribution. This continues until

$$df = \infty \text{ (infinity)},$$

and at that point the two distributions (the *t* and the normal) are identical.

As I stated earlier, the *t*-distribution will generally be associated with the distribution of the means of samples. So was the normal distribution, and this is a reason for our association between the normal and t-distribution. When we worked with a normal distribution we *assumed* that we knew what the population variance or standard deviation was—and we assumed it was known without error (it does not vary). Now we are not going to state that we have knowledge of the population standard deviation. Instead, we are going to use the standard deviation *of the sample* to estimate the population standard deviation. Because we are now working with an estimation we need to recognize that this will create inexactness in whatever other comparisons we work with. Since the t-distribution (as we use it) will involve our estimation of a standard deviation, the poorer or less reliable that estimation is, the less exact we can consider our distribution to be.

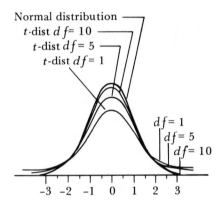

Figure 12.12 Several t-distributions and their relationship to the normal distribution.

The way we will observe that the distribution is less exact will be by the greater spread of the distribution. By now you should realize that the larger the sample, or the larger the number of degrees of freedom (since here $df = n - 1$), the closer the t-distribution is to the normal distribution. The extreme case of using a very large sample is equivalent to finding the *true* population standard deviation almost without error, and thus as the degrees of freedom (or sample size) approach about 100, the two distributions become almost identical.

As with the normal distribution, we will work with a table of values (Appendix Table A.2). Since we have such a large variety of possible t-distributions (one for *each* value of degrees of freedom) the table contains only selected critical values. In addition, just as we "normalized" the many possible normal distributions, we do a similar operation with the t-distribution. All t-distributions in the table have a center of 0 and all values along the X-axis are considered as t-values (just as we had called them z-values for the normal distribution). Thus, for 5 degrees of freedom we have Figure 12.13. The $t_{0.95}$ says we are looking at the t-value below which 95% of the possible cases occur.

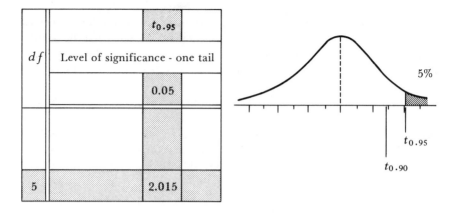

Figure 12.13 t-distribution with 5 df. Finding the t-value for which 5% are above (in one tail). This is the same point that 95% are below and therefore is labeled as $t_{0.95}$.

Similarly, 5% are above this point (called $t = 2.015$) and we say that there is a probability of 0.05 for a value to be in this tail. In Appendix Table A.2 you will find this value under the "level of significance for one-tailed tests" of 0.05. Two-tailed cases are those cases where we are interested in the area between $\pm t$. In these cases we use the line labeled *level of significance for two-tailed test* to determine what are the limits. Thus, if $t = 2.015$, then 5% of the values are above $+t$ and 5% are below $-t$ (see Figure 12.14) or a total of 10% are outside the limits, and 90% are inside the limits.

(2) The χ^2-distribution (pronounced chi-square distribution) is the distribution we will encounter when we deal with collections of values that have been squared. One kind of quantity that we have been working with extensively that involves squares is our variance. Thus, variances of samples have distributions that are related to χ^2-distributions.

If we take each one of a collection of sample variances, divide them by the known (or theoretical) population variance and multiply these quotients by

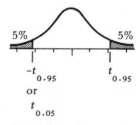

Figure 12.14 t-distribution showing how two-tailed limits are determined. Here the total area that is outside the limits is 10%.

$n - 1$ (where n is the sample size) we will obtain a χ^2-distribution. Thus,

$$\frac{s^2}{\sigma^2}(n-1) = \frac{s^2}{\sigma^2}(df)$$

would have the same distribution as a χ^2 with $n - 1$ degrees of freedom.

Since the larger the sample the closer the sample variance and the true variance would become, we would expect that the χ^2-distribution would change with sample size. Again we relate sample size and degrees of freedom. However, since there are other applications of the χ^2 where degrees of freedom are not connected to sample size, the degrees of freedom notation is even more important with the χ^2-distribution than it was with the t-distribution.

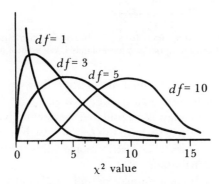

Figure 12.15 Some χ^2-distributions and the relationship of degrees of freedom (df).

The χ^2-distribution is not symmetrical and all the values are positive. Again we need to know the degrees of freedom, since for different degrees of freedom we have different curves. The smaller the number of degrees of freedom the more skewed the distribution is (see Figure 12.15). Appendix Table A.3 gives selected critical values of χ^2 for a number of different numbers of degrees of freedom. Note that the χ^2-values are the quantities recorded on the X-axis, and the values in the table are *areas below that value*.

For example, if $df = 8$, then χ^2-value below which are 10% of the values are found under the column labeled 0.100 and is equal to 1.61 (see Figure 12.16). The value which 5% are above is the $\chi^2_{0.95}$ (remember that if 5% are to be above a point then 95% are below the same point). This is 15.51.

To find the middle 95% we must find two limits. We need to split the area outside of our limits into two 2.5% areas and determine the two limits separately. For $df = 5$ we would find these two limits to be $\chi^2_{0.25}$ (2.5% below) and $\chi^2_{0.975}$ (2.5% above). Figure 12.16 shows these limits and the appropriate entries in the table. We find that

$$\chi^2_{0.025} = 0.831 \quad \text{and} \quad \chi^2_{0.975} = 12.83$$

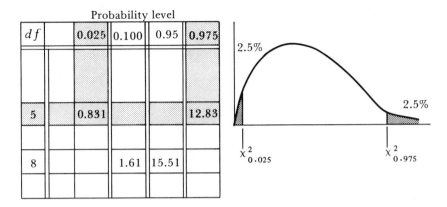

Figure 12.16 Finding limits for a χ^2 (chi-square) distribution. Here we must look up both the upper *and* lower limits if we want a two-tailed test.

(3) The *F*-distribution. The final distribution that we will consider is the *F*-distribution. We will use this distribution when we need to determine the chance that two *samples* may be considered as essentially having the same spreads or variances. We will perform this comparison by calculating a ratio of the variances of two samples. The distribution of all possible ratios of this type will be called the *F*-distribution.

Since this is a ratio, it has a numerator and a denominator where both the numerator and denominator vary separately, and in fact are the variances of our two samples.

Since we will have two samples, and they both have sizes associated with them, we will have *two* values of freedom to operate with. We will therefore designate the value of *F* as

$$F(f_1, f_2) = \frac{\text{numerator with } f_1 \, df}{\text{denominator with } f_2 \, df}$$

(The f_1 means degrees of freedom for 1, the numerator, and f_2 means degrees of freedom for 2, the denominator. Other symbols such as v_1 and v_2 (Greek letters) or *m* and *n* are also used.)

Appendix Table A.5 is a set of *F*-tables. Each table is for a different probability or significance level. Thus one table is for $F_{0.90}$ (90% of the values are less than these *F*-values), another is for $F_{0.95}$, and so forth. To find a particular value for *F* for which 95% of the values are below ($F_{0.95}$), we must also state the two values of the degrees of freedom, f_1 and f_2, for the data. We will now examine several examples of finding *F*-values from the tables. In later chapters we will identify the types of problems involving *F*-values.

Find $F_{0.95}$ (12, 6). Here we are saying $f_1 = 12$ (*df* for numerator) and $f_2 = 6$ (*df* for denominator).

We look up in the table labelled $F_{0.95}$ (see Figure 12.17) and find $F = 4.00$. This can be interpreted as: We expect less than a 5% chance that given

2 samples, one of 12 and the other of 6, the ratio of the variances of these two samples will exceed 4.00 (that the 12 item sample will have a variance 4 times as large as the 6 item variance).

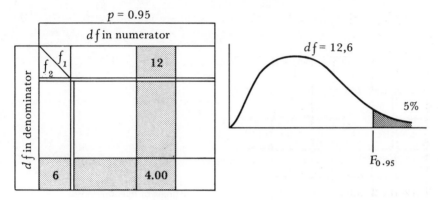

Figure 12.17 Finding a limit for an F-distribution. $F_{0.95}$ (12, 6).

Find $F_{0.99}$ (24, 10). (Then $f_1 = 24$ and $f_2 = 10$.)

Looking in Table A.5 labeled $F_{0.99}$ we find the value of 4.33 (see Figure 12.18).

Find the pair of limits outside of which are 10% of the values (for 12 and 6 degrees of freedom). In our first example we found the upper limit. To find the lower limit you must either consult other tables or use a conversion technique. The conversion for going from one limit to the other is

$$F_p(f_1, f_2) = \frac{1}{F_{1-p}(f_2, f_1)} \qquad (12.3)$$

where p is the probability limit you desire.

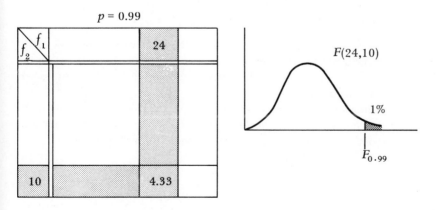

Figure 12.18 Finding a limit for an F-distribution. $F_{0.99}$ (24, 10).

For this example we desire $F_{0.05}(12, 6)$; then $p = 0.05, f_1 = 12$, and $f_2 = 6$. Substituting these into Equation (12.3) we have

$$F_{0.05}(12, 6) = \frac{1}{F_{1-p}(f_2, f_1)} = \frac{1}{F_{1-0.05}(6, 12)} = \frac{1}{F_{0.95}(6, 12)}.$$

Looking up in Table A.5 (see Figure 12.19) we find $F_{0.95}(6, 12) = 3.00$.

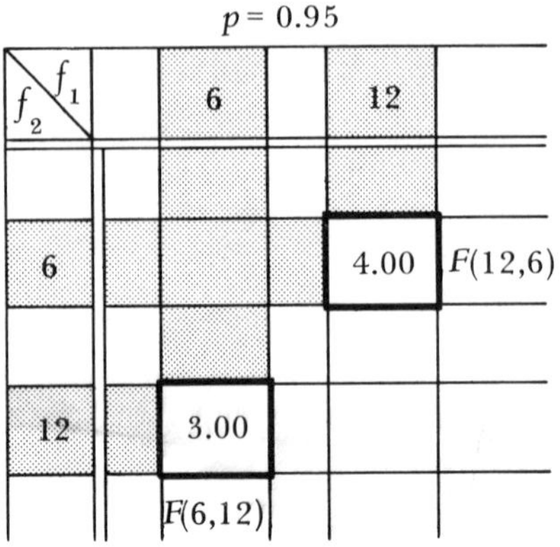

Figure 12.19 The order of the degrees of freedom in the F-distribution is important. Here the values for (6, 12) and (12, 6) degrees of freedom are shown.

Therefore,

$$F_{0.05}(12, 6) = \frac{1}{F_{0.95}(6, 12)} = \frac{1}{3.00} = 0.33.$$

We now have two limits, a 5% and a 95% limit.

Generally, however, we do not bother finding lower limits. If we are interested in *both tails* then instead of *randomly* deciding which of the values is to be placed in the numerator, we decide ahead of time to place the *larger* of the two in the numerator. Thus we will compute the actual F as

$$F = \frac{\text{Larger value (with } df = f_1)}{\text{Smaller value (with } df = f_2)}.$$

This will *always* be greater than 1. In these cases, when we now look up an $F_{0.95}$, a 5% critical limit, *we will consider it as a 10% limit*. For this reason the tables contain two headings—"when used as a one-tailed table" and "when used as a two-tailed table."

One final important observation should be made. Refer back to Figure 12.19 and note that $F(12, 6)$ *does not equal* $F(6, 12)$. The order of degrees of freedom is very important in determining critical values of the F-distribution.

REVIEW

Sample · Population · Distribution of sample means · Bias ·

Estimate · Transformation · Central Limit Theorem ·

Standard error of the mean · Efficiency · Sampling distribution ·
t χ^2 F · Degrees of freedom

PROJECTS

1. The population of numbers on a die represent a rectangular or flat distribution from 1 to 6. Take 5 dice and roll them. Find the average values or the sum (you can also use 1 die that you roll five times). Do this about twenty times (twenty samples of 5), and plot a histogram for the average and the sums. Compare the distribution of the sample values to the population values. You can also use different size samples.

PROBLEMS

Section 12.1

12.1 Take an additional 5 samples of 4 values from the sampling distribution in the Appendix Table A.18. What are the means and the variances of these samples?

Section 12.3

12.2 Refer to Figure 12.5. This contains the distribution of samples of 2, 4, 8, and 20 from the population on the bottom (and in Appendix Table A.18). If you were to plot these distributions on normal probability paper how close do you think they would be to normal distributions? How would the lines differ?

12.3 Using the random number table, Appendix Table A.16, calculate the mean of 4 digits. Do this for about 10 sets of 4. Draw a histogram for them. How close to a normal distribution is it? What is the mean and standard deviation of this distribution? What would you expect the mean should be? What would you expect the standard deviation to be? (*Hint:* 10% of the values should be in each category, so if you choose just 10 values the frequency of each would be expected to be 1. Also this is a theoretical distribution so use n not $n - 1$.) What should the standard error be? How close is the standard error of the distribution to the theoretical value?

12.4 If one pound bars of butter have been averaging 0.98 pounds with $\sigma = 0.04$ (see Problem 10.8), then what is the chance that an inspector from the Weights and Measures Department would find that the average weight of 8 pounds of butter would be less than 0.96 pounds?

12.5 If a barber takes an average of 14 minutes to give a haircut with $\sigma = 3$ minutes
   a. What would he expect the total time for giving 24 haircuts to be?
   b. What would the standard error for the average time for giving 24 haircuts be?
   c. 95% of the time the average time for giving 24 haircuts would fall within what limits?

d. If you multiply the limits for the average by n (or 24) you will obtain limits for the sum (the total time for 24 haircuts). Within what limits would 95% of the total times for 24 haircuts fall?

12.6 Baking potatoes were weighed and found to average 0.50 pounds, with standard deviation of 0.08 pounds. A banquet of 100 people was planned. How many pounds of potatoes should be purchased to provide 95% assurance that there will be enough potatoes, if each person gets 1 potato?

Section 12.4

12.7 Appendix Table A.9a gives efficiencies for the median. How does the efficiency vary according to sample size? Can you explain the peculiar behavior? (*Hint:* Look at the efficiency of the midrange for samples of 2, 3, and 4).

12.8 What happens to the efficiencies of the midrange as n gets very large? Explain.

12.9 Look at the last column in Appendix Table A.9a. Why does that value increase its efficiency as the sample size gets larger?

12.10 For estimating σ we have found that we can use R (range) or s. Appendix Table A.9b and c provide the constants needed to multiply R and s by in order to obtain unbiased estimates of σ. Also included are constants needed to multiply the population variances by to obtain variances of our possible distributions of estimates (the variances of our distribution of estimated variances). By comparing variances verify the efficiency of the range for samples of 5 and 10.

12.11 Assume you are interested in estimating the population standard deviation and you are willing to have a variance of that estimate as large as $0.060\sigma^2$, thus permitting the use of a sample of 10 (since the variance for 10 is $0.057\sigma^2$, which is smaller than $0.060\sigma^2$). How large a sample would be required to have the same precision except by using the range?

12.12 If the cost of taking extra samples is negligible, then it does not matter if you use a sample of 10 or 12. If you were to do the calculations by hand how much time would you estimate it would take to calculate a variance of 10? A range of 12? Which do you prefer? How long would the extra measuring time, or how much would the extra cost of samples, need to be for you to decide to use standard deviation instead of range? How would a desk calculator change your opinion?

Section 12.5

12.13 From the tables find
  a. $t_{0.95}$ for 10 df.
  b. t for which 10% of the distribution values are below ($df = 16$).
  c. t-values for which 90% of the distribution values are between ($df = 22$).
  d. 5% upper limit of t for 8 df.

e. Two-tailed t-values for which 20% of the distribution values are in the tail and $df = 30$.

12.14 From the tables find
   a. $\chi^2_{0.95}$ for 10 df.
   b. $\chi^2_{0.05}$ for 5 df.
   c. χ^2 for which 10% of the distribution values are beyond ($df = 12$).
   d. χ^2-values for which 80% of the distribution values are beyond ($df = 25$).
   e. 10% upper limit of χ^2 for 200 df.

12.15 From the tables find
   a. $F_{0.95}(3, 5)$.
   b. F-value for which 10% of the distribution values are above, df in numerator = 5, df in denominator = 20.
   c. F-value for which 10% of the distribution values are above, df in numerator = 20, df in denominator = 5.
   d. 1% upper critical value for $f_1 = 30, f_2 = 20$.
   e. $F_{0.10}(6, 10)$.
   f. $F_{0.01}(1, 5)$.
   g. A pair of F-values that enclose 99% of the values with $f_1 = 3$, $f_2 = 15$.

12.16 Using the sampling pieces (Appendix Table A.18) compute a correlation coefficient for 3 pairs of successive chips. Compute another correlation coefficient for the next 3 pairs of chips. Work about 10 correlation coefficients in this manner. What type of distribution of correlation coefficients do you get?

12.17 Repeat Problem 12.16 except now use three chips and do a correlation of front to back (call front X and back Y). Do this for 10 sets of 3 values. What does the distribution look like? (These chips have a correlation of approximately 0.8.)

12.18 Refer to Table 12.1. For each successive pair of samples of 4 compute the ratio of the variances (s^2/s^2). Thus the first ratio is 4.33/2.67. You can calculate 25 ratios. Plot these and compare the 90% and 95% percentages of this distribution to the theoretical 90% and 95% limits of an F-distribution (3, 3 df).

13

Estimation and Hypothesis

In Chapter 11 we discussed hypothesis testing and in the last chapter estimation was covered in some detail. I will now attempt to show you some of the similarities between these two concepts.

Let us look at our Mystery Can samples and discuss several ways of describing the relationships of point estimates, interval estimates, and tests of hypothesis (three topics that will be discussed at length in this chapter). In each case we will use the results of our first house and then our first sample of four houses. Recall that the first four values were 6, 8, 3, and 5. For now we will only concern ourselves with the average number of people, and we will recall that the overall population had a mean of 5 and a variance of 2.94 (or approximately 3).

13.1 Point Estimation

In the last chapter we devoted considerable time to discussing the goal of estimation. We noted that we are actually interested in guessing the parameters of the parent universe or population. Remember that these parameters are fixed constants for the particular population, although the sample is not constant. Generally we want to choose what we believe to be the *best* possible guess for the mean, μ, and the variance, σ^2, of the parent distribution. We found that the *best* estimators for these quantities are the *sample mean* and the *sample variance*. These are *point* estimators because they only establish a single value or point out of all the possible values that could arise.

Look at the first individual value. It is a 6. At this point it would be our best estimate of the actual population mean.

After we look at the sample of 4 we can say that we now estimate the mean of the population to be equal to the mean of this new sample. We say:

$$\text{Estimated } \mu = \hat{\mu} \text{ (read this as mu-hat)}$$

$$= \bar{X} = \frac{6 + 8 + 3 + 5}{4} = 5.50.$$

We therefore have refined our estimate of the mean of the population. We still do not expect that this will be a perfect estimate. The problem is how imperfect is the estimate. The most logical way of expressing our unsureness is to place an interval around that single point. Thus we come to interval estimates, or confidence intervals.

13.2 Interval Estimates or Confidence Intervals

You should be well aware that variation is to be expected from estimate to estimate. The two we just obtained were different. So whenever we arrive at a point estimate we realize that it is only an estimate, and that the true population parameter (the mean in this case) falls in some interval around the sample value we obtained.

For this example, it will be assumed that we know that the variance is 2.94. This now permits us to draw a distribution around the individual or the sample. Figure 13.1a shows a normal distribution (with $\sigma = \sqrt{2.94} = 1.7$) drawn around the sampled individual. We already know that approximately

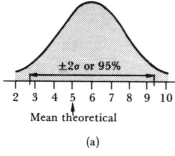

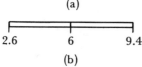

Figure 13.1 Distribution of individuals (a) distribution drawn around first *individual*, (b) confidence interval drawn around first individual (95% *CI*).

95% of the distribution lies between ±1.96 standard deviations (or ±3.4 in this case).* If we draw ±1.96 sigma limits around the individual we will obtain an interval. Since these limits were designated in a way such that 95% of

---

* You may readily substitute ±2σ limits. However, since we will be discussing different probability levels and we have already discussed the normal distribution I will use 1.96σ limits. They are more appropriate.

the values should be between the limits, we say this is a 95% confidence interval (abbreviated as 95% CI). We then interpret the interval to imply that there is a 95% chance that the *true* population mean lies inside this interval. So we now state that there is a 95% chance of being correct (or a 5% chance of being wrong) when we declare that the population mean lies between

$$6 - 1.96\sigma \quad \text{and} \quad 6 + 1.96\sigma$$

or between

$$6 - 3.4 = 2.6 \quad \text{and} \quad 6 + 3.4 = 9.4.$$

We have done two extremely important and new things—we have specified a *range of possible values* that the actual population could be centered at *and* we have assessed the chance that the population is not inside that interval (the 5% chance).

A second way of approaching this concept would be by looking at the actual population (which in real cases we do not know). We can compute an interval within which we would expect 95% of the individual samples to fall. Thus we have the distribution of Figure 13.2, which shows where we expect 95% of the individuals to be (between $5 - 1.96\sigma$ and $5 + 1.96\sigma$ or from 1.6 to 8.4). We can consider this as a confidence interval around the population mean.

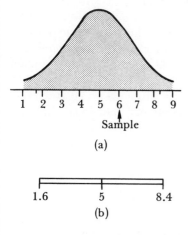

Figure 13.2 Distribution of individuals (a) distribution drawn around *true mean*. This is the theoretical population. (b) Confidence interval drawn around true population mean (95% *CI*).

Recall that the first four individuals were 6, 8, 3, and 5. They all fall in this 95% confidence interval of the population.

Now let us return to placing intervals around the individual samples. If we draw an interval for each of the first 4 individuals, we observe that all of the intervals cross the mean of the population (Figure 13.3). In effect, we observe a drifting of confidence intervals around the true population. In addition, 95% of them will vary back and forth along a line and produce distributions that can be anywhere between two extremes—the left dotted and the right dotted histograms of Figure 13.4. A mere 5% will fall outside of this interval. Thus we expect that, with many different samples, 95% of the confidence intervals we would draw will cover the true mean.

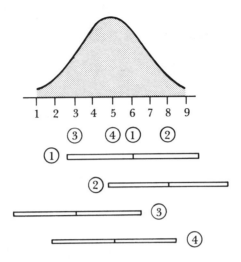

Figure 13.3 95% confidence intervals drawn around each of the first four individuals. Observe how all four intervals cross the population mean.

Now what if I want to be more sure of myself? What if I want to draw an interval such that I will be 99% sure that it will include the true population mean? We must now find the z-values for which 99% of the individual samples fall between (instead of the 95% limits as before). We find that $z = \pm 2.58$ standard deviations, or for this example the limits will be ± 4.4. This is a larger interval than before. We then say that for our first individual the 99% confidence level is from

$$6 - 4.4 = 1.6 \quad \text{to} \quad 6 + 4.4 = 10.4.$$

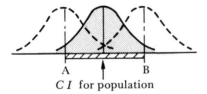

Figure 13.4 95% of the time we expect the distribution we draw around the sample to wander only as far to the left as point A and as far to the right as point B.

In Figure 13.5 we look at the two intervals, 95% and 99%. Observe that by *increasing our probability or chance of including the true population mean, we decrease our discrimination ability*—we have enlarged the collection of possible values to which the true mean may belong. Similarly, if we were willing to be only 90% sure that our interval contained the true population mean, we would create a smaller range.

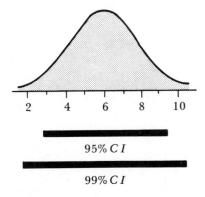

Figure 13.5 The effect of increasing "confidence" is to increase the limits.

Our next step is to look at a sample. The Central Limit Theorem of the last chapter introduced the concept of the distribution of sample means. We found that samples tend to be closer together than individuals. So we should expect that a confidence interval for samples should be smaller than one for individuals. Instead of using σ to find the end points, now we will use $\sigma_{\bar{X}}$, the standard deviation of the sample means or the *standard error* of the means. We find $\sigma_{\bar{X}}$ from Equation (12.1) where

$$\sigma_{\bar{X}} = \frac{\sigma}{\sqrt{n}}.$$

Then a 95% confidence interval would span approximately $\pm 1.96\sigma_{\bar{X}}$. Thus for our samples of 4

$$\sigma_{\bar{X}} = \frac{\sigma}{\sqrt{4}} = \frac{1.71}{2} = 0.855,$$

or only *half* the size of the confidence interval for individuals. (You can also find $\sigma_{\bar{X}}$ from Appendix Table A.6.)

The limits would be drawn at approximately

$$\pm 1.96\sigma_{\bar{X}} = \pm 1.96(0.855) = \pm 1.71$$

from the mean. For our first sample we had found that $\bar{X} = 5.50$, so we would draw confidence limits at

$$\bar{X} - 1.71 \quad \text{and} \quad \bar{X} + 1.71$$

or between

$$5.50 - 1.71 = 3.79 \quad \text{and} \quad 5.50 + 1.71 = 7.21.$$

The most important thing to observe is how much the *width of the interval decreases as the sample size increases*. By taking larger and larger samples we can get as small a confidence interval as we might desire. This is illustrated in Figure 13.6 where confidence intervals are drawn for individuals and samples

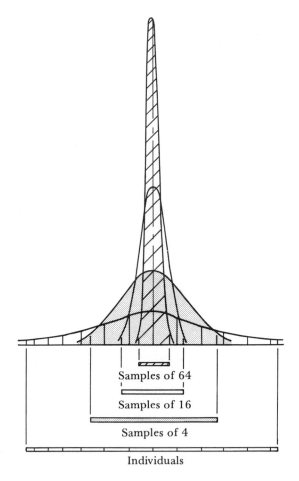

Figure 13.6 The effect of sample size on confidence intervals. Distributions of individuals, samples of 4, 16, and 64 are drawn with the respective 95% confidence intervals.

of 4, 16, and 64, all from the same population. For samples the main statement is still the same as with individuals, only slightly modified. We now say that 95% of the confidence intervals for *samples* will cross the true population center. In Figure 13.7 I have drawn the intervals for the first 10 samples of 4. One of these, the 4th, does not reach the true population mean. Bear in mind that if this was the only sample we had taken, then these would have been the *only limits* (the only confidence intervals) that we could have drawn, and we would *not* have included the true center in our interval. We would have erred. We must expect to make errors sometimes, and our choice of 95% limits forces us to be prepared for about 1 wrong interval out of every 20 intervals we calculate. As we previously said, we can change this chance of error by changing the size of the interval. There is always a compromise. In addition, you must always remember that in real practice we are generally dealing with *only one sample* and we are drawing *only one* confidence interval.

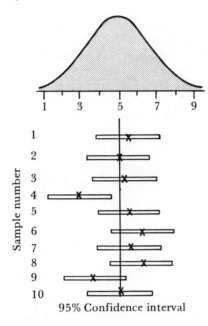

Figure 13.7 95% confidence intervals drawn for 10 sample means of 4 from a given population. Observe how one interval, the fourth, does not cross the true mean.

13.3 Hypothesis Testing

With estimation most of our work was done with the sample. We looked at the sample and then drew an interval around it. We did all our analysis or study after we looked at the sample. This is not so with hypothesis testing. Here, much of the work will be done *before* the sample is even taken. We begin by stating what our main hypothesis and our alternatives are (see Chapter 11 for review).

We might feel or hypothesize that the average number of people is 5. We might also *know* that the standard deviation of occupants is 2.94. Here we are questioning an average, but assuming that the population standard deviation is known. We also will asume that the number of occupants approximates a normal distribution. We might now say:

Hypothesis: The population mean is equal to 5.0 ($\mu = 5.0$)

Alternate Hypothesis: The mean is different than 5.0, it is larger *or* it is smaller than 5.0 ($\mu > 5.0$ or $\mu < 5.0$)★

Next we must decide upon a set of limits. Since we are working with a normal distribution, we can choose a pair of z-values to be our limits. The limits are chosen by stating where we desire to set our α error. (Recall that this α error was the chance of rejecting the hypothesis when the population really has *not* changed—here this is how often we would say the average is *not* 5.0 when it actually is 5.0). If we choose $\alpha = 0.05$ (saying that 5% of the time we will be willing to be wrong) then we look for z-values such that 5% are divided amongst the two tails. We would pick $z = 1.96$ (see Figure 13.8a). For our sample of 4 these two z-values then correspond to

$$\mu + z \frac{\sigma}{\sqrt{n}} = 5.0 + 1.96 \frac{(1.71)}{\sqrt{4}}$$
$$= 5.00 + 1.67 = 6.67$$

and

$$\mu - z \frac{\sigma}{\sqrt{n}} = 5.0 - 1.96 \frac{(1.71)}{\sqrt{4}}$$
$$= 5.00 - 1.67 = 3.33.$$

In this case we are essentially drawing a confidence interval around the hypothesized or guessed mean. We then decide to accept or reject the hypothesis depending upon which region the sample falls in. If you study Figure 13.8b you will see the relationship between the accept and reject regions and the distribution of samples.

The underlying reason for stating that there is a 5% chance of possible error rests with the fact that *if the distribution actually has a mean of 5.0 we would still expect to end up with approximately 5% of the samples in the shaded region. For these 5 out of 100 times, our decision would be "reject the mean of 5.0"—an obviously incorrect decision.* This concept is present in all hypothesis testing.

You might observe that in spite of the seemingly different approach, all we have done is put confidence limits around our hypothesized mean and then checked to see if the sample was inside or outside this particular confidence interval. When we begin to delve into the massive variety of statistical tests, of which hypothesis testing is one technique, we will use a slightly different approach. We will state what type of statistic or value we will calculate from the data, that is, we will be calculating an average or mean. There is a relationship between this value and the hypothesized value, that is, the mean of the sample and the mean of the population are related to a z-value. We then find

---

★ From now on I will begin to use the symbols $<$ and $>$ for less than and greater than. An easy way to distinguish them apart is to realize that the large opening goes with the larger value. So $5 < 8$, the opening going with 8, the small point with the 5. Similarly $3 > 2$. Thus $\mu < 5.0$ means that μ could be 4.9, 4.0, 3.5, 0, 1.5, or anything else less than 5.0.

limits for that new value—the z-value in this case. In other words we find limits of z (with a confidence level or α that is the same as before). We will then find for α = 0.05 that the two z-values would be z = −1.96 and z = +1.96. You then calculate the same statistic or value for the sample and you compare the new calculated value against the limits. Hence for this example you would look at the sample mean of 5.50 and calculate its z-value, which would be

$$z = \frac{\bar{X} - \mu}{\sigma/\sqrt{n}} = \frac{5.50 - 5.00}{1.71/\sqrt{4}} = \frac{0.50}{0.855} = 0.585.$$

(If you can't derive this formula just accept it for now.) You then compare this 0.585 to the limits. Since 0.585 is *not* smaller than −1.96 nor larger than +1.96 it is in *the accept* region for z and we may say that "we do not have sufficient evidence to state that the mean is not 5.00."

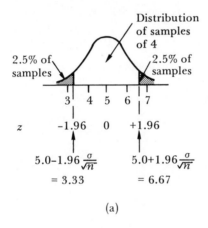

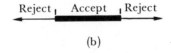

Figure 13.8 (a) Relationship to confidence interval for samples from a given population. (b) Relationship of accept–reject regions.

13.4 Introducing β Error

One of the primary advantages of these statistical methods is to permit us to identify and perhaps to limit our error. We have already discussed how the α error is determined and how we can adjust or change it. But we have not seen how a value is applied to a β error. (Recall that the β error is the chance that we will accept our hypothesis when it is not correct.)

Suddenly we are going to find that the alternates that have crept up in the previous tests of hypotheses are going to be important. Beta error is entirely dependent upon the alternate we have chosen. In addition, we can only talk about one alternate at a time. When we used as an alternate the case $\mu < 5.0$ or $\mu > 5.0$ in the previous section, we actually had an infinity of alternates. This occurred because 5.1 or 5.2 or 6.0 or 3.7 or any other value except 5.00 fit the general description we used for alternates. We actually have a multi-alternate problem.

Let's look at the general way in which α and β errors are related to the distributions. In Figure 13.9 we again have our hypothesized distribution of occupants with limits for an α error of 0.05. Now let us suppose that we really have reason to suspect that as an alternate there is an average of 7 people living

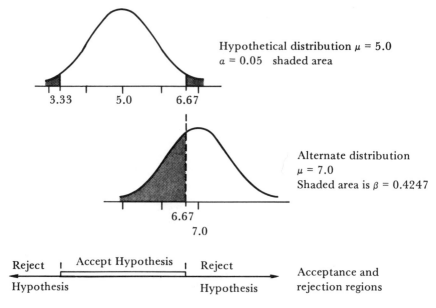

Figure 13.9 The relation of α and β errors for one alternate distribution. Samples of 4 for distribution, with $\sigma = 1.71$.

in the houses (perhaps we are interviewing in an area of large homes). We can draw this alternate distribution, and in Figure 13.9 this distribution is shown below the hypothesized distribution.

Now, what can happen?

We already covered the cases where the *population really is at 5.0*. The upper distribution holds and roughly 5% of the time we will make a mistake (the α error).

For the alternate situation, the *population* would now be at 7.0. Our sample would still wander and:

1. If the particular sample we selected is *above 6.67* (say 7.5) we will say

reject the hypothesis because we are in the shaded area of the upper distribution and this is the reject region. Thus we accept the alternate—and we are correct.

2. If the sample is *less than* 6.67 (say 6.1) we will say *accept* the hypothesis since it is in the range between 3.33 and 6.67 and we will be *wrong* (because the population is actually 7.0). This is where our β error is.

The overall or total β error is that area of the alternate distribution that overlaps the accept region. In this example it is the area of the alternate distribution that is below 6.67 (the shaded area). So we must compute that area (assuming that we have a normal distribution).

$$z = \frac{6.67 - 7.00}{1.71/\sqrt{4}} = \frac{-0.33}{0.855} = -0.38.$$

The β error would then be equal to the area of the portion of the distribution with a z-value of less than 0.38. This area is 0.3520. Thus, there is about a 35% chance (1 in 3) of not observing or not detecting a change in the distribution from 5 to 7.

We have found a single value for β, but this was because we selected a single alternate. As an example of a multiple alternate problem let us examine a control chart.

Essentially with a control chart we have been working with a hypothesis and a corresponding interval. We haven't stated it as such, but the limits we have used are identical to confidence limits for a theoretical center. Basically they are 3 sigma limits for means.* It has been found for most cases involving control charts that these limits provide a good trade-off between the possible errors.

Now what kind of errors am I talking about? I am referring again to the α and β errors. In what way? Well, assuming that our process has shown itself to be stable (I will say what effect this has in a few moments), then in two possible ways errors can occur:

1. There is the rare case of a point being outside the limit, when *no change* in the population has taken place (an α error).

The consequences of making an α error generally involves an unnecessary investigation with time wasted in an effort to locate the causes of a change in the system when there actually is no change in the system. If we assume the

---

★ To see how this has taken place, observe that we have estimated the population variance in some way, either with $\bar{R}$ or σ, or it was given to us from past information. I have already shown that we can convert from say $\bar{R}$ to σ using d_2 and $\sigma_{\bar{x}} = \sigma/\sqrt{n}$. Since $3\sigma_{\bar{x}}$ is what we want, then

$$3\sigma_{\bar{x}} = 3\bar{R}/(d_2 \cdot 1/\sqrt{n}),$$

or

$$[3/(d_2\sqrt{n})]\bar{R}$$

Thus for $n = 4$, $\sqrt{n} = 2$, $d_2 = 2.059$ and $3/(2.052 \cdot 2) = 0.7285 = 0.73$. We have found a constant to multiply $\bar{R}$ by to get $3\sigma_{\bar{x}}$ limits, and it is the same as A_2.

parent population is normally distributed, then we have less than $\frac{1}{4}$ of 1% chance of samples falling beyond the limit, and we thereby limit the chance of a wild goose chase to about once every 370 times we take samples.

The other possibility of error occurs when:

2. The population shifted, but the sample remained within the control limits (β error). This case depends upon how much of a shift we believe may have occurred or what differences concern us.

We often are interested in how much of a chance we have of detecting all different amounts of change in the population, so we will end up with a complete curve of many different values of β. For each possible difference of actual population and control chart center we will have a different β value. The complete curve is called an *operating characteristic curve* and it will depend on sample size. (If we plot $1 - \beta$ instead of β it is called a *power curve*.)

Look at Figure 13.10. This is the control chart for the means of our sample houses. If we go back to Table 12.1 we will find that the $\bar{R} = 3.72$. What are the control limits (for the means chart note that $\bar{X} = 5.00$)?* The population

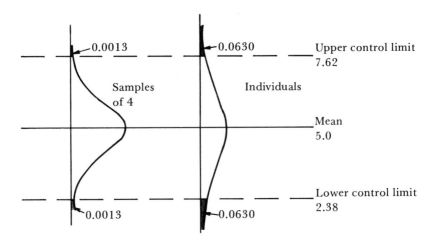

Figure 13.10 Using control chart for test limits we find α for samples to be 0.0026 and for individuals $\alpha = 0.1260$.

distribution is also shown with our distribution of samples of size 4. For this latter distribution (samples of 4) we set up the control limits (in effect this is a confidence limit). There is only a 0.23% chance of a random sample falling outside this set of limits, thus establishing our α error. What is the α error for *individuals* using these same limits?

---

* ANSWER: $\text{UCL}_{\bar{X}} = 5.00 + 0.73(3.72) = 5.00 + 2.62 = 7.62.$
$\text{LCL}_{\bar{X}} = 5.00 - 0.73(3.72) = 5.00 - 2.62 = 2.38.$

We must find the z-values in the distribution of individuals that corresponds to the control limits of 2.38 and 7.62. These are

$$z = \frac{7.62 - 5.00}{1.71} = 1.53,$$

and

$$z = \frac{2.38 - 5.00}{1.71} = -1.53.$$

Then above the z of 1.53 we would have 0.0630 (6.8%) and below $z = -1.53$ we have 0.0630. This totals 0.1260 or $\alpha = 0.1260$. Then 12.6% of the time a single individual will go beyond the limits (whereas 0.23% of the *samples* will do so).

Now let us assume that a shift in the population mean from 5.00 to 6.00 takes place. In Figure 13.11a note how both the population and the sample distributions cross the control limits. The shaded area within the control

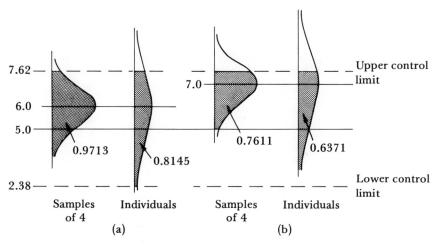

Figure 13.11 Effect of shift of mean on chance of detecting it (β error). (a) A shift of 1.0 or $1.0/1.71 = 0.565\sigma$ in the population. The effect is to change the distribution location. The shaded areas represent the chances of *not* observing the shift and are β errors. (b) Same as (a) except a shift of 2.0 or $2.0/1.71 = 1.71\sigma$ has occurred.

limits represents the amount of the time that we will *not* recognize that a change has taken place. This area is equal to the probability of making a β error. This occurs because the only way we *detect a change* is when a point is *outside* the limits. But the shaded area is the region inside the limits, and, therefore, this shaded area represents the proportion of the time that we will *not* detect the change. So for a change of 1.00 units or 0.565σ, the distribution of averages will have a β error of 0.9713, or 97% of the time we will not be made aware of a change of 1.00 individuals in the population center. Let's see how I found that.

To find the value for β we must return to the assumption that the distributions are normal—or at least the distribution of samples is normal. We must then determine the value(s) corresponding to the control limit(s) for the particular distribution. Hence, for the samples of 4,

$$\sigma_{\bar{x}} = \sigma/\sqrt{n} = 1.71/\sqrt{4} = 0.855.$$

The new center is regarded as μ, and the control limits are our two X-values. Thus, the upper value has a z-value of

$$z = \frac{\text{UCL}_{\bar{x}} - \mu}{\sigma_{\bar{x}}} = \frac{7.62 - 6.00}{0.855} = \frac{1.62}{0.855} = 1.90,$$

and the lower value is

$$z = \frac{\text{LCL}_{\bar{x}} - \mu}{\sigma_{\bar{x}}} = \frac{2.38 - 6.00}{0.855} = \frac{-3.62}{0.855} = -4.23.$$

Looking up the area or probability for a normal distribution between $z = -4.23$ and $z = +1.90$ we obtain a probability of 0.9713 or about 0.97. This is the probability of our β error.

There is also a β error associated with the distribution of individuals. It too is the shaded area between the control limits we have drawn. Again we must find the z-values (assuming we have a normal distribution). What are they?

$$\text{Upper } z\text{-value} = \frac{\text{UCL} - \mu}{\sigma} = \frac{7.62 - 6.00}{1.71} = 0.96,$$

$$\text{Lower } z\text{-value} = \frac{\text{LCL} - \mu}{\sigma} = \frac{2.38 - 6.00}{1.71} = -2.12.$$

Now that you have found the z-values, what is the area? This will be equal to the probability of making a β error.

The two areas are 0.3315 and 0.4830, which total 0.8145, our probability of making a β error.

You should now be aware that given the same set of limits, but on different distributions, we will obtain different β error values just as we found different α-values.

Now another shift takes place, this time to 7.00. Figure 13.11b shows this change for both the distribution of individuals and the samples of 4.

Again we want to find the β error values and again we must determine the z-values for the limits. For the samples of 4 we have

$$\text{Upper } z\text{-value} = \frac{7.62 - 7.00}{0.855} = \frac{0.61}{0.855} = +0.71,$$

$$\text{Lower } z\text{-value} = \frac{2.38 - 7.00}{0.855} = \frac{-4.62}{0.855} = -5.5.$$

The area between these limits is 0.7611 and the resulting probability of β error would be 0.7611. Thus, even if the population mean shifted by as much as 2 people—from 5 to 7 (a 40% increase), we would only be made suspicious about 24% of the time. Roughly 76% of the time we would not have a point go out of the limits when this change takes place.

The natural extension of this idea is to explore an entire set of values, and in Figure 13.12a this is done. Eleven different values were chosen as possible centers with each progressively larger. The shaded area of each distribution represents the probability of β error for each new average (note that for a center of 5.00 we are observing that no change is taking place, and here $\beta = 1 - \alpha$). Directly below I have drawn a scale with each different value positioned below the corresponding distribution mean. I then plotted the β for the distributions. If we connect the points we obtain a smooth curve of all β-values. Here we can see just how the *difference between the centers of the two distributions* affects the probability of a β error.

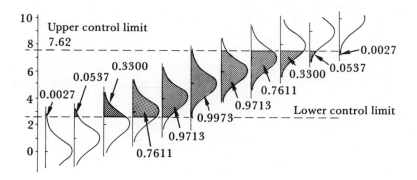

Relation of distribution to control limits
as center shifts

(a)

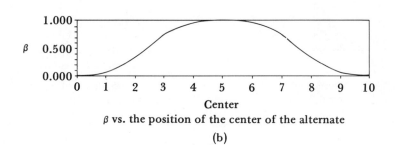

β vs. the position of the center of the alternate

(b)

Figure 13.12 β and distributions of samples of 4 with different centers.

Another observation we can make is how α and β errors are related to each other. In Figure 13.9 we observed a β value for samples of 4 where the population and alternate centers were at 5.0 and 7.0. We did the same in Figure 13.11b. The first case had an $\alpha = 0.05$ and the resulting β was 0.3520. In the latter case we used an α of 0.0023 and β ended up as 0.7611.

So when α changed so did β, and in general *if the sample size remains the same*

When α decreases then β increases.
When α increases then β decreases.

This brings us to the next question—what if the sample size changes?

In Figure 13.13 we look at several pairs of distributions. In each case we have the same means for the distribution and the same α level, 0.05. The first pair is for individuals and we found that $\beta = 0.7852$ (the limits are the same as in Figure 13.2).

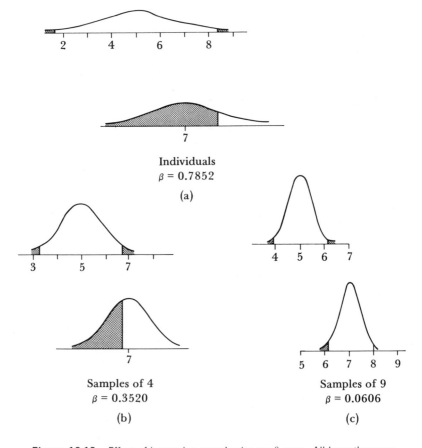

Figure 13.13 Effect of increasing sample size on β error. All have the same $\alpha = 0.05$. All have the same hypothesized mean of 5.0 and alternate mean of 7.0.

Increasing the sample size to 4 we obtain the pair of curves in Figure 13.9, which are reproduced in Figure 13.13b—here, the β equals 0.3520. If we further increase to a sample of 9 we could find that β has decreased to a mere 0.0606. The amount of times we can expect to detect a shift of two units has increased from about 22% of the time using a single individual as the sample, to nearly 94% of the time with a sample of 9. As you can see if we further increased the sample size we could decrease the possibility of a β error even more.

As a result we are able to specify, predetermine, or select values for both α and β by adjusting the sample size. As we take larger and larger samples we decrease the overlap to whatever degree we desire, thus permitting us to control the amount of error we may be willing to accept.

Of course we are not always able to obtain the sample sizes that might be desirable and often we must compromise somewhere. The important issue is that by certain statistical techniques we are able to plan, judge, and manipulate testing procedures and experiments in such a way as to provide a much better picture of the implication and interrelationship of sample size and of certain errors we may incur in our decisions, as well as of the interrelationships of the types of errors themselves.

One final note: If we use a large enough sample we can say that almost any difference is going to be unusual. In other words, with a sample of perhaps 1000 we would be able to say that a difference of perhaps 0.01 is unusual and the possibility of α and β errors would be very small—but if we really did not care about such a small difference, then the statistical significance is meaningless. So, often you may find things that may differ significantly, but only because of the numbers of samples taken. The net difference, however, may be absurd and useless.

REVIEW

Point estimate   ·   Interval estimate   ·   Confidence interval—*CI*   ·   < >   ·   Hypothesis test   ·   Beta (β) error   ·   Alpha (α) error   ·   Significance

PROBLEMS

Section 13.1

13.1 Compare the mean and the median of a sample as possible point estimators of the population mean. Do this for two kinds of population distributions—a normal distribution and a highly skewed distribution. Then discuss how point estimation is related to the bias of the estimator.

13.2 In what way is efficiency related to the selection of a point estimator?

Section 13.2

13.3 Estimate the number of days between germination and the first pickable cucumbers using the following sample of cucumber plants:

Date of germination	First fruit
May 1	June 17
May 4	June 18
May 8	June 21
May 5	June 16
May 12	June 28
May 18	July 3
May 11	June 25
May 9	June 26

What is the 95% confidence interval assuming $\sigma = 2$ days?

13.4 Light bulbs average 100 hours of life with a variance 36 hours (see Problem 10.7). If 25 light bulbs are tested, then 90% of the time the mean should fall between what limits?

13.5 Verify the confidence intervals given in Chapter 1 for egg lengths (see (page 16).

13.6 You are designing a key case for men and are interested in determining the keys most men carry. You sample 25 men and arrive at an average of 6.5 keys (with $s = 4.2$). What would you estimate is the true average number of keys carried by men? What is the 95% confidence interval for the true average? Is this confidence interval of any use in deciding how many keys to provide for (what does s tell you about the distribution)? (You should also note that s was used, therefore z is not actually the proper distribution to work with; you should use a t-value).

13.7 If the monthly weight loss at "weightwatchers" averages 4 pounds with a standard deviation of 2 pounds, what would you expect the 90% confidence interval for the yearly average loss to be? (*Hint:* treat this as a sample of 12 months to obtain the limits, then multiply by 12.)

13.8 A discouraged train rider decided to keep a record of the times between trains. He waited for 10 different trains and recorded an average waiting time of 27 minutes. (His standard deviation of 4.8 was not significantly different from the σ of 4 that the train company claims). What is the 99% confidence interval for the mean waiting time? How does this compare to the company's claim of an average of 19 minutes?

Section 13.3

13.9 How are the number of limits used related to the hypothesis? When would you only be concerned with one limit and when would you need two limits?

13.10 In Chapter 11 we discussed alpha and beta errors. How are these two errors related to the hypothesis testing based on the normal distribution?

13.11 If we believe textbooks cost an average of $10, with a standard deviation of $2, what limits should the average of the cost of your 6 textbooks fall within for you not to reject the hypothesis that textbooks cost is no different than an average of $10? Start by stating the hypothesis and the alternate hypothesis, then provide the test limits (use $\alpha = 0.10$).

13.12 A church official believes that the members of his church in Executive Township have an average income of $15,000 with $\sigma = \$4000$ (see Problem 10.10). What is his hypothesis? What is his alternate hypothesis? He will sample 15 families. What are his 95% test limits? If the average income of the sample of 15 families is $16,500, what is his decision?

13.13 The manufacturer of a stomach antacid (Antiseltzer) claims it relieves stomach discomfort in 5 minutes (with a standard deviation of 2 minutes). At a party 10 people volunteer to take part in an experiment and tell how long it takes for them to get relief. State the hypothesis and alternate hypothesis. Determine the test limit(s) for an α of 0.10. If the average time to get relief was $7\frac{1}{2}$ minutes do you accept or reject the hypothesis?

Section 13.4

13.14 When operating properly a machine that fills soda cans dispenses exactly 16 ounces with a standard deviation of 0.9 ounces. If a sample of 5 cans is taken and α is set at 0.05 what are the test limits? Now if the machine goes out of adjustment and begins to dispense 17 ounces, what is the probability that you will *not* detect this change with the sample of 5? (This is the beta error for an alternate of 17 ounces). What is the consequence to the company? If the machine begins to dispense 14.5 ounces, what is the chance you *will* detect this change? What is the consequence of not detecting this change?

13.15 Fire drills have shown a mean time for evacuation of 4 minutes 15 seconds, with standard deviation of 40 seconds (see Problem 10.6). A new procedure for evacuating is instituted and 6 fire drills are run. If α is set at 0.10, what are the limits for declaring a significant change in time to evacuate has occurred? Now if the average time to evacuate has actually decreased to $3\frac{1}{2}$ minutes, what is the probability of *not* detecting the change (the β error)? What is the consequence? If the new procedure actually increases the time to $4\frac{1}{2}$ minutes, what is the chance we will end up with a sample that shows a significant decrease?

13.16 Refer to Problem 13.11. If the average cost of textbooks actually increases to $12, what is the chance you will not reject the hypothesis?

13.17 Refer to Problem 13.12. The church has tried to reach the lower income group in the community. If the church membership actually averages $14,000, what is the chance of rejecting the original hypothesis? Find the chance of rejecting the hypothesis for true averages of 13,000; 12,000; 11,000; and 10,000. Plot these values. Are they β or $1 - \beta$?

13.18 If the stomach antacid (Problem 13.13) actually takes an average of 6 minutes to give relief, what is the chance that we will detect a significant difference? How about 7 minutes?

13.19 The manufacturer knows its antacid really averages relief in 10 minutes, but it wants to advertise with a shorter time period. The company would like to state that if 4 people take Antiseltzer the average time for relief would be in a range of "*ave.*" ± 2 minutes (where "*ave.*" is the value the company wants to use in its advertisement and the ± 2 minutes are the 2 sigma or 0.05 limits). The company would like to pick the smallest possible value of "*ave.*" However, it realizes that a value too small will be rejected too often. It decides to pick a value of "*ave.*" that will be rejected only 25% of the time. What value should that be?

PART V

Statistical Tests

14

Univariate Tests: General Discussion and Tests of General Assumptions

Before embarking upon the detailed study of the many statistical tests, you might find it extremely advantageous to review Chapter 2—"Questions About the Data." Much of the work of this and subsequent chapters will rely heavily upon your understanding and awareness of the questions and descriptions presented there.

This chapter will begin the work of what most experimenters regard as the crux of statistics—the performance of statistical tests. We have already carried through a few simple statistical tests. We tested to see if a correlation coefficient was large enough for use to consider some data as truly correlated. We tested to determine if points fell outside of control limits and we declared such occurrences as significant. Now we will begin a systematic approach to evaluating and comparing certain characteristics of data to determine whether there are peculiarities or differences between our sample data and some hypothetical populations or population characteristics. We will attempt to make these decisions based upon particular statistical procedures.

In this and the next two chapters we will concern ourselves strictly with univariate data, data that describes a single characteristic and of which there is only one set. The sets of data we looked at in Chapters 3 and 4 were of such a type. There is much we can do in the way of testing a single set of data. We may possibly be interested in the shape, spread, or central value of the distribution that the data came from, in the randomness of the data, or whether a time-dependency is present. We may be concerned about the stability, independence, normality, or the effect of an extreme value. In all of these cases we will need to recognize the assumptions that we make and the effect that these assumptions will have on our analysis and tests.

In Chapters 17 and 18 we will look at pairs of samples to make determinations about their relationships to each other or to decide if they have certain common properties. Chapter 19 will explore those cases where several sets of data are to be compared. The final chapter will expand our interpretation and delve more deeply into the concepts of correlation and regression.

14.1 Statistics and Significance

First we should review what is meant by a statistic and a statistical test. *A statistic is a number derived or calculated from a sample.*

We have encountered a number of different quantities that were computed from a sample. We have found means, medians, and standard deviations, to name a few. These were understandable from just a descriptive nature. Now, however, we will arrive at new quantities (a few of which already have shown up briefly in previous chapters). We will compute values of z or t or χ^2 in *particular ways* and we will call these specific values which describe our sample "statistics" which also describe our sample. We will compare the "statistic" that we have just computed to a table of possible values for the statistic. We may then determine if the statistic we obtained was unusual in any way—unusually large or small. If it is *too large* or *too small* we may then declare that the *characteristic* that we are considering is *also* large, small, changed, or different in the same way that the sample behaves.

If you look back at Chapter 8 you will recall that we discussed how to compare a sample value against some possible population values to determine if our sample could reasonably be expected to have come from a particular population. We are now to do the same, except that the population we will be considering may seem at times to be just a table of values in the back of the book.

Generally, we will regard a *statistical test* as the comparison of a statistic to some limit or limits that are obtained from a table. The limits are associated with the degree of unusualness of the magnitude or size of the statistic. We will normally refer to two degrees of unusualness—probabilities of 0.05 (5% chance) and 0.01 (1% chance). Thus for a 0.05 probability level we mean that there is a 1 in 20 chance that the calculated statistic could randomly have been so large (or small). The implication is that there is a reason for such a large value of the statistic and we declare that the test or the difference is significant. Similarly, a 0.01 level implies that there is a smaller than 1 in 100 chance of obtaining a value so large just randomly and *without cause*. This we consider as highly significant.

14.2 One-sided and Two-sided Tests

With many of the tests that we perform we must initially make another determination. We must declare whether we are interested only if an increase has taken place, only a decrease, or both. The first time this question was posed was in Chapter 2, but no explanation was offered at that time for why we need to state which we are considering. This becomes of paramount concern with certain tests. (We also mentioned it in Chapter 11.)

For *each* type of change we are interested in detecting—either an increase or a decrease—*we need to establish a limit*. Corresponding to *each* limit is an error. The total error, and hence the significance level, is the sum of these errors. If you just have one limit, then the total error is only the error corresponding to that one limit. If you have two limits then the total error is the sum of the two errors. When we dealt with hypothesis and confidence intervals in the last chapter we worked with *two* limits and we essentially had a two-sided test. Our beyond the limits (refer back to Figure 13.8, page 335). In Figure 14.1 I have again shown the relationship of the tails and the errors associated with the tails

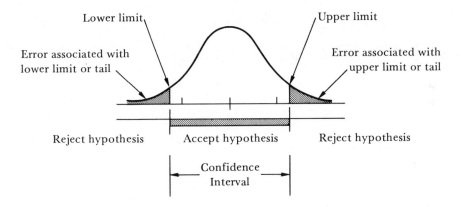

Figure 14.1 Interpreting the limits and errors for a distribution.

to the decisions. Observe that if we are interested in *only one* direction or change, *only when an increase takes place*, then we only use *one* limit. Figure 14.2a shows such a case where *the* limit is the limit for a probability level of 0.05 (just one limit). The *accept* region goes all the way to the left since we are not concerned with a decrease. (For example, if my gas mileage decreases when I try a new brand of gasoline, I will *not* change my brand. I will only change brands if there is an increase.) Note that the *overall error* for the test is at the 0.05 or 5% level. We call this a *one-tailed* or *one-sided* test since the error and rejection regions are in only *one tail* of the distribution of possible statistic values.

Next go to the case when we are interested in both possibilities—increases and decreases. Now we have two limits and now *that same 5% error must be shared between the two tails*. The limits are therefore affected since each limit is to be a $2\frac{1}{2}\%$ limit *not* a 5% limit. If we were to use the same limits as before we would have two 5% limits yielding a total of 10% chance of error (see Figure 14.2b).

Finally, the third case illustrated is that of a lower limit only. Again all the error is associated with one tail and the one limit.

One caution however: Distributions of statistics (which is what we are now

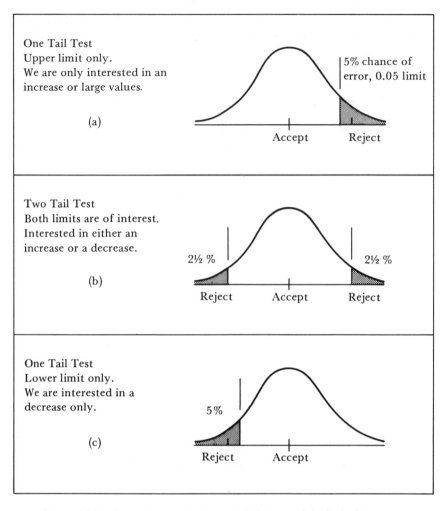

Figure 14.2 Comparing one- and two-tailed tests and the limits for accepting or rejecting a hypothesis. All cases are totals of 5% chance of error. (a) and (c) are one-tail tests, (b) is a two-tailed test.

working with) need not be symmetric. Thus, an upper limit and a lower limit are not necessarily the same *distance* from the center. Just the *areas* are to be the same.

14.3 Parametric and Nonparametric Tests

Another pair of terms that need to be defined at this time are *parametric* and *nonparametric* tests.

When we talked about theoretical distributions, namely the binomial and normal distributions, it was mentioned that by knowing one or two particular quantities (p for the binomial, μ and σ for the normal) we could fully describe the distributions.

A large portion of the major statistical tests that are employed require a major assumption *that the population behaves according to one type of theoretical model or type of distribution.* Thus, we *may* be required to *assume that the population is a normal distribution.* Furthermore, we frequently must use the sample values to estimate the population parameters. We would use the sample mean to estimate population mean, and so forth. These are educated guesses, but still they are guesses and when the guesses are incorrect, particularly guesses about the nature of the distribution type, we may make large errors.

To avoid the need for such assumptions certain types of "distribution-free" or nonparametric techniques have been developed. They often require no specific knowledge of the type of distribution or of the parameters of the distribution that the data comes from. In addition, they may be considerably easier to calculate. However, they may be far less efficient. In most cases, a parametric test will be a more powerful test, one that needs fewer samples to do the same job. But if the assumptions are incorrect, the parametric test may give erroneous results. Hence, as always, a trade-off must occur. Where possible I will treat both types of tests in an interchangeable fashion, identifying and emphasizing the assumptions and clarifying the types of test we are dealing with. In some instances, however, there may not even exist a parametric test. When we deal with data that is merely ordered or categorical certain aspects may be tested only by nonparametric tests.

14.4 Univariate Tests

The following is an outline of the statistical tests for univariate data that we will cover in this and the next two chapters. It is by no means an exhaustive survey, but the major tests generally encountered by beginning students are included.

Chapter 14: Univariate tests: tests of general assumptions
  A. A runs test for randomness
  B. A test for extreme values
  C. Some other assumptions
     1. Are the values *independent*?
     2. Is the data *stable*, or does it show *time-dependency*?
     3. Is the data *normally distributed*?

Chapter 15: Tests on categorical data
  A. Tests to compare data to a theoretical distribution
     1. The general χ^2 test for *any* theoretical distribution
     2. The χ^2 test as a test for *normality*
     3. The Kolmogorov–Smirnov one-sample test
  B. Tests for proportions (binomial distributions)

Chapter 16: Tests involving interval or ratio measurements with an assumption of normality
  A. Comparing the *spread* of a sample to a theoretical spread—the χ^2/df test
  B. Comparing the *mean* of a sample to a theoretical mean
     1. If the population variance is *known*—z-test
     2. If the population variance is *unknown*—t-test

14.5 A Runs Test for Randomness

The question of randomness has to do with the sequencing of the values and whether or not they seem to follow some type of pattern. Data that was obtained in a purely random fashion will show no particular pattern. One way we will attempt to detect unusual patterns will be by analyzing a kind of run in the data. We will look at a test which considers the number of *runs above and below the median*.★

In Figure 14.3 we have a set of points in a time sequence. There are 19 points, so the 10th from the bottom is the median. We have 9 values above and 9 values below this median point. We designate these two numbers as $N_1 =$ *number* of points above the median and $N_2 =$ *number* of points below the median. Obviously, N_1 and N_2 are equal.

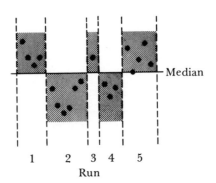

Figure 14.3 Finding the runs above and below the median. The longest run is Run 2 with six points (a length of six). There are a total of five runs.

A run constitutes a set of successive values on one side of the median line. To simplify matters we disregard the median point. Thus, the first 4 points constitute the first run, the fifth point is across the median line and therefore begins the second run. After 5 more points (a total of 6) there is another jump

---

★ The median has been selected because of the ease in working with it and the table associated with these runs. Splitting of the sample into any two categories is possible, and tables are available to work with many other cases. However, this discussion will be limited to equal points above and below a line.

across the median line and a new run begins. This third run is only one point long. Continuing on we find a *total of 5 runs*. Our question now is whether this is an unusual value for the number of runs.

We have three possible questions:

1. We can ask if there are *too few* runs. Generally when we see only a few runs we are probably watching a process that is cycling, shifting, or grouping. If, for example, we suspect a particular trend we would then consider a one-sided or one-tailed test for "small" number of runs. (Again, see Chapter 7 under Section 7.5, Nonrandom Variability.)

$N_1 = N_2$	Significantly small values of u. If the number of runs is less than or equal to the value below, it is significant at the level indicated			Significantly large values of u. If the number of runs is greater than or equal to the value below, it is significant at the level indicated		
	One-tailed test			One-tailed test		
	0.01	0.025	0.05	0.05	0.025	0.01
	Two-tailed test			Two-tailed test		
	0.02	0.05	0.10	0.10	0.05	0.02
9	4	5	6	14	15	16

(a)

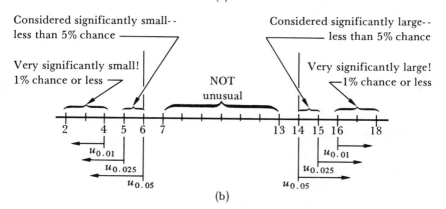

(b)

Figure 14.4 Finding the significant values of number of runs above and below the median. (a) A portion of the Appendix Table A.12 showing the critical values for nine points above and nine points below the median. (b) Scale showing the relationship of the significant values of *u* for nine points above and nine points below the median. (Note that two-tailed tests require adding together the upper *and* lower significance levels from the one-tailed tests.)

2. We may believe that there are *too many* runs. Abnormally large numbers of runs are not encountered very often and may indicate that we are selecting our samples from two different sources. For example, differences between morning and afternoon may occur on a regular basis and if we sample once in the morning and once in the afternoon, a bouncing effect may appear. This then is a test that may act as a signal to look at the way the data was obtained.
3. Finally, we may look at *both* possibilities of too few and too many runs at the same time. Here we would perform a two-tailed test.

Returning to our example, we designate the number of runs by the symbol u (this now is our statistic). In this case $u = 5$. This value of u is to be compared to those values that are listed as unusual in Appendix Table A.12. Figure 14.4a shows the row for $N_1 = N_2 = 9$ from the table. In Figure 14.4b I have located these numbers on a scale and have emphasized the meaning of the various intervals. Thus, values of u that are from 7 to 13 are to be considered typical *random* amounts of number of runs. If you are considering a one-tailed test, then 5 runs (for the few runs case) and 14 (for the too many runs case) are unusual numbers of runs, since they occur less than 5% of the time (or a probability of about 0.05 or a 1 in 20 chance) and are labeled as outside the $u_{0.05}$ limit. Hence they are termed "significant" and indicate a cause for investigation. Runs of 4 or less and 16 or more are considered "very highly significant" since there is less than a 1% chance (0.01 probability or a 1 in 100 chance) that a value should randomly fall outside this limit ($u_{0.01}$). The values in-between (the 5 and the 15) fall between the "significant" and the "highly significant" values. To sum up, as we progress from the center of this scale toward one *or* the other end, we approach critical limits that act as signals to tell us if there is an unusual number of runs.

If for our example we are *only* looking at small numbers of runs then $u = 5$ is beyond the $u_{0.05}$ limit, of the "significant" limit, but is not beyond the $u_{0.01}$, the "very significant" limit. The degree of unusualness is *between* a 1 in 20 and a 1 in 100 chance (or about 0.025). Note that if we thought that the data might have few *or* many runs, then this would be a two-tailed test and the degree of unusualness would only be 0.05.

EXAMPLE 2.1 (Cont.)

In Chapter 7 we identified the fact that the tickets were issued on a number of different days. We averaged the samples of tickets for each day and we plotted a chart, called a control chart, for the samples. We also noted that two of the days were unusual and we then disregarded those two unusually low days. We will now examine the control chart of the 26 samples that are left to see if any pattern might exist. In Figure 14.5 I have reproduced the points of the control chart (omitting the limits).

STEP 1. First we must find the median of these points. Since there are 26 values, the median is between the 13th and the 14th values. The 12th and

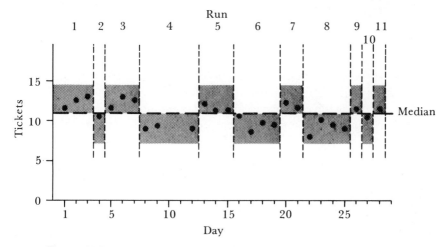

Figure 14.5 The runs above and below the median for the averages of the number of tickets issued on each day (Example 2.1). The points are reproduced from the control chart of Chapter 7 except for the two unusual low values and the control limits.

13th values are 11.00 and the 14th and 15th are 11.25, thus, the median is between 11 and 11.25, or 11.12. I have drawn this line on the chart in Figure 14.5.

STEP 2. Calculate u, the number of runs above and below 11.12. The first 3 points are above the median and make up runs 1, the fourth is below the median (run 2), the 5th through 7th are above it (run 3), and so forth. The 26 points divide up into runs of the following lengths: 3, 1, 3, 3, 3, 4, 2, 4, 1,

$N_1 = N_2$	Significantly small values			Significantly large values		
	Two-tailed test			Two-tailed test		
	0.02	0.05	0.10	0.10	0.05	0.02
13	7	8	9	19	20	21

Figure 14.6 Finding the critical values of number of runs above and below the median for $N_1 = N_2 = 13$, for two-tailed test (Example 2.1).

1, 1. They are set off at the top of Figure 14.5. There are a total of 11 runs, or $u = 11$.

STEP 3. Compare this to the critical limits. For $N_1 = N_2 = 13$, the two-tailed critical $u_{0.05}$ limits from Table A.12 are 8 and 20 (see Figure 14.6). Thus, any value from 9 up to and including 19 are not considered unusual. Since 11 falls in this category, we state that the number of runs is *not* significant and the sample appears to be random (actually it is not even significant at the 0.10 level since the 0.10 limits are 9 and 19).

After following these examples you should be aware of the relative ease of performing a runs test. This is our first example of a nonparametric test. There are other types of runs tests that may be performed, which may tend to detect different kinds of unusualness in the data. However, this test of runs above and below the median is the one most commonly used.

14.6 A Test for Extreme Values

Sometimes one value of a set of data appears to be exceptionally large or small. Frequently this may be the result of a gross error in measurement, such as reading your weight on a scale as 132 instead of 137 because you misinterpreted the various lines on the scale itself. Body temperature may be read as 102.3 degrees instead of 102.6 degrees (since every line stands for 0.2 and not 0.1 degrees as might be expected). An instrument or item to be tested may be dropped, damaged, or bumped and cause an erroneous reading. A distraction during a test may result in a poor performance. In these and many other ways we find that a sample may easily contain values that would be classed as blunders, errors, or extremes. Such an extreme could be particularly serious in making a decision based on a small sample (less than about 20). We will look at one test that provides an indication of whether an end value (the largest or smallest value) of a distribution is unusual.

In this test the data must be placed in an ordered arrangement—in an increasing or decreasing sequence. We then label the values of X_1, X_2, X_3, ... all the way to X_n for the last or nth value in the sample (of size n). The range of this set of data will be $X_n - X_1$, the difference between the extremes. Now consider two possibilities:

1. If there are *no unusual values*, then the difference between the two smallest values, $X_2 - X_1$, will be small compared to the total range. Figure 14.7b shows how these differences will appear on an evenly distributed sample (the total range and the $X_2 - X_1$). These two differences are very different in size.
2. If an *unusual (extreme) value is present*, then the total range will consist mostly of the distance between that single value and the bulk of the distribution. In this case the two ranges (the total $X_n - X_1$ and $X_2 - X_1$) will begin to approach each other (see Figure 14.7a).

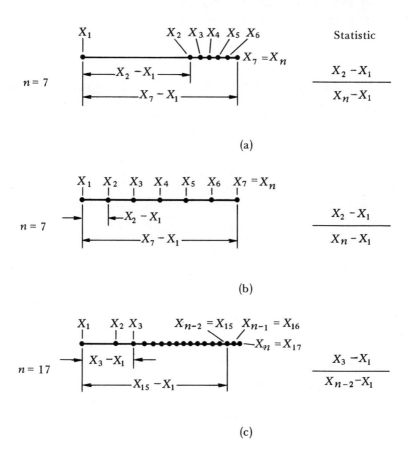

Figure 14.7 Considering whether one of the individuals is an extreme. Essentially, two ranges are being compared: (a) The n value is very small. The distance from the extreme under consideration (X_1) to its nearest neighbor (X_2) is compared to the total range ($X_n - X_1$). The closer this is to 1.00, the more unusual the value. (b) n is the same as in (a) but this time there seems to be no unusual value. Observe how small $X_2 - X_1$ is in comparison to $X_7 - X_1$. (c) The sample size is somewhat larger ($n = 17$). Instead of dealing with the nearest neighbor and the total range, slight changes are made in the distances to compensate for the possibility of several errors.

A table of critical values of the ratio of certain ranges is provided in Appendix Table A.14. You will find some cases when different points are used. This generally occurs when n begins to get large. Figure 14.7c indicates in part why this is true. You should also recognize that the same table can be used for either extreme; if you are comparing values on the large end you just use the second set of equations.

EXAMPLE 2.2 (Cont.)

There are a total of 20 trips recorded in Table 2.2. If they are placed in increasing order they are 13.6, 14.1, 14.3, 14.5, . . ., 16.4, 16.4, 16.5, 16.5, 16.7. Is the 13.6 an extreme?

Consulting Appendix Table A.14 we find that for $n = 20$ we are to compute the statistic called r_{22} which is

$$r_{22} = \frac{X_3 - X_1}{X_{n-2} - X_1}. \quad (14.1)$$

Furthermore, there is less than a 10% chance that r_{22} would be greater than 0.401, a 5% chance of greater than 0.450, and a 1% chance of greater than 0.535. For our data: $X_1 = 13.6$, $X_3 = 14.3$, $X_n = X_{20} = 16.7$, so

$$X_{n-2} = X_{20-2} = X_{18} = 16.5.$$

Therefore

$$r_{22} = \frac{X_3 - X_1}{X_{18} - X_1} = \frac{14.3 - 13.6}{16.5 - 13.6} = \frac{0.7}{2.9} = 0.241.$$

This is smaller than the critical value for the chosen levels, so we declare it is *not* unusual.

EXAMPLE 2.4 (Cont.)

Consider the diameters of the twenty cups listed in Table 2.4. Again $n = 20$ and we can use Equation (14.1). For our data $X_1 = 2.42$, $X_2 = 2.47$, $X_3 = 2.47$, $X_{18} = 2.53$, $X_{19} = 2.53$, $X_{20} = 2.54$. Then

$$r_{22} = \frac{X_3 - X_1}{X_{18} - X_1} = \frac{2.47 - 2.42}{2.53 - 2.42} = \frac{0.05}{0.11} = 0.455.$$

This greater than the 5% critical limit but less than the 1% critical limit. It could be considered a significantly unusual value—perhaps the result of an error in measurement or a damaged cup.

EXAMPLE 14.1

The amount of tread on the original tires on my car were measured after 30,000 miles of driving. The depth of tread on the tires was

0.095, 0.102, 0.098, 0.215, 0.108 inches.

Is the 0.215 an extreme value? If we order these values with the extreme as X_5 we have

X_1	X_2	X_3	X_4	X_5
0.095	0.098	0.102	0.108	0.215

For $n = 5$ the statistic to be calculated (see Table A.14) is

$$r_{10} = \frac{X_2 - X_1}{X_n - X_1}$$

or

$$r_{1.0} = \frac{X_n - X_{n-1}}{X_n - X_1} \qquad (14.2)$$

where $X_n - X_{n-1}$ is the difference between our extreme and its nearest value. For our problem this is

$$r_{10} = \frac{X_5 - X_4}{X_5 - X_1} = \frac{0.215 - 0.108}{0.215 - 0.095} = \frac{0.107}{0.120} = 0.892.$$

Appendix Table A.14 lists the 1% critical limit of 0.780. Since 0.892 exceeds 0.780 we declare the 0.215 is a very significantly extreme value. It happens to correspond to my spare tire, which essentially was never used.

14.7 Some Other Assumptions

Several other assumptions may often be very critical for the proper functioning of certain statistical tests. Of the many possible assumptions we will briefly discuss three—independence, stability, and normality.

Are the Values Independent?

In the univariate case, the testing of independence is a difficult, if not impossible, task. Dependence may be detected or observed as a form of nonrandomness. However, we will find that independence will require a second quantity to measure or relate our values to. One possibility with univariate data could be that the magnitude of one value would depend upon a prior value. To observe this we might have to set up a dummy or artificial second measurement. Perhaps we might attempt to find a relationship between each value and its predecessor and perform a correlation. Generally, however, we just accept the assumption of independence and the possibility of error associated with such an assumption.

Stability or Time-Dependency (refer to Chapters 6 and 7)

With a small sample it is nearly impossible to make a determination of whether a time-dependency exists. One technique would be to plot a graph and observe if there is a significant correlation between the values and the time, or if a regression line can be fit to the data. If the sample is sufficiently large then the data can be separated into several smaller samples and a control chart can be plotted. Finally, in either case, we can look for trends or cycles or shifts (as described in Section 7.5). These may also be treated as cases of nonrandomness (see Section 14.5).

Is the Data Normally Distributed?

This is an extremely important assumption. The decision to use many of the parametric tests depends almost entirely upon whether there is reason to accept the data as coming from a normal distribution. If the source is not normally distributed then certain tests will give erroneous results. We have already discussed this concept (in the beginning of this chapter) and we have also considered one technique for deciding if a distribution appears to be normally distributed. In Chapter 3 we observed how plotting the cumulative frequencies on normal probability graph paper could serve as an inexact but rough estimation method for quick evaluation of a distribution. We will soon encounter a more valuable technique, the chi-square test. This will be done in Chapter 15 (see page 366).

REVIEW

Statistic · Test · Limit · One-sided test · One-tailed test · Two-sided Test · Two-tailed test · Parameter · Parametric test · Nonparametric test · Stability · Independence · Normal distribution · Randomness and runs test · Extreme value · Error · Univariate

PROBLEMS

Section 14.1

14.1 What other "statistics" have we already discussed that could be compared to a theoretical distribution to determine significance?

14.2 If something was declared unusual at the 0.001 level, would it be more or less significant than the levels discussed? How about the 0.10 level?

Section 14.2

14.3 For each of the following provide reasons or a hypothesis that would involve a one-sided test, and separate reasons for a two-sided test:
  a. Comparing two types of light bulbs.
  b. A traffic light is to be installed at an intersection.
  c. A researcher is concerned with the effect of LSD on hormones.

14.4 Refer to Problem 2.1. In what ways would these problem statements lead to one-sided statistical tests and how might they be tested by two-sided tests?

14.5 Examine the following theoretical distributions. Which are symmetrical? For those distributions that are not symmetrical describe the effect of nonsymmetry on the limits.
  a. z b. χ^2
  c. t d. F
  e. r f. χ^2/df

14.6 For each of the distributions in Problem 14.5 compare the limit for a one-sided test to the limits for a two-sided test using the given significance levels (and degrees of freedom where necessary).
   a. z for 10%
   b. χ^2 for 2% $(df = 10)$
   c. t for 1% $(df = 20)$
   d. F for 2% $(f_1 = 10, f_2 = 20)$
   e. r for 5% $(df = 10)$
   f. χ^2/df for 5% $(df = 200)$

Section 14.5 (when not otherwise indicated use the 5% significance level)

14.7 The number of home runs hit by the National League Home Run Champions from 1930 to 1969 were as follows:

   56, 31, 38, 28, 35, 34, 33, 31, 36, 28, 43, 34, 30, 29, 33, 28, 23, 51, 40, 54, 47, 42, 37, 47, 49, 51, 43, 44, 47, 46, 41, 46, 49, 44, 47, 52, 44, 39, 36, 45

   Is the hypothesis that home runs are easier to hit now than in previous years demonstrated by the data? (Use a runs test.)

14.8 Refer back to the control chart for Example 2.3 (Figure 7.7, on page 202). We considered that the average bread weights had a trend-like nature. Does a runs test confirm this suspicion?

14.9 In Problem 7.3 the absorbency of diapers was recorded. The first row is for one type of diaper. Does the absorbency change randomly as the diaper is washed (do only one diaper)?

14.10 Problem 4.1 listed the prices of homes that were sold. If they were listed in the order of sales, has there been an upward trend in the price? (Test this by a runs test.)

Section 14.6

14.11 Aspirin tablets were analyzed in Problem 7.4 and the average amount of aspirin in tablets from 30 bottles was recorded. Is there any unusual variation in the bottles? (Use the extreme values test.)

14.12 Six library shelves were picked at random and the number of books were counted. We found:

   22, 17, 17, 24, 23, 35 books.

   Is the 35 an unusual number of books (test at the 10% level)? What would be an explanation if it is unusual?

14.13 Problem 3.6 listed the number of customers each day of the week. Would you consider certain days as having unusual numbers of customers?

14.14 Problem 4.1 listed prices of homes sold in a development. Are any of the homes unusually high or low priced?

14.15 Refer to Problem 4.2. Twelve women attended "weightwatchers." Did any women have an abnormal amount of weight loss (compared to the rest of the group)?

14.16 In Problem 4.18 donations to a charity were listed. Would you consider any donations as being abnormally large?

15

Univariate Tests of a Categorical Nature

In this chapter we will discuss several tests used to make decisions about data that may be recorded merely in categories. In some cases these same tests may be used with other types of data as well. We will first look at the comparison of an actual distribution of data to a theoretical distribution, and then we will consider tests about proportions.

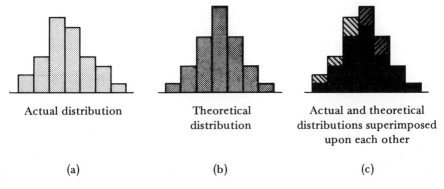

Actual distribution Theoretical distribution Actual and theoretical distributions superimposed upon each other

(a) (b) (c)

Figure 15.1 Comparing an actual and a theoretical distribution. We compare them by superimposing the actual distribution (a) over the theoretical distribution (b) to see the differences (c). In (c) the differences show up as shading. The lines drawn as \ \ \ are for greater frequencies of actuals and the lines drawn as / / / are for less frequencies of actuals.

15.1 Tests to Compare Data to a Theoretical Distribution

The principal way that we can describe categorical data is either by breaking it up into percentages belonging to each category or by listing the actual frequencies of occurrence for each category. We may then draw a histogram or a cumulative frequency diagram. (Of course, a cumulative frequency diagram is only useful if the data is at least ordered, otherwise there is no meaning in successively grouping the various categories.) We will first look at histograms that can deal with *any* type of data. In all cases we will compare the distribution of our data to a theoretical distribution. Thus, in Figure 15.1a we have a distribution or histogram for an actual set of data. Adjacent to it, Figure 15.1b, we have the histogram for what we will hypothesize or believe to be the actual or true population distribution. They are *not* identical. The important question is, "Are they so dissimilar, or so unlike, that there is *a very small chance* that the sample came from the hypothetical distribution?" If we superimpose one of the distributions on top of the other distribution, as in Figure 15.1c, are the differences, the shaded areas, just random differences? Or are they unusually large? We will attempt to gauge these differences in the following tests.

The General χ^2 Test (Chi-square Test)

The general chi-square test looks at the shaded areas in Figure 15.1c and provides us with a way of measuring them. The value we compute is called chi-square (symbolized χ^2) and is found thus

$$\chi^2 = \sum \frac{(\text{Actual frequency} - \text{Theoretical frequency})^2}{\text{Theoretical frequency}}. \quad (15.1)$$

We will examine how to apply this formula and the concept it describes through several examples.

An insurance company is interested in knowing if the distribution of cars used by students on a particular college campus is different from the general distribution of automobiles.

(1) *We take a sample of students* and record the manufacturer of their cars. The results of one such sample is given in Table 15.1.

Table 15.1 Manufacturer of car operated by a group of students (number of students = n = 300)

	Sample
Ford Motor Company	93
General Motors	163
American Motors	4
Chrysler Corp.	40

(2) *We need a theoretical distribution.* One such possibility is the overall passenger car production of all United States manufacturers for a given year. Production for the year 1968 is given in Table 15.2 (percentages are also given).

Table 15.2 Passenger car production for 1968. This to serve as a theoretical population to compare the actual sample of student's cars

Manufacturer	Actual number	Percentage of sales	Proportion
Ford Motor Company	2,397,000	27.1	0.271
General Motors	4,592,000	52.0	0.520
American Motors	269,000	3.0	0.030
Chrysler Corp	1,586,000	17.9	0.179

(3) *We must determine what we would expect the theoretical numbers of each make of car would be for the sample of 300 students.* Since we know what the proportion of each make should be, we merely multiply the proportion by 300 for each make. We then obtain our expected or theoretical frequencies of

$$
\begin{array}{lll}
\text{Ford} & (0.271)(300) & = 81.3 \\
\text{G.M.} & (0.520)(300) & = 156.0 \\
\text{American} & (0.030)(300) & = 9.0 \\
\text{Chrysler} & (0.179)(300) & = 53.7
\end{array}
$$

(4) First we will *compute the differences* between the actual and theoretical frequencies. We then obtain:

Frequency Actual − Frequency Theoretical

Ford Motor Company	93 − 81.3	=	11.7
General Motors	163 − 156.0	=	7.0
American Motors	4 − 9.0	=	−5.0
Chrysler Corp.	40 − 53.7	=	−13.7

(5) We *square each of these differences*. This will get rid of the minus signs and will emphasize large differences.

(Frequency Actual − Frequency Theoretical)^2

Ford Motor Company	$(11.7)^2$	=	136.89
General Motors	$(7.0)^2$	=	49.00
American Motors	$(-5.0)^2$	=	25.00
Chrysler Corp.	$(-13.7)^2$	=	187.69

(6) We must realize another important characteristic of the data. A difference of 5.0 out of an expected 9.0 (for American Motors) is more than 50% of the expected value. It is a very large *relative* difference. The 7.0 out of an expected 156 (for G.M.) is less than a 5% difference. So proportionately this second difference is a much smaller relative difference. Similarly, we desire to work with the relative size of the squared differences and therefore the

squared quantities we have just computed are divided by the *expected frequencies*. Thus, we compute for each difference

$$\frac{(\text{Actual frequency} - \text{Theoretical frequency})^2}{\text{Theoretical frequency}}.$$

We find

$$\text{Ford} \qquad \frac{136.89}{81.3} = 1.68$$

$$\text{G.M.} \qquad \frac{49.00}{156.0} = 0.31$$

$$\text{American} \qquad \frac{25.00}{9.0} = 2.78$$

$$\text{Chrysler} \qquad \frac{187.69}{53.7} = 3.49$$

(7) We sum up these values to obtain the χ^2-value.

$$\chi^2 = 1.68 + 0.31 + 2.78 + 3.49 = 8.26.$$

(8) Finally, we compare this to the critical value of the χ^2 in Table A.3 (under one-sided test). The *degrees of freedom* are equal to *categories minus 1*.★ Since there were 4 categories, the test limit would be χ^2 with $4 - 1 = 3$ degrees of freedom. The 95% value (that critical value which has a 5% chance of error) is 7.81. Our quantity exceeds this, and we therefore state that there is a significant difference between the sample distribution and the theoretical population of cars. Apparently, the students have different buying preferences than the overall population (or some other reason may be proposed).

EXAMPLE 15.2

It is hypothesized by a Chamber of Commerce of a resort area that the weather in its area is clear 60% of the time, partly cloudy 20% of the time, cloudy 15% of the time, and rainy 5% of the time. A potential tourist was skeptical and followed the daily report from the weather bureau for a full year. He obtained the values in Table 15.3.

To perform the comparison of these two distributions and to calculate the χ^2-value we will use a single table to carry out a systematic solution. Table 15.4 contains all of the calculations. The actual values (From Table 15.3) are placed in Column 1 (labeled F_a for *frequency actual*). In Column 2 we have the

---

★ The degrees of freedom has to do with how many of the values can be independently varied. If you look back at the frequencies and recognize that there is a known total, then only 3 of the values can be varied. Once you know that the frequencies of the first three are 93, 163, and 4, you can automatically find that the remainder is 40. Thus, only three categories have the freedom to vary in size.

Table 15.3 Weather for 1 year as reported by the Weather Bureau

Weather	Number of days
Clear	211
Partly cloudy	60
Cloudy	52
Rain	42

corresponding theoretical frequencies, F_t, of rainy, clear, and so forth, days. These are obtained by multiplying the relative times that the particular type of weather is *supposed* to occur by the number of days being compared (365).

Table 15.4 Computations of χ^2 value to compare weather distribution

	1	2	3	4	5
Category of weather	Actual frequency (F_a)	Theoretical frequency (F_t) $(365 \cdot \%)$	$F_a - F_t$	$(F_a - F_t)^2$	$(F_a - F_t)^2/F_t$
Clear	211	219.0	−8	64	0.3
Partly cloudy	60	73.0	−13	169	2.3
Cloudy	52	54.75	−2.75	7.6	0.1
Rain	42	18.25	23.75	569	30.9
Sum	365	365	0		33.6

$\chi^2 = 33.6$. Degrees of freedom = categories − 1 = 4 − 1 = 3.
$\chi^2_{0.95} = 7.81$. $\chi^2_{0.99} = 11.3$.

Column 3 has the differences of the frequencies. These are then squared in Column 4. We then divide each of these squares by the respective theoretical frequency (from Column 2). The quotients are entered in Column 5. The sum of the values in Column 5 is the χ^2-value. Finally, this is compared to the critical value of the χ^2 in Table A.3 and we find the χ^2 for this problem to be very highly significant. It is greater than the χ^2 for a 1% chance of error. Our potential tourist apparently had good reason to be skeptical.

One note should be made. A requirement that is generally assumed to be necessary for the χ^2 test is that all the expected frequencies should be *at least* 5. Thus, in Example 15.1 we see that there is a need for the sample to be at least 169 to obtain a frequency of 5 for American Motors. We have exceeded this requirement, but you should be conscious of it when you consider working with this test.

The χ^2 as a Test for Normality

We have already looked at one way that we can attempt to determine if a distribution appears to be normal, namely using probability paper. However, we were unable to assign a chance of error to any analysis using that technique and therefore the interpretation was very subjective (all we could say was that it looks pretty close to a straight line, or it is far from a straight line). Now we will use the χ^2-test to examine a set of data that we believe to be normally distributed.

The data serves as our *actual* distribution. Next we need a theoretical distribution. If we know the mean and standard deviation, we can fully describe a single complete theoretical normal distribution. Furthermore, we can find limits between which different theoretical percentages of the data should lie. We merely need a normal distribution table and we will utilize the mean and standard deviation from the sample.

Having found actual and theoretical frequencies, we can apply the χ^2-test in a manner similar to the last section. The primary difference involves the number of degrees of freedom. Since we are estimating two things, *both* the mean and the standard deviation, we must decrease the number of degrees of freedom by 1. Thus, our test limit will be a χ^2-value with $k - 3$ degrees of freedom (where $k = $ number of categories).

EXAMPLE 2.1 (Cont.)

In Chapter 3 we drew a histogram for the number of tickets issued at a radar trap (Figure 3.7). I have reproduced it again here (Figure 15.2). By previous calculations we found the *mean* to be *10.11* and the *standard deviation* to be *3.41*. We will now *assume that the theoretical population has the same mean and standard deviation as this sample.*

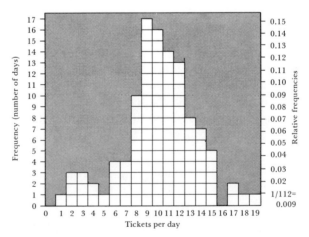

Figure 15.2 Distribution of tickets issued, Example 2.1. This is the actual distribution as shown in Figure 3.7.

Next we must decide how many categories we want to work with. Twenty categories is a fairly large number of categories. Generally, we do not want less than about *five* individuals in *any* of the categories. We therefore choose to use grouped distributions. To simplify our calculations we also select the intervals so that they contain *equal* theoretical percentages.

I chose 8 categories, so that each category selected should contain $\frac{100}{8}\%$ or 12.5% of the total distribution. Figure 15.3 shows how the normal distribution is broken up into sections. The limits have not been indicated yet, except as small letters. We now need to specify just what these limits are

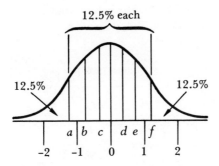

Figure 15.3 A normal distribution split approximately into eight equal sections. Each section is 12.5% of the total area. The z-values are labeled with letters a, b, c, d, e, and f.

Since the normal distribution is symmetrical, *d* has the same magnitude as *c*, *e* has the same magnitude as *b*, and *f* the same magnitude as *a*. So we only need to look up 3 limits in the normal table.

Point *d* is the z-value for which 12.5% (or 0.1250) of the area, lie between it and the mean. This is a *z of about 0.32*.

Point *e* is the z-value for which $0.1250 + 0.1250 = 0.250$ of the area is between it and the mean, or $z = 0.67$.

Point *f* is the z-value for which 0.3750 of the area is between it and the mean, or $z = 1.15$.

Similarly, the limits *a*, *b*, and *c*, have z-values of -1.15, -0.67, and -0.32, respectively.

Since all of these z-values correspond to numbers of standard deviations, to compute the theoretical limits for our distribution we must multiply each limit by 3.41 (the standard deviation) and add the limits to our mean of 10.11 (thus we calculate the limits as $\bar{X} + z\sigma$). The new distribution would be as in Figure 15.4, where the 7 limits are calculated as follows:

1. -1.15 standard deviations or $10.11 - 3.93 = 6.18$.
2. -0.67 standard deviations or $10.11 - 2.29 = 7.82$.

3. -0.32 standard deviations or $10.11 - 1.09 = 9.02$.
4. 0.00 standard deviations or $10.11 + 0 = 10.11$.
5. $+0.32$ standard deviations or $10.11 + 1.09 = 11.20$.
6. $+0.67$ standard deviations or $10.11 + 2.29 = 12.40$.
7. $+1.15$ standard deviations or $10.11 + 3.39 = 14.04$.

Since the total number of individuals in the sample was 112, then 12.5% would be 14 individuals. So we should *expect* 14 individuals in each of the above categories.

Table 15.5 contains the calculations to find the chi-square value. The *actual* frequencies are taken off the histogram of Figure 15.2, so that the first category includes all of the frequencies for the values from 1 through 6, the second includes just the times that 7 occurred, the next includes the frequencies of 8 and 9, and so forth.

Table 15.5

1	2	3	4	5	6
Intervals	Frequency theoretical F_t	Frequency actual F_a	$F_a - F_t$	$(F_a - F_t)^2$	$(F_a - F_t)^2/F_t$
Over 14.04	14	9	-5	25	1.78
12.40–14.04	14	15	1	1	0.07
11.20–12.39	14	13	-1	1	0.07
10.11–11.19	14	14	0	0	0.00
9.02–10.10	14	16	2	4	0.29
7.02– 9.01	14	27	13	169	12.07
6.18– 7.81	14	4	-10	100	7.15
Under 6.18	14	14	0	0	0.00
				Sum	21.43

Degrees of freedom = categories $- 3 = 8 - 3 = 5$.
$\chi^2_{0.95} = 11.1$. $\chi^2_{0.99} = 15.1$.
Since 21.43 exceeds $\chi^2_{0.99}$, we have a very significant difference.

We then compute the differences for each category (Column 4), square these differences (Column 5), and divide by the theoretical value (Column 6). Summing up the values in Column 6 we obtain the χ^2 statistic, which is 21.43. This is to be compared to the critical value of the χ^2 with degrees of freedom equal to categories minus 3. The number of categories is 8, therefore degrees of freedom = 5. From Table A.3 we find there is less than a 1% chance of having a χ^2 value as large as 15.1. (Note that there is the implication that there is a less than a 1% chance of a χ^2 being as large as this *if* the sample came from a normal distribution.) The sample χ^2 of 21.43 is therefore very highly significant and we declare that the distribution is *not* normal. This confirms the suspicion we

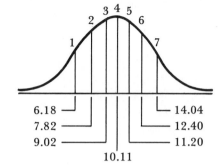

Figure 15.4 A theoretical distribution for Example 2.1. The limits for a set of eight equal areas have been located. The points are also labeled with numbers. The areas are all 12.5% of the total area.

had in Chapter 3 when we found the graph was not a straight line on probability paper.

The Kolmogorov-Smirnov One-Sample Test

The Kolmogorov-Smirnov one-sample test works with *cumulative frequencies* and therefore you will have to have *at least* ordered data (so that you can talk about "less than or equal to" categories).

For each category (or interval) we will determine the *relative* cumulative frequency. We are then to compare this cumulative distribution against some given *theoretical* cumulative distribution. Thus, for each category we will also compute the relative cumulative frequencies for the theoretical distribution.

We then calculate the *differences* between *each* of the pairs of relative cumulative frequencies. We look for the largest difference, positive or negative, which we call D. This value, D, is our test statistic. In Appendix Table A.15 you will find a list of critical limits of D for different significance levels. We only need to know the sample size, n.

If you look at Table A.15 you will observe that if n is greater than 35 you must perform a calculation. For example, if $n = 50$ and we desire to know the limit for an α of 0.10 (a 10% chance of error) we must calculate

$$1.22/\sqrt{50} = 0.173.$$

This then is our limit.

If D is larger than the limit, we declare that there is a significant difference (at the α level) between the sample distribution and the theoretical distribution that we are using as our comparison.

EXAMPLE 15.1 (Cont.)

Since the data for distribution of automobile manufacturers is *not* ordered, and is merely categorical, we *cannot* apply the Kolmogorov-Smirnov test. There is no valid way to create a cumulative distribution out of the data given.

EXAMPLE 15.3

A college official questioned whether students withdrew or left the college in equal amounts each year. The number of students in each class withdrawing during one year, before completing the year, were found to be as follows:

Freshmen	64
Sophomores	52
Juniors	41
Seniors	43

The total withdrawals were 200. If these were in equal amounts we would expect that 50 would leave each year.

Table 15.6 Calculating the maximum difference or D of the Kolmogorov-Smirnov one-sample test. Data is for Example 15.3

	1	2	3	4	5	6
Category	Actual frequency	Actual cumulative frequency	Actual relative cumulative frequency	Theoretical cumulative frequency	Theoretical relative cumulative frequency	Difference between Column 3 and Column 5
Seniors	43	200	1.00	200	1.00	0.00
Juniors	41	157	0.785	150	0.75	0.025
Sophomores	52	116	0.58	100	0.50	0.08
Freshmen	64	64	0.32	50	0.25	0.07

The maximum difference is the largest value in Column 6, and is 0.08.
The 5% limit for D would be $1.36/\sqrt{n} = 1.36/\sqrt{200} = 1.36/14.14 = 0.096$.

Table 15.6 contains the relative cumulative frequencies for the actual values in Column 3, and the expected relative cumulative frequencies for the theoretical distribution in Column 5. The differences for each category are listed in Column 6 and the maximum difference is found to be 0.08. Hence $D = 0.08$. The 5% limit is calculated and has a value of 0.096. Since our test statistic value (the 0.08) is *not* larger than the limit, we *cannot conclude* that the number of students dropping out of college each year is any different from freshman to senior years. There is *no* significant difference between the two distributions (the actual and the theoretical distributions).

15.2 Tests of Proportions

Much of the data that we encounter may be described as coming from a binomial-type source. This is the simplest of categorical data. When we select a sample from such a source we essentially obtain a single value. It may either be written as an X-value (the number of times one of the possibilities occurred out of a total sample of size n) or it may be computed as a proportion (the value

X/n). In either case we may freely convert from one form to the other. The types of questions we shall attempt to answer are:

"Is there a significant difference between this sample proportion and some hypothetical, theoretical proportion?" or "Within what interval would we expect the true population proportion to lie (confidence interval for proportion)?"

Before considering these questions we will have to recognize the role that the size of the sample plays in making a decision.

With small samples we should be working with the exact binomial distribution (as described in Chapter 9) and with large samples we can use the normal approximation to the binomial (when $n \cdot p$ and $n \cdot p(1-p)$ are both at least about 10).

Using the Exact Binomial Test (Small sample size)

Basically, Chapter 9 concerned itself with the question of describing the complete set of probabilities for all of the possibilities in a given binomial distribution. It provided a technique for establishing a theoretical distribution to use as a guide to compare with a particular sample. We will now discuss a technique used to decide if the sample could be expected to have come from such a population. The procedure we will follow here will simply involve examining the binomial distribution tables to decide if a sample falls in a tail, as an extreme value. If our *one sample* falls within the *extreme* 5% or 1% of a given theoretical distribution, we will declare that the sample comes from a different population. We will follow the procedure outlined in Chapter 11 (you could review Example 11.3 for an excellent illustration of a binomial test).

We choose the *hypothesis* that the population proportion is a particular value, which we will call P. We are then to decide on an *alternate* hypothesis—this establishes whether we are interested in a one- or two-tailed test. Determining the sample size establishes n and the complete distribution.

The level of significance or α (alpha) error is to be chosen (generally this will have to be approximate). We decide on our *test limit* or *limits*. To do this we have to examine the exact distribution as given in Table A.8. We accumulate all of the probabilities in the tail until we reach the total probability we want to be in that particular tail (for our limit). The X-values associated with the tail constitute our rejection values, or those cases that will lead us to declare the hypothesis to be false.

The *test statistic* will be our actual number of cases, which we will then compare to the limits.

EXAMPLE 9.1f (Cont.)

This was the problem of the number of times a traffic light is red. Our assumption was that the probability or proportion of the time that the light is red is 0.25. We are going to observe this traffic light 8 times and see how many times it *is* red. We randomly choose 8 times and find it red 6 of those times. Our procedure is as follows:

Hypothesis: The probability of a red light is 0.25.

Alternate Hypothesis: The probability of a red light is greater than 0.25.

Level of significance: We will be interested in limiting ourselves to a 5% chance of error.

Limit: Examining Table A.8 for $n = 8$ and $P = 0.25$ we find the probability of 8 red lights is 0.0000, 7 is 0.0004, 6 is 0.0038, and 5 is 0.0231. Adding these together gives us the probability of 5 or more red lights as

$$0.0231 + 0.0038 + 0.0004 + 0.0000 = 0.0273$$

or a 2.73% chance of error.

If we include 4 red lights, this tail increases to a probability of

$$0.0865 + 0.0273 = 0.1138$$

or 11.38% chance of error. Since we desire a 5% limit we must take the closest, which is the 2.7%. We declare the limit to be 5 red lights (any more will be beyond this limit and are included in the rejection area in the tail).

Test Statistics: Our statistic is merely the number of red lights we observed. We found a total of 6 red lights.

Decision: Since the test statistic, 6, is greater than the limit, 5, we declare that the hypothesis is not true. We state that the proportion of time that the light is red is *not* 0.25, it is *greater* than 0.25.

Confidence Limit: It would be desirable to place a confidence limit or interval around the actual proportion that we have observed. In this example the observed proportion, P, was $X/n = \frac{6}{8} = 0.75$. Unfortunately, with small samples confidence intervals are so large that they provide little aid in decisions. In addition, to obtain meaningful limits on the intervals when working from the tables requires a much more extensive collection of tables, as well as a slightly different technique of working with the table. We will defer the discussion of confidence intervals until the next section.

Large Samples: The Normal Approximation

With larger samples we will use the normal distribution as an approximation to the binomial. This will greatly simplify and at the same time broaden the types of statements we can make. However, we will limit our discussions here to hypothesis and the estimating of *proportions*. If you require estimations regarding X-values or means, then you merely multiply the corresponding proportions, or p-values, by n.

As before, we hypothesize that the population has a particular known value of P associated with the source of our sample. We will then compare the proportion of the sample to this P-value. Again, the procedure is as follows:

Hypothesis: The population proportion is P.

Alternate Hypothesis: The population proportion is not P. (This may require further clarification depending on whether we are dealing with a one- or two-tailed test. We will limit this discussion to a two-tailed test.)
The size of n and the level of significance, α, must be given.

Limits:

$$\text{Upper} \quad z_{1-\alpha/2} \quad \text{(from Table A.1a or b).} \quad (15.2)$$
$$\text{Lower} \quad z_{\alpha/2}$$

Test Statistic:

$$z = \frac{(X/n) - P}{\sqrt{P(1-P)/n}}. \quad (15.3)$$

Decision: If the value of z obtained in Equation 15.3 is between the two limits, we accept the hypothesis that there is *no* significant difference between the sample proportion and our hypothesized P.

Confidence interval: We can draw a confidence interval around the particular *sample* proportion. This provides us with a range of possible values of the population proportion from which we would assume our sample may reasonably have been taken. We will use $\bar{X}$ to stand for the sample proportion, X/n.

$$\bar{X} - z_{\alpha/2}\sqrt{\frac{\bar{X}(1-\bar{X})}{n}} < P < \bar{X} + z_{1-\alpha/2}\sqrt{\frac{\bar{X}(1-\bar{X})}{n}}.$$

EXAMPLE 15.4

A museum purchased a \$1 million painting expecting it would attract 75% of its visitors. To verify this estimate 50 people were randomly chosen as they entered the museum and followed until they left. Of these 50, it was found that 27 looked at the painting. Does this contradict the original estimate?

Hypothesis: The proportion of all visitors looking at the painting, P, is 0.75.

Alternate Hypothesis: The proportion of all visitors is significantly different than 0.75 (larger or smaller).
We will use an α of 0.05 (5% chance of error). Sample size $n = 50$.

Limits:

$$\text{Upper} \quad z_{1-\alpha/2} = z_{1-0.05/2} = z_{0.975} = 1.96.$$
$$\text{Lower} \quad z_{\alpha/2} = z_{0.05/2} = z_{0.025} = 1.96.$$

Test Statistic:

$$z = \frac{(X/n) - P}{\sqrt{P(1-P)/n}} = \frac{(27/50) - 0.75}{\sqrt{(0.75)(0.25)/50}} = \frac{0.54 - 0.75}{\sqrt{0.00375}}$$

$$= \frac{-0.21}{0.061} = -3.44.$$

Decision: Since the test statistic of -3.44 is outside the test limits we *reject* the hypothesis that $P = 0.75$. We declare that p is significantly different from 0.75. It appears to be considerably less.

Confidence interval: Let us determine an estimate of the limit that P probably should lie within.

$$\bar{X} = X/n = 27/50 = 0.54.$$

Then

$$\bar{X} - z_{\alpha/2}\sqrt{\bar{X}(1-\bar{X})/n} < P < \bar{X} + z_{1-\alpha/2}\sqrt{\bar{X}(1-\bar{X})/n}$$

$$0.54 - 1.96\sqrt{(0.54)(0.46)/50} < P < 0.54 + 1.96\sqrt{(0.54)(0.46)/50}$$

$$0.54 - 1.96\sqrt{0.005} < P < 0.54 + 1.96\sqrt{0.005}$$

$$0.54 - 1.96(0.071) < P < 0.54 + 1.96(0.071)$$

$$0.54 - 0.14 < P < 0.54 + 0.14$$

$$0.40 < P < 0.68.$$

Thus we give 19 to 1 odds that the true proportion of people looking at the painting is between 0.40 and 0.68 (or between 40% and 68% of the visitors).

REVIEW

Chi-square test · Goodness of fit ·

$$\chi^2 = \sum \frac{(\text{Frequency}_{\text{actual}} - \text{Frequency}_{\text{theoretical}})^2}{\text{Frequency}_{\text{theoretical}}}$$

Kolmogorov-Smirnov test · D · Proportions · Small sample

Large sample

PROJECTS

1. See the project in Chapter 9. Compare the results you had for the distribution of students having class during various periods to the possible periods. Use the χ^2-test. How close are your results to a binomial distribution?
2. Go to a telephone book. Record the last number of each phone number for about 100 numbers. Assuming that all digits are equally likely, perform a χ^2-test to see if this assumption is valid.

PROBLEMS

Section 15.2

15.1 A teacher claims to give 10% A's, 20% B's, 40% C's, 20% D's, and only 10% F's. A student polled 150 other students who took courses with this instructor and found they actually received

$$16 \text{ A's}, \quad 32 \text{ B's}, \quad 58 \text{ C's}, \quad 24 \text{ D's}, \quad 20 \text{ F's}.$$

Do you agree with the instructor's statement (you want to make a wrong accusation less then 5% of the time)?

15.2 The doctor at a large ski resort declared that the percent of people who broke legs was the same regardless of how good they were. The resort had the following people for a week:

2000 beginner, 1500 novices, 1000 intermediates, 500 advanced.

Of these people the following numbers broke legs:

41 beginners, 24 novices, 18 intermediates, 9 advanced.

Do you agree with the doctor? (Test at the 10% level.)

15.3 If you examine a large collection of cars you would expect the number of dented fenders to be approximately the same for all four fenders. Three hundred cars were examined and we found a total of 560 dented fenders. The distribution was:

passenger front	165
passenger rear	170
driver front	120
driver rear	105

Is there any indication that certain fenders are dented more than others? What is the implication for a manufacturer or distributor of automobile parts?

15.4 The number of suicides might be expected to be constant and random throughout the year. The following numbers of suicides were recorded bimonthly in a large metropolitan area

J–F	M–A	M–J	J–A	S–O	N–D
16	22	28	20	11	17

Is there reason to believe that time of the year might be related to the number of suicides? Test at the 5% level of significance.

15.5 A box of grass seed is labeled as having:
30% Kentucky Blue (very good)
30% Perennial Ryegrass (cheaper)
30% Fescue (okay)
10% Annual Ryegrass (cheapest—only lasts a year)

A curious gardener counted out 1000 seeds and planted them. When all sprouted he counted the different seedlings. There were

260 Kentucky Blue
320 Perennial Ryegrass
240 Fescue
180 Annual Ryegrass

Is there reason to doubt the labeling (is he being gypped)?

15.6 The number of fire alarms experienced in the past in a small town were distributed approximately:

30% false alarms
40% 1-alarm
15% 2-alarm
10% 3-alarm
5% 4-alarm

One hundred and sixty fires reported the next year were as follows:

50 false alarms
62 1-alarm
26 2-alarm
18 3-alarm
4 4-alarm

Is this distribution unusual? (Test at the 0.10 level.)

Section 15.3

15.7 Refer back to the data in Example 2.3 (bread weights). Using about 8 categories, test to determine if this data could be considered a normal distribution (refer back to Problem 3.2 or 3.7 for the histogram).

15.8 In Problem 6.3 we attempted to determine if a correlation existed between the number and weight of 39 football players.
Draw a histogram for the player's numbers. Does it resemble a normal distribution? Calculate the mean and standard deviation (the sums are given with the data). Perform a χ^2-test for normality using about 5 groups. (Use an $\alpha = 0.10$.)

15.9 Determine if the weights of the players in Problem 6.3 are normally distributed (again calculate the mean and standard deviation and perform a χ^2-test, only this time use a 5% critical value).

15.10 Sometimes nonrandomness may show up because the sample means are not normally distributed. We have already observed that the distribution of tickets in Example 2.1 was essentially normal. Thus we would expect that sample means should also be normally distributed. Figure 12.8 (page 309) provides the distribution of sample means where $\bar{X} = 10.68$ and $\sigma_{\bar{x}} = 1.70$. (Note that the means of the sample were given in Table 7.1, page 193). Perform a χ^2-test using about 5 categories to determine if the distribution of samples might not be normal. (Test at the 0.025 level. Note that days 10 and 11 are eliminated.)

15.11 Refer to the number of home runs hit by the National League Home Run Leader (Problem 14.7). Is this a normal distribution? (Test at the 0.10 level.)

Section 15.4

15.12 Grades may be considered ordered. Using the Kolmogorov-Smirnov test, compare the grades given by the teacher in Problem 15.1 to the theoretical grade distribution he claims he follows. (Use the same 5% level.)

15.13 A psychologist was attempting to determine if people's preference for colors was uniform or depended upon wavelength of the light. He asked 30 people to pick out their favorite color from among 6 colors. The number of people who chose the various colors in order of longest wavelength to shortest wavelength were as follows:

```
Longest:   red      3
           orange   2
           yellow   5
           green    7
           blue     7
Shortest:  violet   6
```

Would you consider that wavelength and preference are not related? Test by using the Kolmogorov-Smirnov test and hypothesizing that all the colors would have equal numbers of people favoring them. (Use a 10% level.)

15.14 Redo Problem 15.6 using the Kolmogorov-Smirnov test. Is the distribution of fire alarms unusual?

15.15 During a hot June the following distribution of fire alarms was recorded:

12 false alarms
5 1-alarms
2 2-alarms
1 4-alarm

Is this an unusual distribution of fire alarms? (Refer to Problem 15.6 for the theoretical distribution.)

15.16 The Kolmogorov-Smirnov test can also be used to decide if a distribution appears to be normally distributed. Redo Problem 15.7 using the Kolmogorov-Smirnov test. (*Hint:* Instead of determining areas for each category, you are comparing areas up to and including the particular category.)

15.17 Redo Problem 15.8 using the Kolmogorov-Smirnov test.

15.18 Redo Problem 15.10 using the Kolmogorov-Smirnov test.

Section 15.5

15.19 The management of a large department store has felt that if the number of shoplifters is more than 20% of the customers who enter the store, extreme measures will be necessary. Since the cost of investing in security equipment is very large, a risk of less than 1% is considered necessary.
   a. A group of 150 people were randomly chosen as they entered the store and followed or observed until they left. Of these 150 people 41 were seen to have taken something without paying. Is there sufficient reason for the management to consider the purchase of security equipment?
   b. What would you estimate the proportion of shoplifters to range within? (Assume 1% limits.)
   c. If the number of customers average about 6000, what would you estimate the number of shoplifters to range within?

15.20 If the expected proportion of false alarms has always been 30%, is a month with 12 false alarms out of a total of 20 "reported" fires unusual? (See Problem 15.15.)

15.21 A dishonest cashier believed that only 10% of the customers count their change. She then proceeded to shortchange every customer. By the end of the day she saw 200 people and 28 complained of receiving incorrect change. How good is her estimation of the proportion of people who count their change? (Use an $\alpha = 0.05$.)

15.22 A prospective father decided to estimate how many cigars he would need when the new baby was to be born. He polled 20 of his friends and co-workers asking them if they accept cigars and he found only 15 "takers." If there are a total of 250 men to offer cigars to, how many cigars should he plan on giving out (with 95% assurance of having enough)? (*Hint:* Use the sample to estimate the population proportion and then go to the population.)

15.23 The recreational department of a small township is attempting to determine if the town should invest in two new tennis courts. The town fathers will approve them only if there is reason to expect them to be used 85% of the nights (weather permitting). They decide to observe each night at the old courts to see if there are people waiting for others to finish. After checking on 50 nights they find people waiting on 40 of the nights. Is there reason to believe that there would be people waiting significantly less than 85% of the time (at the 0.10 level)?

15.24 A new pizza parlor believes that if it gives away free pizzas the first day, 75% of the original customers will return in two weeks. On the first day 385 pizzas were served, with each person filling out a card with his name and telephone number. Two weeks later the pizza owner called 30 of the people and found that 23 had already purchased another pizza. Is the owner's assertion correct? (Test at the 10% level.)

15.25 A printer feels that his printing machine is operating satisfactorily when less than 2% of the pages are poorly printed. He examines 100 sheets every hour. After a week (40 hours) he finds a total of 98 bad sheets among those checked. Should he have his machine repaired? What do you estimate the total number of bad sheets to be if he printed 600 pages an hour?

15.26 The president of a large company was persuaded to purchase a $100,000 automatic machine. The chief engineer claimed it would operate 60% of the time. The president randomly selected 15 times to observe if the machine was in operation. Seven of these times it was not being run. Should he complain to the engineer (he wants to be wrong less than 5% of the time)? What is the 95% confidence interval?

★15.27 The president now asks how many times he must sample the machine to make certain his sample values will be off no more than 3%. The formula he is to use is

$$n = P(1 - P)(z_\alpha/\text{Error})^2 \qquad (15.4)$$

where *error* is the maximum amount he wants to be off (it is $\frac{1}{2}$ of the confidence interval). If α is 0.05 (95% assurance) then

$$n = P(1 - P)(1.96/\text{Error})^2.$$

If P is expected to be 0.60 and the error is to be 0.03 (3%) how large a sample must he take?

★15.28 If you have no idea what P should be, then the largest possible value of Equation (15.4) would be when

$$P(1 - P) = \tfrac{1}{2}(1 - \tfrac{1}{2}) = \tfrac{1}{4}.$$

Thus, you can always be safe and use the formula as

$$n = \tfrac{1}{4}(z_\alpha/\text{Error})^2.$$

Redo Problem 15.27. How different is the necessary sample size?

★15.29 The department store owner in Problem 15.19 wanted to be 99% certain that his estimate of the number of shoplifters was good within $\pm 2\%$. If he doesn't want to make a preliminary guess of P, how large a sample should he take? (Note that z_α is not 1.96 but 2.58 because of the 99% significance level.)

★15.30 If the department store manager feels that P is about 0.20, then how large a sample should he take? How does this sample size compare to the one in Problem 15.29?

★15.31 Derive Equation (15.4). Start with Equation (15.3) where X/n is the extreme and $(X/n) - P$ would be the error.

16

Univariate Tests Involving Interval or Ratio Measurements and an Assumption of Normality

This chapter will deal with a collection of statistical methods that are used to test or compare sample values to the hypothetical parameters of a normal distribution (either the hypothetical mean or variance). As a consequence we only consider using these tests when we have reason to believe that it is valid to assume that a normal distribution is present. This requires that the data be more than just ordered. Differences in the values along the scale must be comparable throughout the scale (thus the measurements must be at least of interval type).

In some cases the actual population may not be normally distributed, yet the tests will be acceptable. This situation occurs since we are dealing with samples, and we have already observed (in Chapter 12) that the means of many kinds of samples closely approach normal distributions.

Obviously then, these tests are all parametric tests. In addition, all of these tests will have a common format. It should be noted at this point that all of the tests in this section may be one-sided or two-sided tests. That is, we may concern ourselves with the possibility of a value being *only greater than* a fixed value or *only smaller than* a fixed value, or we may look *simultaneously* at either possible difference. This consideration affects our limits for particular levels of significance and we will treat these as separate problems.

16.1 Comparing the Spread of a Sample to a Theoretical Spread

Two-sided Test

Hypothesis: The population variance is *not* different (larger or smaller) than a theoretical (theo) *given* population variance, σ^2_{theo}. Thus $\sigma^2 = \sigma^2_{\text{theo}}$.

Alternate: The sample may come from a population that has a variance greater than or less than a given theoretical population variance. ($\sigma^2 < \sigma^2_{theo}$ or $\sigma^2 > \sigma^2_{theo}$).*

This is a *two-sided test*. In actual practice it is generally only a one-sided test that is performed. However, for confidence intervals we must discuss two-sided tests. Also, recall in Chapter 12 that we discussed the distribution of sample variances and found it was related to the χ^2-distribution (more specifically the χ^2/df). We will now use this knowledge.

The sample is of size n, σ_{theo} is given. The level of significance, called α, must be provided.

Limits:

$$\text{Upper} \quad \chi^2_{1-\alpha/2}/df = \chi^2_{1-\alpha/2}/(n-1).$$
$$\text{Lower} \quad \chi^2_{\alpha/2}/df = \chi^2_{\alpha/2}/(n-1). \tag{16.1}$$

where $df = n - 1$ and we use Appendix Table A.4† (or use Table A.3 and divide by $n - 1$).

Test Statistic:

$$\frac{s^2}{\sigma^2_{theo}}.$$

Decision: If the calculated s^2/σ^2_{theo} is between the two limits, there is no significant difference between the sample variance and the theoretical variance (σ^2_{theo}) we accept the hypothesis.

If the s^2/σ^2_{theo} is outside either limit, there is a significant difference, at the α-level, between this sample population and the theoretical population. We say "reject the hypothesis and accept the alternate."

Confidence intervals:

$$\frac{s^2}{\chi^2_{1-\alpha/2}/df} < \sigma^2 < \frac{s^2}{\chi^2_{\alpha/2}/df} \tag{16.2}$$

EXAMPLE 16.1

Gravel that is used on a road should have a certain amount of variation, yet not too large a variation. Most of the stone to be used is about $\frac{3}{4}$ inch diameter with some finer and some coarser stone included. The theoretical standard deviation of gravel from the Crushed Stone Corp. quarry is 0.20 inches. Thus, $\sigma^2_{theo} = (0.20)^2 = 0.04$. A sample of 25 stones had a variance of 0.10. Is there a significant difference in the variability of this sample and the typical stones?

Given: Sample size $n = 25$, $df = n - 1 = 24$, $\sigma^2_{theo} = 0.04$, $s^2 = 0.10$. A significant difference generally implies at least an α of 0.05 (a 5% chance of error or a 1 in 20 chance of error).

---

* Roman letters such as s and $\bar{X}$ represent sample values. Greek letters such as σ and μ are to stand for population or theoretical values.
† Table A.4 provides the division—it gives χ^2 divided by degrees of freedom.

Limits:

$$\text{Upper} \quad \frac{\chi^2_{1-\alpha/2}}{df} = \frac{\chi^2_{1-0.05/2}}{df} = \frac{\chi^2_{0.975}}{df} = 1.64.$$

$$\text{Lower} \quad \frac{\chi^2_{\alpha/2}}{df} = \frac{\chi^2_{0.025}}{df} = 0.517.$$

These values are from Table A.4. (See Figure 16.1 for an illustration of obtaining these values.)

df	Probability				
	0.005	0.025		0.975	0.995
24	0.412	0.517		1.64	1.90

Figure 16.1 Finding the critical values of χ^2/df from Table 16.4. Several critical probability levels are shown for degrees of freedom equal to 24. (Generally $df = n - 1$.)

Test Statistic:

$$\frac{s^2}{\sigma^2_{theo}} = \frac{0.10}{0.04} = 2.5.$$

Decision: Since the test statistic is greater than the upper limit (2.5 is greater than 1.64) we declare that there is a significant difference in variability between this gravel and the theoretical variance of gravel from Crushed Stone Corp.

Confidence interval: What is the 99% confidence interval for the true population *for this gravel*? The variance, σ^2, lies between the following two limits:

$$\frac{s^2}{\chi^2_{\alpha/2}/df} \quad \text{and} \quad \frac{s^2}{\chi^2_{1-\alpha/2}/df}$$

where $\alpha = 0.01$.

The limits are

$$\frac{0.10}{\chi^2_{0.995}/df} \quad \text{and} \quad \frac{0.10}{\chi^2_{0.005}/df}$$

or

$$\frac{0.10}{1.90} = 0.053 \quad \text{and} \quad \frac{0.10}{0.412} = 0.243,$$

so that

$$0.053 < \sigma^2 < 0.243.$$

(Observe that this interval also does not include the theoretical σ^2 of 0.04.)

One-sided Test

Hypothesis: The population variance is *not* different from a theoretical given population variance ($s^2 = \sigma_{theo}^2$).

Alternate: The sample comes from a population with variance *greater than* the theoretical variance ($\sigma^2 > \sigma_{theo}^2$).

Sample size is n, σ_{theo} is given, level of significance α must be provided.

Limit:

$$\text{Upper} \quad \frac{\chi_{1-\alpha}^2}{df} \quad df = n - 1 \quad \text{(Use Table A.4.)}$$

Test Statistic:

$$\frac{s^2}{\sigma_{theo}^2}.$$

Decision: If s^2/σ_{theo}^2 is *smaller* than the limit then accept the hypothesis that the sample came from a population with variance σ_{theo}^2.

If s^2/σ_{theo}^2 is *larger* than the limit then accept the *alternate*; state that the sample source has a variance greater than σ_{theo}^2. Note that the difference between this test and the previous test is mainly in the subscript of the χ^2-value that we use. In the first case we are splitting up the α-value and in the second case (the one-sided case) we don't divide the value of α.

EXAMPLE 16.2

It has been found that not speeding, but rather the variability of speeds amongst drivers is the main cause of accidents. A very slow driver may cause an accident as frequently as a very fast driver. A study showed an optimum standard deviation of speeds on a major highway to be 3 miles per hour. The variance would be 9. A sample of the speeds of 16 cars had an s^2 of 14. At the 5% level of significance, is this a greater degree of variability than optimum?

Given: Sample size $n = 16$, $df = n - 1 = 15$, $\sigma_{theo}^2 = 9$, $\alpha = 0.05$.

Limit:

$$\frac{\chi_{1-\alpha}^2}{df} = \frac{\chi_{0.95}^2}{df} = 1.67.$$

Test Statistic:

$$s^2/\sigma^2_{theo} = 14/9 = 1.56.$$

Decision: The test statistic is *not* greater than the limit of 1.67, therefore we conclude that there is no significant difference between the variability of this sample and our theoretical variance of 9.

16.2 Comparing the Mean of a Sample to a Theoretical Population Mean

The following set of tests differ in only one respect, namely by what is used as the value of the population standard deviation. Our previous discussions of confidence intervals and hypotheses concerning the means of samples always required the knowledge of or the inclusion of a value of spread. The main question that now must be answered is how that value of spread will be derived.

We will consider two possibilities:

1. It is *given or known* from prior information.
2. It is *unknown* but we will use the *sample variance to estimate* the population variance.

(A third possibility, using the range, is somewhat easier to handle, but much less efficient.)

The only true difference is in the name of the test statistic and the table we use to find the limit of the test statistic. The general format will be the same. (Again we will look at one- and two-sided tests.)

The Population Variance or Standard Deviation is Given or is Known

This is called a *z-test*. The standard deviation (or variance) of the sample should be checked against the given theoretical value by using the χ^2/df test first. This is to assure that the sample did in fact come from a population having the given variance.

(a) Two-sided test

Hypothesis: The population mean is *not* significantly different than a given theoretical population mean, called μ_{theo} ($\mu = \mu_{theo}$).

Alternate: The sample comes from a population whose mean is *greater than* OR *less than* μ ($\mu > \mu_{theo}$ or $\mu < \mu_{theo}$).

The sample is of size n, sample mean = $\bar{X}$, hypothetical mean is μ_{theo}, level of significance α is to be provided.

Limits:

$$\text{Upper} \quad z_{1-\alpha/2}$$
$$\text{Lower} \quad z_{\alpha/2} \quad (16.3)$$

(These values are from Table A.1a or b.)

Test Statistic:

$$z = \frac{\bar{X} - \mu}{\sigma/\sqrt{n}}. \quad (16.4)$$

Decision: If z calculated is greater than the z_{upper} or less than the z_{lower} we declare that the sample does not come from the population with a mean equal to μ_{theo}. Otherwise, we state that there is no difference between the sample mean and μ_{theo}.

Confidence interval:

$$\bar{X} - z_{\alpha/2} \frac{\sigma}{\sqrt{n}} < \text{True mean} < \bar{X} + z_{1-\alpha/2} \frac{\sigma}{\sqrt{n}} \quad (16.5)$$

EXAMPLE 16.3

An inspector from the Weights and Measures Department of a large city was to check the meat scale in a supermarket. Five readings of a true 2 pound steak were recorded as

$$1.96, \quad 2.02, \quad 1.92, \quad 1.94, \quad 1.98.$$

If this type of scale is *known* to have an error in measurement that is equal to a σ of 0.04 pounds, is there a significant difference between these values and a true mean of 2.00? (The sample standard deviation is 0.038, which is not different than 0.04.) Consider an α of 0.05.

Given:

$$n = 5, \quad \bar{X} = 1.964, \quad \mu_{\text{theo}} = 2.00.$$

Limits:

$$\text{Upper} \quad z_{1-\alpha/2} = z_{0.975} = 1.96$$
$$\text{Lower} \quad z_{\alpha/2} = z_{0.025} = -1.96$$

Test Statistic:

$$z = \frac{\bar{X} - \mu_{\text{theo}}}{\sigma/\sqrt{n}} = \frac{1.964 - 2.00}{0.04/\sqrt{5}} = \frac{-0.036}{0.04/2.24} = -2.01.$$

Decision: The test statistic, z, is outside the lower limit, therefore *we reject the hypothesis* that this scale might give the average weight of 2.00 pounds. The consequence of rejecting this hypothesis might lead to a large fine.

Confidence interval:

$$1.964 - 1.96 \frac{0.04}{\sqrt{5}} < \text{true population mean} < 1.964 + 1.96 \frac{0.04}{\sqrt{5}}$$

or

$$1.964 - 0.035 < \mu < 1.964 + 0.035$$
$$1.929 < \mu < 1.999.$$

(This is rather close to including the hypothetical mean of 2.00.)

(b) One-sided test

Hypothesis: The population mean is *not* different from a given theoretical mean, μ_{theo} ($\mu = \mu_{theo}$).

Alternate Hypothesis: The sample comes from a population whose mean is greater than the given theoretical mean μ_{theo} ($\mu > \mu_{theo}$).

Given: All of the given quantities are the same as in the two-sided test.

Limit:

$$\text{Upper} \quad z_{1-\alpha} \quad \text{(from Table A.1)} \tag{16.6}$$

Test Statistic [this is the same formula as Equation (16.4)]:

$$z = \frac{\bar{X} - \mu_{theo}}{\sigma/\sqrt{n}}.$$

Decision: If the calculated z is *less than* the $z_{1-\alpha}$ limit, we declare that there is no difference between the population mean and the given theoretical mean.

If the calculated z-value is *greater than* $z_{1-\alpha}$ we declare that the sample comes from a population whose mean is greater than μ.

EXAMPLE 16.4

A professional bowler has been very consistent in his bowling. His long-time average score has been 230, with a standard deviation of 10. He has been experimenting with a new delivery (method of running and dropping the ball onto the alley). As a result of this change he averaged 242 for 4 trial games. However, the bowler does not want to change his method unless he can truly say there has been an increase. He is willing to take only a 1 in 20 chance of changing his method if there is really no improvement.

Since we are *only* concerned with an increase this is a one-sided test.

Hypothesis: There is no change in bowling average (the true population mean = 230).

Alternate: There is an *increase* in bowling average (the population mean that this sample came from is greater than 230).

Given:

$n = 4, \quad \sigma = 10, \quad \bar{X} = 242, \quad \alpha = 0.05 \text{ (1 in 20 chance)}.$

Limit:

$$z_{1-\alpha} = z_{0.95} = 1.645.$$

Test Statistic:

$$z = \frac{\bar{X} - \mu_{\text{theo}}}{\sigma/\sqrt{n}} = \frac{242 - 230}{10/\sqrt{4}} = \frac{12}{10/2} = 2.4.$$

Decision: The test statistic is greater than the limit, therefore we reject the hypothesis and accept the alternate. Our bowler has made a significant improvement in his bowling score.

Observe also that if we increased the α value to 0.01 (a 1 in 100 chance of error) and thereby made this a "very significant" level, the limit would only be 2.326. The test statistic even exceeds this value and we could therefore declare the increase as *very highly* significant.

(c) We will now treat the *less than* case.

Hypothesis: The population mean is *not* different from a given theoretical mean, μ_{theo} ($\mu = \mu_{\text{theo}}$).

Alternate: The population mean is *less than* a given theoretical mean, μ_{theo} ($\mu < \mu_{\text{theo}}$).

Given: Same as previous test. The only difference between this and the previous test is the limit and the decision.

Limit:

Lower z_α.

Test Statistic: Same as before, Equation (16.4).

Decision: If

$$z = \frac{\bar{X} - \mu_{\text{theo}}}{\sigma/\sqrt{n}}$$

is greater than z_α, then accept the hypothesis that there is no difference between the sample mean and the theoretical mean.

If z is *less than* z_α then accept the alternate that the population mean is less than μ_{theo}.

EXAMPLE 16.5

A pollution control device for automobiles was being tested for long term use. It has been known from previous tests that a particular type of pollutant is present in automobile exhaust with an average of 90 parts per million and a standard deviation of 25. Ten tests with the device had produced an average exhaust level of 75 parts per million. The company is not ready to begin marketing the device until it is 99% sure that an improvement in lowering the emission level is evident.

This is a one-sided test and σ is known.

Hypothesis: The population mean is 90.

Alternate: The population mean is *less than* 90.

Given:

$$n = 10, \quad \sigma = 25, \quad \bar{X} = 75, \quad \alpha = 0.01.$$

Limit:

$$z_\alpha = z_{0.01} = -2.326.$$

Test Statistic:

$$z = \frac{\bar{X} - \mu_{\text{theo}}}{\sigma/\sqrt{n}} = \frac{75 - 90}{25/\sqrt{10}} = \frac{-15}{25/3.16} = -1.89.$$

Decision: Since the test statistic of -1.89 is larger than the limit of -2.326, we declare that there is *not* a significant decrease in emission level. We accept the hypothesis that the mean is still 90 and the company will keep working on the device.

If the Population Variance or Standard Deviation is Unknown—t-Test

In this case the population standard deviation is unknown and it must be calculated from the sample (as s) and used as our *estimate* of the population spread.

(a) Two-sided test

Hypothesis: The population mean is *not* different from a given theoretical mean, called μ_{theo} ($\mu = \mu_{\text{theo}}$).

Alternate: The sample mean comes from a population whose mean is greater than *or* less than the given theoretical mean, μ_{theo} ($\mu > \mu_{\text{theo}}$ or $\mu < \mu_{\text{theo}}$).

Sample size $= n$, s is calculated from the sample, $\bar{X}$ is the sample mean, degrees of freedom $= df = n - 1$, μ_{theo} is given, α must be specified.

Limits:

$$\begin{array}{lll} \text{Upper} & t_{1-\alpha/2} & \text{(with } n - 1 \text{ degrees of freedom)} \\ \text{Lower} & t_{\alpha/2} & \text{(with } n - 1 \text{ degrees of freedom)} \end{array} \quad (16.7)$$

Use Table A.2 to find these limits (refer back to Chapter 12 for details on using the *t*-table).

Test statistic:

$$t = \frac{\bar{X} - \mu_{\text{theo}}}{s/\sqrt{n}}. \qquad (16.8)$$

Decision: If the *t*-value is between the two limits then we accept the hypothesis, and declare that the population mean that the sample came from is not different from the theoretical mean, μ_{theo}.

If the *t*-value is greater than the upper limit or smaller than the lower limit we declare that the sample mean is *not* from a population whose mean is μ_{theo}.

Confidence interval:

$$\bar{X} - t_{\alpha/2}\frac{s}{\sqrt{n}} < \mu < \bar{X} + t_{1-\alpha/2}\frac{s}{\sqrt{n}} \qquad (16.9)$$

EXAMPLE 16.3 (Cont.)

This time we are to consider the σ-value as unknown. We still have $\bar{X} = 1.964$, $\mu_{\text{theo}} = 2.00$, and $\alpha = 0.05$.

We calculate s from the sample and find that $s = 0.038$. Degrees of freedom $= n - 1 = 4$.

Limits:

Upper $\quad t_{1-\alpha/2} = t_{0.975} = 2.776$.

Lower $\quad t_{\alpha/2} \;\;= t_{0.025} = -2.776$.

Test Statistic:

$$t = \frac{\bar{X} - \mu_{\text{theo}}}{s/\sqrt{n}} = \frac{1.964 - 2.00}{0.038/\sqrt{5}} = \frac{-0.036}{0.038/2.24} = -2.12.$$

Decision: The *t*-value is between the two limits. Therefore, we accept the hypothesis that there is no difference between the sample mean and a theoretical mean of 2.00.

Confidence interval:

$$1.964 - 2.276\frac{0.038}{\sqrt{5}} < \mu < 1.964 + 2.276\frac{0.038}{\sqrt{5}},$$

$$1.964 - 0.039 < \mu < 1.964 + 0.039,$$

$$1.925 < \mu < 2.003.$$

Observe that even though the standard deviation is smaller than the previously given theoretical standard deviation (it was stated as being 0.04), the confidence interval and the test limits are larger than in the *z*-test. As the sample size gets large, however, the *t*-values and *z*-values become more and more alike (mathematically we say that the *t*-values approach the *z*-values) and s becomes more like σ, so that the tests become essentially identical.

(b) One-sided test

Hypothesis: The population mean is not different from a given theoretical mean, μ_{theo} ($\mu = \mu_{theo}$).

Alternate: The population mean is *greater* than a given theoretical mean, μ_{theo} ($\mu > \mu_{theo}$).

Standard deviation is not given. Level of significance, α, must be given.

Limit:

$$\text{Upper } t_{1-\alpha} \quad (df = n - 1).$$

Test Statistic:

$$t = \frac{\bar{X} - \mu_{theo}}{s/\sqrt{n}}.$$

Decision: If the calculated t is less than $t_{1-\alpha}$ then accept the hypothesis that there is no difference between the population mean and μ_{theo}.

If t is greater than the limit, we reject the hypothesis and accept the alternative. We declare that the true mean is *greater* than this given theoretical mean, μ_{theo}.

EXAMPLE 16.4 (Cont.)

If we now assume the standard deviation of the population is unknown, the standard deviation of the bowling scores must be calculated from the sample. It was found to be equal to 12.

Given:

$n = 4$, degrees of freedom $= n - 1 = 3$, $\bar{X} = 242$, $\mu_{theo} = 230$, $\alpha = 0.05$.

Limit:

$$\text{Upper } t_{1-\alpha} = t_{0.95} = 2.353$$

Test Statistic:

$$t = \frac{\bar{X} - \mu_{theo}}{s/\sqrt{n}} = \frac{242 - 230}{12/\sqrt{4}} = \frac{12}{12/2} = 2.00.$$

Decision: Since this test statistic (2.00) is smaller than the limit (2.353) we cannot say that there is an increase in bowling score.

Here again, because we do not have the population standard deviation and must work with an estimate, the limit becomes correspondingly larger. The estimate of the standard deviation is also larger, causing the test statistic to be smaller. The combination of these two situations changes our decision from saying there is an increase in bowling score (when the standard deviation was given) to saying there is no increase.

(c) One-sided test, *less than* case

Hypothesis: Population mean is not different from a given theoretical mean. Standard deviation is unknown.

Alternate: Population mean is smaller than the given theoretical mean.

Given: The only difference between this and the previous test is the limit.

Limit:

$$\text{Lower } t_\alpha \quad (df = n - 1)$$

Test Statistic:

$$t = \frac{\bar{X} - \mu_{theo}}{s/\sqrt{n}}.$$

Decision: If t is greater than t_α we accept the hypothesis.

If t is less than t_α we accept the alternate that the mean is smaller than the theoretical mean.

EXAMPLE 16.5 (Cont.)

Our pollution control manufacturer decided to take another 15 readings. Further investigation also presented some doubt on the reliability of the standard deviation that was given before. It was decided to use all 20 test readings and use the data to also estimate variance. The mean of the 25 tests was found to be 78 with a standard deviation of 21.

Hypothesis: The mean pollution level is 90.

Alternate: The mean pollution level is less than 90.

Given:

$$n = 25, \quad df = 24, \quad \bar{X} = 78, \quad s = 21, \quad \alpha = 0.01.$$

Limit:

$$\text{Lower } t_\alpha = t_{0.01} = -2.492.$$

Test Statistic:

$$t = \frac{\bar{X} - \mu_{theo}}{s/\sqrt{n}} = \frac{78 - 90}{21/\sqrt{25}} = \frac{-12}{21/5} = -2.88.$$

Decision: Since the test statistic of -2.88 is less than the limit of -2.492 we accept the alternate that there is a decrease in pollution level.

Comment on Sample Size

This last example points up something that was mentioned at the end of Chapter 13. If we keep increasing the sample, eventually we will end up with a significant difference. To demonstrate why this is true let me rewrite the test statistic

$$\frac{\bar{X} - \mu_{theo}}{s/\sqrt{n}} \quad \text{as} \quad \frac{\bar{X} - \mu_{theo}}{s}\sqrt{n}$$

Now as long as $\bar{X} - \mu_{theo}$ is not zero and if we keep both $\bar{X}$ and s constant, then as n increases so does the test statistic. Eventually the test statistic would become larger than the limit *no matter how small the difference is*. This leads us to the question of β error and the need for a *specific* alternate to be testing against. Only by presenting an alternative, which implies that you are concerned about some minimum or particular difference or change, can you reconcile this problem. However, we will not pursue this problem in this text.

REVIEW

Interval data · Ratio data · *Limits:* Larger · Smaller · Both ·

Test for spreads: · χ^2/df test ·

Test for means: Variance given or known: z-test ·

Variance unknown: t-test ·

In general:

$$\text{Statistic} = \frac{\text{Sample mean} - \text{Theoretical mean}}{\text{Standard error of the mean}} \cdot$$

$$z = \frac{\bar{X} - \mu}{\sigma/\sqrt{n}} \quad \cdot \quad t = \frac{\bar{X} - \mu}{s/\sqrt{n}}$$

PROBLEMS

Section 16.1

16.1 I have 4 light bulbs in a ceiling light fixture that is inconvenient to reach. I have decided that if the variance of the life of light bulbs is small it pays to change all 4 when one burns out. I have decided that if σ is less than 50 hours I will change all the bulbs at one time. I tested a group of 12 light bulbs to determine the variation among them. The variance of these light bulbs was 1600. Should I change all light bulbs? (First decide if this is a one- or two-sided test. Then test at the 5% level.) What are the possible errors and the consequence of these errors?

16.2 An oven that is regulated must not vary much for consistent baking. A particular food requires the variance of the temperature not to be more than 30. The following eight readings were taken in the oven:

350, 370, 360, 355, 360, 365, 355, 365   ($\bar{X} = 360$).

Is the variation in the oven excessive? (Test at the 5% level.)

16.3 A machinery manufacturer claims that his new machine will reduce the variation in the size of parts. The company that is interested in purchasing the machine has him make 10 parts that usually have a size of 1.000 inches with variance of 0.000005. The variance in the size of the 10 parts was 0.000002. Is this a significant reduction in variation? (Test at 1% level since the machine is very expensive.)

16.4 The precision of an instrument is related to the variance of repeated readings. A new type of laser beam transit is claimed to repeat its distance measurements so that 95% of the time it is within ± 1.2 inch (over a distance up to several miles). (These are 2σ limits.)

Two points were chosen approximately a mile apart and 16 measurements were taken. The variance of these measurements was found to be 2.43. Do you challenge the manufacturer's claim? What would you give as 99% confidence intervals that the variance of his readings are?

16.5 A psychologist has been attempting to determine if marijuana affects response time to a stimulus. The variance of previous tests of response time for a volunteer has been 0.5 seconds. The volunteer then smoked a cigarette and tried the test 5 times. His times in seconds were

$$2.6,\ 4.3,\ 2.2,\ 4.4,\ 5.5.$$

Has the variation in response increased? (Use $\alpha = 0.05$.)

16.6 With certain types of plants harvesting is made easier if all the fruits or vegetables on a tree or in a field mature at the same time. A certain type of corn has matured in 45 days with a standard deviation of 3 days. Twenty-five plants of a new strain of corn plant were grown. They matured in 48 days with a standard deviation of 1 day. Is there a significant change in the variation? (Test at the 10% level.)

16.7 If the width of floor tiles varies too greatly from tile to tile then the pattern does not match when the tiles are set down on the floor. A tile manufacturer asserts that the 12-inch tiles have a standard deviation of only 0.05 inches. I measured the tiles in one carton of 24 tiles and found the variance of the tiles was 0.01. Is this significant (with $\alpha = 0.05$)? What interval would the true variance probably be within (95% CI)? Under the worst condition (largest variance) try to picture the possible effect such a variation would have if you were setting down tiles of this size. Would they match poorly?

Section 16.2

16.8 A class of 25 sixth graders are given a standardized reading test. The national norm on the test is a mean of 100 with standard deviation of 10. The class averaged 108. Is this class unusual? Remember that the limits would have been decided upon before the test was given—they are interested in knowing if the class is good or bad. (Use a 5% level.)

16.9 A special automobile tire has been developed that is believed to reduce the distance it takes for your car to stop. On a special surface the old tires have required an average of 110 feet (with $\sigma = 15$ feet) to stop at 60 miles per hour. Ten sets of new tires were tested and they averaged 98 feet. You want at least 99% assurance they are significantly better before marketing will take place. Are they?

16.10 A newspaper has been selling an average of 150,000 copies every weekday (with $\sigma = 20,000$). As a gimmick to increase circulation the paper ran a contest. Circulation for the first week then was

153,000; 158,000; 161,000; 160,000; 167,000; 161,000.

Has the contest been a success?

★If the cost of the contest was $30,000 and the paper sells for 10 cents has it been a success monetarily?

16.11 An airline claims its departures average only 8 minutes delay with a standard deviation of 4 minutes. One day while waiting for a flight I checked 9 flights and found they averaged 12 minutes delay. Is the claim correct? (Test at $\alpha = 0.05$.)

16.12 A chemist has known that 95% of the time a certain polymer reacts between 18 and 24 seconds after adding the standard ingredients. After adding a new catalyst reactions averaged 19.5 seconds for 6 tests. Has the catalyst changed the average reaction time? (Use the 0.05 level.)

16.13 An olive jar contains the statement "net weight 12 oz.—average of 50 olives." Ten jars were picked and the number of olives were counted. They contain the following number of olives:

43, 43, 47, 44, 48, 51, 46, 48, 47, 53.

Do you agree that the olives could be averaging 50? Within what limits do you expect the true mean to be? (Use 95% limits.)

16.14 Refer to the airline statement in Problem 16.11. Typically they would state that flights averaged only 8 minutes delay time. What would the decision be if the sample of 9 flights had a variance of 21?

16.15 A large orchard has averaged a yield of 140 pounds of apples per tree. A new fertilizer is tested to try to increase the yield. Twelve trees are selected and the following yields in pounds were recorded:

155, 145, 140, 165, 150, 145, 145, 160, 150, 140, 135, 155.

Is there a significant increase? (Use a 0.10 level.)

16.16 Refer back to the 12 women attending "weightwatchers" (Problem 4.2). Is the average weight loss significantly different from zero? (Use $\alpha = 0.05$.)

16.17 Refer back to the new houses of Problem 4.1. Does this data contradict the publicity statement of:

"Our homes' average cost is significantly less than $29,500."

How do you test this?

16.18 A neighbor claims it should only take 14 minutes to get to school. Does the time it takes me (Example 2.2) indicate otherwise? (I don't talk back to this neighbor unless I am 99% sure.)

16.19 A common question that is asked is how large a sample must I take. For example, the Weights and Measures inspector in Example 16.3 must decide beforehand how many readings to take. He must declare several things. He must approximate the value of σ, state α and decide how small a confidence interval or difference is permissible. This last value is essentially the maximum amount he is willing to accept as an error in the location of the true population. The formula for sample size, n, is then

$$n = \left(\frac{z_{1-\alpha/2}\sigma}{\text{Error}}\right)^2, \qquad (16.10)$$

where *error* is the maximum difference he will accept.

Thus if the inspector desired to be off in his estimate of true weight no more than 0.02 pounds, error = 0.02. Since $\alpha = 0.05$, $z_{1-\alpha/2} = z_{0.975} = 1.96$; σ was 0.04. Then

$$n = \left(\frac{(1.96)(0.04)}{0.02}\right)^2 = (3.92)^2 = 15.37.$$

So he must take a sample of 16.

a. How large a sample must he take if he wants to be off by no more than 0.01?

b. How large a sample must he take if he wants to have 99% assurance of being off by no more than 0.02 pounds?

16.20 If the bowler in Example 16.4 wants to be certain that he will detect a 5 point shift in his bowling average in either direction (with $\alpha = 0.05$), how large a sample must he take?

16.21 Referring to Problem 16.9, how large a sample of tires must be used if we want 99% assurance of detecting a decrease of 5 feet? (Note that you should be using $z_{1-\alpha}$ not $z_{1-\alpha/2}$.)

16.22 Referring to Problem 16.10, if the newspaper raises its price to 15 cents, how long will it take to be 99% sure that a decrease in circulation of 5000 has or has not taken place? How long will it take to detect a 10,000 decrease?

17

Comparing Two Independent Samples

We will now explore a collection of statistical tests that are used to decide if the difference between a pair of samples is significant. Of course, as always, it is not the difference in samples that we are interested in, but whether or not the populations that these samples came from are different. This chapter will be restricted to those cases in which samples are completely independent or at least assumed to be independent. Also, one of the tests will aid in determining if two sample conditions *are* independent.

17.1 Statistical Tests for Comparing Two Independent Samples

The tests we will consider in this chapter all differ either in the characteristic they are measuring or in the type of numbers they work with—categorical, ordered, interval, or ratio. Table 17.1 lists the tests, the characteristics they are testing, and their requirements or restrictions.

It should be stressed that the samples chosen are assumed to be independent. Again, except for the chi-square test, whose function is to determine if two characteristics are independent of each other, all cases involve two independent *sets* of measurements. Some examples of the types of independent samples we could consider are:

1. We compare the effectiveness of two different kinds of birth control devices as given to two completely different groups of people. We randomly choose two groups of people and request them to use one particular method. (Dependent samples would involve some or all of the people using both of the techniques.)
2. The effectiveness of teaching by computer or by classroom approach is to be measured. Two independent samples would be involved when we take

two groups of students who register for the course and randomly assign them to one or the other method.
3. We poll people for their attitudes toward the President and the economy. Independent samples would require administering a *separate* poll about the President to one group of people and another poll to another group of people regarding the economy. There may be a strong likelihood (as we will explore under the chi-square test) that people who express strong feelings on one question will tend to regard the second question in much the same way.

Table 17.1 List of statistical tests in Chapter 17

Name	To test	Requirements	
		Minimum type of numbers	Restrictions
1. χ^2 (chi-square) two-way	Independence	Categorical	Should expect at least 5 in each category
2. Difference in proportions	Difference in proportions	Categorical	Depends on size of sample
3. Mann-Whitney	Difference in populations	Ordered	Continuous scale
4. F-test	Difference in Variances	Interval	Normality
5. Means tests			
a. z-test	Difference in means	Interval	Normality population variances known
b. t-test	Difference in means	Interval	Normality

17.2 Test for Independence of Two Sets of Data

This test is an extension of the previous work we did with the chi-square test. At that time we compared the breakdown of a set of measurements that were in a collection of categories against a theoretical or expected set of values for the same categories. We will now do essentially the same thing except this time we will *simultaneously classify each individual observation* or frequency according to two *separate* categories.

We will also calculate a collection of expected frequencies based on the assumption that the two characteristics we are working with are independent. We then calculate a χ^2-value in the same way as before.

First we should use a simple example to illustrate the concept and then I will broaden the scope of the categories. I should emphasize at this point that the chi-square test is generally not considered very good for comparisons with only two possibilities for each of the characteristics. There are correction factors (one called the Yates correction) that make the results more realistic. However, a two-by-two arrangement is excellent for illustrating the technique.

EXAMPLE 17.1

There is a continual argument by the "older" generation that long hair and "hippie" style is related to smoking marijuana—that is, long-haired "hippies" smoke marijuana, clean-cut students don't. A survey of 75 students provided the following two pairs of breakdowns:

| Long hair | 25 | Smoke marijuana | 33 |
| Clean-cut | 50 | Don't smoke marijuana | 42 |

If you now recall our discussion of correlation (Chapter 6) you should recognize the analogy to that situation. The listing above is not a satisfactory arrangement for observing a possible relationship between marijuana and hair length. The question of whether two things that are being measured are independent is really a question of relationship. Thus, we must again have pairs of observations for each individual. In this case we have the breakdown in Table 17.2. Here we find, for example, that of the 33 students who smoke marijuana, 13 have long hair and 20 are clean-cut. Of the 50 who are classified as clean-cut, 20 smoke marijuana and 30 do not, and so forth.

Table 17.2 Breakdown and relation between marijuana smoking and life style (as expressed by hair length)

Style	Marijuana		Total for hair
	Smoke	Don't smoke	
Long hair	13	12	25
Clean-cut	20	30	50
Total for marijuana	33	42	Overall total 75

The problem is that just looking at Table 17.2 does not really give you an indication of whether or not there is any relationship between these characteristics. We need to compare the values in the table against some kind of reference. We will now attempt to determine what values to expect in *each* box.

We start off by agreeing on the totals. They are the values we listed before and are listed along the sides of the chart in Table 17.2 (see also Table 17.3a). Then

If the two characteristics are independent then each subgroup will be divided up into the same proportions as the totals are divided.

Table 17.3 Arriving at the set of theoretical frequencies assuming independence

	Marijuana		
	Smoke	Don't smoke	
Long hair		25	
Clean-cut		50	
	33	42	75

a. Identify the totals for each row and each column.

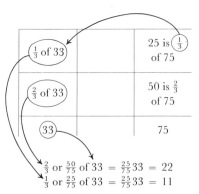

$\tfrac{2}{3}$ or $\tfrac{50}{75}$ of $33 = \tfrac{25}{75} 33 = 22$
$\tfrac{1}{3}$ or $\tfrac{25}{75}$ of $33 = \tfrac{25}{75} 33 = 11$

b. Find the proper proportions for one column (or row) by calculating the proportion of the row total and multiplying the proportion times the total in the particular column you are splitting.

11	14	25
22	28	50

c. Final theoretical values can be found by subtraction. The 14 is found as 25 − 11.

	Also $\tfrac{33}{75}$ of 25		25
			50
	33	42	75

d. We can also find the same 11 in the upper left corner as $\tfrac{33}{75}$ths of the 25 in the row total.

Thus, we find that the total of 75 students are split into two groups according to hair length with $\tfrac{25}{75}$ or $\tfrac{1}{3}$ having long hair and the other $\tfrac{50}{75}$ or $\tfrac{2}{3}$ labeled as "clean-cut". We then reason that *if there is no relation* between hair length and smoking marijuana then each of the two categories under marijuana will *also* be subdivided $\tfrac{1}{3}$ long hair and $\tfrac{2}{3}$ clean-cut. Therefore, we expect $\tfrac{25}{75}$ of 33 or 11 of the 33 students who smoke to have long hair and $\tfrac{50}{75}$ of 33 or 22 students who smoke to be clean-cut (see Table 17.3b).

In a similar manner we can subdivide the 42 students who don't smoke into two categories of $(\tfrac{25}{75}) \cdot 42 = 14$ *with long hair* and $(\tfrac{50}{75}) \cdot 42 = 28$ *clean-cut*. We finally arrive at Table 17.3c.

Two additional important observations should also be made:

1. It does not matter which order we use to arrive at the theoretical values. We can also calculate the upper left corner as $\left(\frac{33}{75}\right) \cdot 25 = 11$. Here we are finding the proportion of long-hair students who also smoke marijuana. The net result, however, is the same.
2. In this problem, no matter what value you start with in the table, once you have found *one* value all of the others are established. Once you have found the 11 you can immediately find the 22 and the 14 by subtraction. After you find either the 22 or the 14 you can also find the 28 by subtraction. So only one value can be freely varied, or there is only 1 degree of freedom.

We have now found a collection of theoretical values to complement the actual observations. In Table 17.4 we have both sets of values entered in the same boxes. Next we will compare the differences by computing a value of χ^2

Table 17.4 Final table and calculations for determining if there is a relation between marijuana and hair type. The values in the shaded small boxes are the theoretical values (assuming independence)

Style	Marijuana		Total
	Smoke	Don't smoke	
Long hair	11 / 13	14 / 12	25
Clean-cut	22 / 20	28 / 30	50
Total	33	42	75

$$\chi^2 = \sum \frac{(\text{Frequency}_{\text{actual}} - \text{Frequency}_{\text{theo}})^2}{\text{Frequency}_{\text{theo}}} = \frac{(13-11)^2}{11} + \frac{(12-14)^2}{14} + \frac{(20-22)^2}{22} + \frac{(30-28)^2}{28}$$

$$= \frac{(2)^2}{11} = \frac{(-2)^2}{14} + \frac{(-2)^2}{22} + \frac{(-2)^2}{28} = \frac{4}{11} + \frac{4}{14} + \frac{4}{22} + \frac{4}{28}$$

$$= 0.36 + 0.29 + 0.18 + 0.13 = 0.90$$

For 1 degree of freedom $\chi^2_{0.95} = 3.84$

essentially in the same way as we did in Chapter 15. We use Equation (15.1), which was

$$\chi^2 = \sum \frac{(\text{Actual frequencies} - \text{Theoretical frequencies})^2}{\text{Theoretical frequencies}}.$$

The sum here is to include the calculations for *all* of the boxes. For this example, the computed χ^2 is 0.90 (see Table 17.4). It is to be compared to a χ^2-value in Appendix Table A.3 having 1 degree of freedom. We find the 95% χ^2-limit to be 3.84. As before, if our computed statistic is less than the limit we would state that there is no significant difference or these two general categories are independent of each other. The computed statistic is much less than the limit and is therefore *not* significant. Therefore, we state there is no relation between smoking marijuana and style, they are independent.

EXAMPLE 17.2

Two hundred people were polled for their opinion of the economic situation and the President's overall job performance. The complete breakdown for the responses are given in Table 17.5.

Table 17.5 Comparison of attitudes of 200 people to President's overall job performance and the economic situation

Economic Situation	President's performance				Total
	Very good	O.K.	Poor	No opinion	
Good	31	9	4	6	50
Poor	10	21	60	9	100
No opinion	9	16	10	15	50
Total	50	46	74	30	200

Observe the similarity between this table and Table 17.2 for the previous example. The only difference is its size. As in the previous example we must find a set of theoretical or expected frequencies given the assumption that the responses on the two questions are unrelated. If they are unrelated then when we look at the responses to the question on economic situation we observe the distribution of these responses ($\frac{1}{4}$ good, $\frac{1}{2}$ poor, and $\frac{1}{4}$ no opinion). We then assume that the 50 very good responses to the *second* question *will be distributed in the same* manner ($\frac{1}{4}$ good, $\frac{1}{2}$ poor, $\frac{1}{4}$ no opinion). Thus we would expect 12.5 good, 25 poor, and 12.5 no opinion responses. Table 17.6 includes all of the

Table 17.6 Calculating the expected frequencies (theoretical) for poll on attitude on President's performance and economic situation

Economic Situation	President's performance				
	Very good	O.K.	Poor	No opinion	
Good	12.5	11.5	18.5	7.5	
Poor	25.0	23.0	37.0	15.0	100
No opinion	12.5	11.5	18.5	7.5	
		46			200

As one example of the calculations the box under row "Poor" and column "O.K." is emphasized. It is found by

$$\frac{\text{(Row total for row of box) (Column total for column of box)}}{\text{Grand total}} = \frac{(100)(46)}{200} = 23.0.$$

Table 17.7 Calculations to determine the χ^2 value for independence of attitudes on Presidential performance and economic situation

Actual	Expected	(Actual − Expected)	(Actual − Expected)^2	$\dfrac{(Actual - Expected)^2}{Expected}$
31	12.5	18.5	342.25	27.3
9	11.5	− 2.5	6.25	0.5
4	18.5	−14.5	210.25	11.4
6	7.5	− 1.5	2.25	0.3
10	25.0	−15.0	225.00	9.0
21	23.0	− 2.0	4.00	0.2
60	37.0	23.0	529.00	14.3
9	15.0	− 6.0	36.00	2.4
9	12.5	− 3.5	12.25	1.0
16	11.5	4.5	20.25	1.8
10	18.5	− 8.5	72.25	3.9
15	7.5	7.5	56.25	7.5
				$\Sigma = 79.6$

$\chi^2_{0.99} = 16.8$
$df = 6.$

expected responses. Note that in general the expected frequency for any given box is

$$\frac{\text{(Row total for the row of that box) (Column total for column of that box)}}{\text{(Grand total)}} \tag{17.1}$$

We compute the χ^2 as before and Table 17.7 provides the computations. If you prefer, you could also do all the calculations in the boxes themselves. The total is found to be 79.6. This is to be compared to a χ^2-value in the Appendix Table A.3. The only question remaining is degrees of freedom.

Degrees of Freedom

A single formula is available to calculate the number of degrees of freedom for any table of the form we have been using (commonly called a χ^2 contingency table).

$$\text{Degrees of freedom} = (\text{Number of rows} - 1)(\text{Number of columns} - 1)$$
$$= (r - 1)(c - 1) \tag{17.2}$$

where r = the number of rows, and c = the number of columns.
For this example $r = 3$, and $c = 4$. Therefore,

$$(r - 1)(c - 1) = (3 - 1)(4 - 1) = (2)(3) = 6 \, df.$$

Another way of arriving at the number of degrees of freedom is to find how many numbers you can freely choose to place in the body of the table (remember that the outside values are already established). For example, starting in the upper left corner you can put any of a variety of values in each of the first *3* columns of the top row. After you pick 3 values the 4th can be only one number. You can then do the same thing in the second row, now having a total of six values that you can choose. But that is where it ends. In the third row there is no longer any choice—try it. Thus we have six degrees of freedom.

Now that we have determined the number of degrees of freedom, we can find the χ^2-limit. Looking up in Appendix Table A.3 we find under $df = 6$ that the 1% limit, $\chi^2_{0.99}$, is 16.8. Our χ^2 of 79.6 is much larger than this limit and therefore we say that there is much less than a 1% chance of error in declaring that these two questions have responses that are *not* independent. We conclude that the responses *are dependent*—the attitude toward Presidential Performance and the economic situation are related to each other in some way.

17.3 Difference in Proportions Test

Quite often we have two independent pairs of proportions to compare (two samples from binomial distributions). In Chapter 9 we had a collection of examples of types of measurements that could be reported as proportions of the two categories. We now will look at how some of these examples could be extended to cases involving the questions of differences or changes in proportions.

EXAMPLE 19.1 (Cont.)
- b. A poll is taken before and after a major speech. We question if there has been a change in the percent of the people in favor of the candidate.
- d. Are the proportion of defective automobiles different for two different models or between different parts of the automobile?
- f. Is there a difference between the amount of time two different traffic lights are red?
- h. Is there a significant difference in the proportion of students who fail the same course taught by two different instructors?
- i. Is there a change in the proportion of empty taxicabs after a fare increase takes place?

In all these cases we are to take two samples and determine two proportions, which we will then designate as $X_1/n_1 = \bar{X}_1$ and $X_2/n_2 = \bar{X}_2$ (where the subscripts 1 and 2 designate the values for the two samples). The hypothesis that we will consider is

Hypothesis: There is no difference between the population proportions for the two samples. That is, P_1, population proportion for first sample, $= P_2$, population proportion for the second sample.

Alternate Hypothesis: There is a difference; P_1 does not equal P_2 (two-tailed test).

Assumption: Both sample sizes are large (all $n \cdot p$'s and $n \cdot (1-p)$'s are larger than about 5).

Limits:

$$\text{Upper} \quad z_{1-\alpha/2}$$
$$\text{Lower} \quad z_{\alpha/2}$$

(These values are obtained from Table 1a or 1b.)

Test Statistic:

$$z = \frac{\bar{X}_1 - \bar{X}_2}{\sqrt{P(1-P)(1/n_1 + 1/n_2)}} \tag{17.3}$$

where

$$P = \frac{X_1 + X_2}{n_1 + n_2}.$$

Decision: If z is greater than $z_{1-\alpha/2}$ or smaller than $z_{\alpha/2}$ we declare that these proportions are different; otherwise we accept the hypothesis that there is no difference.

Confidence interval: Since for large n_1 and n_2 the standard error or the difference in two proportions is

$$\sqrt{\frac{\bar{X}_1(1-\bar{X}_1)}{n_1} + \frac{\bar{X}_2(1-\bar{X}_2)}{n_2}}$$

and is also normally distributed, then the confidence interval is

$$\bar{X}_1 - \bar{X}_2 + z_{\alpha/2}\sqrt{\frac{\bar{X}_1(1-\bar{X}_1)}{n_1} + \frac{\bar{X}_2(1-\bar{X}_2)}{n_2}} < P_1 - P_2$$

$$< \bar{X}_1 - \bar{X}_2 + z_{1-\alpha/2}\sqrt{\frac{\bar{X}_1(1-\bar{X}_1)}{n_1} + \frac{\bar{X}_2(1-\bar{X}_2)}{n_2}} \quad (17.4)$$

EXAMPLE 17.3

The flu vaccine was given to 125 employees while the other 75 did not receive it. Of the employees who lost days because of illness 30 had received the vaccine and 25 did not. Thus $\bar{X}_1 = X_1/n_1 = 30/125 = 0.24$ (the proportion of those that received the vaccine and became ill), and $\bar{X}_2 = X_2/n_2 = 25/75 = 0.33$ (the proportion of those that did not receive the vaccine but became ill).

Hypothesis: The proportion of people who became ill after receiving the vaccine is not different from the proportion of people who became ill but did not receive the vaccine.

Alternate Hypothesis: There is a difference in proportions.
We will use a 5% level of significance ($\alpha = 0.05$).

Limits:

$$\text{Upper} \quad z_{1-\alpha/2} = z_{0.975} = 1.96.$$
$$\text{Lower} \quad z_{\alpha/2} = z_{0.025} = -1.96.$$

Test Statistic: First we calculate the overall average P:

$$P = \frac{X_1 + X_2}{n_1 + n_2} = \frac{30 + 25}{125 + 75} = \frac{55}{200} = 0.275.$$

Then

$$z = \frac{\bar{X}_1 - \bar{X}_2}{\sqrt{P(1-P)(1/n_1 + 1/n_2)}} = \frac{0.24 - 0.33}{\sqrt{(0.275)(1-0.275)(1/125 + 1/75)}}$$

$$= \frac{-0.09}{\sqrt{(0.275)(0.725)(0.008 + 0.013)}}$$

$$= \frac{-0.09}{\sqrt{(0.275)(0.725)(0.021)}}$$

$$= \frac{-0.09}{\sqrt{0.0042}} = \frac{-0.09}{0.065} = -1.39.$$

Decision: Since -1.39 is not larger than 1.96 nor smaller than -1.96, we accept the hypothesis that there is no difference. We decide that our chance of becoming ill is essentially the same whether or not we receive the vaccine.

Confidence interval:

$$\bar{X}_1 - \bar{X}_2 + z_{\alpha/2}\sqrt{\frac{\bar{X}_1(1-\bar{X}_1)}{n_1} + \frac{\bar{X}_2(1-\bar{X}_2)}{n_2}} < P_1 - P_2$$

$$< \bar{X}_1 - \bar{X}_2 + z_{1-\alpha/2}\sqrt{\frac{\bar{X}_1(1-\bar{X}_1)}{n_1} + \frac{\bar{X}_2(1-\bar{X}_2)}{n_2}}$$

$$0.24 - 0.33 + (-1.96)\sqrt{\frac{(0.24)(0.76)}{125} + \frac{(0.33)(0.67)}{75}} < P_1 - P_2$$

$$< 0.24 - 0.33 + 1.96\sqrt{\frac{(0.24)(0.76)}{125} + \frac{(0.33)(0.67)}{75}}$$

$$-0.09 - 1.96\sqrt{0.00146 + 0.00295} < P_1 - P_2$$

$$< -0.09 + 1.96\sqrt{0.00146 + 0.00295}$$

$$-0.09 - 1.96\sqrt{0.00441} < P_1 - P_2 < -0.09 + 1.96\sqrt{0.00441}$$

$$-0.09 - 0.13 < P_1 - P_2 < -0.09 + 0.13$$

$$-0.22 < P_1 - P_2 < 0.04$$

Thus we say the difference in proportions (taking vaccine versus not taking vaccine) can be anywhere from $+0.04$ (a 4% increase in illness) to -0.22 (a 22% decrease in illness). Since this straddles 0% there is a possibility of *no* change.

17.4 Mann-Whitney U-test

This is a nonparametric test that will permit us to compare two distributions to determine if they are different. It is sensitive to differences in mean, variance, or shape, although we may not be able to identify which of these differences is the cause for a significant result in the test.

The requirements of the test are that the data must be at least ordered and that the population values actually cover a continuous range or scale. The measurements that we observe may be discrete such as "I like the cafeteria food, I hate the cafeteria food" or your grade is "A, B, C, D, or F." In both of these cases there are obviously many differences within the categories so that we could hypothetically rank the individual observations even further.

When the data is merely categorical or is ordered, we should not use tests involving an assumption of a normal distribution—for example t, z, χ^2/df, tests. In addition, even when the data is measured on an interval scale there are occasions when we do not believe that the population is normally distributed. In such cases, the Mann-Whitney test is quite useful.

If we take two samples from the same population and rank the values in each of the samples we would expect that considerable overlapping between the samples should take place. If we lump all of the values into one ordered sequence we would then expect a random pattern to occur. On the other hand, if the two samples came from different populations, we would expect the two samples to tend toward different ends of the overall group. A statistic called U can be used to gauge the degree of randomness between the two samples.

We are to look at two samples, one of size n_1 and the other of size n_2. We then combine all of the values by placing them in a single sequence in increasing order. In addition, we label or identify in some way the group the values belong to. We then calculate for each and every value of *one* of the samples the number of values of the other sample that is below it. The overall sum is U. The following example should explain this procedure.

EXAMPLE 17.4

Two brands of flashbulb batteries are tested for the number of flashes they can trigger before they are used up. Six "Long-Life" batteries and eight "Brand X" batteries are tested and the number of flashes are recorded in Table 17.8. We can assume that the true population includes the possibility of less than a full flash, which often is the case with a partly dead battery. It is also possible for the distribution to be quite skewed.

Table 17.8 Number of flashes with different batteries before being used up

"Long-Life"			"Brand X"			
64	71	68	69	78	85	80
84	93	76	97	81	104	98
$n_1 = 6$			$n_2 = 8$			

First we rank all of the fourteen batteries in a single sequence. Table 17.9 shows the ranking. Next to each value is the symbol representing the type of battery associated with the number of flashes. LL stands for "Long-Life" and X stands for "Brand X."

Next we are to calculate the value of U. First we select one of the types of values, for example the LL's. For *each LL* we count how many X's come before the particular LL. Thus we observe that there are *no* X's before the first LL and we calculate a zero for the first LL. In a similar fashion there are zero X's before the second LL. For the third LL there is *one* X before it, and we count 1 for this *LL*. The next LL also has one X before it. The fifth LL has a *total* of

Table 17.9 Mann-Whitney test to determine if there is a difference in the number of flashes different brands of batteries can trigger

Number of flashes	Brand of battery	Rank	Number of X's below the LL's	Number of LL's below the X's
104	X	14	–	6
97	X	13	–	6
95	X	12	–	6
93	LL	11	5	–
85	X	10	–	5
84	LL	9	4	–
81	X	8	–	4
80	X	7	–	4
78	X	6	–	4
76	LL	5	1	–
71	LL	4	1	–
69	X	3	–	2
68	LL	2	0	–
64	LL	1	0	–
			Total, U 11	37

4 X's before it and we assign a 4 for this LL. The final LL has 5 X's preceding it and therefore is assigned a 5. We then *add up all of the numbers* associated with the LL's and we obtain an 11. This is the value of U.

Appendix Table A.21 provides a set of critical values for U. Let us consider making our decision as a two-tailed test with $\alpha = 0.05$ (5% level). Since $n_1 = 6$ and $n_2 = 8$ we find the minimum critical value of U to be 8. The value we obtained is larger (11) and is therefore *not* significant.

Although we have satisfactorily completed this example, go back to Table 17.9 and look at the line labeled "number of LL's below the X's." Here you can see that if you had selected the other sample values for comparison you would have obtained a much larger value of U, 37 in this case. We label this larger value of U as U' (called U prime). There is a relationship between these two values of U.

$$U = n_1 n_2 - U'. \qquad (17.5)$$

Thus, when we know that one value of U is 11 we can calculate the other as:

$$U = n_1 n_2 - U' = 6 \cdot 8 - 11 = 48 - 11 = 37,$$

which agrees with the value we found by adding up the values. This is a convenient and simple way of finding the smaller of the two possible values of U or may serve as a check if you have computed both U-values.

Another way of computing U is to work directly with the ranks. This is particularly useful if the n are large. We choose one of the groups and add up the ranks for each of the values. We call this R_1 if we are working with the n_1-values and R_2 if we are working with the n_2-values. Then U is

or
$$U = n_1 n_2 + \frac{n_1(n_1 + 1)}{2} - R_1$$
$$U = n_1 n_2 + \frac{n_2(n_2 + 1)}{2} - R_2.$$
(17.6)

If we choose LL then from Table 17.9 we find that the ranks are 1, 2, 4, 5, 9, and 11. Therefore

$$R_1 = 1 + 2 + 4 + 5 + 9 + 11 = 32.$$

Since $n_1 = 6$ and $n_2 = 8$

$$U = n_1 n_2 + \frac{n_1(n_1 + 1)}{2} - R_1$$

$$= 6 \cdot 8 + \frac{6(6 + 1)}{2} - 32$$

$$= 48 + \frac{6 \cdot 7}{2} - 32$$

$$= 48 + 21 - 32 = 37.$$

To find the smaller value of U we would then substitute into Equation (17.5) and we would obtain $U' = 11$.

Ties: Sometimes we obtain several measurements with the same value. Perhaps several batteries had the same number of flashes. If the ties were with batteries of the same brand, samples from the same group, there is no effect on U. However, if the ties are with values from both groups, then there is a problem. The recommended policy is to use the average rank for all of the tied values and then proceed as usual. Unless there is a very large proportion of ties the effect of having ties is negligible and any correction is generally unnecessary.

When n is greater than 20: Appendix Table A.21 stops with $n = 20$. When either n is larger than 20 (or even somewhat less) the normal distribution can be used to find limits. You calculate the value of U in the same way as before. Then, knowing n_1 and n_2 we consider that U should be distributed with mean

$$\mu = n_1 n_2 / 2,$$

and standard deviation

$$\sigma = \sqrt{\frac{n_1 n_2 (n_1 + n_2 + 1)}{12}}.$$

Then the test statistic would be

$$z = \frac{U - \mu}{\sigma} = \frac{U - n_1 n_2/2}{\sqrt{\dfrac{n_1 n_2 (n_1 + n_2 + 1)}{12}}}. \qquad (17.7)$$

For a two-tailed test an an alpha significance level, the limits would be

Upper $z_{1-\alpha/2}$

Lower $z_{\alpha/2}$.

(Problems 17.12 and 17.13 are examples which require this technique.)

17.5 Difference in Variances: *F*-test

Many times we take two samples and want to know if their spreads are essentially the same. In addition, a number of statistical tests that are used to study other characteristics of data, such as the mean, make the assumption that the two samples came from populations whose spreads are the same. If there is any suspicion that the spreads may not be the same, you should compare the spreads of the two samples. Here we will make one assumption, that the samples *both came from normal distributions*. If this is true, and if the spreads *are* the same, then the ratio of the variances of the two samples will fall into an *F*-distribution. We therefore call this the *F*-test.

We take two independent samples and compute the sample variances. If we label the two samples 1 and 2, then

s_1^2 = variance of the first sample.
n_1 = size or number of values in the first sample.
σ_1^2 = variance of the population that the first sample came from.

and

s_2^2 = variance of the second sample.
n_2 = size of the second sample.
σ_2^2 = variance of the population that the second sample came from.

We will asume that s_1^2 and s_2^2 will undoubtedly be different for any given set of samples, but σ_1^2 and σ_2^2 could be the same. The underlying question is *whether the population variances are the same*.

We will look only at a two-tailed test, but I will comment about the difference between one- and two-tailed tests.

Hypothesis: There is no difference between the population variances of the two samples ($\sigma_1^2 = \sigma_2^2$).

Alternative Hypothesis: There is a difference. One population variance is larger than the other ($\sigma_1^2 < \sigma_2^2$ or $\sigma_1^2 > \sigma_2^2$).

Both sample sizes are known, α is to be provided, σ_1^2 and σ_2^2 are *not* specified in any way. s_1^2 and s_2^2 are computed from the data.

Test Statistic:

$$F = \frac{\text{Larger } s^2}{\text{Smaller } s^2} = \frac{s_1^2}{s_2^2}.$$

Limit:

$F(f_1, f_2)$ (From Appendix Table A.5* listed as two-tailed test. You select the α level.)

where
$$f_1 = df \text{ in numerator} = n_1 - 1.$$
and
$$f_2 = df \text{ in denominator} = n_2 - 1.$$

Decision: If the calculated F is smaller than the limit, accept the hypothesis of no difference between the variances. If the calculated F is larger than $F(f_1, f_2)$ we reject the hypothesis and accept the alternate hypothesis. We declare that σ_1^2 does not equal σ_2^2.

EXAMPLE 17.5

The quality of paper used for printing is very dependent upon the variation in thickness. Two different brands of paper are to be compared. Ten readings of paper taken from Quality Paper Company had $s^2 = 0.0022$ inches, 13 readings taken on paper from Superior Paper Inc. had $s^2 = 0.0031$ inches.

Hypothesis: There is no difference in variability between paper from Quality Paper Company and Superior Paper Co. ($\sigma_1^2 = \sigma_2^2$).

Alternate Hypothesis: There is a difference in variability between the brands of paper.

We will use a 5% level of significance and a two-tailed test.

Test statistic:

$$F = \frac{\text{Larger } s^2}{\text{Smaller } s^2} = \frac{0.0031}{0.0022} = 1.41.$$

Now we can identify which is s_1^2 and which is s_2^2 since s_1^2 is to be associated with the larger value of variance. Therefore:

$$s_1^2 = 0.0031 \qquad s_2^2 = 0.0022$$
$$f_1 = n_1 - 1 \qquad f_2 = n_2 - 1$$
$$= 13 - 1 \qquad \quad = 10 - 1$$
$$= 12. \qquad \qquad = 9.$$

---

* See Chapter 12 to review the method of obtaining values from the F-tables.

Limit: The 5% F-limit or $F_{0.95}$ is

$$F(f_1, f_2) = F(12, 9) = 3.87.$$

Decision: Since the test statistic, 1.41, is less than the limit, we *accept* the hypothesis that there is no significant difference in variability between the two kinds of paper.

When you want to perform one-tailed F-tests, the only difference in the above procedure is that you select which sample goes on top of the test statistic prior to performing the calculation. You select the variance of the sample that you expect should be the larger one to go in the numerator. Then you select the limit with the appropriate significance level from the table labeled One-Tailed Test.

17.6 Difference in Two Means: z- and t-tests

One of the most common types of problems encountered concerns the question of whether there are differences in the means of two populations. This section will present several tests for cases when two independent samples are drawn from distributions that are relatively normal (or else are moderately large samples). Actually, one basic test is being employed here. The only distinction between the several variations to be discussed have to do with what we use for standard deviation. In particular we need to distinguish between situations of the following types:

1. We know or do not know what the standard deviation is, and
2. We consider the standard deviations of the two populations to be equal or not equal.

The first question establishes the limit value and the second provides us with the test statistic.

1. If we *know the standard deviation* the limit is a z-value and we have a z-test. If we do *not* know the population standard deviation, then we must estimate it from the samples. We then use a t-limit, and call this a t-test.

2. If the standard deviation is to be the same for both samples then the z-test is somewhat simplified, as we only have one value of σ. For a t-test we should first check to see if the variations are the same (by using an F-test). If the F-test shows no differences we then average them together into what we call a pooled variance—a single estimate of σ. If they are different we have some complications in the calculations.

Since in Chapter 16 we spent such a large amount of time demonstrating small differences between a variety of z- and t-tests, you should be quite familiar with the consequences of such differences in conducting those statistical tests. Here we have a similar collection of different tests, all following a common format. In this section we will work with an outline of the hypothesis as we have before, but we will use a table to identify the limits and the test statistics for each case. Several examples will then follow.

Hypothesis: There is no difference between the population means (μ_1, mean of the first population $= \mu_2$, mean of the second population).

Alternate Hypothesis: There is a difference between the population means.
1. If this is a two-sided test then we say $\mu_1 < \mu_2$ or $\mu_1 > \mu_2$.
2. If this is a one-sided test then we say $\mu_1 > \mu_2$ and we choose sample 1 to be the sample that we believe might be the larger of the two.

Significance level, α, is to be chosen. $\bar{X}_1$ and $\bar{X}_2$ are the means of the first and the second samples respectively.

Table 17.10 Table of test statistics and limits for a variety of different parametric tests for the difference of two means

What is known	Test statistic	One- or two-sided test	Limit(s)
Standard deviations known	Standard deviations equal $$z = \frac{\bar{X}_1 - \bar{X}_2}{\sigma\sqrt{\dfrac{1}{n_1} + \dfrac{1}{n_2}}}$$	Two	Upper $= z_{1-\alpha/2}$ Lower $= z_{\alpha/2}$
	Standard deviations *not* equal $$z = \frac{\bar{X}_1 - \bar{X}_2}{\sqrt{\dfrac{\sigma_1^2}{n_1} + \dfrac{\sigma_2^2}{n_2}}}$$	One $(\bar{X}_1 > \bar{X}_2)$	$z_{1-\alpha}$
Standard deviations *not* known but assumed equal	$$t = \frac{\bar{X}_1 - \bar{X}_2}{s_p\sqrt{\dfrac{1}{n_1} + \dfrac{1}{n_2}}}$$ where $$s_p^2 = \frac{(n_1 - 1)s_1^2 + (n_2 - 1)s_2^2}{n_1 + n_2 - 2}$$	Two	Upper $= t_{1-\alpha/2}$ Lower $= t_{\alpha/2}$ $df = n_1 + n_2 - 2$
		One	$t_{1-\alpha}$ $df = n_1 + n_2 - 2$
Standard deviations *not* known and *not* equal	$$t = \frac{\bar{X}_1 - \bar{X}_2}{\sqrt{\dfrac{s_1^2}{n_1} + \dfrac{s_2^2}{n_2}}}$$	Two	$t_{1-\alpha/2}$ and $t_{\alpha/2}$ $$df = \frac{[(s_1^2/n_1) + (s_2^2/n_2)]^2}{\dfrac{(s_1^2/n_1)^2}{n_1} + \dfrac{(s_2^2/n_2)^2}{n_2}}$$

Limit: See Table 17.10. The limit depends on whether σ is known and if this is a one- or two-sided test.

Test Statistic: The general case is

$$\frac{\bar{X}_1 - \bar{X}_2}{\text{Standard deviation of the difference}}$$

(The standard deviation of the difference is also called the standard error of the difference.) The particular cases are found in Table 17.10.

Decision: As before, these depend on whether we have a one- or two-tailed test.

1. *Two-tailed or two-sided test.* If our test statistic is between the two limits we accept the hypothesis of no difference between the means. If the test statistic is outside either limit we accept the alternate hypothesis.
2. *One-tailed or one-sided test.* If our test statistic is less than the limit we accept the hypothesis of no difference. If the test statistic is more than the limit we accept the alternate hypothesis that there is a difference.

EXAMPLE 17.6

Students were randomly selected and placed into two classes. Two different teaching techniques, one traditional and one experimental, were then tried with the hope that the students in the experimental group would progress faster.

A standardized test was given to the two classes of students to determine if there was a difference between the groups. The test has a known standard deviation of 25.

The traditional group had 22 students and a mean score of 127 was recorded. The experimental group had 18 students and scored an average of 136. If there is a significant increase we will attempt to use the experimental method on a broader scale. An α level of 0.10 is chosen.

Hypothesis: There is no difference between teaching methods.

Alternate Hypothesis: There is a difference, the experimental group does better ($\mu_{\text{experimental}} > \mu_{\text{traditional}}$).

Limit: Since the standard deviation is known this is a z-test. It is also a one-tailed test, therefore
$$\text{Limit} = z_{1-\alpha} = z_{0.90} = 1.645.$$

Test Statistic: Since the standard deviations are equal for both tests we use the formula
$$z = \frac{\bar{X}_1 - \bar{X}_2}{\sigma\sqrt{\frac{1}{n_1} + \frac{1}{n_2}}} = \frac{136 - 127}{25\sqrt{\frac{1}{18} + \frac{1}{22}}}$$

$$= \frac{9}{25\sqrt{0.10}} = 1.14.$$

Decision: Since the test statistic is smaller than the limit we conclude that there is no significant increase (we accept the hypothesis). The experimental method will be discontinued.

EXAMPLE 17.7

In deciding what type of paving surface to use on a highway one consideration is the stopping distance for cars. A number of tests were conducted on two types of surfaces by measuring the distance it took for cars travelling at 50 miles per hour to come to a complete stop. The results in feet were:

Asphalt A	Hard surface P
127 108 136 119	129 118 144 117
121 126 142 134	119 129 151 130
$\bar{X}_1 = 126.6,\ s_1^2 = 116.6$	$\bar{X}_2 = 129.6,\ s_2^2 = 153.1$

Since n_1 and n_2 are equal we may simply average the two variances together to obtain a common variance of $(116.6 + 153.1)/2 = 134.8$.

Hypothesis: There is no difference in stopping distance between the pavement surfaces.

Alternate Hypothesis: There is a difference between these pavement surfaces. This is a two-sided test ($\mu_1 < \mu_2$ or $\mu_1 > \mu_2$).
Significance level is chosen as 0.05.

Limits: Standard deviations are considered unknown, therefore this is a t-test. We find the variances are essentially equal (if an F-test is performed). Since $df = n_1 + n_2 - 2 = 8 + 8 - 2 = 14$ we have:

$$\text{Upper limit} \quad t_{1-\alpha/2} = t_{0.975} = 2.145.$$
$$\text{Lower limit} \quad t_{\alpha/2} = t_{0.025} = -2.145.$$

Test Statistic:

$$s_p = \sqrt{134.8} = 11.6.$$

$$t = \frac{\bar{X}_1 - \bar{X}_2}{s_p\sqrt{\dfrac{1}{n_1} + \dfrac{1}{n_2}}} = \frac{126.6 - 129.6}{11.6\sqrt{\dfrac{1}{8} + \dfrac{1}{8}}} = \frac{-3.0}{11.6\sqrt{\dfrac{2}{8}}} = -0.517.$$

Decision: Since the test statistic is *between* the limits, we accept the hypothesis that there is *no* difference between the pavements.

REVIEW

Independence   ·   Chi-square test for independence   ·

$$\text{Expected frequency} = \frac{(\text{Row total})(\text{Column total})}{\text{Grand total}} \quad \cdot$$

$$df = (\text{Row} - 1)(\text{Columns} - 1) \quad \cdot$$

Difference in proportions (large samples)   ·   Mann-Whitney test   ·

U-statistic   ·   Difference in spreads   · F-test   ·

Difference in means • Variance known: z-test •
Variance unknown: t-test •

$$\text{Statistic} = \frac{\bar{X}_1 - \bar{X}_2}{\text{Standard deviation of difference}}$$

PROJECTS

1. Refer to Project 1 Chapter 1. Compare the temperature in the refrigerator in two different locations (like top and bottom shelf).
2. Refer to Project 3 Chapter 1. Compare the weights of granular sugar and the hard packed kind of sugar.
3. Refer to Project 4 Chapter 1. Compare the length or weight of boys versus girls.
4. Refer to Projects 5 or 6, Chapter 1. Compare two different radio or TV stations.
5. Refer to Project 9 Chapter 1. Is your right foot different from your left foot?
6. (For Section 17.2.) From a telephone book record the last digit of about 250 phone numbers listed in the white pages and the last digits of about the same number of phone numbers from the yellow pages. Do a χ^2-test to determine if there is a relationship.

PROBLEMS

Section 17.2

17.1 In examining 150 grades given by a teacher (see Problem 15.1) it was found that he taught two courses—a basic one and an advanced course. The grade distribution for the two courses was as follows:

	A	B	C	D	F
Basic	10	20	38	12	10
Advanced	6	12	20	12	12

Does the teacher appear to give different grades for the two courses? (Test at the 5% level.)

17.2 A survey was conducted by a heating contractor to determine preference in method of heating a house. Five hundred homeowners were asked to declare how well they liked the method of heating they were using. The following results were obtained:

	Oil	*Gas*	*Electricity*
Very good	56	79	15
Satisfactory	65	105	30
Adequate	51	46	3
Poor	28	20	2

Do these results indicate that there are different types of preferences for the different heating methods? (Use the 10% level.)

17.3 A sociologist surveyed a group of students graduating from high school. He asked them if their parents had gone to college and if they were planning to go to college. His results were as follows:

Student planning to attend	Only father attended	Only mother attended	Both attended
Yes	36	6	18
No	27	8	5
Maybe	37	11	52

Is there a relationship between parent and child?

17.4 Redo Problem 17:3. First classify the parents into two types: Father went to college and Mother went to college. Is there a significant relationship between these? Compare the two results.

Section 17.3

17.5 A grocery store kept a record of the sales of its own brands of coffee (in pounds). It found the following results for sale days and nonsale days (the sales were on other brands):

	Sale days	Nonsale days
Total coffee	185	153
Its brand	37	51

Is there a difference in the proportion of sales at the different times? (Test at the 10% level.)

17.6 A department store had found that 41 out of 150 customers had been shoplifting (Problem 15.19). A closed circuit TV system was installed with warning notices posted. After a month another study was conducted. Another 150 people were followed and 28 were observed taking unpaid merchandise. Was there any significant decrease in the amount of theft? (Justify at the 1% level.)

17.7 A student stood at the entrance to a large discount shopping center and polled people entering the store. He recorded yes, no, no opinion, and refused to answer. On a day when he dressed up in a suit he had 45 refusals out of a total of 140 persons asked. The second time he had a beard and dirty clothes. This time 60 out of 135 persons refused to answer. Is there a significant difference in proportions?

17.8 A group of students protested to the Dean that two teachers were giving very different numbers of failure grades. Teacher A failed 28 out of 100 students, while Teacher B failed 43 out of 125 students. What would the

Dean's response be if he stated that he would only continue investigation if there was less than a 1% chance that the grades could be essentially from the same population?

17.9 After the telephone company instituted a reduced rate for long distance calls after 6 P.M., a study was conducted to see if the proportion of long distance calls changed. Before the change took place two hundred calls were checked and 37 were found to be long distance calls. After the change another 200 calls were checked and this time 42 were long distance calls. Was there a significant change? (Test at the 5% level.)

Section 17.4

17.10 Ten garments are washed in "X" detergent and 10 other garments are washed in "X" detergent without phosphates. The whiteness was rated on a scale of 1 to 100 as follows (100 was the whitest):

"X" detergent   37, 48, 49, 61, 55, 71, 59, 78, 67, 66.
"X" without phosphate   44, 33, 27, 36, 27, 68, 49, 51, 60, 63.

Is there a significant difference in the whiteness? (Test at the 5% level.)

17.11 It is now revealed that two people did the rating of whiteness (in Problem 17.10). The first 5 readings in each group were by Mr. A. and the second 5 readings were by Mrs. T. Is there a difference in the way these people rate the whiteness? How might this affect the results of the previous comparison? How might you redo Problem 17.10 to account for any difference in rater? (*Hint:* Think in terms of two tests.)

17.12 Refer to the number of home runs hit by the National League Home Run Champions from 1930 to 1969 in Problem 14.7. Below are listed the number of home runs for the American League Champions. Does one league tend to have home run leaders that hit more home runs? (Use the Mann-Whitney test at the 5% level.)

49, 46, 58, 48, 49, 36, 49, 46, 58, 35, 41, 37, 36, 34, 22, 24, 44, 32, 39, 43, 37, 33, 32, 43, 32, 37, 52, 42, 42, 40, 61, 48, 45, 49, 32, 49, 44, 44, 44, 49.

17.13 Further investigation of the employees who were ill (Example 17.3) revealed the following number of days absent for those who had the flu vaccine and those who did not have the flu vaccine:

Had vaccine   3, 2, 1, 1, 2, $3\frac{1}{2}$, 4, 3, 2, 2, 3, 4, 5, 2, $2\frac{1}{2}$, 2, 1, $\frac{1}{2}$, 1, 2, 1, 1, 3, 2, 1, 3, $2\frac{1}{2}$, $1\frac{1}{2}$, 3, 1.

Did not have vaccine   2, 2, 1, 2, 3, 3, 5, 4, 2, 4, 3, 3, $2\frac{1}{2}$, $3\frac{1}{2}$, 5, 7, 3, 2, 1, 2, 4, 6, 3, 2, 3.

Is there a difference in the number of days ill for these two groups of people?

Section 17.5

17.14 A researcher was interested in knowing if there is a difference in the variability of the response for a person taking marijuana versus a person drinking alcohol. Ten similar subjects smoked a test cigarette and 10 others drank alcohol. The variance on the test of those smoking marijuana was 175 and the variance among the drinkers was 150. Is there a difference in variability? (Test at the 5% level.)

17.15 The variation of speeds of cars on a highway is considered at least as important as the actual speeds (see Example 16.2). Two superhighways were compared. Thirty vehicles were clocked on an interstate highway and 25 cars were clocked on a toll superhighway (automobiles only). The variance of speeds was 400 for the interstate highway and 40 for the toll road. Is there a significant difference in variation? Comment on the problem of the sampling. (Was it random?)

17.16 The variation in voltage at the wall socket can affect the performance of many electrical appliances. An inspector checked the voltage in a house and in a factory. He took 15 readings in the house and 12 in the factory. The resulting variances were

$$s^2_{house} = 83 \qquad s^2_{factory} = 18.$$

Is there a significant difference in the variation of voltage for these two locations?

17.17 Two paving contractors were asked to give estimates to pave identical driveways. The estimates were:

	Best paving	Pave-it
Driveway A	450	525
Driveway B	375	500
Driveway C	675	475

Do the two contractors differ in the amount of variation in their quotes? (Test at the 10% level.)

17.18 The degree of variation in the readings of an instrument is very important. A micrometer and a vernier caliper were compared by measuring the same part 6 times with each instrument. The variance among the six readings with the micrometer was 0.00000004 and for the six readings with the vernier caliper it was 0.0000016. Are the readings with one instrument more variable than with the other? (Test at the 5% level.)

17.19 An aspirin manufacturer that does considerable advertising claims its aspirin are more uniform than the very inexpensive brand. Samples from several bottles of both manufacturers were checked for the amount of aspirin in the tablets.

Much advertised brand	Inexpensive brand
$n = 30$	$n = 20$
$s^2 = 0.0009$	$s^2 = 0.0016$
$\bar{X} = 5.01$	$\bar{X} = 4.98$

Is there a significant difference in the brands? Is this a one- or a two-sided test? (Test at the 5% level.)

Section 17.6

17.20 If the average time between arrivals of express buses is not significantly longer than for local buses, it will pay me to wait for an express. The standard deviation between arrivals of buses is 4 minutes. I checked a sample of buses and found that for 6 locals the average time between buses is 14 minutes and for 8 express buses it was 18 minutes. Should I wait for the express? (Test at the 5% level.)

17.21 One Sunday two fathers had an argument over whose teenage daughter talked more on the phone. So they both started keeping a record of how many hours their daughter talked on the phone each night. The following Saturday they compared notes (times in hours):

Mr. C's Daughter   1.4   2.1   1.7   1.9   2.3.
Mr. K's Daughter   1.3   1.6   1.5   0.8   1.8.

Is there a significant difference? (Test at the 10% level.)

17.22 A new form of sole for shoes was being tested for wear. This was an artificial sole that was to be compared to leather soles. After 2 months of use two sets of shoes were measured to determine the amount of wear in thousandths of an inch.

Artificial   12, 13, 12, 9, 13, 12, 14, 10, 11, 8, 13, 11.
Leather   16, 13, 13, 14, 14, 15, 12, 16, 13, 12, 15, 16.

Is there a significant difference (improvement)? (Test at the 5% level.)

17.23 Refer to Problem 17.5. If the variance of the number of pounds of coffee usually sold in the grocery store is 26, there is a significant change in the total number of pounds of coffee sold during the sale days?

17.24 The telephone company did some investigation of the effect of reducing the rate for long distance calls (see Problem 17.9). A record of the number of long distance calls for the 15 minute interval from 6 to 6:15 P.M. was kept for two weeks before and after the change.

Before		After	
47	47	61	50
52	53	60	54
57	56	59	63
48	51	71	65
49	50	49	61

Was there a significant change in the number of long distance calls? (Test at the 1% level.)

17.25 Refer to Problem 17.13. Use a t-test to determine if there is a difference in the average number of days absent for people who have and have not had the vaccine.

17.26 Refer to Problem 17.19. Is there a significant difference in the amount of aspirin in the tablets of the two manufacturers?

17.27 Refer to Problem 13.3. All the seeds were planted on April 18. If the first 4 plants were in fertilized soil and the second 4 were in unfertilized soil, is there a difference in the time to germinate for seeds in fertilized and unfertilized soil? (Test at the 5% level.)

18

Comparison of Two Related Samples

We will now extend the type of two sample cases that may be compared. We will consider those times when each value in one sample has a counterpart in the second sample or is related to a specific value in that second sample. Under such circumstances we have several tests to determine if there is a difference between the population central values that are more sensitive than those provided in the previous chapter.

In Chapter 6, we discussed how a high correlation between two distributions of values permitted us to go from a location in one distribution to a very similar location in the other distribution. We used this notion to create a method of predicting from one type of value to another. You should also be aware of how a large distribution with large variation can mask or hide relatively small changes. We will now see how pairing of values can aid in comparing samples.

18.1 Statistical Tests for Comparing Two Related Samples

If we gave a reading test to a group of 9th graders we would find that the distribution of reading grades would have a large variance. If we then selected another group of 9th graders at a later time and gave them a similar test, we would find that the variance of their reading scores would also be large. In addition, because of these very broad distributions, we would probably be unable to detect any significant change unless the samples were very large. Even then, we might question the practical significance of such a difference.

Table 18.1 Reading grade levels of two groups of 9th grade students. Tests taken at two different times

| Beginning of the year | 9.4 | 10.3 | 8.4 | 6.8 | 7.8 | | 9.8 | 9.2 | 11.2 | 9.4 | 9.0 |
| End of the year | | 9.3 | 10.1| 8.8 | 6.6 | 7.7 | 10.0 | 9.8 | 11.7 | 9.7 | 9.0 |

$\bar{X}_{beg} = 9.13$, $\quad \bar{X}_{end} = 9.27$, $\quad S_p = 1.24$, $\quad t = 0.253$.

In Table 18.1 we have a pair of samples of the reading scores of 10 students given at two different times. A t-value for the difference in two means has been calculated and is not significant. In Figure 18.1 we see the histograms for the two samples. You may detect what appears to be a difference, but remember that these are *small* samples and samples are expected to vary.

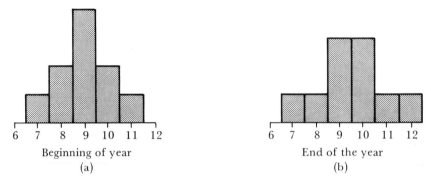

Figure 18.1 Histograms for reading grade test scores at the beginning of the 9th grade and at the end of the 9th grade.

Now let us consider what might happen if we were to match up the students at the two times or test the same individual at the two times in the 9th grades. You might expect that there should be a very high correlation between reading levels at these two times. A student who was a poor reader in the beginning of the 9th grade is probably still a poor reader at the end of the 9th grade, and a good reader will probably stay good relative to the overall class. We no longer have independent samples. Look at Table 18.1 again, only this time compare the score for each student the first time with the score below it at the later time. Is there a pattern? Twice as many students show an increase versus those decreasing. Is the pattern significant? This chapter will explore three statistical tests for deciding if there is a significant difference.

1. Sign test—if we only know which set of values is greater than the other.
2. Wilcoxon matched-pairs test—if we can also rank the differences or say which differences are larger than others.
3. Paired t-test—if we have an interval scale that the values are measured on and the magnitudes of the differences are also on an interval scale.

Before we begin to work with these tests we should discuss a few additional examples of types of cases that are generally treated as related.

Before and after tests, or test-retests, such as: A driving test that measures reflexes is taken both before and after a drug (or alcohol) is administered. Absorbency of towels is measured before and after washing. A group of adults is given a test of attitudes, shown a film, and then retested to see if a change in attitude has occurred. A metal bar is tested for hardness, heat treated, and then retested to see if the hardness has changed.

Matched conditions, such as: Pieces of wood are painted with two different paints to determine which wears best. Two brands of tires are tested on a car at the same time. Differences between sexes may be tested by considering husband and wife, or brother and sister. A control can be chosen with attributes as similar as possible to the subject or individual being tested (this may be important if other factors such as time may also influence the experiment. We assume the control is similarly affected in every way except for the test).

18.2 Sign Test

In any of the examples just mentioned, if there is no real change or difference between the two times that measurements are taken, then there should be an equal chance for a change in either direction. As long as we can identify the direction of change in each case, such as improvement in reflexes, increase in absorbency, positive change in attitude, decrease in hardness, and so forth, *we can apply the sign test*. We can use it when we have numerical differences (I finished the race in 1 minute 20 seconds you finished in 1 minute 12 seconds) or when we have categorical values for difference (you won). We will call one of the possible changes a *plus*, + (possibly increase, or won, or improvement) and the other a *minus*, − (decrease, or lose, or fail).

In all cases we assume that the chance of having a plus is the same as that of a minus. Thus, the probability of either sign of change is $\frac{1}{2}$. If this is true then the possible samples of plus and minus signs will be a binomial distribution with $p = 0.50$ and $n =$ number of differences or pairs of values.

Since we are really working with a binomial test we must consider two cases: *small samples* using the table and *large samples* using the normal approximation. The only difference, however, is in the determination of the limit. The rest of the test procedure is the same.

Hypothesis: There is no difference in population averages.

Alternate Hypothesis: There is a difference in population average. (The sign test actually tests to determine if the medians are different. We may perform a one- or two-tailed test. As always this influences our limit and requires us to decide if we are interested in +, −, or both.)

We must choose α. With small samples it can only be approximate. When using the sign test, it is usually better to calculate the test statistic first. If there is no change in some cases we call the pair tied. Ties will influence the computation of the limit. For each pair of values enter a +, −, or 0 depending on whether there is a + change or difference, a − change, or no change. The total number of plusses *or* minusses constitute our test statistic. (However, the test statistic will depend upon whether we have a one- or two-sided test. If the test is to be *one*-sided then the sign we are to add up must be decided upon first.)

Limit: The total number of plusses and minusses must be computed (this is also the total number of pairs minus the ties). This is n, the sample size.

If n is 20 or less: Use the binomial tables, Appendix Table A.8 with $p = 0.5$. For a one-tailed test add up the probability in one tail until you reach α. For a two-tailed test add up the probability in one tail until you reach $\alpha/2$, then subtract the X from n to find the second limit.

If n is more than 20: Use the normal approximation. To find the X-limits we would choose the appropriate equation(s) below:

One-tailed
$$X = \frac{n}{2} + z_{1-\alpha/2} \tfrac{1}{2}\sqrt{n}.$$

Two-tailed
$$\text{Upper } X = \frac{n}{2} + z_{1-\alpha/2} \tfrac{1}{2}\sqrt{n}. \tag{18.1}$$

$$\text{Lower } X = \frac{n}{2} - z_{\alpha/2} \tfrac{1}{2}\sqrt{n}.$$

(Remember that $p = 0.5$.)

Decision: If the total plusses (or minusses) are within the limits accept the hypothesis of no difference. If the plusses (or minusses) are outside the limits then accept the alternate that there is a difference.

EXAMPLE 18.1

Let us return to the reading scores and apply a sign test to see if there is a general increase. We are to treat this as a one-tailed test with + standing for an increase in reading level from the beginning of grade 9 to the end of the year.

Hypothesis: There is no change in reading level.

Alternate Hypothesis: There is a change, there is an increase between the two times tested.

Use an $\alpha = 0.10\star$.

Test Statistic: Table 18.2 shows the values and signs. There are 6 plusses and 3 minusses out of the 9 values that had changed.

Table 18.2 Computing the test statistic for the sign test. Change in reading from the beginning of the 9th grade to the end of the 9th grade for 10 students

Beginning of year grade	9.4	10.3	8.4	6.8	7.8	9.8	9.2	11.2	9.4	9.0
End of year grade	9.3	10.6	8.8	6.6	7.7	10.0	9.8	11.7	9.7	9.0
Change in grade	−	+	+	−	−	+	+	+	+	0

Total changes: 6 plus changes (+), 3 minus changes (−), 1 no change (0).

★ At this point I will use a significance level of 0.10. This is somewhat of an arbitrary choice. As we proceed I will be changing this level. Quite frequently statisticians also work a reverse statement by determining the level at which a test value *is* significant. In such cases the 10% level is about the least significant level that is used.

Limit: Since the sum of the plusses and minusses was 9, (7 + 2), then $n = 9$. From Table A.8 for $n = 9$, $P = 0.50$ we have: $X = 9$, $p = 0.0020$; $X = 8$, $p = 0.2176$ (sum = 0.0196); $X = 7, p = 0.0703$ (sum = 0.0899). This is the closest to 0.10 so we chose $X = 7$.

Decision: Since the test statistic is *not* in the tail or critical region we accept the hypothesis that there is no change in reading level. This would confirm what might appear to you as only a slight change in population level.

18.3 Wilcoxon Matched-pairs Test

If you go back to Table 18.2, and this time examine the magnitudes or sizes of the differences, you may begin to get a different picture. The sign test merely considered if there was a difference of any kind, and in many cases this is all you can really do. But the sign test assumes that the differences are roughly the same in both directions. If you can at least place the differences in an order from small to large then we can make a more discriminating analysis. We can see if the large differences tend to be in one direction (say increases) and almost all of the small differences in the other direction.

We will redo Example 19.1, but now we will consider the amount of change that has taken place. You should bear in mind, though, that you do not need interval data to apply this test. The only requirement is that the change in the first individual be bigger or smaller than that in the second individual, and so forth. Most of the times, however, you will have numerical values on some sort of scale. We will follow a format almost identical to the one used with the sign test except for the test statistic and the limit.

The *test statistic* is called T. To compute this value we

1. Determine all of the magnitudes or sizes of the differences. If a 0 difference appears we drop it from the sample.
2. We then rank these differences (ignoring the signs). If two or more differences are the same we use the *average* rank for all of these differences.
3. We multiply the ranks by the *sign* of the difference.
4. We observe which sign has the fewer number of ranks and we add them up. This is T.

The *limit* for T is found in Appendix Table A.19.

EXAMPLE 18.1 (Cont.)

Hypothesis: There is no difference in reading level from the beginning of the 9th grade to the end of the 9th grade.

Alternate Hypothesis: There is a difference in reading level—there is an increase (one-sided test).

Significance level $\alpha = 0.05$.*

---

* I have now chosen a more appropriate significance level.

Test Statistic: In Table 18.3, Columns 1 and 2, we have the reading levels for the beginning and the end of the 9th grade.

1. In Column 3 the differences are calculated. Observe that the last student did not change and therefore we only consider 9 students (or $n = 9$).

Table 18.3 Calculating the test statistic in the Wilcoxon matched-pairs test. The data is the change in reading level from the beginning to the end of the 9th grade

Column					
1	2	3	4	5	6
Beginning of the 9th grade	End of the 9th grade	Difference between beginning and end (Column 2 − Column 1)	Rank of difference ignoring sign	Rank with sign included	Ranks with the less frequent sign
9.4	9.3	−0.1	1.5	−1.5	1.5
10.3	10.6	0.3	5.5	5.5	
8.4	8.8	0.4	7	7	
6.8	6.6	−0.2	3.5	−3.5	3.5
7.8	7.7	−0.1	1.5	−1.5	1.5
9.8	10.0	0.2	3.5	3.5	
9.2	9.8	0.6	9	9	
11.2	11.7	0.5	8	8	
9.4	9.7	0.3	5.5	5.5	
9.0	9.0	0	ignore		
					Sum = T = 6.5

2. We rank the differences. They would be

   Difference   0.1   0.1   0.2   0.2   0.3   0.3   0.4   0.5   0.6.
   Rank   1   2   3   4   5   6   7   8   9.

   Since the first and second ranks are both 0.1 we consider these two as a rank of 1.5. We treat the 0.2 and the 0.3 similarly. Therefore, we will use the following ranks.

   Difference   0.1   0.1   0.2   0.2   0.3   0.3   0.4   0.5   0.6.
   Rank   1.5   1.5   3.5   3.5   5.5   5.5   7   8   9.

   These have been placed in Column 4.
3. We now multiply Column 4 by the appropriate sign of the difference. I have put these in Column 5 to avoid confusion. However, you may skip writing out the extra column of numbers.
4. There are fewer minus signs than plus signs, therefore we add the rank for minusses. In Column 6 the ranks for minus values are rewritten. The sum, T, is 6.5.

Limit: From Table A.19 we find for $n = 9$ and a one-sided test with $\alpha = 0.05$.

$$T_{min} = 8.$$

Decision: The value of T for the sample is less than the limit, or it is in the *critical* range. Therefore: We reject the hypothesis and accept the alternate. We declare that there *is* a significant increase in reading level.

Comment: I hope you have recognized that when we used the sign test we were unable to declare that a significant increase had taken place, even though we were willing to take a 10% risk of error. Now we have made some further refinement in the test. By using more information we *are* able to say that there is a significant increase and with only a 5% risk of error.

18.4 Paired *t*-test

We will now consider the case when in addition to being able to recognize that the differences between pairs of values may vary

1. We can calculate actual differences (on an interval scale) for *each* of the pairs of related values and
2. The values (the differences) are also close to being normally distributed.

Under these circumstances we can calculate the mean and standard deviation of the *differences* and then apply a *t*-test. Our assumption is that if there is no *real overall* difference, then the average of the differences should be zero. So we will conduct a *t*-test to determine if there is a significant difference between the mean of these differences and a hypothetical mean of zero.

Hypothesis: The average difference (mean) is equal to zero. There is no difference.

Alternate Hypothesis: The mean difference is *not* zero.

For a one-sided test choose the *order* of subtraction so that the expected difference will be positive.

For a two-sided test we consider both differences.

Significance level α, must be given; n = number of pairs; $\bar{d}$ = mean of the differences; s_d = standard deviation of the differences.

Limit: One-sided test

$$t_{1-\alpha}$$

Two-sided test
Upper $\quad t_{1-\alpha/2}$
Lower $\quad t_{\alpha/2}$ $\quad$ (with $df = n - 1$).

Test Statistic:

$$t = \frac{\bar{d}}{s/\sqrt{n}}. \qquad (18.2)$$

Decision: If t falls below $t_{1-\alpha}$ (one-sided test) or between the two-sided limits *accept the hypothesis* that the difference is zero. If t is outside the limit(s) accept the alternate hypothesis that a significant difference exists.

EXAMPLE 18.1 (Cont.)

We now recognize that the differences we observed are more than just ordered and are also not too non-normal. (I wouldn't exactly call them normally distributed but you must also keep in mind that there are only 10 values.)

The mean of the differences (now we include that 0 difference) is

$$\bar{d} = 19/10 = 1.9,$$

and the standard deviation for the 10 differences is 2.51.

Hypothesis: The mean difference is not different from 0.

Alternate Hypothesis: The mean difference is greater than 0. We will use a one-sided test with significance level $\alpha = 0.025$.

Limit: $df = n - 1 = 9$ (one-sided test).

$$t_{1-\alpha} = t_{0.975} = 2.262.$$

Test Statistic:

$$t = \frac{\bar{d}}{s/\sqrt{n}} = \frac{1.9}{2.51/\sqrt{10}} = \frac{1.9\sqrt{n}}{2.51} = \frac{1.9(3.16)}{2.51} = 2.39.$$

Decision: Since $t = 2.39$ exceeds the limit of 2.262 we accept the alternate hypothesis that there *is* a difference. Observe that this time we reject the hypothesis while assuming only a 0.025 chance of error ($2\frac{1}{2}\%$ level).

18.5 Comparison of These Three Tests: Power

In this chapter I have presented three statistical tests that are all designed to perform the same basic function. They all examine differences and try to determine if these differences are significant. In addition to working with these particular tests I hope that you may have become aware of some even more important concepts that underly the field of statistics. You have seen how three different tests that are used to make a decision about the same characteristic could lead you to arrive at different decisions.

There was a kind of ordering among these three tests; in fact, there were several orderings. The most important observation centers about the degree of assertion that we can make regarding our rejection of the hypothesis. If several tests can all be applied to the same data, then we say that the test that rejects a hypothesis at the *smallest* alpha level is the most powerful of those tests. Obviously, if one test cannot lead to a rejection while another does, then the second must also be more powerful.★

With our three tests we found that the sign test failed to permit us to declare that a difference existed even at the risk of being wrong 1 out of 10 times (0.10 level). The second test, the Wilcoxon, did reject the hypothesis of no difference. It was capable of recognizing the significant differences and it was able to do so with an even smaller chance of error than we asked of the sign test (0.05 level).

The third test also was capable of rejecting the hypothesis, but was capable of being more emphatic about the rejection than was the Wilcoxon test. The t-test was able to reject the hypothesis with about half the chance of error (0.025) that the Wilcoxon test assumed. We therefore state that the t-test is more powerful than the Wilcoxon test.

The degree of effort involved in the calculation of the test statistic as well as the restrictions on the kind of measurements become greater (from categorical to ordered to interval) as we progressed to the more powerful tests. This is generally the case. When we calculated a median we observed that it was easier to compute, but not as efficient as a mean. This was largely because it used less information from the sample. Similarly, the sign test uses much less of the available information in attempting to arrive at a decision, and hence is less efficient and less powerful. These examples should also illustrate that when a parametric test is available and the assumptions are correct, it is usually the most powerful of tests. The first two tests here are nonparametric tests while the t-test is parametric, requiring a mean and a standard deviation to be calculated. Of course, if the assumptions of the t-test and Wilcoxon test are not met, the sign test would be the only one available to you and it would stand as the most powerful test. So power is completely relative to what can be used.

REVIEW

Before–after · Matched pairs · Test–retest ·

Sign test (Binomial distribution) · Wilcoxon matched-pairs · T ·

Paired t-test for differences · Power of a statistical test

PROBLEMS

Section 18.2

18.1 A heating oil company has advertised that by installing a special device on the furnace your heating bills will be reduced substantially. Before I was going to have the device installed, I questioned 7 people who had it

---

★ In statistical theory the *most powerful test* or the test having the greatest power is defined as that test which is more powerful than all other possible alternate statistical tests. In this chapter I am only dealing with three tests and, therefore, we cannot arrive at a statement like "the t-test is *the most powerful test*." We are merely comparing three tests.

installed. Their heating bills in dollars the year before and the year after installation were:

Before	225	195	235	165	260	245	230
After	200	205	215	170	250	255	225

Is there a significant decrease in cost? (Use the sign test at about 5%.)

18.2 The pulse rates of 12 red-blooded American males were recorded before and after looking at the centerfold of *Playboy* magazine.

Before	68	71	84	93	67	74	82	77	71	83	62	66
After	71	70	81	97	73	80	90	76	80	79	80	67

Is there a significant increase in pulse rate? (Test at the 10% level.)

18.3 Two hair sprays were being compared—"The End" and Brand "X". The company that makes "The End" was comparing it to Brand "X" to determine if its brand had longer holding power. Ten women had half their heads sprayed with "The End" and the other half with Brand "X". A strong fan blew in their faces and the number of minutes for their hair to become windblown was recorded as follows

The End	8	6	9	5	7	4	8	8	9	10
Brand "X"	10	5	7	6	4	4	7	7	7	8

Is "The End" a better hair spray? (Test at the 10% level.)

18.4 A botanist was interested in knowing if there was a difference in the time fruits matured on different parts of a plant. He recorded the day of the first fruit on the top and on the bottom for 15 plants. All the fruits came out during July.

Top	3	6	7	5	8	9	10	10	7	8	6	9	10	12	4
Bottom	7	9	5	8	8	10	11	12	6	9	7	13	8	13	8

Is there a significant difference in the time to mature (at the 5% level)?

18.5 Refer to Problem 17.22. If each of the values of artificial soles is listed above the corresponding value for leather sole on the second shoe for pairs of shoes (in other words one sole goes on one shoe, perhaps the left, and the other type of sole goes on the other shoe, the right) then we have matched pairs of shoes to compare and the wear is more readily compared (if people are to do the testing). Using a sign test determine if there is a significant difference between the two types of soles. (Use $\alpha = 0.05$.)

Section 18.3

18.6 Redo Problem 18.1 using the Wilcoxon matched-pairs test.

18.7 Redo Problem 18.2 using the Wilcoxon matched-pairs test.
18.8 Redo Problem 18.3 using the Wilcoxon matched-pairs test. If the values were actually recorded to number of seconds as well as minutes how does this affect the results? Below are the complete times.

The End		Brand X		The End		Brand X	
minutes	seconds	minutes	seconds	minutes	seconds	minutes	seconds
8	15	10	10	4	11	3	48
5	50	5	20	8	25	7	27
8	35	7	15	7	40	6	55
5	35	5	50	9	5	7	20
6	50	4	10	10	15	8	5

18.9 Redo Problem 18.4 using the Wilcoxon matched-pairs test.
18.10 Redo Problem 18.5 using the Wilcoxon matched-pairs test.

Section 18.4

18.11 Redo Problem 18.1 using the paired t-test.
18.12 Redo Problem 18.2 using the paired t-test.
18.13 Redo Problem 18.3 using the paired t-test.
18.14 Refer to the more complete times for the hair sprays listed in Problem 18.8. Compare these using the paired t-test and then compare the results to Problem 18.13.
18.15 Redo Problem 18.4 using the paired t-test.
18.16 Redo Problem 18.5 using the paired t-test.
18.17 Most meats lose weight when they are cooked. A restaurant owner carefully weighed ten 1 pound steaks before and after cooking. He is willing to accept an average of a 10% loss in weight without complaining to the meat supplier. The weights were

Before	1.10 1.08 1.03 1.09 0.93 0.99 1.01 1.07 1.01 0.94 0.98
After	0.92 0.95 0.91 0.96 0.85 0.88 0.89 0.93 0.96 0.87 0.88

Is there a difference in weight significantly greater than 10%? (*Hint*: First decrease the before weights by the 10%, then compare them.)

19

Multiple Sample Cases— Analysis of Variance

We have done considerable testing of pairs of variables to decide if there are significant differences between them. We now want to enlarge this type of problem to include multiple sets of data. In a great many cases we do not want to be limited to only two sets of data or two conditions or two characteristics.

19.1 Investigating Multiple Sets of Data

Let me pose a question which might serve as an area of investigation. A practical inquiry that any automobile driver should be interested in would be "What factors significantly affect the gasoline cost per mile?"

Immediately we should be able to write down a collection of criteria that would appear to be of practical and reasonable importance.

A. Several factors affecting cost per mile depend on the gasoline:
  1. The cost per gallon;
  2. The brand of gasoline;
  3. The kind of gasoline (premium, regular, or intermediate).
B. A few factors involve the designation of kind of automobile:
  1. Engine (including such things as number of cylinders, engine cubic inches, horsepower);
  2. Make of car;
  3. Length of time since tuning the engine;
  4. Size of tires or air pressure.
C. A third group of factors involve the driving conditions:
  1. The actual driver;
  2. Weather;
  3. Kind of roads (including city versus rural, city versus superhighway);
  4. Speed.

We have now listed about a dozen important considerations and I suppose you might be able to add at least several more. Go back to the previous chapters and try to think about how we might investigate some of these variables. I hope you realize the hopelessness of such a request. At best we can look at a large number of pairs of conditions, each of which would require a great number of tests. So we might hold everything constant by picking one individual car and testing two different brands of premium gasoline. We would need to take a fairly large number of test measurements before we would be able to make a decision. Then we might pick a third brand and test it against each of the others by running many more tests. And then you would *rerun* the previous sets of values for the gasolines already checked. If we wanted to compare a different make car we would again have to run a large number of trials in order to perform a satisfactory statistical test. This technique is often called the "hold everything constant" method. Unfortunately, this method is very common. It requires an excessive number of tests and far more time and effort than should be necessary. In addition, the more times you use the same set of data in calculating a statistic or performing a statistical test, the greater is your chance of making an error. (For example, we may have set alpha at 0.05 and calculated a t-value. If our sample t exceeded the limit, we say there is a difference, and we assume there is only a 5 per cent chance of making an error. Now, if you use one of these same sets of data over again, and compare it to a third set of values by calculating another t-value, then the *same limit* would not be giving you an alpha of 0.05, but alpha would be almost 0.10.* Your chance of making an error has doubled.) If you perform enough pairs of tests you will certainly get a few to be significant even if there are no actual significant differences. This then is one of the serious drawbacks of the "hold everything constant" procedure, unless you perform all of the reruns.

A second problem is that certain relationships may be left unidentified. There often exist subtle interdependencies that we cannot identify without a more elaborate kind of statistical manipulation. I am referring to a relation called an *interaction*. An example of interaction is the kind of situation where the brands of gas may work differently with two cars. Perhaps Brand A does well in my car while Brand X does well in your car. Is it the brands that are different, or the cars? Or is it just that there is a peculiar interaction or interrelation between the cars and the brands? This interaction confuses our ability to arrive at a straightforward decision. Interdependencies become even more difficult to recognize when we look at numerous sets of data or broaden the problem to include several more of the variables.

Thus, you should be aware of some of the major limitations of the previous approaches to statistical analysis. For practical problems and true research work we must expand our horizons. The general category or heading that covers the cases of looking at more than two samples is called the *design of experiments*. The most common approach to statistical design of experiments is

---

* This is found as $1 - (0.95)^2 = 1 - 0.902 = 0.098$.

the application of a technique called the *Analysis of Variance* (abbreviated as ANOVA). As the name implies, we will be doing some kind of analysis, determination, or comparison of the variances associated with some sets of data.

A second method called the *Analysis of Means* (ANOM) has already been introduced in Chapter 7. If you recall, this technique looked at the means of a collection of samples.

19.2 The Basic Principle of Analysis of Variance (ANOVA)

We originally set the stage for the analysis of variance in Chapter 3 and again in Chapter 7. In Chapter 3 it was demonstrated how you could add together two different distributions to obtain a third, more variable distribution. We later examined (in Chapter 7) in some detail how we can use the variation within small, instantaneous-type samples as a measure of the smallest or inherent variation in a process. We used this small variation to help us establish a gauge of how samples should vary and how an overall distribution should look if there are no differences in sample means. We will now extend these ideas.

EXAMPLE 2.1 (Cont.)

Our policeman and his tickets will serve as an example of the analysis of variance. In Chapter 7 we observed that the readings taken each day also were classified into 4 different categories.*

Tickets in eastbound lane, morning.
Tickets in westbound lane, morning.
Tickets in westbound lane, afternoon.
Tickets in eastbound lane, afternoon.

At that time we directly compared the means of each sample. Now let us look at these sets of values from a different point of view. In Figure 19.1a we have four histograms, one for each of the four different conditions listed above. Observe that *each of these* small samples has a mean *and* a spread or variance—and we can calculate both of these quantities for *each* sample.

The mean, variance, and standard deviation of the samples are as follows:

	Mean	Variance	Standard deviation
East —morning	12.42	6.41	2.53
West—morning	9.85	5.02	2.24
West—afternoon	12.08	6.07	2.46
East —afternoon	8.38	3.13	1.77

---

* I am referring to Section 7.6, A Type of Experimental Design. However, covering that portion of Chapter 7 is not essential to an understanding of this chapter.

Observe that the 4 variances are quite close to each other. Even more important, however, is the fact that each of these variances represent variation that we can consider only as random variation. These values of variance are the best quantities that we have to judge pure or unidentifiable error.

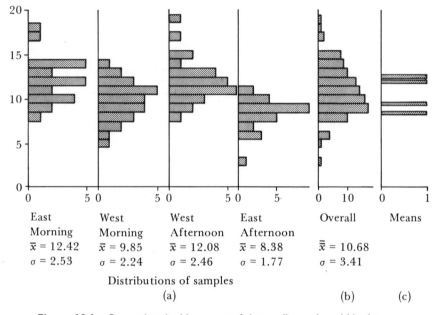

Distributions of samples
(a) (b) (c)

Figure 19.1 Comparing the histograms of the small samples within the overall distribution of tickets issued. We are to determine if the means of the smaller samples differ by comparing the spreads within the samples (a) to the spread of the sample means (c) and the overall distribution (b).

We frequently identify the different categories as treatments. We then talk about the variances of the samples as the *within treatments variance*. Generally, we will assume that the variances of the different samples are effectively the same. If the variances are to be considered essentially identical,* then we should average them together to obtain an overall *within treatments* or *within samples* variance. This then becomes one important measure of variability that we will use. The averaged within sample variance is 5.15 (and is computed by weighting the variances according to the size of the sample used in computing the respective variances).

Another histogram that we can draw is for the overall distribution. It has a mean of 10.68† and a variance of 11.64. Now comes a key question. If the small distributions all had the *same mean*—were centered at the same location—

---

* Remember though, they are not really identical. We are assuming that they are from sources that all have the same variance.
† In Chapter 7 we also observed that 2 days were unusual and we eliminated the 8 values associated with them. Thus we are working with 104 values.

then how would the variance of the overall distribution compare to the variance of the small samples?

The two variances would essentially be the same (again, refer back to Chapter 7). If there was a difference in the locations of each of the small samples, the *variance* of the *overall*, combined distribution would be *larger than* the variance of the small distributions. Then the two estimates of the true population variance (the one based upon the small samples and the other based upon the overall distribution) would differ significantly.

Now, how do we test the variances? We compute an F-ratio. In this case we are concerned only if the variance of the overall population is significantly larger than our estimation based on the little groups. Thus we are looking at *one-tailed* tests.

The value we compute is

$$F = \frac{\text{Variance of overall distribution}}{\text{Pooled or averaged variance of samples}}, \quad (19.1)$$

or for this example

$$F = \frac{10.68}{5.15} = 2.07.$$

This quantity is to be compared to the F-values in Table A.5. The degrees of freedom are to be 103 for the numerator (total number of individuals minus 1) and 100 for the denominator (each of the four samples has $26 - 1$ or 25 degrees of freedom, therefore there is a total of 100). For a one-tailed test with $p = 0.99$ (1% chance of error) the table gives the following F-values:

$$F(60, 60) = 1.84 \quad \text{and} \quad F(120, 120) = 1.53.$$

Thus, $F(103, 100)$ should lie somewhere between these values. But 2.07 exceeds both of these so we can confidently say (with less than a 1% chance of being wrong) that there *is* a difference in the variances—and we attribute that difference to the difference in locations or values of the means.

We are not finished yet. There is still another way to estimate the true population variance and to look at the data. If you refer to Figure 19.1c you will find a third type of distribution. This is a histogram of the *means* of the samples. Observe that they are not all located at the same point.

Should they be?

Of course not!

But how should they be distributed? Well, first let us look at the samples themselves. We can calculate the variance of the *means* as differences from the overall grand mean ($\bar{\bar{X}}$ or X double bar) and we find

$$s_{\bar{X}}^2 = \frac{\sum(\bar{X} - \bar{\bar{X}})^2}{k - 1}$$

$$= \frac{(12.42 - 10.68)^2 + (9.85 - 10.68)^2 + (12.08 - 10.68)^2 + (8.38 - 10.68)^2}{4 - 1}$$

$$= \frac{10.98}{3} = 3.66 \qquad \text{where } k = \text{categories or treatments.}$$

This value is called the *between samples variance*. It is a measure of just how much the samples themselves vary. Now an important question: If you were to rely on the variance *between the samples*, what would you estimate that the *population variance* would be? We must refer back to the Central Limit Theorem and recall that

$$\sigma_{\bar{X}}^2 = \frac{\sigma^2}{n} \quad \text{or} \quad \sigma^2 = n\sigma_{\bar{X}}^2.$$

The *between samples variance* is $\sigma_{\bar{X}}^2$ and n is the sample size (observe that things are not so simple if the samples have different sizes). We would now find our estimation of the *population* variance as

$$\sigma^2 = n\sigma_{\bar{X}}^2 = 25(3.66) = 91.5.$$

Again we will have to test variances. We now have two different estimates of the overall population variance and we will perform an F-test to compare the within samples variance and the between samples variance.

Let's try to reason out what would be unusual. If the four populations were actually located at the same place, then the sample means would be located within a narrow region. We would feel that either of the variances we would calculate should serve as a good estimator of the true population variance.

But what if the samples were actually spread apart and came from different populations? We now have a different situation. The variances of each of the samples might remain the same (thus having the same pooled or averaged variance). However, the means could be spread apart much more than by chance alone. So the variance *between* the samples would be *large*. If there was a significant shift in the means, then the between samples variance would be *very much larger* than the within samples variance. So again we would have a one-tailed test, since we will only be concerned if the between samples estimate of variation is too great in comparison to the within samples variation. We do this with an F-test of

$$F = \frac{\text{Estimate of population variance based on between samples variance}}{\text{Within samples variance}}.$$

For our example we have

$$F = \frac{n\sigma_{\bar{X}}^2}{\sigma^2} = \frac{25(3.66)}{5.15} = \frac{91.5}{5.15} = 17.8.$$

This is to be compared to an F-limit. Once more we must determine the degrees of freedom. The denominator we found before to contain 100 degrees of freedom and the numerator 3 degrees of freedom (categories or treatments minus 1). Thus the critical limit with $p = 0.99$ is between

$$F(3, 60) = 4.13 \quad \text{and} \quad F(3, 120) = 3.95.$$

Our F-value far exceeds these critical values. We again can declare that there are significant differences between the sample means.

The net effect of all of this manipulation is that we have found a very useful and valuable method of comparing the means of several samples. The standard approach, however, will consist of some slight modifications and a formalization of the technique.

19.3 One-way or Single-factor ANOVA

We will begin the study of the ANOVA with the simplest type of "many sample" problem. We will consider only one factor. The goal at this point may seem to be no more than an extension of the t-test for two samples, and in many respects this is true. Later, we will look at some other more complex types of problems that can be evaluated by the ANOVA.

Refer back to the list of factors that might affect gasoline cost per mile. We can select *one of these factors* and then observe that the reason this factor may be important is that several possible levels or kinds or types of samples can occur within that factor. We are then to determine if there are differences within that factor. Thus, we may look at 4 different brands of gasoline; 3 kinds of gasoline; 5 different drivers; 4 brands or models of cars; and so forth. We will also take several values or repeated measurements to create samples in each of the levels or categories of the factor, but we will only look at one overall classification.

In order to state in a precise fashion what is to be done I will resort briefly to the summation notation. The example that will follow should demonstrate that the computations are actually quite straightforward and it will emphasize exactly what you must do in order to perform the analysis.

Table 19.1 Table for presenting data in the ANOVA, including some comments and showing some of the calculations

		Category				Overall values	Comment
		1	2	3	4		
Repeats	1	X_{11}	X_{12}	X_{13}	X_{14}		$X_{ij} = X_{24}$
	2	X_{21}	X_{22}	X_{23}	X_{24}		The first subscript, i, is the repeat
	3	X_{31}	X_{32}	X_{33}	X_{34}		and the second, the j, stands for category.
Totals		T_1	T_2	T_3	T_4	T	Sometimes written as $T_{.4}$ where . stands for "summed over that subscript"
Sample size		n_1	n_2	n_3	n_4	n	T also written as $T_{..}$
Means		$\bar{X}_1$	$\bar{X}_2$	$\bar{X}_3$	$\bar{X}_4$	$\bar{\bar{X}}$	Also written as $\bar{X}_{.4}$ or $\bar{\bar{X}}_{..}$

Note that $T = T_1 + T_2 + T_3 + T_4 = X_{11} + X_{21} + X_{31} \times X_{12} + \cdots + X_{34}$

$n = n_1 + n_2 + n_3 + n_4$

$\bar{X}_1 = \dfrac{T_1}{n_1}, \quad \bar{X}_2 = \dfrac{T_2}{n_2}, \text{ and so forth.} \quad \bar{\bar{X}} = \dfrac{T}{n}$

We will approach the general technique by looking at 4 categories or samples, each with a sample size of 3 (3 replications or repeated measurements). In Table 19.1 we have an arrangement showing the subscript designation for each of the individuals, and some of the calculations we will need for the $\bar{X}$-values. Note that the table need not be restricted to the number of values listed. For example, the sample sizes could be any amount (and the samples may not necessarily be all the same size).

In addition to the individual values, the important quantities that are to be used in further calculations are the totals or the T and the n. We will use these values and the X to calculate variances. Think back to Chapter 4 and our discussions of various ways of computing variances, and try to reconsider how we might interpret a variance. As a starter let us consider a variance as the average of a collection of *sums of squares*. An average requires a division and we will divide by the number of degrees of freedom used in calculating the squares. Thus, we can state that

$$\text{Variance} = \frac{\text{Sum of squares}}{\text{Degrees of freedom}} = \frac{SS}{df}. \qquad (19.2)$$

We will now try to work with these two quantities, the sum of squares and the degrees of freedom, separately.

In Chapter 4 we found that instead of calculating and squaring many differences, especially from the mean, we could expand the numerator into the form

$$\Sigma X^2 - \frac{(\Sigma X)^2}{n}.$$

In addition, it did not matter if we coded. The only consequence of coding is that the numbers become easier to work with. In fact, if we happened to use the mean as the constant then the right term, $(\Sigma X)^2$, would disappear. So actually that right term is nothing but a *correction factor* that is needed because we are not actually subtracting from the mean. We now need to find three different variances, so we will compute three different sums of squares:

(1) The sum of squares for *all* of the values, which is called the *total corrected sum of squares* or just *total SS*. It corresponds to the overall variance and is calculated as

$$\text{Total corrected sum of squares} = \Sigma \Sigma X_{ij}^2 - CF$$
$$= \Sigma X^2 - \frac{(\Sigma X)^2}{n}$$
$$= \Sigma X^2 - \frac{(T)^2}{n}, \qquad (19.3)$$

where

$$\Sigma \Sigma X_{ij}^2 = X_{11}^2 + X_{21}^2 + X_{31}^2 + X_{12}^2 + \cdots + X_{34}^2,$$

and

$$CF = \text{Correction factor} = \frac{(X_{11} + X_{12} + \cdots + X_{34})^2}{n}$$

$$= \frac{(T_1 + T_2 + T_3 + T_4)^2}{n}. \quad (19.4)$$

Observe that since there are n-values used here, the *degrees of freedom is* $n - 1$.

(2) The sum of squares *between samples* or *between categories*. This corresponds to the variance between or among the means of the categories and we calculate this as

$$\text{Between categories sum of squares} = \Sigma \left(\frac{T_j^2}{n_j} \right) - CF$$

$$= \Sigma \left(\frac{T_j^2}{n_j} \right) - \frac{(T)^2}{n}$$

$$= \frac{T_1^2}{n_1} + \frac{T_2^2}{n_2} + \frac{T_3^2}{n_3} + \frac{T_4^2}{n_4}$$

$$- \frac{(T_1 + T_2 + T_3 + T_4)^2}{n}. \quad (19.5)$$

We identify the number of samples or categories as k to distinguish them from n. Therefore, we have $k - 1$ degrees of freedom or $4 - 1 = 3$ degrees of freedom in this case.

(3) The sum of squares for variation *within the samples*. This corresponds to our smallest measure of variance. Here we actually have a separate correction factor for each sample but since they are all of similar form we can write the sum of squares as

$$\text{Within sample sum of squares} = \Sigma\Sigma X_{ij}^2 - \Sigma \frac{(T_j)^2}{n_j}$$

$$= \Sigma\Sigma X^2 - \left(\frac{T_1^2}{n_1} + \frac{T_2^2}{n_2} + \frac{T_3^2}{n_3} + \frac{T_4^2}{n_4} \right). \quad (19.6)$$

Since we lose one degree of freedom for each category, we would have $n - k$ degrees of freedom (there are k categories) or $12 - 4 = 8$ df. You can verify this by examining each sample and calculating the number of differences for each of them. You would get 2 df for each sample or a total of 8 df.

In order to present the computations in a simplified way, a fairly universal approach has been agreed upon. This is called an *analysis of variance table*. The general format with the computations is presented in Table 19.2. Although this table may look extremely complicated, if you examine it carefully you will find it is not as difficult as it might look. Observe:

Table 19.2 General analysis of variance table for one-way or single-category ANOVA

Source of variation: you identify what is meant by treatment or category	Degrees of freedom df	Sum of squares (SS) (these are never negative)	Mean square (MS) (this is SS divided by df and is an estimate of variance to be used in F-ratio)	F-ratio
Between categories or treatments— between means	$k - 1$	$\sum \dfrac{T_j^2}{n_j} - \dfrac{(T)^2}{n}$	$\dfrac{SS \text{ Between}}{k - 1}$	$\dfrac{MS \text{ Between}}{MS \text{ Within}}$
Within categories or samples, sometimes called *error*	$n - k$	$\sum\sum X_{ij}^2 - \sum \dfrac{T_j^2}{n_j}$	$\dfrac{SS \text{ Within}}{n - k}$	
Totals	$n - 1$	$\sum\sum X_{ij}^2 - \dfrac{(T)^2}{n}$		

(1) The degrees of freedom for *between* and *within* add up to the total. You can quickly find one of them by subtracting. Since n is known, you should immediately be able to write down the total *df*, and since you know how many categories you are looking at, the *df* for categories should also be quickly figured.

(2) The *total sum of squares* is equal to the *sum of squares for between* plus *sum of squares for within*. Again, if you calculate two of these you can find the other by subtracting (or you can use the third as a check). They are *never* negative numbers. If the correction factor is larger than the left term then there is an error in your computation. At most there are only three types or groups of calculations represented in the sum of squares column and they are all repeated in the different rows.

Let us now do an example with data to see how easy the computations really are.

EXAMPLE 19.1

(We will follow this example for most of the next two sections.)

We are to determine if there are differences between 4 brands of regular gasoline. They all cost the same per gallon and therefore differences in cost per mile are related to miles per gallon (except inversely, as miles per gallon goes up cost per mile decreases). We will record miles per gallon.

We choose the order of purchasing the gasoline randomly by using a random number table. Thus, we use gasoline 3 first, then 1, then 4, and then 1 again, then 2, and so forth in accordance with random digits chosen from the table. In Table 19.3 we have the miles per gallon for three samples of each gasoline as well as some of the calculations.

Table 19.3 Miles per gallon for four different brands of regular gasoline (recorded to nearest mile). Some computations are also included. (Three separate readings taken with each brand)

	Brand of gasoline				
	1	2	3	4	
	13	11	12	11	
	12	10	11	9	
	14	11	13	10	Overall values
Total	39	32	36	30	137
Numbers (n)	3	3	3	3	12
Means	13	10.7	12	10	11.5

Degrees of freedom: Since there are a total of 12 values, the total number of $df = 12 - 1 = 11$. There are 4 categories or brands being considered; therefore, there are $4 - 1$ or 3 df for brands. This leaves $11 - 3 = 8$ df within brands. (You can also verify the within brands degrees of freedom by adding up the degrees of freedom in each sample. There are 2 df for each of the 4 samples. For example there are 2 df for the 3 numbers 13, 12, and 14. We therefore have $4 \cdot 2 = 8$ df.)

Sum of squares:

$$CF = \text{Correction factor} = \frac{(T)^2}{n}$$

$$= \frac{(137)^2}{12}$$

$$= \frac{18{,}769}{12}$$

$$= 1564.1.$$

Sum of all $X^2 = \Sigma \Sigma X^2$

$$= (13)^2 + (12)^2 + (14)^2 + (11)^2 + (10)^2 + (11)^2$$
$$+ (13)^2 + (11)^2 + (12)^2 + (9)^2 + (11)^2 + (10)^2$$
$$= 1587.0.$$

454 Statistical Tests

$$\text{Sum of squares for totals} = \sum \left(\frac{T_j^2}{n} \right)$$

$$= \frac{39^2}{3} + \frac{32^2}{3} + \frac{36^2}{3} + \frac{30^2}{3}$$

$$= 507 + 341.3 + 432 + 300 = 1580.3.$$

$$\text{Total sum of squares} = \sum \sum X^2 - \frac{(T)^2}{n}$$

$$= 1587 - CF$$

$$= 1587 - 1564.1$$

$$= 22.9.$$

$$\text{Within samples sums of squares} = \sum \sum X^2 - \sum \frac{T_j^2}{n_j}$$

$$= 1587 - \left(\frac{(39)^2}{3} + \frac{(32)^2}{3} + \frac{(36)^2}{3} + \frac{(30)^2}{3} \right)$$

$$= 1587 - 1580.3$$

$$= 6.7.$$

$$\text{Between sample or between brands sum of squares} = \sum \frac{T_j^2}{n_j} - \frac{T^2}{n}$$

$$= 1580.3 - CF$$

$$= 1580.3 - 1564.1$$

$$= 16.2.$$

We can now fill in the first 3 columns of the analysis of variance or ANOVA table. This is done in Table 19.4. Observe that the totals for degrees of freedom and sum of squares are actual totals for the values in those columns.

Table 19.4   Analysis of variance table for Example 19.1

Source of variation	df	Sum of squares	Mean square	F-ratio
Between brands	3	16.2	$\frac{16.2}{3} = 5.4$	$\frac{5.4}{0.84} = 6.44$
Repeated values within brands	8	6.7	$\frac{6.7}{8} = 0.84$	
Total	11	22.9		

5% test limit for F is $F_{0.95}(3, 8) = 4.07$

Next we calculate the mean squares. Since we are not interested in a mean square for the total we only calculate a mean square (a type of estimation of variance) for the brands and the repeated (within brands) values. We divide the sum of squares by the degrees of freedom to arrive at a mean square of 5.4 for variation between brands, and 0.84 for mean square within brands.

Finally we perform an *F*-test to determine if the mean square *between* brands is significantly larger than the mean square *within* brands. We are to use a *one-tailed* test with the following hypothesis.

Hypothesis: There is *no* difference between the category means (between brands).

Alternate Hypothesis: There is a difference between means of the brands. Choose significance level, α. Let $\alpha = 0.05$.

Limit: $F_{1-\alpha}(f_1, f_2)$—One-tailed test, $p = 0.95$

$$f_1 = df \text{ for between brands}$$
$$f_2 = df \text{ within brands}$$

or

$$F(f_1, f_2) = F_{0.95}(3, 8) = 4.07.$$

Test Statistic: *F*-value calculated in the *F*-ratio column; here $F = 6.44$.

Decision: If the test value of *F* is smaller than the *F*-limit, we accept the hypothesis of no difference. If the test statistic is larger than the limit we accept the *alternate* hypothesis that there *is* a difference in means.

Since $F = 6.44$ is greater than the limit of 4.07, we declare that there is a significant difference among these brands of gasoline and that the means are *not* all the same.

19.4 Comparisons for Significant Differences

Whenever we obtain a significant *F*-ratio for the ANOVA our next concern generally is to determine in what way the means are significantly different. The first approach to this problem is to *plot the data*. As a minimum you should plot the means. A quick glance *may* show some type of obvious distinction between the samples. In some cases just looking at the graph of means may not reveal anything of real importance.

Quite often it is desirable to arrive at some measure of just how large a difference between *pairs of means* would be considered exceptional. If you have decided in advance that one and only one pair is of particular interest then you can perform a *t*-test to see if that pair is significantly different. You use the mean square for "within samples" as your value of *s*.

A more common problem would involve making all possible comparisons between means. We will follow one procedure called Tukey's test, which only works if all of the samples have the same size. This will involve determining a maximum distance by which we would expect *any* two means to be separated.

The formula for the distance is

$$\text{Maximum distance} = q_{1-\alpha}\sqrt{\frac{\text{Mean square for within samples}}{n_j}}. \quad (19.7)$$

The alpha level is to be chosen. The value of q is then found in Appendix Table A.20. To find q requires knowing the number of degrees of freedom in the mean square (labeled df) and the number of means to be compared (k). You locate the particular values in the table similarly to the method employed with the F-tables.

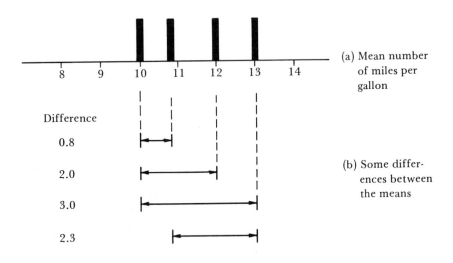

Figure 19.2 For Example 19.1 (a) a plot of the average miles per gallon for all four samples and (b) an illustration of several of the differences between the samples.

EXAMPLE 19.1 (Cont.)

Let us determine if any significant differences can be identified using a 95% confidence.

Limit: Since $df = 8$ (from Table 19.4), and the number of categories are 4 ($k = 4$) we find from Table A.20 for $p = 0.95$ ($\alpha = 0.05$) $df = 8, k = 4$, that $q = 4.53$.

From Table 19.4, the mean square for within brands equals 0.84 and n for each sample is 3. Substituting into Equation (19.3) we find that the maximum distance between samples should be

$$4.53\sqrt{\frac{0.84}{3}} = 4.53(0.53) = 2.39.$$

Examining all of the differences between means (see Figure 19.2b) we find that only one pair—the 10.0 (Brand 4) and the 13.0 (Brand 1)—differ by more than the limit of 2.39. We can therefore at least assert that Brand 1 gives significantly more miles per gallon than Brand 4.

19.5 Two-way Design and Interaction

One extension of the previous problem, Example 19.1, is to look at a second factor at the same time. We will look at the same data as before except we will now recognize that the three readings were not all taken under identical conditions. The readings were with three different cars (called A, B, and C). Look in Table 19.5 and observe that the total and averages for these three cars differ. This tells us that we can go back to the variation that we called "repeated

Table 19.5 Same data as in Table 19.3 (Example 19.1) except the three values are now identified as different cars. This is a two-way table

		Brands of gasoline				Totals for cars	Means for cars
		1	2	3	4		
	A	13	11	12	11	47	11.75
Cars	B	12	10	11	9	42	10.5
	C	14	11	13	10	48	12
Total for brands		39	32	36	30	137	
Means for brands		13	10.7	12	10		11.5

values within brands" and divide it up. We can now assign some of that previously unidentifiable "error" to variations between cars. Thus, we will assume that there is another major classification—between cars. We can calculate the sum of squares for this category in the same manner as we calculated the sum of squares for "between brands." We find:

$$\text{Sum of squares between cars} = \sum \left(\frac{T_i^2}{n_i}\right) - CF,$$

where the totals to be squared are the totals for cars. Thus we have

$$SS \text{ between cars} = \frac{47^2}{4} + \frac{42^2}{4} + \frac{48^2}{4} - 1564.1$$

$$= 552.5 + 441.0 + 576 - 1564.1$$

$$= 1569.2 - 1564.1 = 5.1.$$

The degrees of freedom are 2 since there are 3 categories ($df = k - 1 = 3 - 1 = 2$).

Because we no longer have repeated values we can no longer directly compute the "within samples sums of squares." We calculate this residual or error variation by subtraction. Table 19.6 is the summary table for this ANOVA problem. The F-values for determining if these classifications are significant are both computed using the mean square for the "error." Both of these F-values are significant at the 0.05 level.

Table 19.6 ANOVA table for two-way analysis. Cars are also considered

Source of variation	df	SS	MS	F-ratio	5% F-limit
Between brands	3	16.2	5.4	$\frac{5.4}{0.26} = 20.8$	$F(3, 6) = 4.76$
Between cars	2	5.1	2.55	$\frac{2.55}{0.26} = 9.8$	$F(2, 6) = 5.14$
Residual or error	6 *	1.6 *	0.26		
Total	11	22.9			

* By subtraction: For df the 3 and 2 are subtracted from 11 to give 6. For SS subtract 16.2 and 5.1 from 22.9.
Both F-ratios are significant.

If we were to take repeated measurements for all of the categories we could again obtain a separate independent measure of inherent or smallest variations. We would calculate the sums of squares and degrees of freedom in a way similar to the approach we used in determining the "within samples" variation for one-way analysis. We then find leftover sums of squares and leftover degrees of freedom. These are for another kind of comparison that often is of critical importance. This is a measure of the interrelationship or interdependence among the two different classifications. It is called *interaction*. We will not consider how to perform the calculations for a two-way ANOVA when repeated measurements are taken and an interaction variation would be computed. However, we will look at and attempt to interpret a few ANOVA tables and the corresponding graphs for which these values have been calculated. We will rely upon a typical computer printout for the calculations (there are a number of library programs available that can provide the ANOVA table for you).

Table 19.7 Two-way analysis of variance with repeated measurements of each car and brand. We are able to identify if there is an interaction between car and brand

a. The Data

		Brand			
		1	2	3	4
Car	A	13	11	12	11
		12	11	11	10
	B	12	10	11	9
		12	9	11	8
	C	14	11	13	10
		13	12	13	12

b. ANOVA Table

Source	df	SS	MS	F
Brands	3	25.460	8.486	18.5
Cars	2	16.080	8.042	17.5
Interaction	6	1.917	0.319	0.7
Within	12	5.500	0.458	
Total	23	48.957		

Another complete set of gasoline tests were run and the data is presented in Table 19.7a. The ANOVA table for these values is presented in Table 19.7b. Using an F-test by comparing MS interaction to MS within we would find that the interaction is *not* considered a significantly large value. We can then proceed to test the main effects, brands and cars, as before.

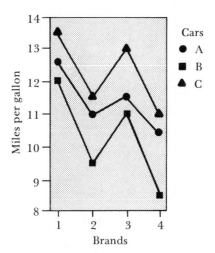

Figure 19.3 Graph of the averages of the miles per gallon for different brands of gasoline and several cars. There is no interaction so that the lines for the different cars are essentially parallel.

The important information that the interaction term tells us is that the cars act in a similar fashion from brand-to-brand (or correspondingly the brands act in the same way from car-to-car). We will plot the averages for all of the samples. We will select *one* of the factors to be used as the X-axis. In order to distinguish the cars we will use different marks to show the averages for each *car* and we then connect the averages for each car. This is done in Figure 19.3. Observe *that all three connecting lines* are essentially identical. When there is NO interaction you should expect to see a set of similar lines (usually with no lines crossing over each other).

Table 19.8 Two-way analysis of variance with repeated measurements of each car and brand. There is a different set of repeated measurements this time and there *is* also a significant interaction term this time

a. Data

		Brand			
		1	2	3	4
Car	A	13	12	12	11
		11	12	11	12
	B	12	10	11	9
		12	10	12	8
	C	14	11	13	10
		13	11	12	10

b. ANOVA table

Source	df	SS	MS	F
Brands	3	21.000	7.000	16.8
Cars	2	8.333	4.166	10.0
Interaction	6	9.000	1.500	3.6
Within	12	5.000	0.4167	
Total	23	43.333		

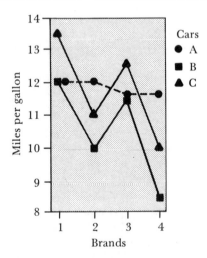

Figure 19.4 Graph of the averages of the miles per gallon for different brands and cars listed in Table 19.8. There is a significant interaction which shows up as different lines for cars.

Table 19.8 shows a different collection of sample values, with the interaction term being significant. Now if we examine the graph (see Figure 19.4), we find that the cars act differently. Car A appears not to do any differently with different gasoline, while the other two cars act very similarly. Cars B and C get higher gas mileage with Brand 1 and 3 and get very low mileage with Brand 4. There is a definite interrelation between the cars and brands and you cannot make any strong statements about gasoline unless you also qualify your conclusions by stating with which car you are dealing. (You can do the same sort of comparison with cars along the X-axis and lines for each brand, and you will also get a set of nonparallel lines). Thus, performing additional F-tests is somewhat meaningless.

Table 19.9 College Board score for students of different sex, on the math and English portions. A significant interaction is present. The math and English averages for opposite sexes are almost the same, and the sex averages are identical

a. The Data

Sex	Test		
	Math	English	Average
Male	550	480	
	570	460	515
	540	490	
Female	480	540	
	490	540	515
	470	570	
Average	516	514	

b. ANOVA Table

Source	df	SS	MS	F
Sex	1	0.0	0.0	
Test	1	33.3	33.3	
Interaction	1	16133.0	16133.0	74.8
Within	8	1733.0	216.7	
	11	17899.3		

EXAMPLE 19.2

Math and English College Board scores for a group of male and female students are given in Table 19.9. The averages are also provided. Obviously there are no differences between males and females and essentially there are no differences in math and English scores. Correct? Not really. The interrelationship between any kind of test and sex is completely masking any differences. The ANOVA table, Table 19.9b, indicates that almost all of the variation is in interaction. Figure 19.5 shows why.

Here we see that there is a complete crisscrossing of the scores from one sex to the other for math and English (or vice versa). *Because of the interaction it is absolutely pointless to talk about* differences *between sexes or differences between tests.* One must discuss tests for the particular sex, or sexes for the particular test.

19.6 General Kinds of Design Problems

Now that we have some idea of the basic principles and operations involved in the analysis of variance it is appropriate to discuss the broader aspects of the kinds of problems and types of interpretations that may be encountered. I will briefly describe some of the different considerations that one must take into account before embarking upon a realistic analysis or setting up a valid study. However, you should realize that to delve into these considerations in depth is beyond the scope of this text and is generally reserved for separate courses in the design and interpretation of experiments. The following outline will give you some idea of the range of some of the questions that are possible. We will then discuss them in a little more detail.

1. What are the factors or classifications?
2. What are the types of classifications—categorical or measurement?
3. Are you interested in just these classifications or are these classifications samples of a larger group of possibilities?
4. Are you interested in estimating variances (error analysis)?
5. How well can you randomize—are there blocks?

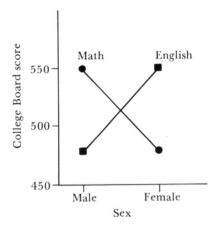

Figure 19.5 Graph of sex and College Board score for the different tests (math and English). There are obvious differences which do not show up as differences in the overall average. The curves for math and English cross over each other indicating a major interaction.

1. *What are the number of factors?*

We have already seen how to consider simultaneously a number of samples in one or two classifications. This can be expanded further to include three or four or even more categories. Of course, the computational procedure becomes an immense job although computer programs are available for some cases.

2. What are the types of classifications?

So far we have only considered categorical values for classifying the things to be compared. Instead of comparing "brands of gasoline" we might have compared octane ratings or we could have replaced "cars" with engine horsepower. Then we would obtain a scale. This scale brings more meaning to the graphs that we looked at under the two-way analysis. Essentially, we would be attempting to see if there are different lines for the second kind of category. This is one of the places where analysis of variance and regression techniques begin to overlay. A technique called *analysis of covariance* also falls into this area.

3. Are you interested in just these classifications?

One of the most important questions we must ask is whether or not we are restricting our classifications to those for which we have data. Let us go back to Example 19.1 and inquire further about the brands and then the cars.

Are we interested in only those *four* brands of gasoline or are we interested in brands of gasoline in general? If we are only interested in these particular brands and none other, then we call this a *fixed* model (sometimes called Model I). It is called *fixed* because you have all of the possible categories included in your analysis, and there is no choice of which sources to sample. Our decision is limited to these four brands and we can truly make statements about the means of just these brands.

If these are only a sample of brands out of a larger group of perhaps fifteen brands of gasoline, then your conclusions and the experimental design are different. You are now attempting to declare if there are significant differences among the larger population of different brands. Often there is a desire to estimate the amount of variability associated with the different classifications.

Several other examples might be:

Cars: Fixed—my two cars and your car are all to be considered;
    Random—we select three cars at random and are to judge about all cars.
Drivers: Fixed—only two drivers are ever to use the car and we sample both;
    Random—the college owned cars may be driven by thirty different people and we randomly select two of them for the study.

The analysis of the random model case, as well as the mixed cases where one of the classifications is fixed and the other is random, are not discussed any further in this text.

4. Are you interested in estimating variances (error analysis)?

Sometimes the main interest is in determining the degree of variability in various types of samples. In taking a blood count only a small sample from a drop of blood is generally used. However, there is a hierarchy of blood sampling that is possible and we would expect that there would be different degrees of

variability associated with each level in the hierarchy. For example:
We would take blood samples from 6 different people
Take 3 vials of blood from each person
From each vial take 2 full eyedroppers
From each eyedropper take 4 drops
For each drop make 3 counts of red blood cells

Each of the lower items is "nested" within the upper levels and a total of 432 readings would be made. Ultimately we would be able to state which variations are the greatest and what collection of samples (at which levels) would give us a certain degree of precision. We can also consider cost for the various types of analysis and arrive at ways of performing the least expensive set of analyses to arrive at a particular degree of precision.

5. *How well can you randomize—are there blocks?*

In many problems you are not able to completely randomize a testing problem. Many items are made in batches or groups and from batch to batch or block to block (as we will call the groups) there may be differences. You can only put four tires on a car at one time. In comparing tire wear, for two sets of tires, you must also recognize that the car is not driven over the exact same roads or at the same speeds in the same way for both sets. In baking cupcakes you are limited to the number you can put in an oven at one time, but the oven temperature may change slightly from one time to the next. Eggs are delivered in cases that may vary. Students are often blocked into different tracks or different classes and may or may not end up with different characteristics or different programs.

What we may be able to do is not to fully randomize our measurements, but rather randomize them over the different blocks. If we cannot take all the tests in one morning we will do half in the morning and the other half in the afternoon. But we will split every category in half and then randomize the experiments in the two parts of the day. We can then treat the blocks as another classification. If the blocks are not significant we can disregard them. This is listed as a *randomized block design*.

REVIEW

Factors · "Hold everything constant" method ·

Analysis of variance—ANOVA · Treatments · Samples ·

Categories · Variances · Within samples · Between samples ·

Total · *F*-tests · One-way ANOVA—Single factor ·

Sums of squares—*SS* · Mean squares—*MS* ·

Degrees of freedom—*df* · Correction factor—*CF* ·

Significant differences · Two-way ANOVA—Two factors · Residual · Error · Interaction · Graph · Factors · Classifications · Fixed · Random · Estimating variance · Randomized blocks

PROBLEMS

Section 19.3

19.1 Three different groups of preschool children were tested to determine their level of reading preparation. After a year they were tested again and the increases in the test were recorded as follows:

Nursery School group	Sesame Street watchers	Control group
6	9	5
7	6	4
6	7	5
8	8	3

Are there significant differences in the amount of improvement that the children have received under the different conditions?

19.2 The number of teenagers per 10,000 who were arrested for possession of drugs were compared for three different types of communities.

Suburban area	Large city	Rural area
12	24	9
16	20	7
15	18	12
22	30	7
10	18	10
6	25	3

Are the means significantly different?

19.3 The level of automotive exhaust emission was measured after using different "low lead" or no-lead gasolines. The results in test units were

Brand E	Brand S	Brand M	Brand A
6	8	10	12
4	6	8	10
10	8	6	4
5	7	9	11

Do the different brands give different average pollution levels?

19.4 Several drugs are tested on a group of people to determine their effectiveness in reducing blood pressure. The amount of blood pressure reduction in mm of Hg was found as

Drug X	Drug Y	Drug Z
14	10	11
15	9	11
12	7	10
11	8	11
10	11	8
11	11	7

Do the drugs act differently?

19.5 The average rate of earnings for several companies in different industries were reported in the following percentages:

Chemicals	Electrical	Metals	Transportation
8.6	8.1	6.4	7.3
6.2	7.3	6.4	5.7

Are the earnings different for different industries?

19.6 Refer to Problem 7.10. Do an analysis of variance for the amount of aspirin in the different tablets.

Section 19.4

For the following problems, if the analysis of variance was significant, determine if any of the sample means are significantly different than the others by using the q-statistic.

19.7 Refer to Problem 19.1.
19.8 Refer to Problem 19.2.
19.9 Refer to Problem 19.3.
19.10 Refer to Problem 19.4.
19.11 Refer to Problem 19.5.
19.12 Refer to Problem 19.6.

Section 19.5

19.13 Refer to Problem 19.2. Each of the rows was for a different month (for 6 months). Are there significant differences among months? (Extend the one-way analysis into a two-way analysis.) How does the difference in areas change (regarding significance)? You can plot the points to see if an interaction would be somewhat apparent.

19.14 Refer to Problem 19.3. The tests were taken on four different cars. Extend the one-way analysis to a two-way analysis. Plot the values. Is anything unusual about the graph?

19.15 Refer to Problem 19.4. The six readings for each drug were actually pairs of values for three different people. The sum of squares for *within* are 3.500, and for people are 14.78. By subtraction calculate the sum of squares for interaction. Is the interaction term significant? Plot the averages.

19.16 Refer to Problem 19.5. If the first row of earnings was for 1970 and the second row for 1971, is there a difference for years?

19.17 Refer to Problem 7.13. Try to do an analysis of variance using only the first reading at each point. You should also plot the values to see if an interaction is apparent. Is either direction significantly variable? (Do the depths vary significantly?)

20

Correlation and Regression Continued

Chapter 6 was devoted to a general introduction to several concepts and statistical approaches involved in determining the degree of correlation and the best straight line equation. In this chapter I will expand or elaborate upon some of that material and introduce some additional concepts.

20.1 Correlation: Nonparametric Tests

In many cases, particularly for work in the behavioral and social sciences or in education, the data that is accumulated is not being measured on an interval scale. It was mentioned in Chapter 6 that a very important assumption behind calculating and interpreting the correlation coefficient was that both sets of values being correlated were normally distributed. In order to have a normal distribution, the data must be measured on at least an interval scale. However, as I just said, there are many kinds of situations in which the measurements are not interval, but are categorical or perhaps ranked. In these cases the correlation coefficient r would be of no value.

If the data is merely categorical we have already used a technique to decide whether or not there is a relationship between the two sets of values. We have used a χ^2 (chi-square) test for independence (Chapter 17). If we find that two classifications are not independent, they must be considered related, and we can say that there is a *significant correlation* of some sort. If we also desire to have a measure of correlation, a value called the *contingency coefficient*, C, is sometimes computed. It is

$$C = \sqrt{\frac{\chi^2}{n + \chi^2}}, \qquad (20.1)$$

Spearman Rank Order Correlation Coefficient

Quite often, situations are encountered where the data can at least be placed on an ordered scale. We may not be able to accurately weigh or measure a set of individuals, but we can tell which ones are larger or heavier, and we can do that for every individual. We can tell that a grade of sixty on a test is better than a fifty, but we may not be able to say that this difference is as large a real difference as from an eighty to ninety (even though both are 10 point differences). But we can say that the 50 is less than the 60 and both are less than the 80, and so forth. We can rank all of the grades.

We are given pairs of measurements. Since the individuals in each pair are associated with two different characteristics, we can separate all of the values into the two categories. We then rank the values within each set. Next we compare the ranks for each pair by calculating the differences in rank (calling them d_i-values). We will square these differences (obtaining d_i^2's), sum up all the squares (yielding a term $\sum d_i^2$) and substitute into Equation (20.2), to obtain a different correlation coefficient called r_{rho} (r sub rho).

$$r_{rho} = 1 - \frac{6 \sum d_i^2}{n^3 - n} = 1 - \frac{6 \sum d_i^2}{n(n^2 - 1)}, \qquad (20.2)$$

where n = number of pairs of values.

Critical values for testing the significance of r_{rho} are found in Appendix Table A.22.

EXAMPLE 20.1

A finance company was interested in knowing if there was a relationship between the amount of a payment and the time it was paid (after the bill was sent). Since neither of these two values would necessarily be normally distributed,* the correlation coefficient, r, was not considered a reasonable quantity to calculate. Instead, a rank order correlation coefficient was computed. Twelve people were selected. The amount of their bills and the order of the receipt of payment was recorded in Table 20.1 (in Columns 2 and 3). These were converted to ranks (the order of receipt is already a ranking). The difference in rank is found for each of the people in Column 6. These differences are then squared in Column 7 and the sum is found. The sum is 180 (observe that the sum of the differences *before* squaring is zero and this serves as a convenient check). Substituting into Equation (20.2) we find a correlation coefficient of

---

* They both would probably be skewed, most amounts would be small and most payments on time but some amounts would be large and some times very long until payed.

Table 20.1 Amount and order of receipt of payment to a finance company, for twelve people

1	2	3	4	5	6	7
Person billed	Amount	Order of receipt	Ranks Amount	order	d (Amount–Order)	d^2
Mr. A. W.	10.57	1	1	1	0	0
Mr. J. N.	19.38	4	8	4	4	16
Mr. T. R.	67.21	10	12	10	2	4
Mrs. S. T.	42.67	12	11	12	−1	1
Miss S. B.	18.31	2	7	2	5	25
Mr. M. A.	23.16	3	9	3	6	36
Mr. A. F.	12.88	7	3	7	−4	16
Mrs. T. N.	17.61	11	6	11	−5	25
Mr. B. N.	37.63	8	10	8	2	4
Mr. D. F.	16.43	5	5	5	0	0
Mrs. L. A.	12.18	9	2	9	−7	49
Miss N. P.	14.39	6	4	6	−2	4
					Sum 0 (check)	180

$n = 12$

$$r_{\text{rho}} = 1 - \frac{6 \sum d^2}{n(n^2 - 1)} = 1 - \frac{6(180)}{12(143)} = 1 - 0.63 = 0.37$$

For $n = 12$, the 5% significant value is 0.506.

0.37. This is considerably less than the 5% limit of 0.506 from Appendix Table A.22, and we therefore conclude that there is not a significant correlation.

20.2 How Good is the Regression Line?

When we discussed correlation we also considered the question of significance. We compared the magnitude of our correlation coefficient to values that would be considered unusually large if there was actually no correlation. However, we did not attempt to do similar analysis for our regression equation or for the coefficients of the regression equation. You also may have observed that we could have a *significant* correlation with data so scattered that the relationship seemed purely coincidental. It would be helpful for us to have some additional gauges of how *good* is the straight line or the correlation.

Look back to the height and weight data in Chapter 6. For now we will assume we want to predict from height to weight. First let us look at the total variation in the weights, or the Y-values. We can do this by calculating the sum of the squares for the difference between the Y-values and $\bar{Y}$ (we have already performed this during the calculation of r). We will call this quantity our *total sum of squares*. If you also recall that when there is no correlation the "best line" is $\bar{Y}$ (refer back to Figure 6.11, page 165), then we can interpret this total sum of squares as one measure of variation around a prediction line. If there actually is a relationship, however, this measure of variation will be the *largest* one we can calculate.

In contrast to the line $\bar{Y}$, the regression line was defined as that line which would leave the *smallest* collection of squared differences along the Y-axis.

To illustrate how to go further we will work with just 6 points. I have plotted one possible situation in Figure 20.1. The first graph, Figure 20.1a, shows the

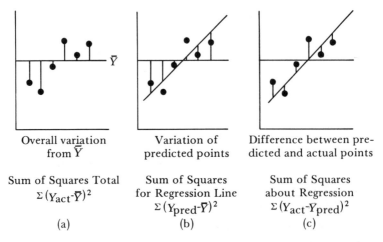

Overall variation from $\bar{Y}$

Variation of predicted points

Difference between predicted and actual points

Sum of Squares Total
$\Sigma(Y_{act}-\bar{Y})^2$
(a)

Sum of Squares for Regression Line
$\Sigma(Y_{pred}-\bar{Y})^2$
(b)

Sum of Squares about Regression
$\Sigma(Y_{act}-Y_{pred})^2$
(c)

Figure 20.1 Different ways of looking at the deviations and the squared differences.

differences between each point and the line $Y = \bar{Y}$. These differences when squared give us our *total sum of squares*. Try to make a rough picture in your mind of the magnitudes of these differences, remembering that we will be emphasizing the larger differences when we square them.

The next illustration, Figure 20.1b, shows the regression line and the predicted values that fall on that line. We can also compute the squared differences for these points and we call the sum of these squares the *sum of squares for regression*. Observe that if the regression line fits the points perfectly, then these differences would be identical to the overall variation. This is illustrated in Figure 20.2.

Returning to Figure 20.1 we find that the actual values and the predicted values, the points on the line, do not coincide. In each case there is something left over. These are often called *residuals* and they constitute a form of error. The final illustration, Figure 20.1c, shows the sizes of these errors. If we have found a good line then the sum of squares of these errors will be considerably smaller than the overall sum of squares.

In Figure 20.2 we have a perfect fit and the errors are zero. In this case *we have explained all of the variation* between the Y-values when we introduced the regression line.

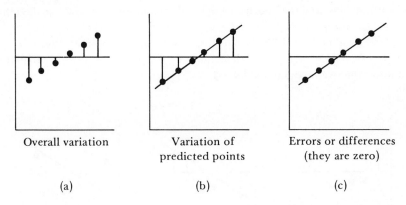

Overall variation	Variation of predicted points	Errors or differences (they are zero)
(a)	(b)	(c)

Figure 20.2 A perfect fit, all the variation among the points is accounted for with the regression line.

At the other extreme we should examine a case where there is essentially *no* correlation. Figure 20.3a shows a set of points and the variation around the line $Y = \bar{Y}$. The overall variation here is not much different than the previous examples. The best regression line is given next (Figure 20.3b) with the variations from $\bar{Y}$ shown. In the final illustration, Figure 20.3c, we have the differences between the points and the line. Even a quick look should indicate to you that the residuals or errors are just about as large as the total variation. In other words, the regression line has not reduced the overall variation by very much.

Interpretation of r^2

If we have calculated the various sums of squares, then we can calculate the ratios of the sum of squares for regression to the total sum of squares, and the error sum of squares to the total sum of squares. These ratios can provide us with statements of the proportion of variation that is explained and unexplained by the regression line. If we calculate:

$$\frac{\text{Sum of squares for regression line}}{\text{Sum of squares total}} = \frac{SS_{\text{reg}}}{SS_{\text{total}}},$$

we have a measure of the proportion of the total spread of Y-values that is explained by the regression line. This ratio is also equal to r^2 or

$$r^2 = \frac{SS_{reg}}{SS_{total}}. \tag{20.3}$$

Thus, the square of the correlation coefficient provides us with an additional indicator of how well the line fits the data. This quantity, r^2, is often called the *index of determination*.

Stated another way, $100r^2$ equals the percent of the variation that is attributed to the regression line, and the remaining $100(1 - r^2)$ is the percent of the variation that we have not explained.

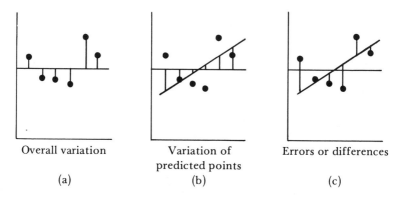

Overall variation
(a)

Variation of predicted points
(b)

Errors or differences
(c)

Figure 20.3 A poor fit. The error variation about the regression line (c) is almost as large as the overall total variation. Very little variation is eliminated or explained away by introducing the regression line.

For the height and weight problem of Chapter 6 we found that $r = 0.849$. Then

$$r^2 = (0.849)^2 = 0.721$$

or

$$100\ r^2 = 100(0.721) = 72.1\%.$$

We observed a distribution of heights and of weights. Now the value of r^2 tells us that 72% of the variation in one of the values (for example, height) can be explained merely by knowing what the other value (weight) is. However, we have no explanation for the remaining 28% of the variation—it could be just random variation.

I hope you can see the obvious ease and importance of this index. For instance, you know that given a sufficiently large sample, perhaps even only 100, a correlation coefficient of 0.20 is significant and a correlation coefficient of 0.25 is very highly significant. But if $r = 0.20$, then $r^2 = 0.04$ and we see that this relationship is only able to give an explanation for 4% of the variation. Even the very highly significant correlation coefficient of 0.25 only increases this percentage to a little over 6% ($r^2 = 0.25^2 = 0.0625$).

You can see that having a correlation coefficient even as large as 0.500 only explains 25% of the variation ($100r^2 = 100(0.5)^2 = 25\%$) *no matter how large the sample might be.*

*Testing the Significance of the Regression Line**

Since we have measures of variation or sums of squares associated with the regression line and the error, it is possible to convert these terms into variances that can be tested by an F-test. We will then be able to conclude whether or not the regression line is significant. (Bear in mind that when you perform a regression analysis you generally do not calculate a correlation coefficient. Regression analysis, though, can also be easily expanded for more complex analyses.)

As we discussed before, we have three values of the sum of squares: total (about the mean or $\bar{Y}$); regression; and error about the regression line. They are also related to each other, since

Total sum of squares = Sum of squares for regression
$$+ \text{ Sum of squares about regression (error).} \quad (20.4)$$

Associated with each of the sum of squares is a value of degrees of freedom. The total number of degrees of freedom (for the total sum of squares) is $n - 1$. The sum of squares for regression has only 1 degree of freedom, and therefore there are $n - 2$ degrees of freedom associated with the error or residual term. (Total degrees of freedom minus regression degrees of freedom $= (n - 1) - 1 = n - 2$.)

Table 20.2 shows the typical method of presenting these calculations and the calculations themselves. The sums of squares for regression and about regression are computed (in Column 3) and then divided by the appropriate number of degrees of freedom. These quotients again are called *mean squares* and as with the ANOVA we finally compare the two mean squares.

We divide the mean square for regression by the error or residual mean square. This is an F-statistic and is to be compared to an F-limit with degrees of freedom 1 and $n - 2$.

Let us return to our example of heights and weights. During Chapter 6 we performed a number of calculations. The results that we now need are

$$b = 7.71$$
$$\sum XY - \frac{(\sum X)(\sum Y)}{n} = 893,$$
$$n = 20$$
$$\sum Y^2 - \frac{(\sum Y)^2}{n} = 9494.$$

---

* This section depends somewhat on previous coverage of analysis of variance. However, it is not necessary to cover Chapter 19.

Table 20.2 General analysis of variance table for regression

1	2	3	4	5
Source of Variation	Degrees of freedom df	Sum of squares SS	Mean square (MS) (sum of squares divided by df)	F-ratio
Regression	1	$b\left[\sum XY - \dfrac{(\sum X)(\sum Y)}{n}\right]$	$MS_{reg} = \dfrac{SS}{1}$	$F = \dfrac{\text{Mean square regression}}{\text{Mean square about regression}}$
About regression (error or residual)	$n-2$	By subtraction	$\dfrac{SS}{n-2} = S_{y.x}^2$	
Total	$n-1$	$\sum y^2 - \dfrac{(\sum y)^2}{n}$		

In Table 20.3 we have the computations leading up to the F-ratio. The sequence of steps involves finding the total sum of squares (9494) first, and then the regression sum of squares $7.71 \cdot 893 = 6885$. The "about regression" sums of squares is then found by subtracting 6885 from 9494. The mean squares are then computed and finally the ratio of mean squares is found. The limit that we are to compare our F-value against is an F-value with 1 and 18 degrees of freedom. The 1% (one-tailed) limit is 7.59. Our computed F-value is 47.5, which is much larger than the limit. We assert that the regression line is significant.

Table 20.3 Analysis of variance table for regression of height and weight

Source	df	Sum of squares	Mean square	F-ratio
Regression	1	$7.71(893) = 6885$	$\dfrac{6885}{1} = 6885$	$\dfrac{6885}{144.9} = 47.5$
About regression	18	2609	$\dfrac{2609}{18} = 144.9$	
Total	$20 - 1 = 19$	9494		

(One tail) 1% limit $F_{0.99}(1, 18) = 7.59$

The natural extension of this technique involves the taking of repeated measurements at each X-value. This provides us with an additional estimate, that of "within groups" variation. This within-group variation then becomes our pure error and the remaining error really serves as a measure of how well or poorly the equation fits the data. This is called *lack of fit* and refers to the fact that we really need to be using a *curve* of some type instead of a straight line.

20.3 Confidence Intervals for the Regression Line

Up until now we have discussed the regression line in general and the general fit of the points. Once we have concluded that the fit is satisfactory and we are willing to use the equation for the purpose of prediction, we need to discuss how good that prediction will be. We need to specify what the error or distribution of possible values would be. Initially we are essentially looking for a way of computing the variance of the residuals. This is often called the *standard error of the estimate* and may be symbolized by $s_{y \cdot x}^2$. It may be found directly from the analysis of variance table for regression. This is the mean square about regression (the error or residual mean square). In our problem of height and weight it is 145.5. This can be computed directly as follows:

$$s_{y \cdot x}^2 = \frac{n-1}{n-2}(s_y^2 - b^2 s_x^2), \qquad (20.5)$$

where

$$s_y^2 = \text{variance of the } Y = \left[\sum Y^2 - \frac{(\sum Y)^2}{n}\right] \bigg/ (n-1)$$

and

$$s_x^2 = \text{variance of the } X = \left[\sum X^2 - \frac{(\sum X)^2}{n}\right] \bigg/ (n-1)$$

Generally, calculating confidence intervals is more valuable than finding the standard error (but we will need the standard error to compute confidence intervals). And here comes a point of confusion. *Two* kinds of confidence intervals are possible.

1. There is a confidence interval for the possible values of the *mean* of the Y for a given X.
2. There is a confidence interval of all possible values of Y for a given X.

There is an important difference between these two intervals that requires a decision on your part and a statement of your objective. Let us briefly examine and compare these two cases.

In the first case we are recognizing an objective that says we are interested in predicting what the average Y-value will be. The prediction is for a long range or average estimate of Y, which is basically what the regression line is. However, we also recognize that the line itself is not perfect, but is just an estimate of the possible lines. Essentially, this confidence interval is an interval containing the possible regression lines. Figure 20.4a shows a typical pair of confidence limits for the distribution of means and a typical distribution of means for an X-value.

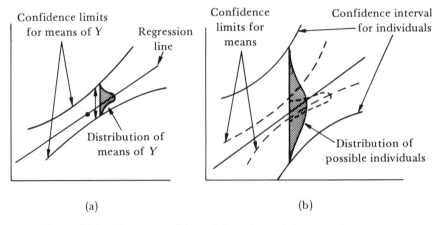

Figure 20.4 The two possible confidence intervals for regression.
(a) Confidence intervals for means. (b) Confidence intervals for individuals.
(*CI* for means is also shown for comparison.)

The second type of confidence limit provides an interval for the distribution of the possible actual Y-values. Obviously this should be a wider distribution and a wider confidence interval since individuals are spread out much more than means. A typical set of confidence limits for individuals is shown in Figure 20.4b. The confidence limits for means are also repeated, this time with dotted lines, for you to see the relative differences in interval sizes.

In both cases you should observe that the confidence limits get broader as you move from the center of the X-values. Previously we noted that the regression line always goes through the point $(\bar{X}, \bar{Y})$. For any of the possible $(\bar{X}, \bar{Y})$ points you could have a number of different regression lines going through that point. Pivot the line to obtain a number of different regression

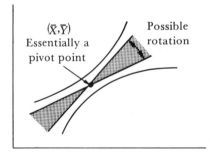

Figure 20.5 A graphical reason for wider estimates the further you get from the center of the X and Y. The regression line may pivot about the overall mean.

lines. As you go further from the center, there becomes a wider spread of possible points that can be on the lines. This is illustrated in Figure 20.5. Of course, this pivoting is true for all of the possible $(\bar{X}, \bar{Y})$ points that might occur and therefore we have a wider spread of limits than would be implied from just the one pivot point.

The actual confidence limits for the two cases are as follows (the significance level or α-value is for one estimate only).

Confidence interval for mean of Y:

$$Y_{\text{pred}} + t_{\alpha/2} s_{y \cdot x} \sqrt{\frac{1}{n} + \frac{(X - \bar{X})^2}{(n-1)s_x^2}} < \text{True mean of } Y$$

$$< Y_{\text{pred}} + t_{1-\alpha/2} s_{y \cdot x} \sqrt{\frac{1}{n} + \frac{(X - \bar{X})^2}{(n-1)s_x^2}}. \quad (20.6)$$

Confidence interval for individual Y:

$$Y_{\text{pred}} + t_{\alpha/2} s_{y \cdot x} \sqrt{1 + \frac{1}{n} + \frac{(X - \bar{X})^2}{(n-1)s_x^2}} < \text{Possible values of } Y$$

$$< Y_{\text{pred}} + t_{1-\alpha/2} s_{y \cdot x} \sqrt{1 + \frac{1}{n} + \frac{(X - \bar{X})^2}{(n-1)s_x^2}} \quad (20.7)$$

where Y_{pred} is the predicted value of Y, and the t-values have $n - 2$ degrees of freedom.

For our example of heights and weights we had calculated one prediction of a Y for a particular X. For $X = \text{height} = 10$ (actually 5 feet 10 inches) we found $Y = \text{weight} = 170.3$ pounds. We will now find 95% confidence intervals for our predicted value ($\alpha = 0.05$).

Since $n = 20$ the degrees of freedom are $n - 2 = 18$, and the t-values are

$$t_{0.975} = 2.101,$$

$$t_{0.025} = -2.101.$$

$$\bar{X} = 7.75,$$

$$s_{y \cdot x} = \sqrt{145.5} = 12.1.$$

$$s_x^2 = \frac{\sum X^2 - (\sum X)/n}{n-1} = \frac{115.75}{19} = 60.7.$$

Confidence interval of the mean of the predicted Y—from Equation (20.6):

$$170.3 + (-2.101)(12.1)\sqrt{\frac{1}{20} + \frac{(10 - 7.75)^2}{19(60.7)}} < \text{Mean of } Y$$

$$< 170.3 + (2.101)(12.1)\sqrt{\frac{1}{20} + \frac{(10 - 7.75)^2}{19(60.7)}}$$

$$170.3 - 25.4\sqrt{0.05 + 0.00048} < \text{Mean of } Y$$
$$< 170.3 + 25.4\sqrt{0.05 + 0.00048}$$
$$170.3 - 25.4(0.224) < \text{Mean of } Y < 170.3 + 25.4(0.224)$$
$$170.3 - 5.67 < \text{Mean of } Y < 170.3 + 5.67$$
$$164.6 < \text{Mean of } Y < 176.0$$

Confidence interval for possible Y-values—from Equation (20.7):

$$170.3 - (2.101)(12.1)\sqrt{1 + \frac{1}{20} + \frac{(10 - 7.75)^2}{19(60.7)}}$$
$$< Y < 170.3 + (2.101)(12.1)\sqrt{1 + \frac{1}{20} + \frac{(10 - 7.75)^2}{19(60.7)}}$$
$$170.3 - 25.4\sqrt{1 + 0.05 + 0.00048}$$
$$< Y < 170.3 + 25.4\sqrt{1 + 0.05 + 0.00048}$$
$$170.3 - 25.4(1.025) < Y < 170.3 + 25.4(1.025)$$
$$170.3 - 26.0 < Y < 170.3 + 26.0$$
$$144.3 < Y < 196.3.$$

You can see the large difference between these two sets of limit.

20.4 Multiple Correlation

Often measurements are taken for a number of different variables and it is desirable to determine which of the many pairs of quantities are correlated. As an example I will discuss one type of situation that should be familiar to all students. At some time or another you have probably taken a battery of achievement or aptitude tests. Many of the broad achievement tests consist of separate tests of reading, arithmetic, English, and so forth. You might also find that the parts of the test appear to be repeated several times. You may have been asked to do a collection of readings and then told to stop. Later you were required to do more reading exercises.

A typical test might have consisted of the following 5 subparts:

1. Reading, Part 1.
2. Reading, Part 2.
3. English.
4. Math achievement.
5. Math aptitude.

Each student who takes this test would receive a separate score for each subpart of the test and therefore we would have 5 scores for *each* individual. If we gave the test to 50 students we would have 50 sets of 5 scores each.

It is possible to determine the correlations between *each* of the pairs of tests (subparts). We can calculate a correlation coefficient for Reading Part 1 and Reading Part 2, for Reading Part 1 and English, for English and Math achievement, and so forth. In all, we can arrive at 20 correlation coefficients. Of course, we only really have 10 different ones since the correlation coefficient of English to Math achievement is the same as Math achievement to English.

After all the correlation coefficients are computed (which is virtually unmanageable without a computer) they are generally recorded in an arrangement that is called a matrix. An example of such a matrix might be the one given in Table 20.4. Along the sides are listed the subparts, which have also been labeled X_1, X_2, X_3, X_4, and X_5. Since we are dealing with a general concept, you should realize that given *any* set of variables we can determine if there are such relationships.

Table 20.4 A matrix of correlation coefficients. All of the correlations between five different subparts of a battery of tests

	X_1 Reading Part 1	X_2 Reading Part 2	X_3 English	X_4 Math achievement	X_5 Math aptitude
X_1 Reading Part 1	1.00	0.862	0.68	0.10	0.20
X_2 Reading Part 2	0.862	1.00	0.57	0.16	0.08
X_3 English	0.68	0.57	1.00	0.19	0.05
X_4 Math achievement	0.10	0.16	0.19	1.00	0.67
X_5 Math aptitude	0.20	0.08	0.05	0.67	1.00

To find the correlation between any pair of subtests we select the column for one of them and the row for the other, and the box in the intersection of row and column gives the correlation coefficient. Thus the correlation between English and Math achievement is 0.19.

Note that it does not matter which you pick for the column and which for the row. However, sometimes a matrix of this type is reported in journals with only half of the values. You then must select the proper order. (If you examine the matrix you will see that the values in the upper right triangle would match the values in the lower left triangle by merely flipping the triangle over. One of the triangles could be omitted with no loss of information.)

You may have observed the set of 1 down the diagonal. These appear because there must be a perfect correlation between any test and itself. For example, the correlation between English and English is 1.00 since these values are identical for each individual. All fifty pairs would be plotted on a perfect straight line.

Now what kind of interpretation can we draw from this collection of correlation coefficients? First, we cannot legitimately use the significance limits for correlation coefficient that we discussed in Chapter 6 because we are actually comparing ten coefficients at one time. The significance *levels* change drastically, but you can use the limits as rough guides (mentally adjusting the levels for now). A high correlation means the particular pair of values are closely related. The two Reading parts show up as having a very high correlation (0.862) and would appear to be measuring the same thing. A low correlation between Reading Part 1 and Math achievement, 0.10, shows that there is little relationship between this math and reading test (and vice-versa). In such a manner we can compare and detect the types of relationships that may exist among a large collection of different measurements.

20.5 Multiple Regression

Multiple correlation permits us to determine the degree of relationship between a collection of different measurements. In many cases, however, the more important analysis involves the ability to predict an outcome based on a number of different variables.

A chemist may desire to predict the density of batches of plastic based on the temperature, pressure, time, and concentration of a catalyst. A biologist might be interested in an equation relating amount of phosphorous, nitrogen, sunlight, and water, to the yield of corn. We might try to predict the amount of rubber to be consumed in the United States as related to the number of cars produced, the amount of gasoline consumed, and the gross national product. Success in college as measured by grade point average might be predicted from a battery of achievement tests. This situation will serve as our example for this section.

As with multiple correlation, multiple regression requires us to observe a collection of variables, all measured on the same individual (or at the same time, or on the same batch, and so forth). *One* of the variables is the quantity we are interested in predicting. In many chemical, physical, biological, and engineering experiments we can *control* the collection of variables and we then measure the thing we are interested in later predicting. Thus, we might legitimately say we have several *independent* variables, and one dependent variable. Other problems mainly involve the simultaneous recording of a variety of variables and also observing the output value. This is the case with a problem like our grade point average.

In our example we would be attempting to find the *best* equation of the form

$$Y = b_0 + b_1 X_1 + b_2 X_2 + b_3 X_3 + b_4 X_4 + b_5 X_5, \qquad (20.8)$$

where the b are constants and the X are the scores on the various parts of the achievement test. For example, if we found that the best equation is

$$Y = 0.71 + 0.01 X_1 + 0.01 X_2 + 0.02 X_3 + 0.02 X_4 + 0.03 X_5,$$

then
$$b_0 = 0.71, b_1 = 0.01, b_2 = 0.01, \ldots, b_5 = 0.03.$$

If a particular student scored a 20 on Reading, Part 1, then $X_1 = 20$; a 25 on Reading, Part 2, then $X_2 = 25$; a 15 on English, then $X_3 = 15$; an 18 on Math achievement, then $X_4 = 18$; and a 20 on Math aptitude, then $X_5 = 20$. We would then predict his grade point average to be

$$Y = 0.71 + 0.01(20) + 0.01(25) + 0.02(15) + 0.02(18) + 0.03(20)$$
$$= 0.71 + 0.2 + 0.25 + 0.3 + 0.36 + 0.6 = 2.42.$$

The calculations involved in arriving at the particular b are generally enormous even for just two independent variables (two X). Fortunately, computer programs are available to solve multiple regression problems. In addition, the programs generally provide ways of determining whether all of the variables need to be retained. Obviously, if we get almost as good a prediction using 4 or 3 of the variables instead of all 5 it is much more convenient. Thus some programs may provide a listing of the index of determination for a variety of possible equations, with different sets of variables (X) included in the equation. You can then select an equation with a high index of determination but with fewer than all of the variables.

Usually the reason it is possible to use fewer than all of the variables is that several of the variables are highly correlated. In this example, since both of the reading tests are so highly correlated it will make little sense to include *both* in the equation. Therefore an equation eliminating either X_1 or X_2 would probably be just as good as one that included both of these.

20.6 Nonlinear Regression

In Chapter 6 I introduced the possibility that a straight line might not be the best line to try to fit a set of data to. Many kinds of problems involve curves and relationships of forms like

$$Y = A + BX^2,$$

$$Y = \frac{1}{A + BX},$$

$$y = Ae^{BX}.$$

In most cases we transform or change the X-values. Often by performing some mathematical manipulation to each X-value we can modify them in such a way that they would fall on a straight line. For example, if we expect that the data falls in the form

$$Y = A + \frac{B}{X},$$

then we can take reciprocals of each of the X-values and plot these "new" or transformed X-values. If the fit is good these points should fall on a straight line and we can test these new points by using a linear regression formula.

As with multiple regression, there are computer programs for nonlinear regression. A typical program would provide the best coefficients for several different types of curves and also give the index of determination for each of the curves.

EXAMPLE 20.2

Let us look at the data in Problem 6.4. The data in this problem was the income tax rate for 1971. If you computed the best linear curve you would have predicted a negative income tax for small incomes, and if you plotted the data you would have found that there was a definite curvature to the graph. In Table 20.5 I have reproduced the data and have also provided the printout showing the "best" fit for 6 types of equations along with the index of determination for each of them.

Table 20.5a The income tax rate schedule for a single person after deductions (1971)

Income	Tax	Income	Tax
1000	145	10,000	2,090
1500	225	14,000	3,210
2000	310	20,000	5,230
4000	690	26,000	7,590
6000	1110	50,000	20,190
8000	1590	100,000	53,090

Table 20.5b The parameters for several different curves

CURVE TYPE	INDEX OF DETERMINATION	A	B
1. $Y = A + (B*X)$	0.977374	-2665.76	0.525605
2. $Y = A*EXP(B*X)$	0.734523	667.285	$5.36886E - 05$
3. $Y = A*(X \uparrow B)$	0.996316	$1.83383E - 02$	1.27607
4. $Y = A + (B/X)$	0.158997	12837.2	$-1.95372E + 07$
5. $Y = 1/(A + B*X)$	0.217701	$2.27484E - 03$	$-3.53437E - 08$
6. $Y = X/(A + B*X)$	0.998658	6.97633	$-1.82448E - 04$

Note that $*$ represents multiplication, the $\uparrow$ means raise to a power.

Curve 1 is of the form $Y = A + BX$ with $A = -2666$ and $B = 0.526$, and we have the equation

$$Y = -2666 + 0.526X.$$

Even though the index of determination is about 0.977 (thus explaining 97.7% of the variation of the values), the computer can provide us with a set of differences and percent differences between the actual Y-values and the ones calculated from the equation (these are the residuals, or errors). For curve 1 we would obtain the following results:

X-ACTUAL	Y-ACTUAL	Y-CALC	DIFFERENCE	PCT DIFFER
1000	145	−2140.16	2285.16	106.7
1500	225	−1877.36	2102.36	111.9
2000	310	−1614.55	1924.55	119.2
4000	690	−563.344	1253.34	222.4
6000	1110	487.866	622.134	127.5
8000	1590	1539.08	50.9247	3.3
10000	2090	2590.28	−500.265	−19.3
14000	3210	4692.7	−1482.7	−31.5
20000	5230	7846.33	−2616.33	−33.3
26000	7590	11000.0	−3409.96	−30.9
50000	20190	23614.5	−3424.48	−14.5
100000	53090	49894.7	3195.29	6.4

Observe the very large differences between the actual and calculated values, in spite of the high index of determination.

Curves 2, 4, and 5 all are much poorer fits, so we might just as well ignore them. Curve 3, though, is a better fit. It is a curve of the type $Y = AX^B$. For this example $Y = 0.0183 X^{1.28}$ (the $E-02$ in the $1.83383 E-02$ under Column A represents "times 10^{-2}" or "times 0.01"; the value of $E+07$ in Column B means "times 10^7" or $-1.95372 \cdot 10^7 = 19{,}537{,}200$). The comparison of the results here is

X-ACTUAL	Y-ACTUAL	Y-CALC	DIFFERENCE	PCT DIFFER
1000	145	123.475	21.5252	17.4
1500	225	207.149	17.8507	8.6
2000	310	299.03	10.9703	3.6
4000	690	724.187	−34.1866	−4.7
6000	1110	1214.94	−104.943	−8.6
8000	1590	1753.83	−163.827	−9.3
10000	2090	2331.58	−241.583	−10.3
14000	3210	3581.96	−371.961	−10.3
20000	5230	5646.6	−416.6	−7.3
26000	7590	7892.0	−301.999	−3.8
50000	20190	18179.7	2010.27	11.0
100000	53090	44027.4	9062.55	20.5

In this case the percent differences are all relatively small (the biggest is $20\frac{1}{2}\%$) even though the largest real difference is over $9000 and is much larger than any difference for the straight line case. However, for most of the points the calculated values are very reasonable.

The best fit (from the standpoint of index of determination) is the last one. This is approximately the equation

$$Y = \frac{X}{7 - 0.0002X}.$$

The results of fitting the actual points is

X-ACTUAL	Y-ACTUAL	Y-CALC	DIFFERENCE	PCT DIFFER
1000	145	147.191	−2.19117	−1.4
1500	225	223.792	1.2083	0.5
2000	310	302.506	7.49392	2.4
4000	690	640.354	49.6455	7.7
6000	1110	1020.12	89.877	8.8
8000	1590	1450.13	139.87	9.6
10000	2090	1941.05	148.95	7.6
14000	3210	3165.95	44.0517	1.3
20000	5230	6010.76	−780.762	−12.9
26000	7590	11645.2	−4055.23	−34.8
50000	20190	−23298.2	43488.2	186.6
100000	53090	−8874.29	61964.3	698.2

Here we have excellent results up to an income (X) of about $20,000, but beyond that point the calculated values are in extreme error.

From this example you should see that the index of determination is not necessarily the most important factor in deciding what is a "best" fit. The future use of the equation and the examination of the differences between the predicted and actual values will generally play extremely vital roles in determining what curve to use. Whatever curve you agree to use, you must be aware of the fact that here again a decision rests upon you, the experimenter. The statistical technique is only to provide you with some tool for exploring and comparing several alternatives.

REVIEW

Contingency coefficient ·

Spearman rank order correlation coefficient—r_{rho} ·

$$r_{rho} = 1 - \frac{6 \sum d^2}{n(n^2 - 1)}$$

Significance of regression line · Index of determination · Residual ·

$100r^2$ = percent variation · Analysis of variance table ·

Sums of squares · Confidence intervals · Means · Individuals ·

Multiple correlation · Matrix · Multiple regression ·

Nonlinear regression · Curve

PROBLEMS

Section 20.1

For Problems 20.1 to 20.5 calculate the contingency coefficient for the designated χ^2-test.

20.1   Refer to Example 17.1.
20.2   Refer to Example 17.2.
20.3   Refer to Problem 17.1.
20.4   Refer to Problem 17.2.
20.5   Refer to Problem 17.3.
20.6   A sociologist studying crime rates in different countries felt that the population density and crime rate were related. He compiled the following information for 12 countries:

People per square mile	Murder rate	People per square mile	Murder rate
415	2.7	195	0.8
300	20.8	708	2.3
606	3.0	5	1.9
42	0.7	166	0.5
82	7.6	311	19.1
454	3.6	238	2.6

Is there a correlation between population density and murder rate? Use a rank order correlation coefficient.

For Problems 20.7 to 20.9 calculate a Spearman rank order coefficient. Determine if the correlation is significant. Compare the results to the correlation coefficient, r, that was calculated in Chapter 6. (Also comment on the difference in effort required.)

20.7   Refer to Problem 6.1 (typing speed).
20.8   Refer to Problem 6.2 (staples).
20.9   Refer to Problem 6.3 (football players).

Section 20.2

For Problems 20.10 to 20.15 calculate the index of determination. Comment on how you can use this index to describe the usefulness of the regression line.

20.10   Refer to Problem 6.16.
20.11   Refer to Problem 6.17 (the same as 6.2).
20.12   Refer to Problem 6.18 (the same as 6.1).
20.13   Refer to Problem 6.19.
20.14   Refer to Problem 6.20 (the same as 6.3).
20.15   Refer to Problem 6.21 (the same as 6.4).

For Problems 20.16 to 20.21 perform an analysis of variance and test the F-ratio.

20.16 Refer to Problem 6.16.
20.17 Refer to Problem 6.17.
20.18 Refer to Problem 6.18.
20.19 Refer to Problem 6.19.
20.20 Refer to Problem 6.20.
20.21 Refer to Problem 6.21.

Section 20.3

20.22 Refer to Problem 6.16. A restaurant would like to know the 95% confidence interval for the average rating of coffee perked 7 minutes. The owner would also like the 95% confidence interval for the rating of all the individuals who would eat in the restaurant.

20.23 See Problem 6.17. What is the 95% confidence interval for the number of staples that would properly staple through 10 sheets (not the average number of staples)?

20.24 See Problem 6.18. What is the 99% confidence interval for the mean number of errors for someone who types 50 words per minute?

20.25 See Problem 6.19. What range of education do you expect for 95% of the people who earn $9000?

20.26 See Problem 6.20. You would expect that there is a 95% chance that the new player, number 50, would be in what weight range?

APPENDICES

Appendix A
Appendix Tables

List of Tables

A.1	Normal distribution	495–498
A.2	t-distribution	501
A.3	χ^2 (chi-square) distribution	503
A.4	χ^2/df distribution (chi-square over degrees of freedom)	504–505
A.5	F-distribution	508–513
A.6	Graph for finding $\sigma/\sqrt{n}$	516
A.7	Critical values for correlation coefficient, r	518
A.8	Binomial distribution	520–524
A.9	Simple estimators for the mean and standard deviation	526–527
A.10	Control chart factors and symbols	528
A.11	Analysis of means limits	530–531
A.12	Runs above and below the median	533
A.13	Ordinates of the normal distribution	534–535
A.14	Test for extreme values: Critical values	536
A.15	Critical values of Kolmogorov–Smirnov one-sample test D	538
A.16	Random numbers	539–543
A.17	Squares, square roots, and reciprocals	545–546
A.18	Sampling pieces	547–550
A.19	Critical values of T in the Wilcoxon matched-pairs test	551
A.20	Critical values of q for determining significant differences	552–553
A.21	Critical values of U and U' for the Mann–Whitney test	554–557
A.22	Critical values of r_{rho} (Spearman rank order correlation coefficient)	558

A.1 Normal Distribution

These tables are used for finding areas of a normal distribution with mean, $\mu = 0$, and standard deviation, $\sigma = 1.00$. Values along the X axis are designated as z values. The entries in Column A of Table A.1a are z values. Column B contains the corresponding areas between 0 and z, and Column C contains the area above z (or from z to infinity). The area between $-z$ and 0 is the same as from 0 to z (because of the symmetry of the distribution).

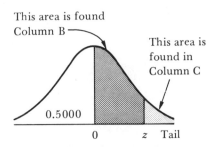

Table A.1b gives the z values for certain commonly encountered areas. Two-tailed and one-tailed values are separated.

EXAMPLE 1

Find the area between the mean and $z = 1.25$. In Column A look for $z = 1.25$. In Column B read 0.3944.

A z	B Area 0 to z
1.25	0.3944

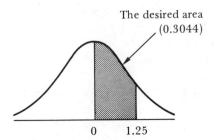

Note that the particular Column A we must look in is actually the 3rd Column A, the seventh column in the overall table. Column A extends for six different columns.

EXAMPLE 2

Find the area below $z = -0.35$.

This is equivalent to the area in the tail or the area beyond $z = 0.35$. Look up in Column C corresponding to $z = 0.35$, and find 0.3632.

This can also be found by observing the area below $z = 0.00$ is 0.5000 and Column B shows that the area from 0.00 to 0.35 is 0.1368. Therefore, the area in the tail is the difference or is 0.3632.

A	B	C
z	Area 0 to z	Area beyond z
0.35	0.1368	0.3632

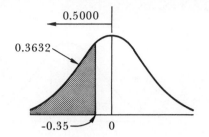

Given a normal distribution with mean μ and standard deviation σ we can convert any X value to a z value by substituting in the formula

$$z = \frac{X - \mu}{\sigma}.$$

EXAMPLE 3

Given a normal distribution with mean $\mu = 50$ and standard deviation $\sigma = 5$, find the area between 40 and 55.

First find the z values corresponding to 40 and 55.

$$55 \text{ corresponds to } z = \frac{55 - 50}{5} = 1.00$$

$$40 \text{ corresponds to } z = \frac{40 - 50}{5} = -2.00.$$

The area between $z = -2.00$ and $z = 1.00$ is the sum of the 2 areas from -2.00 to 0 and 0 to 1.00, or $0.3413 + 0.4772 = 0.8185$.

A	B
z	Area 0 to z
1.00	0.3413
2.00	0.4772

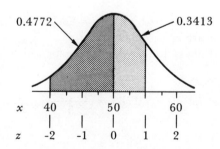

EXAMPLE 4

Below what value of z is 72% of the area?

Either determine z for which 0.2200 (22%) is from 0 to z (because 50% + 22% = 72%) or for which 0.2800 (28%) is beyond (or 100% − 72% = 28%).

In the first case look down Column B to locate 0.2200.

In the second case look down Column C to locate 0.2800.

In either case we need to approximate or *interpolate*.

494   Appendix A

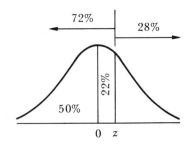

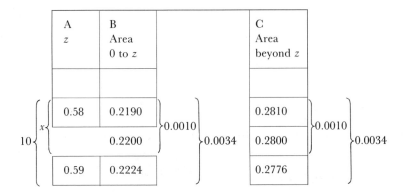

0.2200 is 0.0010/0.0034 or 10/34 or about 3/10 of the way between the two z values. Therefore we go *3/10ths* of the way between 0.58 and 0.59 or we get 0.583.

EXAMPLE 5

In a normal distribution with mean $\mu = 50$ and standard deviation $\sigma = 5$, below what value are 72% of the values?

See Example 4 above to find the z value for which 72% are below 0.583. This is:

$$0.583 = \frac{X - 50}{5},$$

or

$$X = 5(0.583) + 50 = 52.81.$$

Table A.1a Normal distribution

(A) z	(B) Area between mean and z	(C) Area beyond z	(A) z	(B) Area between mean and z	(C) Area beyond z	(A) z	(B) Area between mean and z	(C) Area beyond z
0.00	0.0000	0.5000	0.55	0.2088	0.2912	1.10	0.3643	0.1357
0.01	0.0040	0.4960	0.56	0.2123	0.2877	1.11	0.3665	0.1335
0.02	0.0080	0.4920	0.57	0.2157	0.2843	1.12	0.3686	0.1314
0.03	0.0120	0.4880	0.58	0.2190	0.2810	1.13	0.3708	0.1292
0.04	0.0160	0.4840	0.59	0.2224	0.2776	1.14	0.3729	0.1271
0.05	0.0199	0.4801	0.60	0.2257	0.2743	1.15	0.3749	0.1251
0.06	0.0239	0.4761	0.61	0.2291	0.2709	1.16	0.3770	0.1230
0.07	0.0279	0.4721	0.62	0.2324	0.2676	1.17	0.3790	0.1210
0.08	0.0319	0.4681	0.63	0.2357	0.2643	1.18	0.3810	0.1190
0.09	0.0359	0.4641	0.64	0.2389	0.2611	1.19	0.3830	0.1170
0.10	0.0398	0.4602	0.65	0.2422	0.2578	1.20	0.3849	0.1151
0.11	0.0438	0.4562	0.66	0.2454	0.2546	1.21	0.3869	0.1131
0.12	0.0478	0.4522	0.67	0.2486	0.2514	1.22	0.3888	0.1112
0.13	0.0517	0.4483	0.68	0.2517	0.2483	1.23	0.3907	0.1093
0.14	0.0557	0.4443	0.69	0.2549	0.2451	1.24	0.3925	0.1075
0.15	0.0596	0.4404	0.70	0.2580	0.2420	1.25	0.3944	0.1056
0.16	0.0636	0.4364	0.71	0.2611	0.2389	1.26	0.3962	0.1038
0.17	0.0675	0.4325	0.72	0.2642	0.2358	1.27	0.3980	0.1020
0.18	0.0714	0.4286	0.73	0.2673	0.2327	1.28	0.3997	0.1003
0.19	0.0753	0.4247	0.74	0.2704	0.2296	1.29	0.4015	0.0985
0.20	0.0793	0.4207	0.75	0.2734	0.2266	1.30	0.4032	0.0968
0.21	0.0832	0.4168	0.76	0.2764	0.2236	1.31	0.4049	0.0951
0.22	0.0871	0.4129	0.77	0.2794	0.2206	1.32	0.4066	0.0934
0.23	0.0910	0.4090	0.78	0.2823	0.2177	1.33	0.4082	0.0918
0.24	0.0948	0.4052	0.79	0.2852	0.2148	1.34	0.4099	0.0901
0.25	0.0987	0.4013	0.80	0.2881	0.2119	1.35	0.4115	0.0885
0.26	0.1026	0.3974	0.81	0.2910	0.2090	1.36	0.4131	0.0869
0.27	0.1064	0.3936	0.82	0.2939	0.2061	1.37	0.4147	0.0853
0.28	0.1103	0.3897	0.83	0.2967	0.2033	1.38	0.4162	0.0838
0.29	0.1141	0.3859	0.84	0.2995	0.2005	1.39	0.4177	0.0823
0.30	0.1179	0.3821	0.85	0.3023	0.1977	1.40	0.4192	0.0808
0.31	0.1217	0.3783	0.86	0.3051	0.1949	1.41	0.4207	0.0793
0.32	0.1255	0.3745	0.87	0.3078	0.1922	1.42	0.4222	0.0778
0.33	0.1293	0.3707	0.88	0.3106	0.1894	1.43	0.4236	0.0764
0.34	0.1331	0.3669	0.89	0.3133	0.1867	1.44	0.4251	0.0749
0.35	0.1368	0.3632	0.90	0.3159	0.1841	1.45	0.4265	0.0735
0.36	0.1406	0.3594	0.91	0.3186	0.1814	1.46	0.4279	0.0721
0.37	0.1443	0.3557	0.92	0.3212	0.1788	1.47	0.4292	0.0708
0.38	0.1480	0.3520	0.93	0.3238	0.1762	1.48	0.4306	0.0694
0.39	0.1517	0.3483	0.94	0.3264	0.1736	1.49	0.4319	0.0681
0.40	0.1554	0.3446	0.95	0.3289	0.1711	1.50	0.4332	0.0668
0.41	0.1591	0.3409	0.96	0.3315	0.1685	1.51	0.4345	0.0655
0.42	0.1628	0.3372	0.97	0.3340	0.1660	1.52	0.4357	0.0643
0.43	0.1664	0.3336	0.98	0.3365	0.1635	1.53	0.4370	0.0630
0.44	0.1700	0.3300	0.99	0.3389	0.1611	1.54	0.4382	0.0618
0.45	0.1736	0.3264	1.00	0.3413	0.1587	1.55	0.4394	0.0606
0.46	0.1772	0.3228	1.01	0.3438	0.1562	1.56	0.4406	0.0594
0.47	0.1808	0.3192	1.02	0.3461	0.1539	1.57	0.4418	0.0582
0.48	0.1844	0.3156	1.03	0.3485	0.1515.	1.58	0.4429	0.0571
0.49	0.1879	0.3121	1.04	0.3508	0.1492	1.59	0.4441	0.0559
0.50	0.1915	0.3085	1.05	0.3531	0.1469	1.60	0.4452	0.0548
0.51	0.1950	0.3050	1.06	0.3554	0.1446	1.61	0.4463	0.0537
0.52	0.1985	0.3015	1.07	0.3577	0.1423	1.62	0.4474	0.0526
0.53	0.2019	0.2981	1.08	0.3599	0.1401	1.63	0.4484	0.0516
0.54	0.2054	0.2946	1.09	0.3621	0.1379	1.64	0.4495	0.0505

SOURCE: From Audrey Haber and Richard P. Runyon, *General Statistics*, Addison-Wesley, Reading, Mass., 1969, pp. 290–291.

Table A.1a (Cont.)

(A) z	(B) Area between mean and z	(C) Area beyond z	(A) z	(B) Area between mean and z	(C) Area beyond z	(A) z	(B) Area between mean and z	(C) Area beyond z
1.65	0.4505	0.0495	2.22	0.4868	0.0132	2.79	0.4974	0.0026
1.66	0.4515	0.0485	2.23	0.4871	0.0129	2.80	0.4974	0.0026
1.67	0.4525	0.0475	2.24	0.4875	0.0125	2.81	0.4975	0.0025
1.68	0.4535	0.0465	2.25	0.4878	0.0122	2.82	0.4976	0.0024
1.69	0.4545	0.0455	2.26	0.4881	0.0119	2.83	0.4977	0.0023
1.70	0.4554	0.0446	2.27	0.4884	0.0116	2.84	0.4977	0.0023
1.71	0.4564	0.0436	2.28	0.4887	0.0113	2.85	0.4978	0.0022
1.72	0.4573	0.0427	2.29	0.4890	0.0110	2.86	0.4979	0.0021
1.73	0.4582	0.0418	2.30	0.4893	0.0107	2.87	0.4979	0.0021
1.74	0.4591	0.0409	2.31	0.4896	0.0104	2.88	0.4980	0.0020
1.75	0.4599	0.0401	2.32	0.4898	0.0102	2.89	0.4981	0.0019
1.76	0.4608	0.0392	2.33	0.4901	0.0099	2.90	0.4981	0.0019
1.77	0.4616	0.0384	2.34	0.4904	0.0096	2.91	0.4982	0.0018
1.78	0.4625	0.0375	2.35	0.4906	0.0094	2.92	0.4982	0.0018
1.79	0.4633	0.0367	2.36	0.4909	0.0091	2.93	0.4983	0.0017
1.80	0.4641	0.0359	2.37	0.4911	0.0089	2.94	0.4984	0.0016
1.81	0.4649	0.0351	2.38	0.4913	0.0087	2.95	0.4984	0.0016
1.82	0.4656	0.0344	2.39	0.4916	0.0084	2.96	0.4985	0.0015
1.83	0.4664	0.0336	2.40	0.4918	0.0082	2.97	0.4985	0.0015
1.84	0.4671	0.0329	2.41	0.4920	0.0080	2.98	0.4986	0.0014
1.85	0.4678	0.0322	2.42	0.4922	0.0078	2.99	0.4986	0.0014
1.86	0.4686	0.0314	2.43	0.4925	0.0075	3.00	0.4987	0.0013
1.87	0.4693	0.0307	2.44	0.4927	0.0073	3.01	0.4987	0.0013
1.88	0.4699	0.0301	2.45	0.4929	0.0071	3.02	0.4987	0.0013
1.89	0.4706	0.0294	2.46	0.4931	0.0069	3.03	0.4988	0.0012
1.90	0.4713	0.0287	2.47	0.4932	0.0068	3.04	0.4988	0.0012
1.91	0.4719	0.0281	2.48	0.4934	0.0066	3.05	0.4989	0.0011
1.92	0.4726	0.0274	2.49	0.4936	0.0064	3.06	0.4989	0.0011
1.93	0.4732	0.0268	2.50	0.4938	0.0062	3.07	0.4989	0.0011
1.94	0.4738	0.0262	2.51	0.4940	0.0060	3.08	0.4990	0.0010
1.95	0.4744	0.0256	2.52	0.4941	0.0059	3.09	0.4990	0.0010
1.96	0.4750	0.0250	2.53	0.4943	0.0057	3.10	0.4990	0.0010
1.97	0.4756	0.0244	2.54	0.4945	0.0055	3.11	0.4991	0.0009
1.98	0.4761	0.0239	2.55	0.4946	0.0054	3.12	0.4991	0.0009
1.99	0.4767	0.0233	2.56	0.4948	0.0052	3.13	0.4991	0.0009
2.00	0.4772	0.0228	2.57	0.4949	0.0051	3.14	0.4992	0.0008
2.01	0.4778	0.0222	2.58	0.4951	0.0049	3.15	0.4992	0.0008
2.02	0.4783	0.0217	2.59	0.4952	0.0048	3.16	0.4992	0.0008
2.03	0.4788	0.0212	2.60	0.4953	0.0047	3.17	0.4992	0.0008
2.04	0.4793	0.0207	2.61	0.4955	0.0045	3.18	0.4993	0.0007
2.05	0.4798	0.0202	2.62	0.4956	0.0044	3.19	0.4993	0.0007
2.06	0.4803	0.0197	2.63	0.4957	0.0043	3.20	0.4993	0.0007
2.07	0.4808	0.0192	2.64	0.4959	0.0041	3.21	0.4993	0.0007
2.08	0.4812	0'0188	2.65	0.4960	0.0040	3.22	0.4994	0.0006
2.09	0.4817	0.0183	2.66	0.4961	0.0039	3.23	0.4994	0.0006
2.10	0.4821	0.0179	2.67	0.4962	0.0038	3.24	0.4994	0.0006
2.11	0.4826	0.0174	2.68	0.4963	0.0037	3.25	0.4994	0.0006
2.12	0.4830	0.0170	2.69	0.4964	0.0036	3.30	0.4995	0.0005
2.13	0.4834	0.0166	2.70	0.4965	0.0035	3.35	0.4996	0.0004
2.14	0.4838	0.0162	2.71	0.4966	0.0034	3.40	0.4997	0.0003
2.15	0.4842	0.0158	2.72	0.4967	0.0033	3.45	0.4997	0.0003
2.16	0.4846	0.0154	2.73	0.4968	0.0032	3.50	0.4998	0.0002
2.17	0.4850	0.0150	2.74	0.4969	0.0031	3.60	0.4998	0.0002
2.18	0.4854	0.0146	2.75	0.4970	0.0030	3.70	0.4999	0.0001
2.19	0.4857	0.0143	2.76	0.4971	0.0029	3.80	0.4999	0.0001
2.20	0.4861	0.0139	2.77	0.4972	0.0028	3.90	0.49995	0.00005
2.21	0.4864	0.0136	2.78	0.4973	0.0027	4.00	0.49997	0.00003

Table A.1b Normal distribution (z values for selected areas)

Percent between limits	Area between limits	Total area outside limits	Area in each tail	z
25	0.25	0.75	0.375	0.319
50	0.50	0.50	0.25	0.674
60	0.60	0.40	0.20	0.842
70	0.70	0.30	0.15	1.036
75	0.75	0.25	0.125	1.150
80	0.80	0.20	0.10	1.282
90	0.90	0.10	0.05	1.645
95	0.95	0.05	0.025	1.960
98	0.98	0.02	0.01	2.326
99	0.99	0.01	0.005	2.576
99.5	0.995	0.005	0.0025	2.810
99.9	0.999	0.001	0.0005	3.291
99.99	0.9999	0.0001	0.00005	3.891

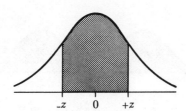

Two-tailed: Areas between the limits of $-z$ and $+z$

Table A.1b (Cont.)

Percent below z	Area below z (shaded area)	Area above z (unshaded areas)	z (or −z)
50	0.50	0.50	0.000
60	0.60	0.40	0.253
70	0.70	0.30	0.524
75	0.75	0.25	0.674
80	0.80	0.20	0.842
90	0.90	0.10	1.282
95	0.95	0.05	1.645
97.5	0.975	0.025	1.960
98	0.98	0.02	2.054
99	0.99	0.01	2.326
99.5	0.995	0.005	2.576
99.9	0.999	0.001	3.090
99.95	0.9995	0.0005	3.291
99.99	0.9999	0.0001	3.719
99.999	0.99999	0.00001	4.265

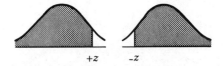

One-tailed: Area is below $+z$ or above $-z$

A.2 *t*-distribution

This is a symmetric distribution with center equal to 0. *df* is *degrees of freedom* and *generally equals n − 1*. For one-tailed tests or one-sided levels of significance you should consult row labeled "level of significance for one-tailed test." The subscript designates the proportion of the area that is below that value.

EXAMPLE 1

$df = 5$. What is the limit below which are *95% of the values*? This is a one-tailed test. It also corresponds to the value of $t_{0.95}$.

df		$t_{0.95}$	
	Level ... one-tailed test		
		0.05	
5		2.015	

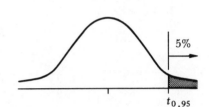

The limit is 2.015.

EXAMPLE 2

$df = 12$. What are the limits between which lie 95% of the values? This is a two-tailed test for which the 5% in the tails are divided between the two tails. Thus $2\frac{1}{2}\%$ or 0.025 are in each tail and the *t*-values are $t_{0.025}$ and $t_{0.975}$.

df		$t_{0.975}$	
	Level ... two-tailed test		
		0.05	
12		2.179	

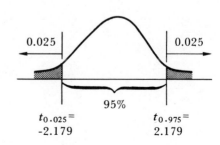

The limits are ±2.179.

Table A.2 t-distribution

df	$t_{0.90}$	$t_{0.95}$	$t_{0.975}$	$t_{0.99}$	$t_{0.995}$	$t_{0.9995}$
	\multicolumn{6}{c}{Level of significance for one-tailed test}					
	0.10	0.05	0.025	0.01	0.005	0.0005
	\multicolumn{6}{c}{Level of significance for two-tailed test}					
	0.20	0.10	0.05	0.02	0.01	0.001
1	3.078	6.314	12.706	31.821	63.657	636.619
2	1.886	2.920	4.303	6.965	9.925	31.598
3	1.638	2.353	3.182	4.541	5.841	12.941
4	1.533	2.132	2.776	3.747	4.604	8.610
5	1.476	2.015	2.571	3.365	4.032	6.859
6	1.440	1.943	2.447	3.143	3.707	5.959
7	1.415	1.895	2.365	2.998	3.499	5.405
8	1.397	1.860	2.306	2.896	3.355	5.041
9	1.383	1.833	2.262	2.821	3.250	4.781
10	1.372	1.812	2.228	2.764	3.169	4.587
11	1.363	1.796	2.201	2.718	3.106	4.437
12	1.356	1.782	2.179	2.681	3.055	4.318
13	1.350	1.771	2.160	2.650	3.012	4.221
14	1.345	1.761	2.145	2.624	2.977	4.140
15	1.341	1.753	2.131	2.602	2.947	4.073
16	1.337	1.746	2.120	2.583	2.921	4.015
17	1.333	1.740	2.110	2.567	2.898	3.965
18	1.330	1.734	2.101	2.552	2.878	3.992
19	1.328	1.729	2.093	2.539	2.861	3.883
20	1.325	1.725	2.086	2.528	2.845	3.850
21	1.323	1.721	2.080	2.518	2.831	3.819
22	1.321	1.717	2.074	2.508	2.819	3.792
23	1.319	1.714	2.069	2.500	2.807	3.767
24	1.318	1.711	2.064	2.492	2.797	3.745
25	1.316	1.708	2.060	2.485	2.787	3.725
26	1.315	1.706	2.056	2.479	2.779	3.707
27	1.314	1.703	2.052	2.473	2.771	3.690
28	1.313	1.701	2.048	2.467	2.763	3.674
29	1.311	1.699	2.045	2.462	2.756	3.659
30	1.310	1.697	2.042	2.457	2.750	3.646
40	1.303	1.684	2.021	2.423	2.704	3.551
60	1.296	1.671	2.000	2.390	2.660	3.460
120	1.289	1.658	1.980	2.358	2.617	3.373
∞	1.282	1.645	1.960	2.326	2.576	3.291

SOURCE: Extracted from Table III of Fisher and Yates, *Statistical Tables for Biological and Medical Research*, published by Oliver and Boyd, Ltd., Edinburgh, by permission of the authors and publishers.

A.3 χ^2 (chi-square) distribution

Generally the *df* (degrees of freedom) is equal to $n - 1$. The probability level indicates what proportion of the particular χ^2 distribution lies below the corresponding entry in the table.

EXAMPLE 1

$df = 5$. We desire a χ^2 for which there is a 10% chance of a value being larger. This is the same as asking what is the value for which 90% are smaller. This is $\chi^2_{0.90}$.

	Probability of value above	
	0.100	
	Probability of value below	
df	0.900	
5	9.24	

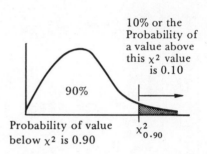

10% or the Probability of a value above this χ^2 value is 0.10

Probability of value below χ^2 is 0.90

$\chi^2_{0.90}$

$\chi^2_{0.90} = 9.24$.

EXAMPLE 2

$df = 8$. We desire χ^2 limits (upper and lower) for which 10% are outside. Then we need two limits—a two-tailed case—with 5% in each tail.

	Probability of value below	
df	0.050	0.950
8	2.73	15.5

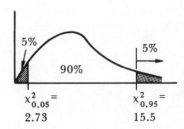

$\chi^2_{0.05} = 2.73$, $\chi^2_{0.95} = 15.5$.

Table A.3 χ^2 (chi-square) distribution

Probability that a value is above the critical value

Degrees of freedom	0.995	0.990	0.975	0.950	0.900	0.750	0.500	0.250	0.100	0.050	0.025	0.010	0.005
	0.005	0.010	0.025	0.050	0.100	0.250	0.500	0.750	0.900	0.950	0.975	0.990	0.995
1	0.0000393	0.000157	0.000982	0.00393	0.0158	0.102	0.455	1.32	2.71	3.84	5.02	6.63	7.88
2	0.0100	0.0201	0.0506	0.103	0.211	0.575	1.39	2.77	4.61	5.99	7.38	9.21	10.6
3	0.0717	0.115	0.216	0.352	0.584	1.21	2.37	4.11	6.24	7.81	9.35	11.3	12.8
4	0.207	0.297	0.484	0.711	1.06	1.92	3.36	5.39	7.78	9.49	11.1	13.3	14.9
5	0.412	0.554	0.831	1.15	1.61	2.67	4.35	6.63	9.24	11.1	12.8	15.1	16.7
6	0.676	0.872	1.24	1.64	2.20	3.45	5.35	7.84	10.6	12.6	14.4	16.8	18.5
7	0.989	1.24	1.69	2.17	2.83	4.25	6.35	9.04	12.0	14.1	16.0	18.5	20.3
8	1.34	1.65	2.18	2.73	3.49	5.07	7.34	10.2	13.4	15.5	17.5	20.1	22.0
9	1.73	2.09	2.70	3.33	4.17	5.90	8.34	11.4	14.7	16.9	19.0	21.7	23.6
10	2.16	2.56	3.25	3.94	4.87	6.74	9.34	12.5	16.0	18.3	20.5	23.2	25.2
11	2.60	3.05	3.82	4.57	5.58	7.58	10.3	13.7	17.3	19.7	21.9	24.7	26.8
12	3.07	3.57	4.40	5.23	6.30	8.44	11.3	14.8	18.5	21.0	23.3	26.2	28.3
13	3.57	4.11	5.01	5.89	7.04	9.30	12.3	16.0	19.8	22.4	24.7	27.7	29.8
14	4.07	4.66	5.63	6.57	7.79	10.2	13.3	17.1	21.1	23.7	26.1	29.1	31.3
15	4.60	5.23	6.26	7.26	8.55	11.0	14.3	18.2	22.3	25.0	27.5	30.6	32.8
16	5.14	5.81	6.91	7.96	9.31	11.9	15.3	19.4	23.5	26.3	28.8	32.0	34.3
17	5.70	6.41	7.56	8.67	10.1	12.8	16.3	20.5	24.8	27.6	30.2	33.4	35.7
18	6.26	7.01	8.23	9.39	10.9	13.7	17.3	21.6	26.0	28.9	31.5	37.2	37.2
19	6.84	7.63	8.91	10.1	11.7	14.6	18.3	22.7	27.2	30.1	32.9	36.2	38.6
20	7.43	8.26	9.59	10.9	12.4	15.5	19.3	23.8	28.4	31.4	34.2	37.6	40.0
21	8.03	8.90	10.3	11.6	13.2	16.3	20.3	24.9	29.6	32.7	35.5	38.9	41.4
22	8.64	9.54	11.0	12.3	14.0	17.2	21.3	26.0	30.8	33.9	36.8	40.3	42.8
23	9.26	10.2	11.7	13.1	14.8	18.1	22.3	27.1	32.0	35.2	38.1	41.6	44.2
24	9.89	10.9	12.4	13.8	15.7	19.0	23.3	28.2	33.2	36.4	39.4	43.0	45.6
25	10.5	11.5	13.1	14.6	16.5	19.9	24.3	29.3	34.4	37.7	40.6	44.3	46.9
26	11.2	12.2	13.8	15.4	17.3	20.8	25.3	30.4	35.6	38.9	41.9	45.6	48.3
27	11.8	12.9	14.6	16.2	18.1	21.7	26.3	31.5	36.7	40.1	43.2	47.0	49.6
28	12.5	13.6	15.3	16.9	18.9	22.7	27.3	32.6	37.9	41.3	44.5	48.3	51.0
29	13.1	14.3	16.0	17.7	19.8	23.6	28.3	33.7	39.1	42.6	45.7	49.6	52.3
30	13.8	15.0	16.8	18.5	20.6	24.5	29.3	34.8	40.3	43.8	47.0	50.9	53.7

Probability that a value is below the critical value

SOURCE: This table is abridged from Catherine M. Thompson's "Table of Percentage Points of the χ^2 Distribution," copyright 1941 by *Biometrika*, **32**, 188–189, and used by permission of the author and editor.

Table A.4 χ^2/df (chi-square over degrees of freedom) distribution

	Probability of a value above the critical value								
	Probability of a value below the critical value								
df	0.005	0.001	0.005	0.01	0.025	0.05	0.10	0.20	0.30	0.40
1	0.0^639*	0.0^5157	0.0^439	0.0^316	0.0^398	0.0^239	0.016	0.064	0.148	0.275
2	0.001	0.001	0.005	0.010	0.025	0.052	0.106	0.223	0.356	0.511
3	0.005	0.008	0.024	0.038	0.072	0.117	0.195	0.335	0.475	0.623
4	0.016	0.023	0.052	0.074	0.121	0.178	0.266	0.412	0.549	0.688
5	0.032	0.042	0.082	0.111	0.166	0.229	0.322	0.469	0.600	0.731
6	0.050	0.064	0.113	0.145	0.206	0.272	0.367	0.512	0.638	0.762
7	0.069	0.085	0.141	0.177	0.241	0.310	0.405	0.546	0.667	0.785
8	0.089	0.107	0.168	0.206	0.272	0.342	0.436	0.574	0.691	0.803
9	0.108	0.128	0.193	0.232	0.300	0.369	0.463	0.598	0.710	0.817
10	0.126	0.148	0.216	0.256	0.325	0.394	0.487	0.618	0.727	0.830
11	0.144	0.167	0.237	0.278	0.347	0.416	0.507	0.635	0.741	0.840
12	0.161	0.184	0.256	0.298	0.367	0.436	0.525	0.651	0.753	0.848
13	0.177	0.201	0.274	0.316	0.385	0.453	0.542	0.664	0.764	0.856
14	0.193	0.217	0.291	0.333	0.402	0.469	0.556	0.676	0.773	0.863
15	0.207	0.232	0.307	0.349	0.418	0.484	0.570	0.687	0.781	0.869
16	0.221	0.246	0.321	0.363	0.432	0.498	0.582	0.697	0.789	0.874
17	0.234	0.260	0.335	0.377	0.445	0.510	0.593	0.706	0.796	0.879
18	0.247	0.272	0.348	0.390	0.457	0.522	0.604	0.714	0.802	0.883
19	0.258	0.285	0.360	0.402	0.469	0.532	0.613	0.722	0.808	0.887
20	0.270	0.296	0.372	0.413	0.480	0.543	0.622	0.729	0.813	0.890
22	0.291	0.317	0.393	0.434	0.499	0.561	0.638	0.742	0.823	0.897
24	0.310	0.337	0.412	0.452	0.517	0.577	0.652	0.753	0.831	0.902
26	0.328	0.355	0.429	0.469	0.532	0.592	0.665	0.762	0.838	0.907
28	0.345	0.371	0.445	0.484	0.547	0.605	0.676	0.771	0.845	0.911
30	0.360	0.386	0.460	0.498	0.560	0.616	0.687	0.779	0.850	0.915
35	0.394	0.420	0.491	0.529	0.588	0.642	0.708	0.795	0.862	0.922
40	0.423	0.448	0.518	0.554	0.611	0.663	0.726	0.809	0.872	0.928
45	0.448	0.472	0.540	0.576	0.630	0.680	0.741	0.820	0.880	0.933
50	0.469	0.494	0.560	0.594	0.647	0.695	0.754	0.829	0.886	0.937
55	0.488	0.512	0.577	0.610	0.662	0.708	0.765	0.837	0.892	0.941
60	0.506	0.529	0.592	0.625	0.675	0.720	0.774	0.844	0.897	0.944
70	0.535	0.558	0.618	0.649	0.697	0.739	0.790	0.856	0.905	0.949
80	0.560	0.582	0.640	0.669	0.714	0.755	0.803	0.865	0.911	0.952
90	0.581	0.602	0.658	0.686	0.729	0.768	0.814	0.873	0.917	0.955
100	0.599	0.619	0.673	0.701	0.742	0.779	0.824	0.879	0.921	0.958
120	0.629	0.648	0.699	0.724	0.763	0.798	0.839	0.890	0.929	0.962
140	0.653	0.671	0.719	0.743	0.780	0.812	0.850	0.898	0.934	0.965
160	0.673	0.690	0.736	0.758	0.793	0.824	0.860	0.905	0.939	0.968
180	0.689	0.706	0.749	0.771	0.804	0.833	0.868	0.910	0.942	0.970
200	0.703	0.719	0.761	0.782	0.814	0.841	0.874	0.915	0.945	0.972
250	0.732	0.746	0.785	0.804	0.832	0.858	0.887	0.924	0.951	0.975
300	0.753	0.767	0.802	0.820	0.846	0.870	0.897	0.931	0.956	0.977
350	0.770	0.783	0.816	0.833	0.857	0.879	0.904	0.936	0.959	0.979
400	0.784	0.796	0.827	0.843	0.866	0.887	0.911	0.940	0.962	0.981
450	0.795	0.807	0.837	0.852	0.874	0.893	0.916	0.944	0.964	0.982
500	0.805	0.816	0.845	0.859	0.880	0.898	0.920	0.946	0.966	0.983
750	0.839	0.848	0.872	0.884	0.901	0.917	0.934	0.956	0.972	0.986
1,000	0.859	0.868	0.889	0.899	0.914	0.928	0.943	0.962	0.976	0.988
5,000	0.936	0.939	0.949	0.954	0.961	0.967	0.974	0.983	0.989	0.995
∞	1	1	1	1	1	1	1	1	1	1

SOURCE: From A. Hald, *Statistical Tables and Formulas*, John Wiley & Sons, Inc., New York, 1952.

* Read 0.0^316 as 0.00016, and so on.

Table A.4 (continued)

	Probability of a value above the critical value										
	0.50	0.40	0.30	0.20	0.10	0.05	0.025	0.010	0.005	0.001	0.0005
	Probability of a value below the critical value										
df	0.50	0.60	0.70	0.80	0.90	0.95	0.975	0.990	0.995	0.999	0.9995
1	0.455	0.708	1.07	1.64	2.71	3.84	5.02	6.64	7.88	10.83	12.12
2	0.693	0.916	1.20	1.61	2.30	3.00	3.69	4.61	5.30	6.91	7.60
3	0.789	0.982	1.22	1.55	2.08	2.60	3.12	3.78	4.28	5.42	5.91
4	0.839	1.011	1.22	1.50	1.94	2.37	2.79	3.32	3.72	4.62	5.00
5	0.870	1.03	1.21	1.46	1.85	2.21	2.57	3.02	3.35	4.10	4.42
6	0.891	1.04	1.21	1.43	1.77	2.10	2.41	2.80	3.09	3.74	4.02
7	0.907	1.04	1.20	1.40	1.72	2.01	2.29	2.64	2.90	3.47	3.72
8	0.918	1.04	1.19	1.38	1.67	1.94	2.19	2.51	2.74	3.27	3.48
9	0.927	1.05	1.18	1.36	1.63	1.88	2.11	2.41	2.62	3.10	3.30
10	0.934	1.05	1.18	1.34	1.60	1.83	2.05	2.32	2.52	2.96	3.14
11	0.940	1.05	1.17	1.33	1.57	1.79	1.99	2.25	2.43	2.84	3.01
12	0.945	1.05	1.17	1.32	1.55	1.75	1.94	2.18	2.36	2.74	2.90
13	0.949	1.05	1.16	1.31	1.52	1.72	1.90	2.13	2.29	2.66	2.81
14	0.953	1.05	1.16	1.30	1.50	1.69	1.87	2.08	2.24	2.58	2.72
15	0.956	1.05	1.15	1.29	1.49	1.67	1.83	2.04	2.19	2.51	2.65
16	0.959	1.05	1.15	1.28	1.47	1.64	1.80	2.00	2.14	2.45	2.58
17	0.961	1.05	1.15	1.27	1.46	1.62	1.78	1.97	2.10	2.40	2.52
18	0.963	1.05	1.14	1.26	1.44	1.60	1.75	1.93	2.06	2.35	2.47
19	0.965	1.05	1.14	1.26	1.43	1.59	1.73	1.90	2.03	2.31	2.42
20	0.967	1.05	1.14	1.25	1.42	1.57	1.71	1.88	2.00	2.27	2.37
22	0.970	1.05	1.13	1.24	1.40	1.54	1.67	1.83	1.95	2.19	2.30
24	0.972	1.05	1.13	1.23	1.38	1.52	1.64	1.79	1.90	2.13	2.23
26	0.974	1.05	1.12	1.22	1.37	1.50	1.61	1.76	1.86	2,08	2.17
28	0.976	1.04	1.12	1.22	1.35	1.48	1.59	1.72	1.82	2.03	2.12
30	0.978	1.04	1.12	1.21	1.34	1.46	1.57	1.70	1.79	1.99	2.07
35	0.981	1.04	1.11	1.19	1.32	1.42	1.52	1.64	1.72	1.90	1.98
40	0.983	1.04	1.10	1.18	1.30	1.39	1.48	1.59	1.67	1.84	1.90
45	0.985	1.04	1.10	1.17	1.28	1.37	1.45	1.55	1.63	1.78	1.84
50	0.987	1.04	1.09	1.16	1.26	1.35	1.43	1.52	1.59	1.73	1.79
55	0.988	1.04	1.09	1.16	1.25	1.33	1.41	1.50	1.56	1.69	1.75
60	0.989	1.04	1.09	1.15	1.24	1.32	1.39	1.47	1.53	1.66	1.71
70	0.990	1.03	1.08	1.14	1.22	1.29	1.36	1.43	1.49	1.60	1.65
80	0.992	1.03	1.08	1.13	1.21	1.27	1.33	1.40	1.45	1.56	1.60
90	0.993	1.03	1.07	1.12	1.20	1.26	1.31	1.38	1.43	1.52	1.56
100	0.993	1.03	1.07	1.12	1.18	1.24	1.30	1.36	1.40	1.49	1.53
120	0.994	1.03	1.06	1.11	1.17	1.22	1.27	1.32	1.36	1.45	1.48
140	0.995	1.03	1.06	1.10	1.16	1.20	1.25	1.30	1.33	1.41	1.44
160	0.996	1.02	1.06	1.09	1.15	1.19	1.23	1.28	1.31	1.38	1.41
180	0.996	1.02	1.05	1.09	1.14	1.18	1.22	1.26	1.29	1.36	1.38
200	0.997	1.02	1.05	1.08	1.13	1.17	1.21	1.25	1.28	1.34	1.36
250	0.997	1.02	1.04	1.07	1.12	1.15	1.18	1.22	1.25	1.30	1.32
300	0.998	1.02	1.04	1.07	1.11	1.14	1.17	1.20	1.22	1.27	1.29
350	0.998	1.02	1.04	1.06	1.10	1.13	1.15	1.18	1.21	1.25	1.27
400	0.998	1.02	1.04	1.06	1.09	1.12	1.14	1.17	1.19	1.24	1.25
450	0.999	1.02	1.03	1.06	1.09	1.11	1.13	1.16	1.18	1.22	1.23
500	0.999	1.01	1.03	1.05	1.08	1.11	1.13	1.15	1.17	1.21	1.22
750	0.999	1.01	1.03	1.04	1.07	1.09	1.10	1.12	1.14	1.17	1.18
1,000	0.999	1.01	1.02	1.04	1.06	1.07	1.09	1.11	1.12	1.14	1.15
5,000	1.00	1.00	1.01	1.02	1.02	1.03	1.04	1.05	1.05	1.06	1.07
∞	1	1	1	1	1	1	1	1	1	1	1

df
Degrees of freedom

A.4 χ^2/df distribution

This table can be developed from Table A.3 by the appropriate division. Similarly, the χ^2-values can be obtained from this table by multiplication. For example, for $10\,df$ Table A.3 gives the 0.100 (10%) value of χ^2 as 4.87. Table A.4 gives the 0.10 value of χ^2/df as 0.487, which is 4.87/10. Generally $df = n - 1$. Entry is the same as in Table A.3.

EXAMPLE 1

Degrees of freedom $df = 18$. What are the 1% and 99% limits? A 1% limit is the same as a probability level of 0.01, and a 99% limit is the same as a probability level of 0.99.

	Probability of value below	
df	0.01	0.99
18	0.390	1.93

Thus, the 0.01 limit of χ^2/df is 0.390 (or only 1% of the possible values are below this value of 0.390); the 0.99 limit or the value of χ^2/df for which 99% are below is 1.93. 98% of the values lie between 0.390 and 1.93 (or a 0.98 probability of a value between these limits.

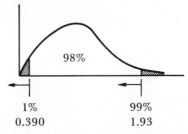

A.5 F-distribution

These are all *one-tailed tables*. The way they are used as two-tailed tables is discussed below.

All F-values involve a *division*. Symbolically,

$$F(f_1, f_2) = \frac{\text{A numerator with } df = f_1}{\text{A denominator with } df = f_2}.$$

$F_{0.90}$
(90% are below)

Thus, $F(8, 6)$ means that two numbers were involved in finding the distribution. The number that is (or was) in the numerator has 8 degrees of freedom and the number in the denominator has 6 degrees of freedom. Generally, these are variances where each number of degrees of freedom is referring to a sample size. Typically $df = n - 1$ so that for 2 samples, one with 9 individuals, and the other with 7 individuals, the df would be 8 and 6 respectively.

EXAMPLE 1

Find the F-value below which 90% of the values are found. You are given that the degrees of freedom are $f_1 = 8$ and $f_2 = 6$.

This is to be a one-tailed test with probability = 0.90, therefore consult the table which reads:

One-tailed test $P = 0.90$

f_2 \ f_1	8
6	2.98

Two-tailed tests: To determine the limit(s) for a two-tailed test there are two choices.

1. The preferred method (because of its ease). Perform the calculation of the F-value as the following:

$$F = \frac{\text{Larger value with } df = f_1}{\text{Smaller value with } df = f_2}.$$

The limit is found by referring to the table labeled "when used as two-tailed test, $P = 0.XX$."

Thus, only *one limit* is found.

2. To find the two limits first find the upper limit as a one-tailed test limit. Then find the lower limit by first looking up the F-value with the reversed degrees of freedon. Then calculate the reciprocal of this value.

EXAMPLE 2

Find the limits for a two-sided test between which 90% of the F-values lie. The degrees of freedom are $f_1 = 8$ and $f_2 = 6$.

(1) The *preferred method*. We will use the table labeled "when used as two-tailed test, $P = 0.90$." (This is the one-tailed table for $P = 0.95$.)

To find the sample value of F you calculate

$$F = \frac{\text{Larger value}}{\text{Smaller value}}.$$

We then look up the appropriate F-value and compare the two. If the calculated quantity is larger than the F-value from the table we would say it has a probability of less than 0.90.

Thus, if the two values were 18.4 with 8 degrees of freedom and 6.4 with 6 degrees of freedom, we would have

$$F = \frac{\text{Larger}}{\text{Smaller}} = \frac{18.4}{6.4} = 2.875.$$

The corresponding F would be $F(8, 6)$ and would be 4.15 (see (a) below).

Note that if the two quantities were reversed with the 18.4 having 6 degrees of freedom (instead of 8) and the 6.4 having 8 degrees of freedom then the ratio of larger/smaller still is 2.875, but the F-value in the table would be found for $f_1 = 6$ and $f_2 = 8$. (These are reversed.) The critical value would be 3.58 (see (b) below).

	f_1	df in numerator
f_2		8
df in denom.	6	4.15

	f_1	df in numerator
f_2		6
df in denom.	8	3.58

(a) $F(8, 6) = 4.15$ (b) $F(6, 8) = 3.58$

(2) To obtain *two limits* (this is necessary if you are finding confidence intervals), refer to the table—one-tailed test, $P = 0.95$ (since 5% are in each tail).

Upper limit: $F_{0.95}$ (one-tailed test 0.95) with 8 and 6 degrees of freedom. This is the 4.15 as in (a) of the figure above.

Lower limit: We find $F_{0.05}$ using the same table as before except *reversing* the degrees of freedom (as in (b) of the figure above). We obtain the 3.58. Then take the reciprocal of this F-value and obtain:

$$\text{Lower limit} = \frac{1}{F(f_1, f_2)} = \frac{1}{3.58} = 0.279.$$

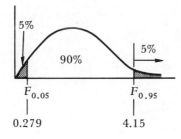

Table A.5 F-distribution

a. One-tailed test, $P = 0.90$ (0.10 or 10% in tail). When used as two-tailed test, $P = 0.80$.

Degrees of freedom in numerator

f_2 \ f_1	1	2	3	4	5	6	7	8	9	10	12	15	20	24	30	40	60	120	∞
1	39.86	49.50	53.59	55.83	57.24	58.20	58.91	59.44	59.86	60.19	60.71	61.22	61.74	62.00	62.26	62.53	62.79	63.06	63.33
2	8.53	9.00	9.16	9.24	9.29	9.33	9.35	9.37	9.38	9.39	9.41	9.42	9.44	9.45	9.46	9.47	9.47	9.48	9.49
3	5.54	5.46	5.39	5.34	5.31	5.28	5.27	5.25	5.24	5.23	5.22	5.20	5.18	5.18	5.17	5.16	5.15	5.14	5.13
4	4.54	4.32	4.19	4.11	4.05	4.01	3.98	3.95	3.94	3.92	3.90	3.87	3.84	3.83	3.82	3.80	3.79	3.78	3.76
5	4.06	3.78	3.62	3.52	3.45	3.40	3.37	3.34	3.32	3.30	3.27	3.24	3.21	3.19	3.17	3.16	3.14	3.12	3.10
6	3.78	3.46	3.29	3.18	3.11	3.05	3.01	2.98	2.96	2.94	2.90	2.87	2.84	2.82	2.80	2.78	2.76	2.74	2.72
7	3.59	3.26	3.07	2.96	2.88	2.83	2.78	2.75	2.72	2.70	2.67	2.63	2.59	2.58	2.56	2.54	2.51	2.49	2.47
8	3.46	3.11	2.92	2.81	2.73	2.67	2.62	2.59	2.56	2.54	2.50	2.46	2.42	2.40	2.38	2.36	2.34	2.32	2.29
9	3.36	3.01	2.81	2.69	2.61	2.55	2.51	2.47	2.44	2.42	2.38	2.34	2.30	2.28	2.25	2.23	2.21	2.18	2.16
10	3.29	2.92	2.73	2.61	2.52	2.46	2.41	2.38	2.35	2.32	2.28	2.24	2.20	2.18	2.16	2.13	2.11	2.08	2.06
11	3.23	2.86	2.66	2.54	2.45	2.39	2.34	2.30	2.27	2.25	2.21	2.17	2.12	2.10	2.08	2.05	2.03	2.00	1.97
12	3.18	2.81	2.61	2.48	2.39	2.33	2.28	2.24	2.21	2.19	2.15	2.10	2.06	2.04	2.01	1.99	1.96	1.93	1.90
13	3.14	2.76	2.56	2.43	2.35	2.28	2.23	2.20	2.16	2.14	2.10	2.05	2.01	1.98	1.96	1.93	1.90	1.88	1.85
14	3.10	2.73	2.52	2.39	2.31	2.24	2.19	2.15	2.12	2.10	2.05	2.01	1.96	1.94	1.91	1.89	1.86	1.83	1.80
15	3.07	2.70	2.49	2.36	2.27	2.21	2.16	2.12	2.09	2.06	2.02	1.97	1.92	1.90	1.87	1.85	1.82	1.79	1.76
16	3.05	2.67	2.46	2.33	2.24	2.18	2.13	2.09	2.06	2.03	1.99	1.94	1.89	1.87	1.84	1.81	1.78	1.75	1.72
17	3.03	2.64	2.44	2.31	2.22	2.15	2.10	2.06	2.03	2.00	1.96	1.91	1.86	1.84	1.81	1.78	1.75	1.72	1.69
18	3.01	2.62	2.42	2.29	2.20	2.13	2.08	2.04	2.00	1.98	1.93	1.89	1.84	1.81	1.78	1.75	1.72	1.69	1.66
19	2.99	2.61	2.40	2.27	2.18	2.11	2.06	2.02	1.98	1.96	1.91	1.86	1.81	1.79	1.76	1.73	1.70	1.67	1.63
20	2.97	2.59	2.38	2.25	2.16	2.09	2.04	2.00	1.96	1.94	1.89	1.84	1.79	1.77	1.74	1.71	1.68	1.64	1.61
21	2.96	2.57	2.36	2.23	2.14	2.08	2.02	1.98	1.95	1.92	1.87	1.83	1.78	1.75	1.72	1.69	1.66	1.62	1.59
22	2.95	2.56	2.35	2.22	2.13	2.06	2.01	1.97	1.93	1.90	1.86	1.81	1.76	1.73	1.70	1.67	1.64	1.60	1.57
23	2.94	2.55	2.34	2.21	2.11	2.05	1.99	1.95	1.92	1.89	1.84	1.80	1.74	1.72	1.69	1.66	1.62	1.59	1.55
24	2.93	2.54	2.33	2.19	2.10	2.04	1.98	1.94	1.91	1.88	1.83	1.78	1.73	1.70	1.67	1.64	1.61	1.57	1.53
25	2.92	2.53	2.32	2.18	2.09	2.02	1.97	1.93	1.89	1.87	1.82	1.77	1.72	1.69	1.66	1.63	1.59	1.56	1.52
26	2.91	2.52	2.31	2.17	2.08	2.01	1.96	1.92	1.88	1.86	1.81	1.76	1.71	1.68	1.65	1.61	1.58	1.54	1.50
27	2.90	2.51	2.30	2.17	2.07	2.00	1.95	1.91	1.87	1.85	1.80	1.75	1.70	1.67	1.64	1.60	1.57	1.53	1.49
28	2.89	2.50	2.29	2.16	2.06	2.00	1.94	1.90	1.87	1.84	1.79	1.74	1.69	1.66	1.63	1.59	1.56	1.52	1.48
29	2.89	2.50	2.28	2.15	2.06	1.99	1.93	1.89	1.86	1.83	1.78	1.73	1.68	1.65	1.62	1.58	1.55	1.51	1.47
30	2.88	2.49	2.28	2.14	2.05	1.98	1.93	1.88	1.85	1.82	1.77	1.72	1.67	1.64	1.61	1.57	1.54	1.50	1.46
40	2.84	2.44	2.23	2.09	2.00	1.93	1.87	1.83	1.79	1.76	1.71	1.66	1.61	1.57	1.54	1.51	1.47	1.42	1.38
60	2.79	2.39	2.18	2.04	1.95	1.87	1.82	1.77	1.74	1.71	1.66	1.60	1.54	1.51	1.48	1.44	1.40	1.35	1.29
120	2.75	2.35	2.13	1.99	1.90	1.82	1.77	1.72	1.68	1.65	1.60	1.55	1.48	1.45	1.41	1.37	1.32	1.26	1.19
∞	2.71	2.30	2.08	1.94	1.85	1.77	1.72	1.67	1.63	1.60	1.55	1.49	1.42	1.38	1.34	1.30	1.24	1.17	1.00

Degrees of freedom in denominator

Table A.5 (Cont.)

b. One-tailed test, $P = 0.95$ (0.05 or 5% in tail). When used as two-tailed test, $P = 0.90$.

Degrees of freedom in numerator

f_2 \ f_1	1	2	3	4	5	6	7	8	9	10	12	15	20	24	30	40	60	120	∞
1	161.4	199.5	215.7	224.6	230.2	234.0	236.8	238.9	240.5	241.9	243.9	245.9	248.0	249.1	250.1	251.1	252.2	253.3	254.3
2	18.51	19.00	19.16	19.25	19.30	19.33	19.35	19.37	19.38	19.40	19.41	19.43	19.45	19.45	19.46	19.47	19.48	19.49	19.50
3	10.13	9.55	9.28	9.12	9.01	8.94	8.89	8.85	8.81	8.79	8.74	8.70	8.66	8.64	8.62	8.59	8.57	8.55	8.53
4	7.71	6.94	6.59	6.39	6.26	6.16	6.09	6.04	6.00	5.96	5.91	5.86	5.80	5.77	5.75	5.72	5.69	5.66	5.63
5	6.61	5.79	5.41	5.19	5.05	4.95	4.88	4.82	4.77	4.74	4.68	4.62	4.56	4.53	4.50	4.46	4.43	4.40	4.36
6	5.99	5.14	4.76	4.53	4.39	4.28	4.21	4.15	4.10	4.06	4.00	3.94	3.87	3.84	3.81	3.77	3.74	3.70	3.67
7	5.59	4.74	4.35	4.12	3.97	3.87	3.79	3.73	3.68	3.64	3.57	3.51	3.44	3.41	3.38	3.34	3.30	3.27	3.23
8	5.32	4.46	4.07	3.84	3.69	3.58	3.50	3.44	3.39	3.35	3.28	3.22	3.15	3.12	3.08	3.04	3.01	2.97	2.93
9	5.12	4.26	3.86	3.63	3.48	3.37	3.29	3.23	3.18	3.14	3.07	3.01	2.94	2.90	2.86	2.83	2.79	2.75	2.71
10	4.96	4.10	3.71	3.48	3.33	3.22	3.14	3.07	3.02	2.98	2.91	2.85	2.77	2.74	2.70	2.66	2.62	2.58	2.54
11	4.84	3.98	3.59	3.36	3.20	3.09	3.01	2.95	2.90	2.85	2.79	2.72	2.65	2.61	2.57	2.53	2.49	2.45	2.40
12	4.75	3.89	3.49	3.26	3.11	3.00	2.91	2.85	2.80	2.75	2.69	2.62	2.54	2.51	2.47	2.43	2.38	2.34	2.30
13	4.67	3.81	3.41	3.18	3.03	2.92	2.83	2.77	2.71	2.67	2.60	2.53	2.46	2.42	2.38	2.34	2.30	2.25	2.21
14	4.60	3.74	3.34	3.11	2.96	2.85	2.76	2.70	2.65	2.60	2.53	2.46	2.39	2.35	2.31	2.27	2.22	2.18	2.13
15	4.54	3.68	3.29	3.06	2.90	2.79	2.71	2.64	2.59	2.54	2.48	2.40	2.33	2.29	2.25	2.20	2.16	2.11	2.07
16	4.49	3.63	3.24	3.01	2.85	2.74	2.66	2.59	2.54	2.49	2.42	2.35	2.28	2.24	2.19	2.15	2.11	2.06	2.01
17	4.45	3.59	3.20	2.96	2.81	2.70	2.61	2.55	2.49	2.45	2.38	2.31	2.23	2.19	2.15	2.10	2.06	2.01	1.96
18	4.41	3.55	3.16	2.93	2.77	2.66	2.58	2.51	2.46	2.41	2.34	2.27	2.19	2.15	2.11	2.06	2.02	1.97	1.92
19	4.38	3.52	3.13	2.90	2.74	2.63	2.54	2.48	2.42	2.38	2.31	2.23	2.16	2.11	2.07	2.03	1.98	1.93	1.88
20	4.35	3.49	3.10	2.87	2.71	2.60	2.51	2.45	2.39	2.35	2.28	2.20	2.12	2.08	2.04	1.99	1.95	1.90	1.84
21	4.32	3.47	3.07	2.84	2.68	2.57	2.49	2.42	2.37	2.32	2.25	2.18	2.10	2.05	2.01	1.96	1.92	1.87	1.81
22	4.30	3.44	3.05	2.82	2.66	2.55	2.46	2.40	2.34	2.30	2.23	2.15	2.07	2.03	1.98	1.94	1.89	1.84	1.78
23	4.28	3.42	3.03	2.80	2.64	2.53	2.44	2.37	2.32	2.27	2.20	2.13	2.05	2.01	1.96	1.91	1.86	1.81	1.76
24	4.26	3.40	3.01	2.78	2.62	2.51	2.42	2.36	2.30	2.25	2.18	2.11	2.03	1.98	1.94	1.89	1.84	1.79	1.73
25	4.24	3.39	2.99	2.76	2.60	2.49	2.40	2.34	2.28	2.24	2.16	2.09	2.01	1.96	1.92	1.87	1.82	1.77	1.71
26	4.23	3.37	2.98	2.74	2.59	2.47	2.39	2.32	2.27	2.22	2.15	2.07	1.99	1.95	1.90	1.85	1.80	1.75	1.69
27	4.21	3.35	2.96	2.73	2.57	2.46	2.37	2.31	2.25	2.20	2.13	2.06	1.97	1.93	1.88	1.84	1.79	1.73	1.67
28	4.20	3.34	2.95	2.71	2.56	2.45	2.36	2.29	2.24	2.19	2.12	2.04	1.96	1.91	1.87	1.82	1.77	1.71	1.65
29	4.18	3.33	2.93	2.70	2.55	2.43	2.35	2.28	2.22	2.18	2.10	2.03	1.94	1.90	1.85	1.81	1.75	1.70	1.64
30	4.17	3.32	2.92	2.69	2.53	2.42	2.33	2.27	2.21	2.16	2.09	2.01	1.93	1.89	1.84	1.79	1.74	1.68	1.62
40	4.08	3.23	2.84	2.61	2.45	2.34	2.25	2.18	2.12	2.08	2.00	1.92	1.84	1.79	1.74	1.69	1.64	1.58	1.51
60	4.00	3.15	2.76	2.53	2.37	2.25	2.17	2.10	2.04	1.99	1.92	1.84	1.75	1.70	1.65	1.59	1.53	1.47	1.39
120	3.92	3.07	2.68	2.45	2.29	2.17	2.09	2.02	1.96	1.91	1.83	1.75	1.66	1.61	1.55	1.50	1.43	1.35	1.25
∞	3.84	3.00	2.60	2.37	2.21	2.10	2.01	1.94	1.88	1.83	1.75	1.67	1.57	1.52	1.46	1.39	1.32	1.22	1.00

Degrees of freedom in denominator

Table A.5 (Cont.)

c. One-tailed test, $P = 0.975$ (0.025 or 2.5%). When used as two-tailed test, $P = 0.95$.

Degrees of freedom in numerator

$f_2 \backslash f_1$	1	2	3	4	5	6	7	8	9	10	12	15	20	24	30	40	60	120	∞
1	647.8	799.5	864.2	899.6	921.8	937.1	948.2	956.7	963.3	968.6	976.7	984.9	993.1	997.2	1001	1006	1010	1014	1018
2	38.51	39.00	39.17	39.25	39.30	39.33	39.36	39.37	39.39	39.40	39.41	39.43	39.45	39.46	39.46	39.47	39.48	39.49	39.50
3	17.44	16.04	15.44	15.10	14.88	14.73	14.62	14.54	14.47	14.42	14.34	14.25	14.17	14.12	14.08	14.04	13.99	13.95	13.90
4	12.22	10.65	9.98	9.60	9.36	9.20	9.07	8.98	8.90	8.84	8.75	8.66	8.56	8.51	8.46	8.41	8.36	8.31	8.26
5	10.01	8.43	7.76	7.39	7.15	6.98	6.85	6.76	6.68	6.62	6.52	6.43	6.33	6.28	6.23	6.18	6.12	6.07	6.02
6	8.81	7.26	6.60	6.23	5.99	5.82	5.70	5.60	5.52	5.46	5.37	5.27	5.17	5.12	5.07	5.01	4.96	4.90	4.85
7	8.07	6.54	5.89	5.52	5.29	5.12	4.99	4.90	4.82	4.76	4.67	4.57	4.47	4.42	4.36	4.31	4.25	4.20	4.14
8	7.57	6.06	5.42	5.05	4.82	4.65	4.53	4.43	4.36	4.30	4.20	4.10	4.00	3.95	3.89	3.84	3.78	3.73	3.67
9	7.21	5.71	5.08	4.72	4.48	4.32	4.20	4.10	4.03	3.96	3.87	3.77	3.67	3.61	3.56	3.51	3.45	3.39	3.33
10	6.94	5.46	4.83	4.47	4.24	4.07	3.95	3.85	3.78	3.72	3.62	3.52	3.42	3.37	3.31	3.26	3.20	3.14	3.08
11	6.72	5.26	4.63	4.28	4.04	3.88	3.76	3.66	3.59	3.53	3.43	3.33	3.23	3.17	3.12	3.06	3.00	2.94	2.88
12	6.55	5.10	4.47	4.12	3.89	3.73	3.61	3.51	3.44	3.37	3.28	3.18	3.07	3.02	2.96	2.91	2.85	2.79	2.72
13	6.41	4.97	4.35	4.00	3.77	3.60	3.48	3.39	3.31	3.25	3.15	3.05	2.95	2.89	2.84	2.78	2.72	2.66	2.60
14	6.30	4.86	4.24	3.89	3.66	3.50	3.38	3.29	3.21	3.15	3.05	2.95	2.84	2.79	2.73	2.67	2.61	2.55	2.49
15	6.20	4.77	4.15	3.80	3.58	3.41	3.29	3.20	3.12	3.06	2.96	2.86	2.76	2.70	2.64	2.59	2.52	2.46	2.40
16	6.12	4.69	4.08	3.73	3.50	3.34	3.22	3.12	3.05	2.99	2.89	2.79	2.68	2.63	2.57	2.51	2.45	2.38	2.32
17	6.04	4.62	4.01	3.66	3.44	3.28	3.16	3.06	2.98	2.92	2.82	2.72	2.62	2.56	2.50	2.44	2.38	2.32	2.25
18	5.98	4.56	3.95	3.61	3.38	3.22	3.10	3.01	2.93	2.87	2.77	2.67	2.56	2.50	2.44	2.38	2.32	2.26	2.19
19	5.92	4.51	3.90	3.56	3.33	3.17	3.05	2.96	2.88	2.82	2.72	2.62	2.51	2.45	2.39	2.33	2.27	2.20	2.13
20	5.87	4.46	3.86	3.51	3.29	3.13	3.01	2.91	2.84	2.77	2.68	2.57	2.46	2.41	2.35	2.29	2.22	2.16	2.09
21	5.83	4.42	3.82	3.48	3.25	3.09	2.97	2.87	2.80	2.73	2.64	2.53	2.42	2.37	2.31	2.25	2.18	2.11	2.04
22	5.79	4.38	3.78	3.44	3.22	3.05	2.93	2.84	2.76	2.70	2.60	2.50	2.39	2.33	2.27	2.21	2.14	2.08	2.00
23	5.75	4.35	3.75	3.41	3.18	3.02	2.90	2.81	2.73	2.67	2.57	2.47	2.36	2.30	2.24	2.18	2.11	2.04	1.97
24	5.72	4.32	3.72	3.38	3.15	2.99	2.87	2.78	2.70	2.64	2.54	2.44	2.33	2.27	2.21	2.15	2.08	2.01	1.94
25	5.69	4.29	3.69	3.35	3.13	2.97	2.85	2.75	2.68	2.61	2.51	2.41	2.30	2.24	2.18	2.12	2.05	1.98	1.91
26	5.66	4.27	3.67	3.33	3.10	2.94	2.82	2.73	2.65	2.59	2.49	2.39	2.28	2.22	2.16	2.09	2.03	1.95	1.88
27	5.63	4.24	3.65	3.31	3.08	2.92	2.80	2.71	2.63	2.57	2.47	2.36	2.25	2.19	2.13	2.07	2.00	1.93	1.85
28	5.61	4.22	3.63	3.29	3.06	2.90	2.78	2.69	2.61	2.55	2.45	2.34	2.23	2.17	2.11	2.05	1.98	1.91	1.83
29	5.59	4.20	3.61	3.27	3.04	2.88	2.76	2.67	2.59	2.53	2.43	2.32	2.21	2.15	2.09	2.03	1.96	1.89	1.81
30	5.57	4.18	3.59	3.25	3.03	2.87	2.75	2.65	2.57	2.51	2.41	2.31	2.20	2.14	2.07	2.01	1.94	1.87	1.79
40	5.42	4.05	3.46	3.13	2.90	2.74	2.62	2.53	2.45	2.39	2.29	2.18	2.07	2.01	1.94	1.88	1.80	1.72	1.64
60	5.29	3.93	3.34	3.01	2.79	2.63	2.51	2.41	2.33	2.27	2.17	2.06	1.94	1.88	1.82	1.74	1.67	1.58	1.48
120	5.15	3.80	3.23	2.89	2.67	2.52	2.39	2.30	2.22	2.16	2.05	1.94	1.82	1.76	1.69	1.61	1.53	1.43	1.31
∞	5.02	3.69	3.12	2.79	2.57	2.41	2.29	2.19	2.11	2.05	1.94	1.83	1.71	1.64	1.57	1.48	1.39	1.27	1.00

Degrees of freedom in denominator

Table A.5 (Cont.)

d. One-tailed test, $P = 0.99$ (0.01 or 1%). When used as two-tailed test, $P = 0.98$.

f_2 \ f_1	1	2	3	4	5	6	7	8	9	10	12	15	20	24	30	40	60	120	∞
1	4052	4999.5	5403	5625	5764	5859	5928	5982	6022	6056	6106	6157	6209	6235	6261	6287	6313	6339	6366
2	98.50	99.00	99.17	99.25	99.30	99.33	99.36	99.37	99.39	99.40	99.42	99.43	99.45	99.46	99.47	99.47	99.48	99.49	99.50
3	34.12	30.82	29.46	28.71	28.24	27.91	27.67	27.49	27.35	27.23	27.05	26.87	26.69	26.60	26.50	26.41	26.32	26.22	26.13
4	21.20	18.00	16.69	15.98	15.52	15.21	14.98	14.80	14.66	14.55	14.37	14.20	14.02	13.93	13.84	13.75	13.65	13.56	13.46
5	16.26	13.27	12.06	11.39	10.97	10.67	10.46	10.29	10.16	10.05	9.89	9.72	9.55	9.47	9.38	9.29	9.20	9.11	9.02
6	13.75	10.92	9.78	9.15	8.75	8.47	8.26	8.10	7.98	7.87	7.72	7.56	7.40	7.31	7.23	7.14	7.06	6.97	6.88
7	12.25	9.55	8.45	7.85	7.46	7.19	6.99	6.84	6.72	6.62	6.47	6.31	6.16	6.07	5.99	5.91	5.82	5.74	5.65
8	11.26	8.65	7.59	7.01	6.63	6.37	6.18	6.03	5.91	5.81	5.67	5.52	5.36	5.28	5.20	5.12	5.03	4.95	4.86
9	10.56	8.02	6.99	6.42	6.06	5.80	5.61	5.47	5.35	5.26	5.11	4.96	4.81	4.73	4.65	4.57	4.48	4.40	4.31
10	10.04	7.56	6.55	5.99	5.64	5.39	5.20	5.06	4.94	4.85	4.71	4.56	4.41	4.33	4.25	4.17	4.08	4.00	3.91
11	9.65	7.21	6.22	5.67	5.32	5.07	4.89	4.74	4.63	4.54	4.40	4.25	4.10	4.02	3.94	3.86	3.78	3.69	3.60
12	9.33	6.93	5.95	5.41	5.06	4.82	4.64	4.50	4.39	4.30	4.16	4.01	3.86	3.78	3.70	3.62	3.54	3.45	3.36
13	9.07	6.70	5.74	5.21	4.86	4.62	4.44	4.30	4.19	4.10	3.96	3.82	3.66	3.59	3.51	3.43	3.34	3.25	3.17
14	8.86	6.51	5.56	5.04	4.69	4.46	4.28	4.14	4.03	3.94	3.80	3.66	3.51	3.43	3.35	3.27	3.18	3.09	3.00
15	8.68	6.36	5.42	4.89	4.56	4.32	4.14	4.00	3.89	3.80	3.67	3.52	3.37	3.29	3.21	3.13	3.05	2.96	2.87
16	8.53	6.23	5.29	4.77	4.44	4.20	4.03	3.89	3.78	3.69	3.55	3.41	3.26	3.18	3.10	3.02	2.93	2.84	2.75
17	8.40	6.11	5.18	4.67	4.34	4.10	3.93	3.79	3.68	3.59	3.46	3.31	3.16	3.08	3.00	2.92	2.83	2.75	2.65
18	8.29	6.01	5.09	4.58	4.25	4.01	3.84	3.71	3.60	3.51	3.37	3.23	3.08	3.00	2.92	2.84	2.75	2.66	2.57
19	8.18	5.93	5.01	4.50	4.17	3.94	3.77	3.63	3.52	3.43	3.30	3.15	3.00	2.92	2.84	2.76	2.67	2.58	2.49
20	8.10	5.85	4.94	4.43	4.10	3.87	3.70	3.56	3.46	3.37	3.23	3.09	2.94	2.86	2.78	2.69	2.61	2.52	2.42
21	8.02	5.78	4.87	4.37	4.04	3.81	3.64	3.51	3.40	3.31	3.17	3.03	2.88	2.80	2.72	2.64	2.55	2.46	2.36
22	7.95	5.72	4.82	4.31	3.99	3.76	3.59	3.45	3.35	3.26	3.12	2.98	2.83	2.75	2.67	2.58	2.50	2.40	2.31
23	7.88	5.66	4.76	4.26	3.94	3.71	3.54	3.41	3.30	3.21	3.07	2.93	2.78	2.70	2.62	2.54	2.45	2.35	2.26
24	7.82	5.61	4.72	4.22	3.90	3.67	3.50	3.36	3.26	3.17	3.03	2.89	2.74	2.66	2.58	2.49	2.40	2.31	2.21
25	7.77	5.57	4.68	4.18	3.85	3.63	3.46	3.32	3.22	3.13	2.99	2.85	2.70	2.62	2.54	2.45	2.36	2.27	2.17
26	7.72	5.53	4.64	4.14	3.82	3.59	3.42	3.29	3.18	3.09	2.96	2.81	2.66	2.58	2.50	2.42	2.33	2.23	2.13
27	7.68	5.49	4.60	4.11	3.78	3.56	3.39	3.26	3.15	3.06	2.93	2.78	2.63	2.55	2.47	2.38	2.29	2.20	2.10
28	7.64	5.45	4.57	4.07	3.75	3.53	3.36	3.23	3.12	3.03	2.90	2.75	2.60	2.52	2.44	2.35	2.26	2.17	2.06
29	7.60	5.42	4.54	4.04	3.73	3.50	3.33	3.20	3.09	3.00	2.87	2.73	2.57	2.49	2.41	2.33	2.23	2.14	2.03
30	7.56	5.39	4.51	4.02	3.70	3.47	3.30	3.17	3.07	2.98	2.84	2.70	2.55	2.47	2.39	2.30	2.21	2.11	2.01
40	7.31	5.18	4.31	3.83	3.51	3.29	3.12	2.99	2.89	2.80	2.66	2.52	2.37	2.29	2.20	2.11	2.02	1.92	1.80
60	7.08	4.98	4.13	3.65	3.34	3.12	2.95	2.82	2.72	2.63	2.50	2.35	2.20	2.12	2.03	1.94	1.84	1.73	1.60
120	6.85	4.79	3.95	3.48	3.17	2.96	2.79	2.66	2.56	2.47	2.34	2.19	2.03	1.95	1.86	1.76	1.66	1.53	1.38
∞	6.63	4.61	3.78	3.32	3.02	2.80	2.64	2.51	2.41	2.32	2.18	2.04	1.88	1.79	1.70	1.59	1.47	1.32	1.00

Degrees of freedom in numerator (f_1); Degrees of freedom in denominator (f_2)

Table A.5 (*continued*)

e. One-tailed test, $P = 0.995$ (0.005 or 0.5%). When used as two-tailed test, $P = 0.99$.

Degrees of freedom in numerator

f_2 \ f_1	1	2	3	4	5	6	7	8	9	10	12	15	20	24	30	40	60	120	∞
1	16211	20000	21615	22500	23056	23437	23715	23925	24091	24224	24426	24630	24836	24940	25044	25148	25253	25359	25465
2	198.5	199.0	199.2	199.2	199.3	199.3	199.4	199.4	199.4	199.4	199.4	199.4	199.4	199.5	199.5	199.5	199.5	199.5	199.5
3	55.55	49.80	47.47	46.19	45.39	44.84	44.43	44.13	43.88	43.69	43.39	43.08	42.78	46.62	42.47	42.31	42.15	41.99	41.83
4	31.33	26.28	24.26	23.15	22.46	21.97	21.62	21.35	21.14	20.97	20.70	20.44	20.17	20.03	19.89	19.75	19.75	19.61	19.47	19.32
5	22.78	18.31	16.53	15.56	14.94	14.51	14.20	13.96	13.77	13.62	13.38	13.15	12.90	12.78	12.66	12.53	12.40	12.27	12.14
6	18.63	14.54	12.92	12.03	11.46	11.07	10.79	10.57	10.39	10.25	10.03	9.81	9.59	9.47	9.36	9.24	9.12	9.00	8.88
7	16.24	12.40	10.88	10.05	9.52	9.16	8.89	8.68	8.51	8.38	8.18	7.97	7.75	7.65	7.53	7.42	7.31	7.19	7.08
8	14.69	11.04	9.60	8.81	8.30	7.95	7.69	7.50	7.34	7.21	7.01	6.81	6.61	6.50	6.40	6.29	6.18	6.06	5.95
9	13.61	10.11	8.72	7.96	7.47	7.13	6.88	6.69	6.54	6.42	6.23	6.03	5.83	5.73	5.62	5.52	5.41	5.30	5.19
10	12.83	9.43	8.08	7.34	6.87	6.54	6.30	6.12	5.97	5.85	5.66	5.47	5.27	5.17	5.07	4.97	4.86	4.75	4.64
11	12.23	8.91	7.60	6.88	6.42	6.10	5.86	5.68	5.54	5.42	5.24	5.05	4.86	4.76	4.65	4.55	4.44	4.34	4.23
12	11.75	8.51	7.23	6.52	6.07	5.76	5.52	5.35	5.20	5.09	4.91	4.72	4.53	4.43	4.33	4.23	4.12	4.01	3.90
13	11.37	8.19	6.93	6.23	5.79	5.48	5.25	5.08	4.94	4.82	4.64	4.46	4.27	4.17	4.07	3.97	3.87	3.76	3.65
14	11.06	7.92	6.68	6.00	5.56	5.26	5.03	4.86	4.72	4.60	4.43	4.25	4.06	3.96	3.86	3.76	3.66	3.55	3.44
15	10.80	7.70	6.48	5.80	5.37	5.07	4.85	4.67	4.54	4.42	4.25	4.07	3.88	3.79	3.69	3.58	3.48	3.37	3.26
16	10.58	7.51	6.30	5.64	5.21	4.91	4.69	4.52	4.38	4.27	4.10	3.92	3.73	3.64	3.54	3.44	3.33	3.22	3.11
17	10.38	7.35	6.16	5.50	5.07	4.78	4.56	4.39	4.25	4.14	3.97	3.79	3.61	3.51	3.41	3.31	3.21	3.10	2.98
18	10.22	7.21	6.03	5.37	4.96	4.66	4.44	4.28	4.14	4.03	3.86	3.68	3.50	3.40	3.30	3.20	3.10	2.99	2.87
19	10.07	7.09	5.92	5.27	4.85	4.56	4.34	4.18	4.04	3.93	3.76	3.59	3.40	3.31	3.21	3.11	3.00	2.89	2.78
20	9.94	6.99	5.82	5.17	4.76	4.47	4.26	4.09	3.96	3.85	3.68	3.50	3.32	3.22	3.12	3.02	2.92	2.81	2.69
21	9.83	6.89	5.73	5.09	4.68	4.39	4.18	4.01	3.88	3.77	3.60	3.43	3.24	3.15	3.05	2.95	2.84	2.73	2.61
22	9.73	6.81	5.65	5.02	4.61	4.32	4.11	3.94	3.81	3.70	3.54	3.36	3.18	3.08	2.98	2.88	2.77	2.66	2.55
23	9.63	6.73	5.58	4.95	4.54	4.26	4.05	3.88	3.75	3.64	3.47	3.30	3.12	3.02	2.92	2.82	2.71	2.60	2.48
24	9.55	6.66	5.52	4.89	4.49	4.20	3.99	3.83	3.69	3.59	3.42	3.25	3.06	2.97	2.87	2.77	2.66	2.55	2.43
25	9.48	6.60	5.46	4.84	4.43	4.15	3.94	3.78	3.64	3.54	3.37	3.20	3.01	2.92	2.82	2.72	2.61	2.50	2.38
26	9.41	6.54	5.41	4.79	4.38	4.10	3.89	3.73	3.60	3.49	3.33	3.15	2.97	2.87	2.77	2.67	2.56	2.45	2.33
27	9.34	6.49	5.36	4.74	4.34	4.06	3.85	3.69	3.56	3.45	3.28	3.11	2.93	2.83	2.73	2.63	2.52	2.41	2.25
28	9.28	6.44	5.32	4.70	4.30	4.02	3.81	3.65	3.52	3.41	3.25	3.07	2.89	2.79	2.69	2.59	2.48	2.37	2.29
29	9.23	6.40	5.28	4.66	4.26	3.98	3.77	3.61	3.48	3.38	3.21	3.04	2.86	2.76	2.66	2.56	2.45	2.33	2.24
30	9.18	6.35	5.24	4.62	4.23	3.95	3.74	3.58	3.45	3.34	3.18	3.01	2.82	2.73	2.63	2.52	2.42	2.30	2.18
40	8.83	6.07	4.98	4.37	3.99	3.71	3.51	3.35	3.22	3.12	2.95	2.78	2.60	2.50	2.40	2.30	2.18	2.06	1.93
60	8.49	5.79	4.73	4.14	3.76	3.49	3.29	3.13	3.01	2.90	2.74	2.57	2.39	2.29	2.19	2.08	1.96	1.83	1.69
120	8.18	5.54	4.50	3.92	3.55	3.28	3.09	2.93	2.81	2.71	2.54	2.37	2.19	2.09	1.98	1.87	1.75	1.61	1.43
∞	7.88	5.30	4.28	3.72	3.35	3.09	2.90	2.74	2.62	2.52	2.36	2.19	2.00	1.90	1.79	1.67	1.53	1.36	1.00

Degrees of freedom in denominator

Appendix A 513

f. One-tailed test, $P = 0.999$ (0.001 or 0.1%). When used as two-tailed test, $P = 0.998$.

Degrees of freedom in numerator

f_1 / f_2	1	2	3	4	5	6	7	8	9	10	12	15	20	24	30	40	60	120	∞
1	4053*	5000*	5404*	5625*	5764*	5859*	5929*	5981*	6023*	6056*	6107*	6158*	6209*	6235*	6261*	6287*	6313*	6340*	6366*
2	998.5	999.0	999.2	999.2	999.3	999.3	999.4	999.4	999.4	999.4	999.4	999.4	999.4	999.5	999.5	999.5	999.5	999.5	999.5
3	167.0	148.5	141.1	137.1	134.6	132.8	131.6	130.6	129.9	129.2	128.3	127.4	126.4	125.9	125.9	125.0	124.5	124.0	123.5
4	74.14	61.25	56.18	53.44	51.71	50.53	49.66	49.00	48.47	48.05	47.41	46.76	46.10	45.77	45.43	45.09	44.75	44.40	44.05
5	47.18	37.12	33.20	31.09	29.75	28.84	28.16	27.64	27.24	26.92	26.42	25.91	25.39	25.14	24.87	24.60	24.33	24.06	23.79
6	35.51	27.00	23.70	21.92	20.81	20.03	19.46	19.03	18.69	18.41	17.99	17.56	17.12	16.89	16.67	16.44	16.21	15.99	15.75
7	29.25	21.69	18.77	17.19	16.21	15.52	15.02	14.63	14.33	14.08	13.71	13.32	12.93	12.73	12.53	12.33	12.12	11.91	11.70
8	25.42	18.49	15.83	14.39	13.49	12.86	12.40	12.04	11.77	11.54	11.19	10.84	10.48	10.30	10.11	9.92	9.73	9.53	9.33
9	22.86	16.39	13.90	12.56	11.71	11.13	10.70	10.37	10.11	9.89	9.57	9.24	8.90	8.72	8.55	8.37	8.19	8.00	7.81
10	21.04	14.91	12.55	11.28	10.48	9.92	9.52	9.20	8.96	8.75	8.45	8.13	7.80	7.64	7.47	7.30	7.12	6.94	6.76
11	19.69	13.81	11.56	10.35	9.58	9.05	8.66	8.35	8.12	7.92	7.63	7.32	7.01	6.85	6.68	6.52	6.35	6.17	6.00
12	18.64	12.97	10.80	9.63	8.89	8.38	8.00	7.71	7.48	7.29	7.00	6.71	6.40	6.25	6.09	5.93	5.76	5.59	5.42
13	17.81	12.31	10.21	9.07	8.35	7.86	7.49	7.21	6.98	6.80	6.52	6.23	5.93	5.78	5.63	5.47	5.30	5.14	4.97
14	17.14	11.78	9.73	8.62	7.92	7.43	7.08	6.80	6.58	6.40	6.13	5.85	5.56	5.41	5.25	5.10	4.94	4.77	4.60
15	16.59	11.34	9.34	8.25	7.57	7.09	6.74	6.47	6.26	6.08	5.81	5.54	5.25	5.10	4.95	4.80	4.64	4.47	4.31
16	16.12	10.97	9.00	7.94	7.27	6.81	6.46	6.19	5.98	5.81	5.55	5.27	4.99	4.85	4.70	4.54	4.39	4.23	4.06
17	15.72	10.66	8.73	7.68	7.02	6.56	6.22	5.96	5.75	5.58	5.32	5.05	4.78	4.63	4.48	4.33	4.18	4.02	3.85
18	15.38	10.39	8.49	7.46	6.81	6.35	6.02	5.76	5.56	5.39	5.13	4.87	4.59	4.45	4.30	4.15	4.00	3.84	3.67
19	15.08	10.16	8.28	7.26	6.62	6.18	5.85	5.59	5.39	5.22	4.97	4.70	4.43	4.29	4.14	3.99	3.84	3.68	3.51
20	14.82	9.95	8.10	7.10	6.46	6.02	5.69	5.44	5.24	5.08	4.82	4.56	4.29	4.15	4.00	3.86	3.70	3.54	3.38
21	14.59	9.77	7.94	6.95	6.32	5.88	5.56	5.31	5.11	4.95	4.70	4.44	4.17	4.03	3.88	3.74	3.58	3.42	3.26
22	14.38	9.61	7.80	6.81	6.19	5.76	5.44	5.19	4.99	4.83	4.58	4.33	4.06	3.92	3.78	3.63	3.48	3.32	3.15
23	14.19	9.47	7.67	6.69	6.08	5.65	5.33	5.09	4.89	4.73	4.48	4.23	3.96	3.82	3.68	3.53	3.38	3.22	3.05
24	14.03	9.34	7.55	6.59	5.98	5.55	5.23	4.99	4.80	4.64	4.39	4.14	3.87	3.74	3.59	3.45	3.29	3.14	2.97
25	13.88	9.22	7.45	6.49	5.88	5.46	5.15	4.91	4.71	4.56	4.31	4.06	3.79	3.66	3.52	3.37	3.22	3.06	2.89
26	13.74	9.12	7.36	6.41	5.80	5.38	5.07	4.83	4.64	4.48	4.24	3.99	3.72	3.59	3.44	3.30	3.15	2.99	2.82
27	13.61	9.02	7.27	6.33	5.73	5.31	5.00	4.76	4.57	4.41	4.17	3.92	3.66	3.52	3.38	3.23	3.08	2.92	2.75
28	13.50	8.93	7.19	6.25	5.66	5.24	4.93	4.69	4.50	4.35	4.11	3.86	3.60	3.46	3.32	3.18	3.02	2.86	2.69
29	13.39	8.85	7.12	6.19	5.59	5.18	4.87	4.64	4.45	4.29	4.05	3.80	3.54	3.41	3.27	3.12	2.97	2.81	2.64
30	13.29	8.77	7.05	6.12	5.53	5.12	4.82	4.58	4.39	4.24	4.00	3.75	3.49	3.36	3.22	3.07	2.92	2.76	2.59
40	12.61	8.25	6.60	5.70	5.13	4.73	4.44	4.21	4.02	3.87	3.64	3.40	3.15	3.01	2.87	2.73	2.57	2.41	2.23
60	11.97	7.76	6.17	5.31	4.76	4.37	4.09	3.87	3.69	3.54	3.31	3.08	2.83	2.69	2.55	2.41	2.25	2.08	1.89
120	11.38	7.32	5.79	4.95	4.42	4.04	3.77	3.55	3.38	3.24	3.02	2.78	2.53	2.40	2.26	2.11	1.95	1.76	1.54
∞	10.83	6.91	5.42	4.62	4.10	3.74	3.47	3.27	3.10	2.96	2.74	2.51	2.27	2.13	1.99	1.84	1.66	1.45	1.00

Degrees of freedom in denominator

SOURCE: Abridged with kind permission from Biometrika Tables for Statisticians, vol. 1, edited by E. S. Pearson and H. O. Hartley (Cambridge, Cambridge University Press, 1962).

* Multiply these entries by 100.

A.6 Graph for Finding $\sigma/\sqrt{n}$

The accompanying graph is for finding the standard error, $\sigma/\sqrt{n}$, given σ and n or for finding the necessary sample size n given σ and the desired standard error, $\sigma/\sqrt{n}$.

To find $\sigma/\sqrt{n}$ given σ and n: Locate σ along the right axis (labeled σ). Find and follow the slanted line going to the lower left from that point. Locate the value of n on the bottom axis. Go straight up from that point (n) until you intersect the slanted line from σ. Note the point of intersection (a). From the intersection (point a) proceed horizontally to the $\sigma/\sqrt{n}$ axis (the left axis) and read the standard error $(\sigma/\sqrt{n})$ directly.

EXAMPLE 1

If $\sigma = 10$, $n = 4$, find the standard error.

Figure A.6.1 shows how you follow the line emanating from $\sigma = 10$ down to the left. The second line from $n = 4$ going up intersects at point a. Going to the left from that point we find $\sigma/\sqrt{n} = 5$.

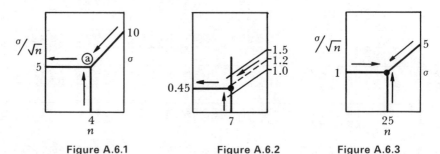

Figure A.6.1   Figure A.6.2   Figure A.6.3

EXAMPLE 2

If $\sigma = 1.2$, $n = 7$, find the standard error.

No line comes from the point $\sigma = 1.2$, so we must approximate the line. It should go between the lines for $\sigma = 1.0$ and 1.5 (if you have a ruler you can move up or down parallel to the slanted lines until it reaches the 1.2 point on the σ axis.) You follow up the line for $n = 7$ until you are the proper distance above the slanted line from $\sigma = 1.0$ (or until you reach your ruler). This corresponds to point a. Follow this point horizontally until you reach $\sigma/\sqrt{n}$ at approximately 0.45 (see Figure A.6.2).

To find n, the necessary sample size, when σ is known.

EXAMPLE 3

Given that $\sigma = 5$ and we desire a standard error of 1, how large a sample is necessary?

Follow the slanted line from $\sigma = 5$ until it intersects the horizontal line from the point $\sigma/\sqrt{n} = 1$. From this point drop down to the n scale and read $n = 25$ (see Figure A.6.3).

Table A.6 Graph for finding $\sigma/\sqrt{n}$ or n

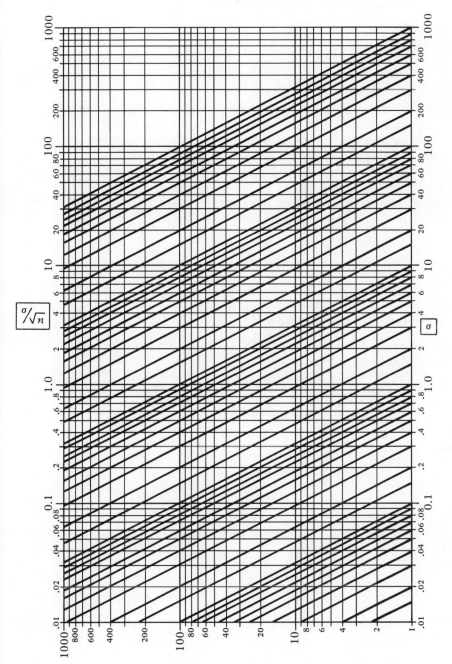

A.7 Critical Values for Correlation Coefficient, r. Pearson product moment correlation coefficient

This table gives critically large values of the correlation coefficient for various sample sizes (degrees of freedom). Several levels of significance are given.

$$n = \text{number of } pairs \text{ of values.}$$

Generally: *Degrees of freedom* $df = n - 2$.

Critical values are given for both one- and two-tailed tests. A *one-tailed* test would mean that we are *only* interested in a positive correlation or we are *only* interested in a negative correlation. A two-tailed test means we are simultaneously considering *either* a positive or a negative correlation.

EXAMPLE 1

Given a sample of 22 what are the significantly large values or limits for the correlation coefficient? Consider a two-tailed test and 5% and 1% limits.

$$n = 22,$$

therefore

$$df = n - 2 = 20.$$

From the table for $df = 20$ we read:

1. The 5% limit or 0.05 limit is 0.4227 (there is less than a 5% chance that r should randomly be as large as 0.4227), This is also labeled as $r_{0.95}$.
2. The 1% limit or 0.01 limit is 0.5368 (there is less than a 1% chance that r should randomly be as large as 0.5368). This is also labeled as $r_{0.99}$.

df	Level of significance for two-tailed test	
	0.05	0.01
20	0.4227	0.5368

Table A.7 Critical values of the Pearson product moment correlation coefficient

	Level of significance for one-tailed test				
	0.05	0.025	0.01	0.005	0.0005
	Level of significance for two-tailed test				
$df = n - 2$	0.10	0.05	0.02	0.01	0.001
1	0.9877	0.9969	0.9995	0.9999	1.0000
2	0.9000	0.9500	0.9800	0.9900	0.9990
3	0.8054	0.8783	0.9343	0.9587	0.9912
4	0.7293	0.8114	0.8822	0.9172	0.9741
5	0.6694	0.7545	0.8329	0.8745	0.9507
6	0.6215	0.7067	0.7887	0.8343	0.9249
7	0.5822	0.6664	0.7498	0.7977	0.8982
8	0.5494	0.6319	0.7155	0.7646	0.8721
9	0.5214	0.6021	0.6851	0.7348	0.8471
10	0.4973	0.5760	0.6581	0.7079	0.8233
11	0.4762	0.5529	0.6339	0.6835	0.8010
12	0.4575	0.5324	0.6120	0.6614	0.7800
13	0.4409	0.5139	0.5923	0.6411	0.7603
14	0.4259	0.4973	0.5742	0.6226	0.7420
15	0.4124	0.4821	0.5577	0.6055	0.7246
16	0.4000	0.4683	0.5425	0.5897	0.7084
17	0.3887	0.4555	0.5285	0.5751	0.6932
18	0.3783	0.4438	0.5155	0.5614	0.6787
19	0.3687	0.4329	0.5034	0.5487	0.6652
20	0.3598	0.4227	0.4921	0.5368	0.6524
25	0.3233	0.3809	0.4451	0.4869	0.5974
30	0.2960	0.3494	0.4093	0.4487	0.5541
35	0.2746	0.3246	0.3810	0.4182	0.5189
40	0.2573	0.3044	0.3578	0.3932	0.4896
45	0.2428	0.2875	0.3384	0.3721	0.4648
50	0.2306	0.2732	0.3218	0.3541	0.4433
60	0.2108	0.2500	0.2948	0.3248	0.4078
70	0.1954	0.2319	0.2737	0.3017	0.3799
80	0.1829	0.2172	0.2565	0.2830	0.3568
90	0.1726	0.2050	0.2422	0.2673	0.3375
100	0.1638	0.1946	0.2301	0.2540	0.3211

SOURCE: Table A.7 is taken from Table VII of Fisher and Yates, *Statistical Tables for Biological, Agricultural, and Medical Research*, published by Oliver and Boyd, Ltd., Edinburgh, and by permission of the authors and publishers.

A.8 Binomial Distribution

To find the probabilities for selected binomial distributions given n and p.

EXAMPLE 1

Determine the probabilities for each X-value in the binomial distribution with $n = 6$ and $p = 0.35$.

Note that
$$q = 1 - p = 1 - 0.35 = 0.65.$$

Under the column for n locate $n = 6$. All of the values from 0 to 6 in the adjacent column (labeled X column) correspond to possibilities when $n = 6$.

Under the $p = 0.35$ column you will find the probabilities for each X-value. Thus the probability of $X = 0$ is 0.0754, the probability that $X = 1$ is 0.2437, and so forth.

n	X	p 0.35
5		
→ 6	0	0.0754
	1	0.2437
	2	0.3280
	3	0.2355
	4	0.0951
	5	0.0205
	6	0.0018

Note that when the probability of a particular value of X is recorded as 0.0000 this is *not* to be interpreted as impossible but just less than a probability of 0.00005.

EXAMPLE 2

For a binomial distribution with $n = 6$ and $p = 0.35$ determine the probability of a value equal to or greater than 5.

Add up the probabilities for *5 and 6* or $0.0205 + 0.0018 = 0.0223$.

EXAMPLE 3

For a binomial distribution with $n = 25$ and $p = 0.30$ determine the limit above which 5% of the values lie.

Add up the various probabilities until you obtain approximately 0.0500.

$X = 15, p = 0.0000$; $X = 14, p = 0.0000$; $X = 13, p = 0.0003$;
$X = 12, p = 0.0016$ and the sum for all p from 12 to 15 is 0.0019;
$X = 11, p = 0.0074$ and the sum $= 0.0093$;
$X = 10, p = 0.0245$ and the sum $= 0.0338$;
$X = 9, p = 0.0612$ and the sum $= 0.0950$.

The closest value to a sum of 5% or 0.05 is the $X = 10$.

Table A.8 Individual terms, binomial distribution

n	x	0.05	0.10	0.15	0.20	p 0.25	0.30	0.35	0.40	0.50	x
1	0	0.9500	0.9000	0.8500	0.8000	0.7500	0.7000	0.6500	0.6000	0.5000	1
	1	0.0500	0.1000	0.1500	0.2000	0.2500	0.3000	0.3500	0.4000	0.5000	0
2	0	0.9025	0.8100	0.7225	0.6400	0.5625	0.4900	0.4225	0.3600	0.2500	2
	1	0.0950	0.1800	0.2550	0.3200	0.3750	0.4200	0.4550	0.4800	0.5000	1
	2	0.0025	0.0100	0.0225	0.0400	0.0625	0.0900	0.1225	0.1600	0.2500	0
3	0	0.8574	0.7290	0.6141	0.5120	0.4219	0.3430	0.2746	0.2160	0.1250	3
	1	0.1354	0.2430	0.3251	0.3840	0.4219	0.4410	0.4436	0.4320	0.3750	2
	2	0.0071	0.0270	0.0574	0.0960	0.1406	0.1890	0.2389	0.2880	0.3750	1
	3	0.0001	0.0010	0.0034	0.0080	0.0156	0.0270	0.0429	0.0640	0.1250	0
4	0	0.8145	0.6561	0.5220	0.4096	0.3164	0.2401	0.1785	0.1296	0.0625	4
	1	0.1715	0.2916	0.3685	0.4096	0.4219	0.4116	0.3845	0.3456	0.2500	3
	2	0.0135	0.0486	0.0975	0.1536	0.2109	0.2646	0.3105	0.3456	0.3750	2
	3	0.0005	0.0036	0.0115	0.0256	0.0469	0.0756	0.1115	0.1536	0.2500	1
	4	0.0000	0.0001	0.0005	0.0016	0.0039	0.0081	0.0150	0.0256	0.0625	0
5	0	0.7738	0.5905	0.4437	0.3277	0.2373	0.1681	0.1160	0.0778	0.0312	5
	1	0.2036	0.3280	0.3915	0.4096	0.3955	0.3602	0.3124	0.2592	0.1562	4
	2	0.0214	0.0729	0.1382	0.2048	0.2637	0.3087	0.3364	0.3456	0.3125	3
	3	0.0011	0.0081	0.0244	0.0512	0.0879	0.1323	0.1811	0.2304	0.3125	2
	4	0.0000	0.0004	0.0022	0.0064	0.0146	0.0284	0.0488	0.0768	0.1562	1
	5	0.0000	0.0000	0.0001	0.0003	0.0010	0.0024	0.0053	0.0102	0.0312	0
6	0	0.7351	0.5314	0.3771	0.2621	0.1780	0.1176	0.0754	0.0467	0.0156	6
	1	0.2321	0.3543	0.3993	0.3932	0.3560	0.3025	0.2437	0.1866	0.0938	5
	2	0.0305	0.0984	0.1762	0.2458	0.2966	0.3241	0.3280	0.3110	0.2344	4
	3	0.0021	0.0146	0.0415	0.0819	0.1318	0.1852	0.2355	0.2765	0.3125	3
	4	0.0001	0.0012	0.0055	0.0154	0.0330	0.0595	0.0951	0.1382	0.2344	2
	5	0.0000	0.0001	0.0004	0.0015	0.0044	0.0102	0.0205	0.0369	0.0938	1
	6	0.0000	0.0000	0.0000	0.0001	0.0002	0.0007	0.0018	0.0041	0.0156	0
7	0	0.6983	0.4783	0.3206	0.2097	0.1335	0.0824	0.0490	0.0280	0.0078	7
	1	0.2573	0.3720	0.3960	0.3670	0.3115	0.2471	0.1848	0.1306	0.0547	6
	2	0.0406	0.1240	0.2097	0.2753	0.3115	0.3177	0.2985	0.2613	0.1641	5
	3	0.0036	0.0230	0.0617	0.1147	0.1730	0.2269	0.2679	0.2903	0.2734	4
	4	0.0002	0.0026	0.0109	0.0287	0.0577	0.0972	0.1442	0.1935	0.2734	3
	5	0.0000	0.0002	0.0012	0.0043	0.0115	0.0250	0.0466	0.0774	0.1641	2
	6	0.0000	0.0000	0.0001	0.0004	0.0013	0.0036	0.0084	0.0172	0.0547	1
	7	0.0000	0.0000	0.0000	0.0000	0.0001	0.0002	0.0006	0.0016	0.0078	0
8	0	0.6634	0.4305	0.2725	0.1678	0.1001	0.0576	0.0319	0.0168	0.0039	8
	1	0.2793	0.3826	0.3847	0.3355	0.2670	0.1977	0.1373	0.0896	0.0312	7
	2	0.0515	0.1488	0.2376	0.2936	0.3115	0.2965	0.2587	0.2090	0.1094	6
	3	0.0054	0.0331	0.0839	0.1468	0.2076	0.2541	0.2786	0.2787	0.2188	5
	4	0.0004	0.0046	0.0185	0.0459	0.0865	0.1361	0.1875	0.2322	0.2734	4
	5	0.0000	0.0004	0.0026	0.0092	0.0231	0.0467	0.0808	0.1239	0.2188	3
	6	0.0000	0.0000	0.0002	0.0011	0.0038	0.0100	0.0217	0.0413	0.1094	2
	7	0.0000	0.0000	0.0000	0.0001	0.0004	0.0012	0.0033	0.0079	0.0312	1
	8	0.0000	0.0000	0.0000	0.0000	0.0000	0.0001	0.0002	0.0007	0.0039	0
		0.95	0.90	0.85	0.80	0.75	0.70	0.65	0.60	0.50	
						p					

Table A.8 *(continued)*

n	x	0.05	0.10	0.15	0.20	p 0.25	0.30	0.35	0.40	0.50	x
9	0	0.6302	0.3874	0.2316	0.1342	0.0751	0.0404	0.0207	0.0101	0.0020	9
	1	0.2985	0.3874	0.3679	0.3020	0.2253	0.1556	0.1004	0.0605	0.0176	8
	2	0.0629	0.1722	0.2597	0.3020	0.3003	0.2668	0.2162	0.1612	0.0703	7
	3	0.0077	0.0446	0.1069	0.1762	0.2336	0.2668	0.2716	0.2508	0.1641	6
	4	0.0006	0.0074	0.0283	0.0661	0.1168	0.1715	0.2194	0.2508	0.2461	5
	5	0.0000	0.0008	0.0050	0.0165	0.0380	0.0735	0.1181	0.1672	0.2461	4
	6	0.0000	0.0001	0.0006	0.0028	0.0087	0.0210	0.0424	0.0743	0.1641	3
	7	0.0000	0.0000	0.0000	0.0003	0.0012	0.0039	0.0098	0.0212	0.0703	2
	8	0.0000	0.0000	0.0000	0.0000	0.0001	0.0004	0.0013	0.0035	0.0176	1
	9	0.0000	0.0000	0.0000	0.0000	0.0000	0.0000	0.0001	0.0003	0.0020	0
10	0	0.5987	0.3487	0.1969	0.1074	0.0563	0.0282	0.0135	0.0060	0.0010	10
	1	0.3151	0.3874	0.3474	0.2684	0.1877	0.1211	0.0725	0.0403	0.0098	9
	2	0.0746	0.1937	0.2759	0.3020	0.2816	0.2335	0.1757	0.1209	0.0439	8
	3	0.0105	0.0574	0.1298	0.2013	0.2503	0.2668	0.2522	0.2150	0.1172	7
	4	0.0010	0.0112	0.0401	0.0881	0.1460	0.2001	0.2377	0.2508	0.2051	6
	5	0.0001	0.0015	0.0085	0.0264	0.0584	0.1029	0.1536	0.2007	0.2461	5
	6	0.0000	0.0001	0.0012	0.0055	0.0162	0.0368	0.0689	0.1115	0.2051	4
	7	0.0000	0.0000	0.0001	0.0008	0.0031	0.0090	0.0212	0.0425	0.1172	3
	8	0.0000	0.0000	0.0000	0.0001	0.0004	0.0014	0.0043	0.0106	0.0439	2
	9	0.0000	0.0000	0.0000	0.0000	0.0000	0.0001	0.0005	0.0016	0.0098	1
	10	0.0000	0.0000	0.0000	0.0000	0.0000	0.0000	0.0000	0.0001	0.0010	0
11	0	0.5688	0.3138	0.1673	0.0859	0.0422	0.0198	0.0088	0.0036	0.0004	11
	1	0.3293	0.3835	0.3248	0.2362	0.1549	0.0932	0.0518	0.0266	0.0055	10
	2	0.0867	0.2131	0.2866	0.2953	0.2581	0.1998	0.1395	0.0887	0.0269	9
	3	0.0137	0.0710	0.1517	0.2215	0.2581	0.2568	0.2254	0.1774	0.0806	8
	4	0.0014	0.0158	0.0536	0.1107	0.1721	0.2201	0.2428	0.2365	0.1611	7
	5	0.0001	0.0025	0.0132	0.0388	0.0803	0.1321	0.1830	0.2207	0.2256	6
	6	0.0000	0.0003	0.0023	0.0097	0.0268	0.0566	0.0985	0.1471	0.2256	5
	7	0.0000	0.0000	0.0003	0.0017	0.0064	0.0173	0.0379	0.0701	0.1611	4
	8	0.0000	0.0000	0.0000	0.0002	0.0011	0.0037	0.0102	0.0234	0.0806	3
	9	0.0000	0.0000	0.0000	0.0000	0.0001	0.0005	0.0018	0.0052	0.0269	2
	10	0.0000	0.0000	0.0000	0.0000	0.0000	0.0000	0.0002	0.0007	0.0054	1
	11	0.0000	0.0000	0.0000	0.0000	0.0000	0.0000	0.0000	0.0000	0.0005	0
12	0	0.5404	0.2824	0.1422	0.0687	0.0317	0.0138	0.0057	0.0022	0.0002	12
	1	0.3413	0.3766	0.3012	0.2062	0.1267	0.0712	0.0368	0.0174	0.0029	11
	2	0.0388	0.2901	0.2924	0.2835	0.2323	0.1678	0.1088	0.0639	0.0161	10
	3	0.0173	0.0852	0.1720	0.2362	0.2581	0.2397	0.1954	0.1419	0.0537	9
	4	0.0021	0.0213	0.0683	0.1329	0.1936	0.2311	0.2367	0.2128	0.1208	8
	5	0.0002	0.0038	0.0193	0.0532	0.1032	0.1585	0.2039	0.2270	0.1934	7
	6	0.0000	0.0005	0.0040	0.0155	0.0401	0.0792	0.1281	0.1766	0.2256	6
	7	0.0000	0.0000	0.0006	0.0033	0.0115	0.0291	0.0591	0.1009	0.1934	5
	8	0.0000	0.0000	0.0001	0.0005	0.0024	0.0078	0.0190	0.0420	0.1208	4
	9	0.0000	0.0000	0.0000	0.0001	0.0004	0.0015	0.0048	0.0125	0.0537	3
	10	0.0000	0.0000	0.0000	0.0000	0.0000	0.0002	0.0008	0.0025	0.0161	2
	11	0.0000	0.0000	0.0000	0.0000	0.0000	0.0000	0.0001	0.0003	0.0029	1
	12	0.0000	0.0000	0.0000	0.0000	0.0000	0.0000	0.0000	0.0000	0.0002	0
		0.95	0.90	0.85	0.80	0.75	0.70	0.65	0.60	0.50	
						p					

Table A.8 (continued)

n	x	0.05	0.10	0.15	0.20	p 0.25	0.30	0.35	0.40	0.50	x
13	0	0.5133	0.2542	0.1209	0.0550	0.0238	0.0097	0.0037	0.0013	0.0001	13
	1	0.3512	0.3672	0.2774	0.1787	0.1029	0.0540	0.0259	0.0113	0.0016	12
	2	0.1109	0.2448	0.2937	0.2680	0.2059	0.1388	0.0836	0.0453	0.0095	11
	3	0.0214	0.0997	0.1900	0.2457	0.2517	0.2181	0.1651	0.1107	0.0349	10
	4	0.0028	0.0277	0.0838	0.1535	0.2097	0.2337	0.2222	0.1845	0.0873	9
	5	0.0003	0.0055	0.0266	0.0691	0.1258	0.1803	0.2154	0.2214	0.1571	8
	6	0.0000	0.0008	0.0063	0.0230	0.0559	0.1030	0.1546	0.1968	0.2095	7
	7	0.0000	0.0001	0.0011	0.0058	0.0186	0.0442	0.0833	0.1312	0.2095	6
	8	0.0000	0.0000	0.0001	0.0011	0.0047	0.0142	0.0336	0.0656	0.1571	5
	9	0.0000	0.0000	0.0000	0.0001	0.0009	0.0034	0.0101	0.0243	0.0873	4
	10	0.0000	0.0000	0.0000	0.0000	0.0001	0.0006	0.0022	0.0065	0.0349	3
	11	0.0000	0.0000	0.0000	0.0000	0.0000	0.0001	0.0003	0.0012	0.0095	2
	12	0.0000	0.0000	0.0000	0.0000	0.0000	0.0000	0.0000	0.0001	0.0016	1
	13	0.0000	0.0000	0.0000	0.0000	0.0000	0.0000	0.0000	0.0000	0.0001	0
14	0	0.4877	0.2288	0.1028	0.0440	0.0178	0.0068	0.0024	0.0008	0.0001	14
	1	0.3593	0.3559	0.2539	0.1539	0.0832	0.0407	0.0181	0.0073	0.0009	13
	2	0.1229	0.2570	0.2912	0.2501	0.1802	0.1134	0.0634	0.0317	0.0056	12
	3	0.0259	0.1142	0.2056	0.2501	0.2402	0.1943	0.1366	0.0845	0.0222	11
	4	0.0037	0.0349	0.0998	0.1720	0.2202	0.2290	0.2022	0.1549	0.0611	10
	5	0.0004	0.0078	0.0352	0.0860	0.1468	0.1963	0.2178	0.2066	0.1222	9
	6	0.0000	0.0013	0.0093	0.0322	0.0734	0.1262	0.1759	0.2066	0.1833	8
	7	0.0000	0.0002	0.0019	0.0092	0.0280	0.0618	0.1082	0.1574	0.2095	7
	8	0.0000	0.0000	0.0003	0.0020	0.0082	0.0232	0.0510	0.0918	0.1833	6
	9	0.0000	0.0000	0.0000	0.0003	0.0018	0.0066	0.0183	0.0408	0.1222	5
	10	0.0000	0.0000	0.0000	0.0000	0.0003	0.0014	0.0049	0.0136	0.0611	4
	11	0.0000	0.0000	0.0000	0.0000	0.0000	0.0002	0.0010	0.0033	0.0222	3
	12	0.0000	0.0000	0.0000	0.0000	0.0000	0.0000	0.0001	0.0005	0.0056	2
	13	0.0000	0.0000	0.0000	0.0000	0.0000	0.0000	0.0000	0.0001	0.0009	1
	14	0.0000	0.0000	0.0000	0.0000	0.0000	0.0000	0.0000	0.0000	0.0001	0
15	0	0.4633	0.2059	0.0874	0.0352	0.0134	0.0047	0.0016	0.0005	0.0000	15
	1	0.3658	0.3432	0.2312	0.1319	0.0668	0.0305	0.0126	0.0047	0.0005	14
	2	0.1348	0.2669	0.2856	0.2309	0.1559	0.0916	0.0476	0.0219	0.0032	13
	3	0.0307	0.1285	0.2184	0.2501	0.2252	0.1700	0.1110	0.0634	0.0139	12
	4	0.0049	0.0428	0.1156	0.1876	0.2252	0.2186	0.1792	0.1268	0.0417	11
	5	0.0006	0.0105	0.0449	0.1032	0.1651	0.2061	0.2123	0.1859	0.0916	10
	6	0.0000	0.0019	0.0132	0.0430	0.0917	0.1472	0.1906	0.2066	0.1527	9
	7	0.0000	0.0003	0.0030	0.0138	0.0393	0.0811	0.1319	0.1771	0.1964	8
	8	0.0000	0.0000	0.0005	0.0035	0.0131	0.0348	0.0710	0.1181	0.1964	7
	9	0.0000	0.0000	0.0001	0.0007	0.0034	0.0116	0.0298	0.0612	0.1527	6
	10	0.0000	0.0000	0.0000	0.0001	0.0007	0.0030	0.0096	0.0245	0.0916	5
	11	0.0000	0.0000	0.0000	0.0000	0.0001	0.0006	0.0024	0.0074	0.0417	4
	12	0.0000	0.0000	0.0000	0.0000	0.0000	0.0001	0.0004	0.0016	0.0139	3
	13	0.0000	0.0000	0.0000	0.0000	0.0000	0.0000	0.0001	0.0003	0.0032	2
	14	0.0000	0.0000	0.0000	0.0000	0.0000	0.0000	0.0000	0.0000	0.0005	1
	15	0.0000	0.0000	0.0000	0.0000	0.0000	0.0000	0.0000	0.0000	0.0000	0
		0.95	0.90	0.85	0.80	0.75	0.70	0.65	0.60	0.50	

p

Table A.8 (continued)

n	x	0.05	0.10	0.15	0.20	p 0.25	0.30	0.35	0.40	0.50	x
16	0	0.4401	0.1853	0.0743	0.0281	0.0100	0.0033	0.0010	0.0003	0.0000	16
	1	0.3706	0.3294	0.2097	0.1126	0.0535	0.0228	0.0087	0.0030	0.0002	15
	2	0.1463	0.2745	0.2775	0.2111	0.1336	0.0732	0.0353	0.0150	0.0018	14
	3	0.0359	0.1423	0.2285	0.2463	0.2079	0.1465	0.0888	0.0468	0.0085	13
	4	0.0061	0.0514	0.1311	0.2001	0.2252	0.2040	0.1553	0.1014	0.0278	12
	5	0.0008	0.0137	0.0555	0.1281	0.1802	0.2099	0.2008	0.1623	0.0667	11
	6	0.0001	0.0028	0.0180	0.0550	0.1101	0.1649	0.1982	0.1983	0.1222	10
	7	0.0000	0.0004	0.0045	0.0197	0.0524	0.1010	0.1524	0.1889	0.1746	9
	8	0.0000	0.0001	0.0009	0.0055	0.0197	0.0487	0.0923	0.1417	0.1964	8
	9	0.0000	0.0000	0.0001	0.0012	0.0058	0.0185	0.0442	0.0840	0.1746	7
	10	0.0000	0.0000	0.0000	0.0002	0.0014	0.0056	0.0167	0.0392	0.1222	6
	11	0.0000	0.0000	0.0000	0.0000	0.0002	0.0013	0.0049	0.0142	0.0667	5
	12	0.0000	0.0000	0.0000	0.0000	0.0000	0.0002	0.0011	0.0040	0.0278	4
	13	0.0000	0.0000	0.0000	0.0000	0.0000	0.0000	0.0002	0.0008	0.0085	3
	14	0.0000	0.0000	0.0000	0.0000	0.0000	0.0000	0.0000	0.0001	0.0018	2
	15	0.0000	0.0000	0.0000	0.0000	0.0000	0.0000	0.0000	0.0000	0.0002	1
	16	0.0000	0.0000	0.0000	0.0000	0.0000	0.0000	0.0000	0.0000	0.0000	0
17	0	0.4181	0.1668	0.0631	0.0225	0.0075	0.0023	0.0007	0.0002	0.0000	17
	1	0.3741	0.3150	0.1893	0.0957	0.0426	0.0169	0.0060	0.0019	0.0001	16
	2	0.1575	0.2800	0.2673	0.1914	0.1136	0.0581	0.0260	0.0102	0.0010	15
	3	0.0415	0.1556	0.2359	0.2393	0.1893	0.1245	0.0701	0.0341	0.0052	14
	4	0.9076	0.0605	0.1457	0.2093	0.2209	0.1868	0.1320	0.0796	0.0182	13
	5	0.0010	0.0175	0.0668	0.1361	0.1914	0.2081	0.1849	0.1379	0.0472	12
	6	0.0001	0.0039	0.0236	0.0680	0.1276	0.1784	0.1991	0.1839	0.0944	11
	7	0.0000	0.0007	0.0065	0.0267	0.0668	0.1201	0.1685	0.1927	0.1484	10
	8	0.0000	0.0001	0.0014	0.0084	0.0279	0.0644	0.1134	0.1606	0.1855	9
	9	0.0000	0.0000	0.0003	0.0021	0.0093	0.0276	0.0611	0.1070	0.1855	8
	10	0.0000	0.0000	0.0000	0.0004	0.0025	0.0095	0.0263	0.0571	0.1484	7
	11	0.0000	0.0000	0.0000	0.0001	0.0005	0.0026	0.0090	0.0242	0.0944	6
	12	0.0000	0.0000	0.0000	0.0000	0.0001	0.0006	0.0024	0.0081	0.0472	5
	13	0.0000	0.0000	0.0000	0.0000	0.0000	0.0001	0.0005	0.0021	0.0182	4
	14	0.0000	0.0000	0.0000	0.0000	0.0000	0.0000	0.0001	0.0004	0.0052	3
	15	0.0000	0.0000	0.0000	0.0000	0.0000	0.0000	0.0000	0.0001	0.0010	2
	16	0.0000	0.0000	0.0000	0.0000	0.0000	0.0000	0.0000	0.0000	0.0001	1
	17	0.0000	0.0000	0.0000	0.0000	0.0000	0.0000	0.0000	0.0000	0.0000	0
18	0	0.3972	0.1501	0.0536	0.0180	0.0056	0.0016	0.0004	0.0001	0.0000	18
	1	0.3763	0.3002	0.1704	0.0811	0.0338	0.0126	0.0042	0.0012	0.0001	17
	2	0.1683	0.2835	0.2556	0.1723	0.0958	0.0458	0.0190	0.0069	0.0006	16
	3	0.0473	0.1680	0.2406	0.2297	0.1704	0.1046	0.0547	0.0246	0.0031	15
	4	0.0093	0.0700	0.1592	0.2153	0.2130	0.1681	0.1104	0.0644	0.0117	14
	5	0.0014	0.0218	0.0787	0.1507	0.1988	0.2017	0.1664	0.1146	0.0327	13
	6	0.0002	0.0052	0.0301	0.0816	0.1436	0.1873	0.1941	0.1655	0.0708	12
	7	0.0000	0.0010	0.0091	0.0350	0.0820	0.1376	0.1792	0.1892	0.1214	11
	8	0.0000	0.0002	0.0022	0.0120	0.0376	0.0811	0.1327	0.1734	0.1669	10
	9	0.0000	0.0000	0.0001	0.0033	0.0139	0.0386	0.0794	0.1284	0.1855	9
	10	0.0000	0.0000	0.0001	0.0008	0.0042	0.0149	0.0385	0.0771	0.1669	8
	11	0.0000	0.0000	0.0000	0.0001	0.0010	0.0046	0.0151	0.0374	0.1214	7
		0.95	0.90	0.85	0.80	0.75	0.70	0.65	0.60	0.50	
						p					

Table A.8 *(continued)*

n	x	0.05	0.10	0.15	0.20	p 0.25	0.30	0.35	0.40	0.50	x
18	12	0.0000	0.0000	0.0000	0.0000	0.0002	0.0012	0.0047	0.0145	0.0708	6
	13	0.0000	0.0000	0.0000	0.0000	0.0000	0.0002	0.0012	0.0045	0.0327	5
	14	0.0000	0.0000	0.0000	0.0000	0.0000	0.0000	0.0002	0.0011	0.0117	4
	15	0.0000	0.0000	0.0000	0.0000	0.0000	0.0000	0.0000	0.0002	0.0031	3
	16	0.0000	0.0000	0.0000	0.0000	0.0000	0.0000	0.0000	0.0000	0.0006	2
	17	0.0000	0.0000	0.0000	0.0000	0.0000	0.0000	0.0000	0.0000	0.0001	1
	18	0.0000	0.0000	0.0000	0.0000	0.0000	0.0000	0.0000	0.0000	0.0000	0
19	0	0.3774	0.1351	0.0456	0.0144	0.0042	0.0011	0.0003	0.0001	0.0000	19
	1	0.3774	0.2852	0.1529	0.0685	0.0268	0.0093	0.0029	0.0008	0.0000	18
	2	0.1787	0.2852	0.2428	0.1540	0.0803	0.0358	0.0138	0.0046	0.0003	17
	3	0.0533	0.1796	0.2428	0.2182	0.1517	0.0869	0.0422	0.0175	0.0018	16
	4	0.0112	0.0798	0.1714	0.2182	0.2023	0.1491	0.0909	0.0467	0.0074	15
	5	0.0018	0.0266	0.0907	0.1636	0.2023	0.1916	0.1468	0.0933	0.0222	14
	6	0.0002	0.0069	0.0374	0.0955	0.1574	0.1916	0.1844	0.1451	0.0518	13
	7	0.0000	0.0014	0.0122	0.0443	0.0974	0.1525	0.1844	0.1797	0.0961	12
	8	0.0000	0.0002	0.0032	0.0166	0.0487	0.0981	0.1489	0.1797	0.1442	11
	9	0.0000	0.0000	0.0007	0.0051	0.0198	0.0514	0.0980	0.1464	0.1762	10
	10	0.0000	0.0000	0.0001	0.0013	0.0066	0.0220	0.0528	0.0976	0.1762	9
	11	0.0000	0.0000	0.0000	0.0003	0.0018	0.0077	0.0233	0.0532	0.1442	8
	12	0.0000	0.0000	0.0000	0.0000	0.0004	0.0022	0.0083	0.0237	0.0961	7
	13	0.0000	0.0000	0.0000	0.0000	0.0001	0.0005	0.0024	0.0085	0.0518	6
	14	0.0000	0.0000	0.0000	0.0000	0.0000	0.0001	0.0006	0.0024	0.0222	5
	15	0.0000	0.0000	0.0000	0.0000	0.0000	0.0000	0.0001	0.0005	0.0074	4
	16	0.0000	0.0000	0.0000	0.0000	0.0000	0.0000	0.0000	0.0001	0.0018	3
	17	0.0000	0.0000	0.0000	0.0000	0.0000	0.0000	0.0000	0.0000	0.0003	2
	18	0.0000	0.0000	0.0000	0.0000	0.0000	0.0000	0.0000	0.0000	0.0000	1
	19	0.0000	0.0000	0.0000	0.0000	0.0000	0.0000	0.0000	0.0000	0.0000	0
20	0	0.3585	0.1216	0.0388	0.0115	0.0032	0.0008	0.0002	0.0000	0.0000	20
	1	0.3774	0.2702	0.1368	0.0576	0.0211	0.0068	0.0020	0.0005	0.0000	19
	2	0.1887	0.2852	0.2293	0.1369	0.0669	0.0278	0.0100	0.0031	0.0002	18
	3	0.0596	0.1901	0.2428	0.2054	0.1339	0.0716	0.0323	0.0123	0.0011	17
	4	0.0133	0.0898	0.1821	0.2182	0.1897	0.1304	0.0738	0.0350	0.0046	16
	5	0.0022	0.0319	0.1028	0.1746	0.2023	0.1789	0.1272	0.0746	0.0148	15
	6	0.0003	0.0089	0.0454	0.1091	0.1686	0.1916	0.1712	0.1244	0.0370	14
	7	0.0000	0.0020	0.0160	0.0545	0.1124	0.1643	0.1844	0.1659	0.0739	13
	8	0.0000	0.0004	0.0046	0.0222	0.0609	0.1144	0.1614	0.1797	0.1201	12
	9	0.0000	0.0001	0.0011	0.0074	0.0271	0.0654	0.1158	0.1597	0.1602	11
	10	0.0000	0.0000	0.0002	0.0020	0.0099	0.0308	0.0686	0.1171	0.1762	10
	11	0.0000	0.0000	0.0000	0.0005	0.0030	0.0120	0.0336	0.0710	0.1602	9
	12	0.0000	0.0000	0.0000	0.0001	0.0008	0.0039	0.0136	0.0355	0.1201	8
	13	0.0000	0.0000	0.0000	0.0000	0.0002	0.0010	0.0045	0.0146	0.0739	7
	14	0.0000	0.0000	0.0000	0.0000	0.0000	0.0002	0.0012	0.0049	0.0370	6
	15	0.0000	0.0000	0.0000	0.0000	0.0000	0.0000	0.0003	0.0013	0.0148	5
	16	0.0000	0.0000	0.0000	0.0000	0.0000	0.0000	0.0000	0.0003	0.0046	4
	17	0.0000	0.0000	0.0000	0.0000	0.0000	0.0000	0.0000	0.0000	0.0011	3
	18	0.0000	0.0000	0.0000	0.0000	0.0000	0.0000	0.0000	0.0000	0.0002	2
	19	0.0000	0.0000	0.0000	0.0000	0.0000	0.0000	0.0000	0.0000	0.0000	1
	20	0.0000	0.0000	0.0000	0.0000	0.0000	0.0000	0.0000	0.0000	0.0000	0
		0.95	0.90	0.85	0.80	0.75	0.70	0.65	0.60	0.50	

p

A.9 Simple Estimators for the Mean and Standard Deviation

These tables give the efficiencies and variances associated with a number of estimators that do not require the use of all of the individuals. Separate tables are given for:

1. Small sample estimators of the mean (the median, midrange and the best of two).
2. Small sample estimators of the standard deviation using the range.
3. Unbiased estimators of the standard deviation using s (standard deviation) of the sample.

(1) *Estimators of the mean* (Table A.9a): For various samples of size n, the efficiencies of the median, midrange, and the best pair of two values are given. The columns labeled variances (var.) give the value to multiply the population variance by in order to obtain the variance of the estimators (that is, the variance of the median, midrange, and average of 2).

EXAMPLE 1

Given the following values:

$$76, 81, 47, 83, 56, 64, 78.$$

The ordering of these is:

$$47, 56, 64, 76, 78, 80, 83.$$

From Table A.9a for $n = 7$

	Median		Midrange		Average of the best two		
	Var.	Eff.	Var.	Eff.	Statistic	Var.	Eff.
7	0.210	0.679	0.218	0.654	$\frac{1}{2}(X_2 + X_6)$	0.168	0.849

The median is 76 with an efficiency of 0.679 (and the variance of the possible sample medians is $0.210\sigma^2$ where σ is the population variance). The midrange is 65 with an efficiency of 0.679 (and the variance of the possible sample midranges is $0.218\sigma^2$).

The best possible average using only 2 values is

$$\tfrac{1}{2}(X_2 + X_6) = \tfrac{1}{2}(56 + 80) = 68$$

with an efficiency of 0.849 (or the variance of possible samples of this type is $0.168\sigma^2$).

Note that the mean is 69.1 (and the variance of possible means is $\sigma^2/n = 0.143\sigma^2$).

(2) *Estimators of the standard deviation—the range* (Table A.9b): For various sample sizes, values are given to multiply the range by in order to give a best estimate of the standard deviation. The variance of these estimators and the efficiency is also given.

$$\text{Range} = 83 - 47 = 35.$$

From Table A.9b

Sample size	Constant to multiply range by	Variance of estimate	Efficiency
7	0.370	0.0949	0.911

The estimate of the standard deviation is $0.370R = 0.370(35) = 12.95$ (and the variance of this estimate of the standard deviation is $0.0949\sigma^2$). The efficiency of this estimate (compared to using s) is 0.911.

(3) *Unbiased estimate of the standard deviation using s* (Table A.9c): We found s^2 to be an unbiased estimate of σ^2 but a correction factor must be used to make s an unbiased estimator of σ.

EXAMPLE 2

The variance of a sample of 5 is found to be 16.00, therefore, $s = 4.00$. The unbiased estimate of σ is found from Table A.9c.

Sample size	Estimate	Variance of the estimate
5	1.064 s	$0.132\,\sigma^2$

Thus $\sigma = 1.064, s = 1.064(4.00) = 4.256.$
The variance of this *estimate* of σ is $0.132\sigma^2 = 0.132(16) = 2.11$.

Table A.9a Several estimates of the mean (variance to be multiplied by σ^2)

n	Median Var.	Median Eff.	Midrange Var.	Midrange Eff.	Average of best two Statistic	Average of best two Var.	Average of best two Eff.	$(x_2 + x_3 + \cdots + x_{n-1})/(n-2)$ Var.	$(x_2 + x_3 + \cdots + x_{n-1})/(n-2)$ Eff.
2	0.500	1.000	0.500	1.000	$\frac{1}{2}(x_1 + x_2)$	0.500	1.000		
3	0.449	0.743	0.362	0.920	$\frac{1}{2}(x_1 + x_3)$	0.362	0.920	0.449	0.743
4	0.298	0.838	0.298	0.838	$\frac{1}{2}(x_2 + x_3)$	0.298	0.838	0.298	0.838
5	0.287	0.697	0.261	0.767	$\frac{1}{2}(x_2 + x_4)$	0.231	0.867	0.227	0.881
6	0.215	0.776	0.236	0.706	$\frac{1}{2}(x_2 + x_5)$	0.193	0.865	0.184	0.906
7	0.210	0.679	0.218	0.654	$\frac{1}{2}(x_2 + x_6)$	0.168	0.849	0.155	0.922
8	0.168	0.743	0.205	0.610	$\frac{1}{2}(x_3 + x_6)$	0.149	0.837	0.134	0.934
9	0.166	0.669	0.194	0.572	$\frac{1}{2}(x_3 + x_7)$	0.132	0.843	0.118	0.942
10	0.138	0.723	0.186	0.539	$\frac{1}{2}(x_3 + x_8)$	0.119	0.840	0.105	0.949
11	0.137	0.663	0.178	0.510	$\frac{1}{2}(x_3 + x_9)$	0.109	0.832	0.0952	0.955
12	0.118	0.709	0.172	0.484	$\frac{1}{2}(x_4 + x_9)$	0.100	0.831	0.0869	0.959
13	0.117	0.659	0.167	0.461	$\frac{1}{2}(x_4 + x_{10})$	0.0924	0.833	0.0799	0.963
14	0.102	0.699	0.162	0.440	$\frac{1}{2}(x_4 + x_{11})$	0.0860	0.830	0.0739	0.966
15	0.102	0.656	0.158	0.422	$\frac{1}{2}(x_4 + x_{12})$	0.0808	0.825	0.0688	0.969
16	0.0904	0.692	0.154	0.392	$\frac{1}{2}(x_5 + x_{12})$	0.0756	0.827	0.0644	0.971
17	0.0901	0.653	0.151	0.389	$\frac{1}{2}(x_5 + x_{13})$	0.0711	0.827	0.0605	0.973
18	0.0810	0.686	0.148	0.375	$\frac{1}{2}(x_5 + x_{14})$	0.0673	0.825	0.0570	0.975
19	0.0808	0.651	0.145	0.362	$\frac{1}{2}(x_6 + x_{14})$	0.0640	0.823	0.0539	0.976
20	0.0734	0.681	0.143	0.350	$\frac{1}{2}(x_6 + x_{15})$	0.0607	0.824	0.0511	0.978
∞	$\dfrac{1.57}{n}$	0.637		0.000	$\frac{1}{2}(P_{25} + P_{75})$	$\dfrac{1.24}{n}$	0.808		1.000

Source: From *Introduction to Statistical Analysis*, 3rd ed., by Dixon & Massey, Copyright © 1969 by McGraw-Hill, Inc. Used by permission of McGraw-Hill Book Co.

Table A.9b Unbiased estimate of σ using range (multiply variance by σ^2)

Sample size	Constant to multiply range	Variance of estimate	Eff.	Sample size	Constant to multiply range	Variance of estimate	Eff.
2	0.886	0.571	1.000	11	0.315	0.0616	0.831
3	0.591	0.275	0.992	12	0.307	0.0571	0.814
4	0.486	0.183	0.975	13	0.300	0.0533	0.797
5	0.430	0.138	0.955	14	0.294	0.0502	0.781
6	0.395	0.112	0.993	15	0.288	0.0474	0.766
7	0.370	0.0949	0.911	16	0.283	0.0451	0.751
8	0.351	0.0829	0.890	17	0.279	0.0430	0.738
9	0.337	0.0740	0.869	18	0.275	0.0412	0.725
10	0.325	0.0671	0.850	19	0.271	0.0395	0.712
				20	0.268	0.0381	0.700

Table A.9c Unbiased estimate of σ based on s

n	Estimate	Variance	n	Estimate	Variance
2	$1.253\,s$	$0.571\,\sigma^2$	7	$1.042\,s$	$0.0865\,\sigma^2$
3	$1.128\,s$	$0.273\,\sigma^2$	8	$1.036\,s$	$0.0738\,\sigma^2$
4	$1.085\,s$	$0.178\,\sigma^2$	9	$1.032\,s$	$0.0643\,\sigma^2$
5	$1.064\,s$	$0.132\,\sigma^2$	10	$1.028\,s$	$0.0570\,\sigma^2$
6	$1.051\,s$	$0.104\,\sigma^2$	∞	$\left[1 + \dfrac{1}{4(n-1)}\right]s$	$\sigma^2/2n$

Source: Reproduced with permission from E. S. Pearson "The Probability Integral of the Range in Samples of n Observations From a Normal Population", *Biometrika* (1942) 301.

A.10 Control Chart Factors and Symbols

Symbols

Individuals	X
Mean of a set of individuals	$\bar{X}$
Mean of a set of means	$\bar{\bar{X}}$
Range of a set of individuals	R
Mean of a set of ranges	$\bar{R}$

Control Limits for Range and Means Control Charts

	Control chart limits	
	Upper control limit	Lower control limit
Range chart	$D_4\bar{R}$	$D_3\bar{R}$
Means chart	$\bar{X} + A_2\bar{R}$	$\bar{X} - A_2\bar{R}$

The D and A are constants which depend on the sample size. Consult the accompanying table for the appropriate values.

To estimate σ (the standard deviation) from $\bar{R}$, use the following formula:

$$\sigma = \bar{R}/d_2.$$

To establish a control chart (for a set of at least about 20 samples all of size n), follow these steps:

1. Calculate $\bar{R}$.
2. Look up D_3 and D_4.
3. Calculate the upper and lower limits.
4. Plot the values of R to see if any are outside the limits.
5. If no points are outside the limits, the range is in control. If any points are outside the limits, continue to investigate.
6. Calculate $\bar{X}$.
7. Look up A_2.

8. Calculate the limits for $\bar{X}$.
9. Plot the points on to see if any are outside the limits.
10. If no means are outside the limits then conclude the means are within control, otherwise investigate.

Table A.10 Control chart factors

Number of individuals in each sample, n	$\bar{R}$-chart		Means chart	Estimator of σ
	D_3	D_4	A_2	d_2
2	0	3.267	1.880	1.128
3	0	2.575	1.023	1.693
4	0	2.282	0.729	2.059
5	0	2.115	0.577	2.326
6	0	2.004	0.483	2.534
7	0.076	1.924	0.419	2.704
8	0.136	1.864	0.373	2.847
9	0.184	1.816	0.337	2.970
10	0.223	1.777	0.308	3.078

SOURCE: These values were extracted from ASTM. *Manual on Quality Control of Material*, STP 15C, Philadelphia, Pa., 1951; with permission.

A.11 Analysis of Means Limits

These tables are for finding the limits for comparing sets of means in the analysis of means. Two probability levels 0.05 and 0.01 are provided.

Two cases are considered:

1. Standard deviation is known (Table A.11a).
2. Standard deviation is unknown and must be calculated from the samples (the average range is to be used) (Table A.11b).

k is the number of samples being considered.
n is the size of each of the samples (all samples of the same size).
$\bar{\bar{X}}$ is the mean of all of the sample means.

(1) *Standard deviation is known and is equal to σ*: To compare a collection of k means we calculate limits as

$$\bar{\bar{X}} + Z_\alpha \frac{\sigma}{\sqrt{n}} \quad \text{and} \quad \bar{\bar{X}} - Z_\alpha \frac{\sigma}{\sqrt{n}}.$$

Z_α-values are found in Table A.11a (for either $Z_{0.05}$ or $Z_{0.01}$).

EXAMPLE 1

Given 5 samples of 4 individuals each, with an overall mean of 10.00 and $\sigma = 2.5$, what are the limits that the means should fall within? From Table

A.11a for $k = 5$

k	Z-values for	
	0.05 limits	0.01 limits
5	2.57	3.09

$$\bar{\bar{X}} = 10.00, \quad \sigma = 2.5, \quad n = 4.$$

The 0.05 (5% level of significance) limits are

$$\bar{\bar{X}} + Z_{0.05}\frac{\sigma}{\sqrt{n}} = 10 + 2.57\frac{(2.5)}{\sqrt{4}} = 10 + 3.21 = 13.21$$

$$\bar{\bar{X}} - Z_{0.05}\frac{\sigma}{\sqrt{n}} = 10 - 2.57\frac{(2.5)}{\sqrt{4}} = 10 - 3.21 = 6.79$$

(2) *Standard deviation is unknown:* Calculate the estimate of the standard deviation from the average range of the samples being compared ($\bar{R}$).

The limits are

$$\bar{\bar{X}} + H_\alpha \frac{\bar{R}}{d_2\sqrt{n}} \quad \text{and} \quad \bar{\bar{X}} - H_\alpha \frac{\bar{R}}{d_2\sqrt{n}},$$

where d_2 is found in Table A.10.★

H_α-values are found in Table A.11b (0.05 or 0.01 limits). Degrees of freedom are approximately equal to $0.90(k)(n-1)$.

EXAMPLE 1

For the 5 samples of 4 values the $\bar{R}$ was found to be 4.8. To find H_α we need the approximate degrees of freedom.

$$df = 0.9(k)(n-1) = 0.9(5)(4-1) = 0.9(5)(3) = 13.5.$$

For the $\alpha = 0.05$ (5% level) we find in Table A.11b, for 0.05:

		k
		5
df	10	2.78
	15	2.60

★ It should be noted that d_2 should be modified for small numbers of samples. However, the use of d_2 will give conservative, larger limits and actually provide smaller values of alpha error.

Since $df = 13.5$ is between the 10 and 15 we must interpolate between the 2.78 and 2.60. (The actual H_α-value would be $3.5/5.0 = 7/10$ths of the way between these values. The difference would be

$$(7/10)(2.78 - 2.60) = (7/10)(0.18) = 0.13.$$

Therefore

$$H_\alpha = 2.78 - 0.13 = 2.65).$$

From Table A.10, $d_2 = 2.059$.

Table A.11a Standard deviation known

k Number of categories	Z-values (two-tailed)	
	0.05 limits	0.01 limits
2	2.24	2.81
3	2.39	2.93
4	2.49	3.02
5	2.57	3.09
6	2.63	3.14
7	2.68	3.19
8	2.73	3.22
9	2.77	3.26
10	2.80	3.29
15	2.93	3.40
20	3.02	3.48
25	3.08	3.54
30	3.14	3.58
50	3.28	3.72
120	3.52	3.94

The limits are

$$\bar{\bar{X}} + H_\alpha \frac{\bar{R}}{d_2\sqrt{n}} = 10 + 2.65 \frac{(4.8)}{(2.059)(\sqrt{4})} = 10 + 3.07 = 13.07$$

$$\bar{\bar{X}} - H_\alpha \frac{\bar{R}}{d_2\sqrt{n}} = 10 - 2.65 \frac{(4.8)}{(2.059)(\sqrt{4})} = 10 - 3.07 = 6.93$$

Table A.11b Analysis of means limits

df \ k	2*	3	4	5	6	8	10	15	20	30	40	60
				H_α for $\alpha = 0.05$								
5	1.82											
6	1.73	2.59	2.94	3.19	3.37							
8	1.63	2.39	2.71	2.92	3.09	3.33						
10	1.58	2.29	2.58	2.78	2.93	3.15	3.31					
15	1.51	2.16	2.42	2.60	2.74	2.93	3.07	3.32				
20	1.48	2.10	2.35	2.52	2.64	2.83	2.96	3.18	3.33			
30	1.44	2.04	2.28	2.44	2.56	2.73	2.86	3.06	3.19	3.37		
40	1.43	2.01	2.25	2.40	2.52	2.69	2.80	3.00	3.13	3.29		
60	1.41	1.98	2.21	2.36	2.48	2.64	2.76	2.94	3.06	3.22		
120	1.40	1.95	2.18	2.33	2.44	2.60	2.71	2.88	3.00	3.15		
∞	1.39	1.93	2.15	2.29	2.40	2.55	2.65	2.82	2.94	3.08		

df \ k	2*	3	4	5	6	8	10	15	20	30	40	60
				H_α for $\alpha = 0.01$								
5	2.85											
6	2.62	3.74	4.21	4.53	4.78							
8	2.37	3.31	3.70	4.97	4.17	4.47						
10	2.24	3.08	3.43	3.67	3.86	4.11	4.29					
15	2.08	2.81	3.12	3.32	3.47	3.69	3.84	4.11				
20	2.01	2.70	3.98	3.17	3.30	3.50	3.63	3.87	4.02			
30	1.94	2.58	2.85	3.02	3.15	3.33	3.45	3.66	3.79	3.96		
40	1.91	2.53	2.79	2.95	3.07	3.24	3.36	3.56	3.68	3.84	3.96	
60	1.88	2.48	2.73	2.88	3.00	3.16	3.27	3.46	3.58	3.73	3.84	3.97
120	1.85	2.43	2.67	2.82	2.93	3.09	3.20	3.37	3.48	3.62	3.72	3.86
∞	1.82	2.39	2.61	2.76	2.87	3.02	3.12	3.29	3.39	3.53	3.62	3.73

SOURCE: Adapted by averaging the upper and lower bounds from "Tables of Percentage Points for the Studentized Maximum Absolute Deviate in Normal Samples," by Halpern, Greenhouse, Cornfield, Zalohar, *J. Amer. Statist. Ass.*, **50** (1955), p. 185, with the kind permission of the authors and publisher.

* The numbers corresponding to $k = 2$ are appropriate modifications of entries in Student t-tables.

A.12 Runs Above and Below the Median

A *run* consists of a set of consecutive points or individuals having values either all greater than or all less than the median value. We define N_1 as the total number of points above the median, which equals N_2, total number of points below the median. Given a total number of individuals n, then if n is even

$$N_1 = N_2 = n/2;$$

if n is odd

$$N_1 = N_2 = (n - 1)/2.$$

One- and two-tailed tests are possible.

For *one-tailed tests* you must decide before testing if you are interested only in detecting either small numbers of runs or large numbers of runs. Critical values for three significance levels are given—only one limit is to be found.

For *two-tailed tests* consider both possibilities simultaneously *and two* limits must be selected from the table.

EXAMPLE

For the following values perform a runs test at 0.05 level:

$$1, 4, 5, 4, 6, 0, 1, 5, 3, 2, 1, 2, 2, 5, 4.$$

The median is 3. The runs are shown as:

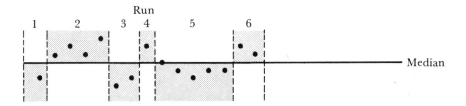

We therefore have 6 runs or $u = 6$. (Note that since there are 15 values $N_1 = N_2 = (15 - 1)/2 = 14/2 = 7$.)

One-tailed test: Test for too few runs at 0.05 level ($\alpha = 0.05$). From the table (see Table A.12) we find the limit $u_{0.05} = 4$. Since $u = 6$ is above this limit we find this is not unusual.

Two-tailed test: Test at 0.05 level ($\alpha = 0.05$). From the table we find:

$$\text{Lower limit} = 3$$

and

$$\text{Upper limit} = 13$$

Since $u = 6$ is greater than 3 and less than 13, it is *not* unusual.

	Significantly small values			Significantly large values		
	One-tailed test			One-tailed test		
$N_1 = N_2$	0.01	0.025	0.05	0.05	0.025	0.01
	Two-tailed test			Two-tailed test		
	0.02	0.05	0.10	0.10	0.05	0.02
7	3	3	4	12	13	13

Table A.12

Number of values above (N_1) and number of values below (N_2) are equal.	Significantly small values of u. If the number of runs is less than or equal to the value below, it is significant at the level indicated.			Significantly large values of u. If the number of runs is greater than or equal to the value below, it is significant at the level indicated.		
	One-tailed test			One-tailed test		
$N_1 = N_2$	0.01	0.025	0.05	0.05	0.025	0.01
	Two-tailed test			Two-tailed test		
	0.02	0.05	0.10	0.10	0.05	0.02
5	2	2	3	9	10	10
6	2	3	3	11	11	12
7	3	3	4	12	13	13
8	4	4	5	13	14	14
9	4	5	6	14	15	16
10	5	6	6	16	16	17
11	6	7	7	17	17	18
12	6	7	8	18	19	19
13	7	8	9	19	20	21
14	8	9	10	20	21	22
15	9	10	11	21	22	23
16	10	11	11	23	23	24
17	10	11	12	24	25	26
18	11	12	13	25	26	27
19	12	13	14	26	27	28
20	13	14	15	27	28	29
21	14	15	16	28	29	30
22	14	16	17	29	30	32
23	15	16	17	31	32	33
24	16	17	18	32	33	34
25	17	18	19	33	34	35
26	18	19	20	34	35	36
27	19	20	21	35	36	37
28	19	21	22	36	37	39
29	20	22	23	37	38	39
30	21	22	24	38	40	41
35	25	27	28	44	45	47
40	30	31	33	49	51	52
50	38	40	42	60	62	64
60	47	49	51	71	73	75
70	56	58	60	82	84	86
80	65	68	70	92	94	97
90	74	77	79	103	105	108
100	84	86	88	114	116	118

Source: Adapted from "Tables for Testing Randomness of Grouping in a Sequence of Alternatives" by Frieda S. Swed and C. Eisenhart, *Annals of Mathematical Statistics*, Vol. 14 (1943), pp. 67–87, with the kind permission of the authors and publishers.

A.13 Ordinates of the Normal Distribution

The ordinates or heights of the standard normal distribution for selected values of z are contained in this table. Since the normal distribution is symmetric, heights of all positive z-values are the same as heights of all negative z-values (see points $+1$ and -1 in Figure A.13, both have a height of 0.242). The table lists z as $\pm z$ to stand for both the $+$ and $-$ values. Figure A.13 shows heights for several examples.

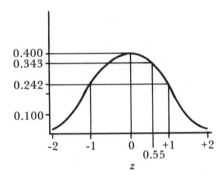

Table A.13 Ordinates of the normal distribution

z	Y	z	Y
0	0.399	±1.00	0.242
±0.05	0.398	±1.05	0.230
±0.10	0.397	±1.10	0.218
±0.15	0.394	±1.15	0.206
±0.20	0.391	±1.20	0.194
±0.25	0.387	±1.25	0.183
±0.30	0.381	±1.30	0.171
±0.35	0.375	±1.35	0.160
±0.40	0.368	±1.40	0.150
±0.45	0.361	±1.45	0.139
±0.50	0.352	±1.50	0.1295
±0.55	0.343	±1.55	0.1200
±0.60	0.333	±1.60	0.1109
±0.65	0.323	±1.65	0.1023
±0.70	0.312	±1.70	0.0940
±0.75	0.301	±1.75	0.0863
±0.80	0.290	±1.80	0.0790
±0.85	0.278	±1.85	0.0721
±0.90	0.266	±1.90	0.0656
±0.95	0.254	±1.95	0.0596

Table A.13 *(continued)*

z	Y	z	Y
±2.00	0.0540	±2.50	0.0175
±2.05	0.0488	±2.55	0.0154
±2.10	0.0440	±2.60	0.0136
±2.15	0.0396	±2.65	0.0119
±2.20	0.0355	±2.70	0.0104
±2.25	0.0317	±2.75	0.0091
±2.30	0.0283	±2.80	0.0079
±2.35	0.0252	±2.85	0.0069
±2.40	0.0224	±2.90	0.0060
±2.45	0.0198	±2.95	0.0051
		±3.00	0.0044

A.14 Test for Extreme Values: Critical Values

This table provides test statistics and critical limits for determining if one of the values of a small set of numbers appears excessively large or unusual. *The data must be ordered.*

The particular test statistic depends upon the size of the sample, n. Look down the column for n, the test statistic is given in the left column. Two sets of statistics are given depending on whether the extreme is the largest or the smallest value. Select the appropriate statistic. Critical values are found in the columns to the right.

EXAMPLE 1

The following 8 values of times were recorded

$$26, 31, 29, 49, 30, 34, 27, 29.$$

Is the 49 unusual? Placing them in order we have

X_1	X_2	X_3	X_4	X_5	X_6	X_7	X_8
26	27	29	29	30	31	34	49

From Table A.14, for $n = 8$ the statistic for a *largest* extreme value is

$$r_{11} = \frac{X_n - X_{n-1}}{X_n - X_2}.$$

Since $n = 8$,

$$r_{11} = \frac{X_8 - X_{8-1}}{X_8 - X_2} = \frac{X_8 - X_7}{X_8 - X_2}$$

$$= \frac{49 - 34}{49 - 26} = \frac{15}{23} = 0.653.$$

Consulting the table for $n = 8$ we find this value exceeds the 0.02 critical value of 0.631 but is less than the 0.01 critical value of 0.683. It is significant but not very highly significant.

		Probability level	
Test statistic	n	0.02	0.01
$r_{11} = \dfrac{X_n - X_{n-1}}{X_n - X_2}$	8	0.631	0.683

Table A.14 Test for extreme values: Critical values

Test statistic			Probability of a value as large as the table						
Extreme small	Extreme small	n	0.30	0.20	0.10	0.05	0.02	0.01	0.005
$r_{10} = \dfrac{X_2 - X_1}{X_n - X_1}$	$r_{10} = \dfrac{X_n - X_{n-1}}{X_n - X_1}$	3	0.684	0.781	0.886	0.941	0.976	0.988	0.994
		4	0.471	0.560	0.679	0.765	0.846	0.889	0.926
		5	0.373	0.451	0.557	0.642	0.729	0.780	0.821
		6	0.318	0.386	0.482	0.560	0.644	0.698	0.740
		7	0.281	0.344	0.434	0.507	0.586	0.637	0.680
$r_{11} = \dfrac{X_2 - X_1}{X_{n-1} - X_1}$	$r_{11} = \dfrac{X_n - X_{n-1}}{X_n - X_2}$	8	0.318	0.385	0.479	0.554	0.631	0.683	0.725
		9	0.288	0.352	0.441	0.512	0.587	0.635	0.677
		10	0.265	0.325	0.409	0.477	0.551	0.597	0.639
$r_{21} = \dfrac{X_3 - X_1}{X_{n-1} - X_1}$	$r_{21} = \dfrac{X_n - X_{n-2}}{X_n - X_2}$	11	0.391	0.442	0.517	0.576	0.638	0.679	0.713
		12	0.370	0.419	0.490	0.546	0.605	0.642	0.675
		13	0.351	0.399	0.467	0.521	0.578	0.615	0.649
$r_{22} = \dfrac{X_3 - X_1}{X_{n-2} - X_1}$	$r_{22} = \dfrac{X_n - X_{n-2}}{X_n - X_3}$	14	0.370	0.421	0.492	0.546	0.602	0.641	0.674
		15	0.353	0.402	0.472	0.525	0.579	0.616	0.647
		16	0.338	0.386	0.454	0.507	0.559	0.595	0.624
		17	0.325	0.373	0.438	0.490	0.542	0.577	0.605
		18	0.314	0.361	0.424	0.475	0.527	0.561	0.589
		19	0.304	0.350	0.412	0.462	0.514	0.547	0.575
		20	0.295	0.340	0.401	0.450	0.502	0.535	0.562
		21	0.287	0.331	0.391	0.440	0.491	0.524	0.551
		22	0.280	0.323	0.382	0.430	0.481	0.514	0.541
		23	0.274	0.316	0.374	0.421	0.472	0.505	0.532
		24	0.268	0.310	0.367	0.413	0.464	0.497	0.524
		25	0.262	0.304	0.360	0.406	0.457	0.489	0.516

SOURCE: Entries are extracted from W. J. Dixon, "Processing Data for Outliers," *Biometrics*, Vol. 9 (1952), p. 74, with the kind permission of the author and publishers.

A.15 Critical Values of Kolmogorov-Smirnov One-sample Test—D

This table provides critical values for several levels of significance for the statistic D in the Kolmogorov-Smirnov one-sample test. This test compares the relative cumulative frequencies of a theoretical distribution to the relative cumulative frequencies of an actual distribution. The *largest difference* is called D. For a given sample size, critically large values of D are given in Table A.15.

EXAMPLE 1

A theoretical distribution was believed to be 10%, 1; 20%, 2; 40%, 3; 20%, 4; and 10%, 5. Twenty-five actual values were six 5's, six 6's, eight 3's, three 2's, and two 1's. Test at alpha (α) of 0.10.

	5	4	3	2	1
1. Theoretical relative frequency	0.10	0.20	0.40	0.20	0.10
2. Theoretical relative cumulative frequency	0.10	0.30	0.70	0.90	1.00
3. Actual relative frequency	0.24	0.24	0.32	0.12	0.08
4. Actual relative cumulative frequency	0.24	0.48	0.80	0.92	1.00
5. Difference in relative cumulative frequency 4. − 2.	0.14	0.18	0.10	0.02	0.00

n	Level of significance
	0.10
25	0.24
over 35	$\dfrac{1.22}{\sqrt{n}}$

Largest difference = D = 0.18. This is less than the critical limit of 0.24 and we declare that there is no significant difference.

If n is more than 35: To find the critical limit substitute n in the appropriate formula.

EXAMPLE 2

If the same relative numbers of values were found in a sample of 140, then D still equals 0.18 but the critical limit is

$$\frac{1.22}{\sqrt{n}} = \frac{1.22}{\sqrt{140}} = \frac{1.22}{11.83} = 0.103.$$

We would conclude that there is a significant difference (at 0.10 level).

Table A.15 Critical values of Kolmogorov-Smirnov one-sample test

Sample size (n)	Level of significance for D				
	0.20	0.15	0.10	0.05	0.01
1	0.900	0.925	0.950	0.975	0.995
2	0.684	0.726	0.776	0.842	0.929
3	0.565	0.597	0.642	0.708	0.828
4	0.494	0.525	0.564	0.624	0.733
5	0.446	0.474	0.510	0.565	0.669
6	0.410	0.436	0.470	0.521	0.618
7	0.381	0.405	0.438	0.486	0.577
8	0.358	0.381	0.411	0.457	0.543
9	0.339	0.360	0.388	0.432	0.514
10	0.322	0.342	0.368	0.410	0.490
11	0.307	0.326	0.352	0.391	0.468
12	0.295	0.313	0.338	0.375	0.450
13	0.284	0.302	0.325	0.361	0.433
14	0.274	0.292	0.314	0.349	0.418
15	0.266	0.283	0.304	0.338	0.404
16	0.258	0.274	0.295	0.328	0.392
17	0.250	0.266	0.286	0.318	0.381
18	0.244	0.259	0.278	0.309	0.371
19	0.237	0.252	0.272	0.301	0.363
20	0.231	0.246	0.264	0.294	0.356
25	0.21	0.22	0.24	0.27	0.32
30	0.19	0.20	0.22	0.24	0.29
35	0.18	0.19	0.21	0.23	0.27
Over 35	$\dfrac{1.07}{\sqrt{n}}$	$\dfrac{1.14}{\sqrt{n}}$	$\dfrac{1.22}{\sqrt{n}}$	$\dfrac{1.36}{\sqrt{n}}$	$\dfrac{1.63}{\sqrt{n}}$

Source: Adapted from Massey, F. J., Jr. 1951. The Kolmogorov-Smirnov test for goodness of fit. *J. Amer. Statist. Ass.*, 46, 70, with the kind permission of the author and publisher.

A.16 Random Numbers

The collections of numbers are arranged so that all digits may be considered as random, regardless of the point you begin to select numbers from or the direction in which you continue to select values. Individual numbers or groups of more than one number may be selected at one time and you may go to the right, left, up, down, or diagonally. They are grouped in fives merely for convenience.

Table A.16 Random numbers

11164	36318	75061	37674	26320	75100	10431	20418	19228	91792
21215	91791	76831	58678	87054	31687	93205	43685	19732	08468
10438	44482	66558	37649	08882	90870	12462	41810	01806	02977
36792	26236	33266	66583	60881	97395	20461	36742	02852	50564
73944	04773	12032	51414	82384	38370	00249	80709	72605	67497
49563	12872	14063	93104	78483	72717	68714	18048	25005	04151
64208	48237	41701	73117	33242	42314	83049	21933	92813	04763
51486	72875	38605	29341	80749	80151	33835	52602	79147	08868
99756	26360	64516	17971	48478	09610	04638	17141	09227	10606
71325	55217	13015	79207	00431	45117	33827	92873	02953	85474
65285	97198	12138	53010	94601	15838	16805	61004	43516	17020
17264	57327	38224	29301	31381	38109	34976	65692	98566	29550
95639	99754	31199	92558	68368	04985	51092	37780	40261	14479
61555	76404	86210	11808	12841	45147	97438	60022	12645	62000
78137	98768	04689	87130	79225	08153	84967	64539	79493	74917
62490	99215	84987	28759	19177	14733	24550	28067	68894	38490
24216	63444	21283	07044	92729	37284	13211	37485	10415	36457
16975	95428	33226	55903	31605	43817	22250	03918	46999	98501
59138	39542	71168	57609	91510	77904	74244	50940	31553	62562
29478	59652	50414	31966	87912	87154	12944	49862	96566	48825
96155	95009	27429	72918	08457	78134	48407	26061	58754	05326
29621	66583	62966	12468	20245	14015	04014	35713	03980	03024
12639	75291	71020	17265	41598	64074	64629	63293	53307	48766
14544	37134	54714	02401	63228	26831	19386	15457	17999	18306
83403	88827	09834	11333	68431	31706	26652	04711	34593	22561
67642	05204	30697	44806	96989	68403	85621	45556	35434	09532
64041	99011	14610	40273	09482	62864	01573	82274	81446	32477
17048	94523	97444	59904	16936	39384	97551	09620	63932	03091
93039	89416	52795	10631	09728	68202	20963	02477	55494	39563
82244	34392	96607	17220	51984	10753	76272	50985	97593	34320
96990	55244	70693	25255	40029	23289	48819	07159	60172	81697
09119	74803	97303	88701	51380	73143	98251	78635	27556	20712
57666	41204	47589	78364	38266	94393	70713	53388	79865	92069
46492	61594	26729	58272	81754	14648	77210	12923	53712	87771
08433	19172	08320	20839	13715	10597	17234	39355	74816	03363
10011	75004	86054	41190	10061	19660	03500	68412	57812	57929
92420	65431	16530	05547	10683	88102	30176	84750	10115	69220
35542	55865	07304	47010	43233	57022	52161	82976	47981	46588
86595	26247	18552	29491	33712	32285	64844	69395	41387	87195
72115	34985	58036	99137	47482	06204	24138	24272	16196	04393

SOURCE: From the RAND Corporation, *The Million Random Digits with 100,000 Normal Deviates.* Published 1955 by the Free Press, reprinted with the permission of the RAND Corporation.

Table A.16 *(continued)*

40603	16152	83235	37361	98783	24838	39793	80954	76865	32713
40941	53585	69958	60916	71018	90561	84505	53980	64735	85140
73505	83472	55953	17957	11446	22618	34771	25777	27064	13526
39412	16013	11442	89320	11307	49396	39805	12249	57656	88686
57994	76748	54627	48511	78646	33287	35524	54522	08795	56273
07428	58863	96023	88936	51343	70958	96768	74317	27176	29600
35379	27922	28906	55013	26937	48174	04197	36074	65315	12537
10982	22807	10920	26299	23593	64629	57801	10437	43965	15344
90127	33341	77806	12446	15444	49244	47277	11346	15884	28131
63002	12990	23510	68774	48983	20481	59815	67248	17076	78910
40779	86382	48454	65269	91239	45989	45389	54847	77919	41105
43216	12608	18167	84631	94058	82458	15139	76856	86019	47928
96167	64375	74108	93643	09204	98855	59051	56492	11933	64958
70975	62693	35684	72607	23026	37004	32989	24843	01128	74658
85812	61876	23570	75754	29090	40264	80399	47254	40135	69911
61834	59199	15469	82285	84164	91333	90954	87186	31598	25942
91402	77227	79516	21007	58602	81418	87838	18443	76162	51146
58299	83880	20125	10794	37780	61705	18276	99041	78135	99661
40684	99948	33880	76413	63839	71371	32392	51812	48248	96419
75978	64298	08074	62055	73864	01926	78374	15741	74452	49954
34556	39861	88267	76068	62445	64361	78685	24246	27027	48239
65990	57048	25067	77571	77974	37634	81564	98608	37224	49848
16381	15069	25416	87875	90374	86203	29677	82543	37554	89179
52458	88880	78352	67913	09245	47773	51272	06976	99571	33365
33007	85607	92008	44897	24964	50559	79549	85658	96865	24186
38712	31512	08588	61490	72294	42862	87334	05866	66269	43158
58722	03678	19186	69602	34625	75958	56869	17907	81867	11535
26188	69497	51351	47799	20477	71786	52560	66827	79419	70886
12893	54048	07255	86149	99090	70958	50775	31768	52903	27645
33186	81346	85095	37282	85536	72661	32180	40229	19209	74939
79893	29448	88392	54211	61708	83452	61227	81690	42265	20310
48449	15102	44126	19438	23382	14985	37538	30120	82443	11152
94205	04259	68983	50561	06902	10269	22216	70210	60736	58772
38648	09278	81313	77400	41126	52614	93613	27263	99381	49500
04292	46028	75666	26954	34979	68381	45154	09314	81009	05114
17026	49737	85875	12139	59391	81830	30185	83095	78752	40899
48070	76848	02531	97737	10151	18169	31709	74842	85522	74092
30159	95450	83778	46115	99178	97718	98440	15076	21199	20492
12148	92231	31361	60650	54695	30035	22765	91386	70399	79270
73838	77067	24863	97576	01139	54219	02959	45696	98103	78867

Table A.16 *(continued)*

73547	43759	95632	39555	74391	07579	69491	02647	17050	49869
07277	93217	79421	21769	83572	48019	17327	99638	87035	89300
65128	48334	07493	28098	52087	55519	83718	60904	48721	17522
38716	61380	60212	05099	21210	22052	01780	36813	19528	07727
31921	76458	73720	08657	74922	61335	41690	41967	50691	30508
57238	27464	61487	52329	26150	79991	64398	91273	26824	94827
24219	41090	08531	61578	08236	41140	76335	91189	66312	44000
31309	49387	02330	02476	96074	33256	48554	95401	02642	29119
20750	97024	72619	66628	66509	31206	55293	24249	02266	29010
28537	84395	26654	37851	80590	53446	34385	86893	87713	26842
97929	41220	86431	94485	28778	44997	38802	56594	61363	04206
40568	33222	40486	91122	43294	94541	40988	02929	83190	74247
41483	92935	17061	78252	40498	43164	68646	33023	64333	64083
93040	66476	24990	41099	65135	37641	97613	87282	63693	55299
76869	39300	84978	07504	36835	72748	47644	48542	25076	68626
02982	57991	50765	91930	21375	35604	29963	13738	03155	59914
94479	76500	39170	06629	10031	48724	49822	44021	44335	26474
52291	75822	95966	90947	65031	75913	52654	63377	70664	60082
03684	03600	52831	55381	97013	19993	41295	29118	18710	64851
58939	28366	86765	67465	45421	74228	01095	50987	83833	37276
37100	62492	63642	47638	13925	80113	88067	42575	44078	62703
53406	13855	38519	29500	62479	01036	87964	44498	07793	21599
55172	81556	18856	59043	64315	38270	25677	01965	21310	28115
40353	84807	47767	46890	16053	32415	60259	99788	55924	22077
18899	09612	77541	57675	70153	41179	97535	82889	27214	03482
68141	25340	92551	11326	60939	79355	41544	88926	09111	86431
51559	91159	81310	63251	91799	41215	87412	35317	74271	11603
92214	33386	73459	79359	65867	39269	57527	69551	17495	91456
15089	50557	33166	87094	52425	21211	41876	42525	36625	63964
96461	00604	11120	22254	16763	19206	67790	88362	01880	37911
28177	44111	15705	73835	69399	33602	13660	84342	97667	80847
66953	44737	81127	07493	07861	12666	85077	95972	96556	80108
19712	27263	84575	49820	19837	69985	34931	67935	71903	82560
68756	64757	19987	92222	11691	42502	00952	47981	97579	93408
75022	65332	98606	29451	57349	59219	08585	31502	96936	96356
11323	70069	90269	89266	46413	61615	66447	49751	15836	97343
55208	63470	18158	25283	19335	53893	87746	72531	16826	52605
11474	08786	05594	67045	13231	51186	71500	50498	59487	48677
81422	86842	60997	79669	43804	78690	58358	87639	24427	66799
21771	75963	23151	90274	08275	50677	99384	94022	84888	80139

Table A.16 *(continued)*

42278	12160	32576	14278	34231	20724	27908	02657	19023	07190
17697	60114	63247	32096	32503	04923	17570	73243	76181	99343
05686	30243	34124	02936	71749	03031	72259	26351	77511	00850
52992	46650	89910	57395	39502	49738	87854	71066	84596	33115
94518	93984	81478	67750	89354	01080	25988	84359	31088	13655
00184	72186	78906	75480	71140	15199	69002	08374	22126	23555
87462	63165	79816	61630	50140	95319	79205	79202	67414	60805
88692	58716	12273	48176	86038	78474	76730	82931	51595	20747
20094	42962	41382	16768	13261	13510	04822	96354	72001	68642
60935	81504	50520	82153	27892	18029	79663	44146	72876	67843
51392	85936	43898	50596	81121	98122	69196	54271	12059	62539
54239	41918	79526	46274	24853	67165	12011	04923	20273	89405
57892	73394	07160	90262	48731	46648	70977	58262	78359	50436
02330	74736	53274	44468	53616	35794	54838	39114	68302	26855
76115	29247	55342	51299	79908	36613	68361	18864	13419	34950
63312	81886	29085	20101	38037	34742	78364	39356	40006	49800
27632	21570	34274	56426	00330	07117	86673	46455	66866	76374
06335	62111	44014	52567	79480	45886	92585	87828	17376	35254
64142	87676	21358	88773	10604	62834	63971	03989	21421	76086
28436	25468	75235	75370	63543	76266	27745	31714	04219	00699
09522	83855	85973	15888	29554	17995	37443	11461	42909	32634
93714	15414	93712	02742	34395	21929	38928	31205	01838	60000
15681	53599	58185	73840	88758	10618	98725	23146	13521	47905
77712	23914	08907	43768	10304	61405	53986	61116	76164	54958
78453	54844	61509	01245	91199	07482	02534	08189	62978	55516
24860	68284	19367	29073	93464	06714	45268	60678	58506	23700
37284	06844	78887	57276	42695	03682	83240	09744	63025	60997
35488	52473	37634	32569	39590	27379	23520	29714	03743	08444
51595	59909	35223	44991	29830	56614	59661	83397	38421	17503
90660	35171	30021	91120	78793	16827	89320	08260	09181	53622
54723	56527	53076	38235	42780	22716	36400	48028	78196	92985
84828	81248	25548	34075	43459	44628	21866	90350	82264	20478
65799	01914	81363	05173	23674	41774	25154	73003	87031	94368
87917	38549	48213	71708	92035	92527	55484	32274	87918	22455
26907	88173	71189	28377	13785	87469	35647	19695	33401	51998
68052	65422	88460	06352	42379	55499	60469	76931	83430	24560
42587	68149	88147	99700	56124	53239	38726	63652	36644	50876
97176	55416	67642	05051	89931	19482	80720	48977	70004	03664
53295	87133	38264	94708	00703	35991	76404	82249	22942	49659
23011	94108	29196	65187	69974	01970	31667	54307	40032	30031

Table A.16 *(continued)*

75768	49549	24543	63285	32803	18301	80851	89301	02398	99891
86668	70341	66460	75648	78678	27770	30245	44774	56120	44235
56727	72036	50347	33521	05068	47248	67832	30960	95465	32217
27936	78010	09617	04408	18954	61862	64547	52453	83213	47833
31994	69072	37354	93025	38934	90219	91148	62757	51703	84040
02985	95303	15182	50166	11755	56256	89546	31170	87221	63267
89965	10206	95830	95406	33845	87588	70237	84360	19629	72568
45587	29611	98579	42481	05359	36378	56047	68114	58583	16313
01071	08530	74305	77509	16270	20889	99753	88035	55643	18291
90209	68521	14293	39194	68803	32052	39413	26883	83119	69623
04982	68470	27875	15480	13206	44784	83601	03172	07817	01520
19740	24637	97377	32112	74283	69384	49768	64141	02024	85380
50197	79869	86497	68709	42073	28498	82750	43571	77075	07123
46954	67536	28968	81936	95999	04319	09932	66223	45491	69503
82549	62676	31123	49899	70512	95288	15517	85352	21987	08669
61798	81600	80018	84742	06103	60786	01408	75967	29948	21454
57666	29055	46518	01487	30136	14349	56159	47408	78311	25896
29805	64994	66872	62230	41385	58066	96600	99301	85976	84194
06711	34939	19599	76247	87879	97114	74314	39599	43544	36255
13934	46885	58315	88366	06138	37923	11192	90757	10831	01580
28549	98327	99943	25377	17628	65468	07875	16728	22602	33892
40871	61803	25767	55484	90997	86941	64027	01020	39518	34693
47704	38355	71708	80117	11361	88875	22315	38048	42891	87885
62611	19698	09304	29265	07636	08508	23773	56545	08015	28891
03047	83981	11916	09267	67316	87952	27045	62536	32180	60936
26460	50501	31731	18938	11025	18515	31747	96828	58258	97107
01764	25959	69293	89875	72710	49659	66632	25314	95260	22146
11762	54806	02651	52912	32770	64507	59090	01275	47624	16124
31736	31695	11523	64213	91190	10145	34231	36405	65860	48771
97155	48706	52239	21831	49043	18650	72246	43729	63368	53822
31181	49672	17237	04024	65324	32460	01566	67342	94986	36106
32115	82683	67182	89030	41370	50266	19505	57724	93358	49445
07068	75947	71743	69285	30395	81818	36125	52055	20289	16911
26622	74184	75166	96748	34729	61289	36908	73686	84641	45130
02805	52676	22519	47848	68210	23954	63085	87729	14176	45410
32301	58701	04193	30142	99779	21697	05059	26684	63516	75925
26339	56909	39331	42101	01031	01947	02257	47236	19913	90371
95274	09508	81012	42413	11278	19354	68661	04192	36878	84366
24275	39632	09777	98800	48027	96908	08177	15364	02317	89548
36116	42128	65401	94199	51058	10759	47244	99830	64255	40550

A.17 Squares, Square Roots, and Reciprocals

This table may be used to find the squares, square roots, and reciprocals of any two-digit number. For any value find the first two digits in the column labeled n.

Squares: The square of the number n is found in the column n^2.

EXAMPLE 1
Find the square of 8.3.

From Table A.17,

n	n^2
8.3	68.89

$n^2 = 68.89$.

If the number is not between 1.0 and 9.9, then you must adjust the decimal point. Table A.17a gives you factors by which to multiply the value in column n^2.

EXAMPLE 2
Find the square of 760.

The first two digits are 76, thus look up for $n = 7.6$ to find $n^2 = 57.76$. From Table A.17a for values from 100 to 990 we multiply n^2 by 10.000. Then $(760)^2 = 10,000(57.76) = 577,600$.

n	n^2
7.6	57.76

Square Roots: Two columns are provided. Consult Table A.17a to decide which column to use and by what to multiply the value in the table.

EXAMPLE 1
Find the square root of 8.3.

8.3 is between 1.0 and 9.9, therefore from Table A.17a we find $\sqrt{n}$ and multiply by 1·0. From Table A.17b we obtain 2.8810.

n	$\sqrt{n}$
8.3	2.8810

EXAMPLE 2
Find the square root of 760. This is between 100 and 990. Table A.17a refers you to Column $\sqrt{n}$; multiply by 10. From Table A.17b we find

n	$\sqrt{n}$
7.6	2.7568

Therefore, $\sqrt{760} = 10(2.7568) = 27.568$.

EXAMPLE 3

Find the square root of 0.37. This is between 0.10 and 0.99. From Table A.17a we find that we should use the column labeled $\sqrt{10n}$ and multiply by 0.1. From Table A.17 we find

n	$\sqrt{10n}$
3.7	6.083

therefore

$$\sqrt{0.37} = 0.1(6.083) = 0.6083.$$

Reciprocals: For any value of n we read the reciprocal or $1/n$ in the $1/n$ column. Table A.17a lists multipliers for values not between 1 and 9.9.

Table A.17a

For numbers between	To find squares multiply n^2 by	To find square roots		To find reciprocals multiply by this
		Use column labeled	Multiply by this value	
1 and 9.9	1	$\sqrt{n}$	1	1.0
10 and 99	100	$\sqrt{10n}$	1	0.1
100 and 990	10,000	$\sqrt{n}$	10	0.01
1,000 and 9,900	1,000,000	$\sqrt{10n}$	10	0.001
10,000 and 99,000	100,000,000	$\sqrt{n}$	100	0.0001
0.10 and 0.99	0.01	$\sqrt{10n}$	0.1	10
0.010 and 0.099	0.0001	$\sqrt{n}$	0.01	100
0.0010 and 0.0099	0.000001	$\sqrt{10n}$	0.001	1000

Table A.17b Squares, square roots, and reciprocals

n	n^2	$\sqrt{n}$	$\sqrt{10n}$	$1/n$	n	n^2	$\sqrt{n}$	$\sqrt{10n}$	$1/n$
1.0	1.00	1.0000	3.162	1.000	2.1	4.41	1.4491	4.583	0.476
					2.2	4.84	1.4832	4.690	0.455
1.1	1.21	1.0488	3.317	0.909	2.3	5.29	1.5166	4.796	0.435
1.2	1.44	1.0954	3.464	0.833	2.5	5.76	1.5492	4.899	0.417
1.3	1.69	1.1402	3.606	0.769	2.5	6.25	1.5811	5.000	0.400
1.4	1.96	1.1832	3.742	0.714					
1.5	2.25	1.2247	3.873	0.667					
					2.6	6.76	1.6125	5.099	0.385
					2.7	7.29	1.6432	5.196	0.370
1.6	2.56	1.2649	4.000	0.625	2.8	7.84	1.6733	5.292	0.357
1.7	2.89	1.3038	4.123	0.588	2.9	8.41	1.7029	5.385	0.345
1.8	3.24	1.3416	4.243	0.556	3.0	9.00	1.7321	5.477	0.333
1.9	3.61	1.3784	4.359	0.526					
2.0	4.00	1.4142	4.472	0.500					

Table A.17b (continued)

n	n^2	$\sqrt{n}$	$\sqrt{10n}$	$1/n$	n	n^2	$\sqrt{n}$	$\sqrt{10n}$	$1/n$
3.1	9.61	1.7607	5.568	0.323	6.6	43.56	2.5690	8.124	0.152
3.2	10.24	1.7889	5.657	0.312	6.7	44.89	2.5884	8.185	0.149
3.3	10.89	1.8166	5.745	0.303	6.8	46.24	2.6077	8.246	0.147
3.4	11.56	1.8439	5.831	0.294	6.9	47.61	2.6268	8.307	0.145
3.5	12.25	1.8708	5.916	0.286	7.0	49.00	2.6458	8.367	0.143
3.6	12.96	1.8974	6.000	0.278	7.1	50.41	2.6646	8.426	0.141
3.7	13.69	1.9235	6.083	0.270	7.2	51.84	2.6833	8.485	0.139
3.8	14.44	1.9494	6.164	0.263	7.3	53.29	2.7019	8.544	0.137
3.9	15.21	1.9748	6.245	0.256	7.4	54.76	2.7203	8.602	0.135
4.0	16.00	2.0000	6.325	0.250	7.5	56.25	2.7386	8.660	0.133
4.1	16.81	2.0248	6.043	0.214	7.6	57.76	2.7568	8.718	0.132
4.2	17.64	2.0494	6.481	0.238	7.7	59.29	2.7749	8.775	0.130
4.3	18.49	2.0736	6.557	0.233	7.8	60.84	2.7928	8.832	0.128
4.4	19.36	2.0976	6.633	0.227	7.9	62.41	2.8107	8.888	0.127
4.5	20.25	2.1213	6.708	0.222	8.0	64.00	2.8284	8.944	0.125
4.6	21.16	2.1448	6.782	0.217	8.1	65.61	2.8460	9.000	0.123
4.7	22.09	2.1679	6.856	0.213	8.2	67.24	2.8636	9.055	0.122
4.8	23.04	2.1909	6.928	0.208	8.3	68.89	2.8810	9.110	0.120
4.9	24.01	2.2136	7.000	0.204	8.4	70.56	2.8983	9.165	0.119
5.0	25.00	2.2361	7.071	0.200	8.5	72.25	2.9155	9.220	0.118
5.1	26.01	2.2583	7.141	0.196	8.6	73.96	2.9326	9.274	0.116
5.2	27.04	2.2804	7.211	0.192	8.7	75.69	2.9496	9.327	0.115
5.3	28.09	2.3022	7.280	0.189	8.8	77.44	2.9665	9.381	0.114
5.4	29.16	2.3238	7.348	0.185	8.9	79.21	2.9833	9.434	0.112
5.5	30.25	2.3452	7.416	0.182	9.0	81.00	3.0000	9.487	0.111
5.6	31.36	2.3664	7.483	0.179	9.1	82.81	3.0166	9.539	0.110
5.7	32.49	2.3875	7.550	0.175	9.2	84.64	3.0332	9.592	0.109
5.8	33.64	2.4083	7.616	0.172	9.3	86.49	3.0496	9.644	0.108
5.9	34.81	2.4290	7.681	0.169	9.4	88.36	3.0659	9.695	0.106
6.0	36.00	2.4495	7.746	0.167	9.5	90.25	3.0822	9.747	0.105
6.1	37.21	2.4698	7.810	0.164	9.6	92.16	3.0984	9.798	0.104
6.2	38.44	2.4900	7.874	0.161	9.7	94.09	3.1145	9.849	0.103
6.3	39.69	2.5100	7.937	0.159	9.8	96.04	3.1305	9.899	0.102
6.4	40.06	2.5298	8.000	0.156	9.9	98.01	3.1464	9.950	0.101
6.5	42.25	2.5495	8.062	0.154	10.0	100.00	3.1623	10.000	0.100

A.18 Sampling Pieces

Table A.18 labeled *Sampling pieces I* is a table of the 200 chips referred to in the beginning of Chapter 12. Each block contains a pair of numbers. One number is to go on each side of the chip that the block represents. For example, the block in the upper right corner is $\boxed{\begin{array}{c}6\\2\end{array}}$. The 6 goes on one side of the chip (cardboard, paper, plastic, or whatever you have available). The 2 goes on the other side. It is also necessary to use two different colors to identify the different sets of numbers. Thus, all of the upper numbers might be drawn in red and the lower numbers in black. The two sets of numbers represent two different distributions.

The chips or blocks have been arranged in the order that they were originally sampled and have an order that has been referred to in Chapter 12 (see page 296). This order is from right to left and starts in the upper right corner.

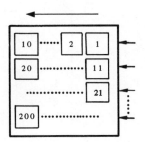

Thus the first chip is $\boxed{\begin{array}{c}6\\2\end{array}}$, the second is $\boxed{\begin{array}{c}8\\5\end{array}}$, the tenth is $\boxed{\begin{array}{c}5\\2\end{array}}$, the eleventh $\boxed{\begin{array}{c}4\\1\end{array}}$, etc., with the very last one $\boxed{\begin{array}{c}5\\3\end{array}}$.

Table A.18 Sampling pieces I

5	6	7	5	3	5	5	3	8	6
2	2	4	2	0	1	3	1	5	2
8	3	7	4	3	3	0	6	6	4
5	1	3	0	1	2	−3	3	4	1
7	9	5	6	5	6	5	8	9	3
3	5	0	1	3	3	2	4	6	0
6	2	7	5	5	4	5	1	4	5
4	0	2	2	4	1	1	−2	2	3

Table A.18 Sampling pieces I *(continued)*

7, 5	5, 2	6, 2	3, 0	7, 4	7, 2	3, 2	7, 4	3, 1	5, 1
4, 1	3, 1	8, 4	4, 0	7, 3	5, 2	7, 4	6, 3	4, 2	6, 3
6, 3	7, 5	6, 1	6, 2	4, 1	5, 2	4, 1	5, 1	5, 2	2, −1
4, 3	3, −1	4, 2	7, 4	8, 5	8, 3	5, 2	3, −1	1, −2	5, 3
5, 1	6, 3	5, 4	4, 0	7, 3	7, 4	4, 2	5, 2	6, 3	6, 4
6, 5	4, 2	5, 2	2, 0	5, 3	6, 3	4, −1	6, 4	6, 2	9, 6
3, 1	5, 1	6, 4	6, 3	4, 0	7, 5	6, 3	8, 5	4, 1	4, 1
5, 2	2, −2	4, 3	3, 0	7, 3	3, −1	4, 2	4, 1	6, 3	6, 2
5, 3	4, 1	6, 2	3, 0	5, 2	5, 1	4, 1	4, 2	8, 4	5, 1
5, 3	4, 1	8, 4	4, 0	6, 1	5, 2	4, 2	6, 3	4, 1	5, 1
5, 2	3, 2	7, 3	7, 2	5, 2	4, 0	3, 0	7, 3	4, 1	8, 6
3, 2	5, 2	3, 0	6, 2	2, −1	3, −1	4, 1	6, 2	2, −1	7, 4
6, 4	4, 2	5, 3	5, 2	6, 3	5, 0	5, 2	7, 3	5, 0	5, 2
6, 4	4, 3	4, 0	3, 0	6, 2	5, 3	6, 3	2, −1	6, 4	10, 7
6, 3	4, 0	7, 4	4, 2	2, 0	5, 1	4, 3	4, 1	7, 5	2, 0
5, 3	5, 1	3, 1	1, −1	5, 4	6, 1	2, 1	4, 2	6, 2	5, 3

A second set of sampling pieces is labeled *Sampling pieces II*. Here too the upper set of numbers represent one distribution and the lower set of numbers represent a second distribution.

The four distributions are as follows:

Sampling Pieces I —upper distribution   Normal   mean = 5 $\sigma = 1.72$
Sampling Pieces I —lower distribution   Normal   mean = 2 $\sigma = 1.72$
Sampling Pieces II—upper distribution   Normal   mean = 0 $\sigma = 1.00$
Sampling Pieces II—lower distribution   Normal   mean = 0 $\sigma = 3.47$

Table A.18 Sampling pieces II

0 3	0 8	2 −1	0 2	0 −3	0 0	0 −2	1 1	0 −5	−1 1
0 0	−2 2	−1 −2	0 4	1 −1	−2 −2	0 3	−1 0	0 1	1 6
1 −6	0 2	−1 −4	1 3	1 −2	1 1	1 −5	−1 2	−1 −1	−2 −1
1 −4	0 0	1 −1	2 2	1 −1	−2 7	0 −3	1 1	−1 5	0 −2
1 −9	−1 4	0 −2	0 0	2 −4	2 1	0 −3	0 −6	0 0	0 1
1 3	−1 0	−1 −2	1 −8	1 6	−1 2	2 2	0 −1	−1 −1	0 0
1 1	−1 5	0 0	1 −2	1 1	1 0	1 10	−1 −4	0 3	−2 5
−2 −3	0 2	0 2	0 −5	1 4	0 −1	0 −3	−1 −1	0 −2	0 4
−1 3	1 5	−1 1	0 −3	−1 4	2 6	0 −2	0 −3	1 −7	0 −4
−1 −2	0 −6	0 −5	−1 −3	−1 1	0 4	−1 2	0 −4	−1 0	−1 −2
−3 2	1 0	0 −4	0 3	2 −2	1 −5	−1 3	−1 −4	0 3	1 −6
1 1	0 3	−1 −1	−1 −7	0 −3	0 −3	1 −1	2 5	−1 2	0 2

Table A.18 Sampling pieces II *(continued)*

-2	1	0	0	1	-1	2	0	1	1
5	-1	2	0	-2	1	1	-7	-5	4
-1	1	0	-2	0	-1	-1	0	0	0
2	1	6	2	4	4	-1	3	2	1
-2	2	1	1	0	-1	0	0	1	3
1	0	-4	0	3	2	9	4	-3	5
0	0	0	1	-1	-1	0	1	-1	0
-1	-5	-2	-1	-6	-1	-4	-3	0	-2
1	1	0	-1	0	0	-1	0	-2	0
-2	0	-1	4	0	3	2	-2	-3	7
0	0	0	-1	-1	-1	0	1	0	-1
0	-10	3	1	-2	1	-5	1	-4	-3
1	-1	1	0	-1	2	0	-1	1	0
-3	2	-1	-3	-1	4	0	3	6	0
0	0	-2	0	1	1	-1	-1	0	1
1	5	1	0	7	-1	-2	3	-2	-4

A.19 Critical Values of T in the Wilcoxon Matched-pairs Test

This table is used for finding limits for the statistic T in the Wilcoxon matched-pairs test. Two samples of matched or related values are compared by observing the differences between each pair of values. The differences are then ranked according to absolute size. The rank of each difference is multiplied by its sign and the ranks for the less frequently observed sign are added up. This sum is T; n = number of pairs.

For any given n, T is significant at a given level if it is equal to or *less than* the value shown in the table.

Table A.19 Critical values of T in the Wilcoxon matched-pairs test.*

	Level of significance for one-tailed test					Level of significance for one-tailed test			
	0.05	0.025	0.01	0.005		0.05	0.025	0.01	0.005
	Level of significance for two-tailed test					Level of significance for two-tailed test			
n	0.10	0.05	0.02	0.01	n	0.10	0.05	0.02	0.01
5	0	—	—	—	28	130	116	101	91
6	2	0	—	—	29	140	126	110	100
7	3	2	0	—	30	151	137	120	109
8	5	3	1	0	31	163	147	130	118
9	8	5	3	1	32	175	159	140	128
10	10	8	5	3	33	187	170	151	138
11	13	10	7	5	34	200	182	162	148
12	17	13	9	7	35	213	195	173	159
13	21	17	12	9	36	227	208	185	171
14	25	21	15	12	37	241	221	198	182
15	30	25	19	15	38	256	235	211	194
16	35	29	23	19	39	271	249	224	207
17	41	34	27	23	40	286	264	238	220
18	47	40	32	27	41	302	279	252	233
19	53	46	37	32	42	319	294	266	247
20	60	52	43	37	43	336	310	281	261
21	67	58	49	42	44	353	327	296	276
22	75	65	55	48	45	371	343	312	291
23	83	73	62	54	46	389	361	328	307
24	91	81	69	61	47	407	378	345	322
25	100	89	76	68	48	426	396	362	339
26	110	98	84	75	49	446	415	379	355
27	119	107	92	83	50	466	434	397	373

SOURCE: From Audrey Haber and Richard P. Runyon, *General Statistics*, Addison-Wesley, Reading, Mass., 1969.

* Slight discrepancies will be found between the critical values appearing in the table above and in Table 2 of the 1964 revision of F. Wilcoxon, and R. A. Wilcox, *Some Rapid Approximate Statistical Procedures*, New York, Lederle Laboratories, 1964. The disparity reflects the latter's policy of selecting the critical value nearest a given significance level, occasionally overstepping that level. For example, for $N = 8$,
  the probability of a T of $3 = 0.0390$ (two-tail)
and
  the probability of a T of $4 = 0.0546$ (two-tail)
Wilcoxon and Wilcox select a T of 4 as the critical value of the 0.05 level of significance (two-tail), whereas Table A.19 reflects a more conservative policy by setting a T of 3 as the critical value at this level.

A.20 Critical Values of q for Determining Significant Differences (Fractional Points of the Studentized Range)

These values are used to determine the maximum difference to be expected among a set of sample means. All the samples must be the same size, which we call n_j; $k = $ the number of *samples*.

Two significance levels are provided: $P = 0.99$ or $P = 0.01$, and $P = 0.95$ or $P = 0.05$.

This is generally used with the analysis of variance where a mean square within samples is calculated. We require the degrees of freedom associated with this mean square. It would be:

$$df = k(n_j - 1).$$

EXAMPLE 1

To compare 5 samples, each with 4 values, find the q for 0.01 level; $k = 5$, $n_j = 4$.

$$df = 5(4 - 1) = 5(3) = 15.$$

From the table the 1% or 0.01 value for q is found to be 5.56.

df \ k	5
15	5.56

Table A.20 Fractional points of the studentized range, $q = (x_{max} - x_{min})/s$

$P = 0.95$ ($\alpha = 0.05$)

df \ k	2	3	4	5	6	7	8	9	10	11	12	13	14	15
1	18.0	27.0	32.8	37.1	40.4	43.1	45.4	47.4	49.1	50.6	52.0	53.2	54.3	55.4
2	6.09	8.3	9.8	10.9	11.7	12.4	13.0	13.5	14.0	14.4	14.7	15.1	15.4	15.7
3	4.50	5.91	6.82	7.50	8.04	8.48	8.85	9.18	9.46	9.72	9.95	10.1	10.3	10.5
4	3.93	5.04	5.76	6.29	6.71	7.05	7.35	7.60	7.83	8.03	8.21	8.37	8.52	8.66
5	3.64	4.60	5.22	5.67	6.03	6.33	6.58	6.80	6.99	7.17	7.32	7.47	7.60	7.72
6	3.46	4.34	4.90	5.31	5.63	5.89	6.12	6.32	6.49	6.65	6.79	6.92	7.03	7.14
7	3.34	4.16	4.68	5.06	5.36	5.61	5.82	6.00	6.16	6.30	6.43	6.55	6.66	6.76
8	3.26	4.04	4.53	4.89	5.17	5.40	5.60	5.77	5.92	6.05	6.18	6.29	6.39	6.48
9	3.20	3.95	4.42	4.76	5.02	5.24	5.43	5.60	5.74	5.87	5.98	6.09	6.19	6.28
10	3.15	3.88	4.33	4.65	4.91	5.12	5.30	5.46	5.60	5.72	5.83	5.93	6.03	6.11
11	3.11	3.82	4.26	4.57	4.82	5.03	5.20	5.35	5.49	5.61	5.71	5.81	5.90	5.99
12	3.08	3.77	4.20	4.51	4.75	4.95	5.12	5.27	5.40	5.51	5.62	5.71	5.80	5.88
13	3.06	3.73	4.15	4.45	4.69	4.88	5.05	5.19	5.32	5.43	5.53	5.63	5.71	5.79
14	3.03	3.70	4.11	4.41	4.64	4.83	4.99	5.13	5.25	5.36	5.46	5.55	5.64	5.72

SOURCE: Abridged from *Biometrica Tables for Statisticians*, Vol. 1, edited by E. S. Pearson and H. O. Hartley, Cambridge University Press, Cambridge, 1962, with the kind permission of the editors and publishers.

Table A.20 *(continued)*

$P = 0.95 \quad (\alpha = 0.05)$

df \ k	2	3	4	5	6	7	8	9	10	11	12	13	14	15
15	3.01	3.67	4.08	4.37	4.60	4.78	4.94	5.08	5.20	5.31	5.40	5.49	5.58	5.65
16	3.00	3.65	4.05	4.33	4.56	4.74	4.90	5.03	5.15	5.26	5.35	5.44	5.52	5.59
17	2.98	3.63	4.02	4.30	4.52	4.71	4.86	4.99	5.11	5.21	5.31	5.39	5.47	5.55
18	2.97	3.61	4.00	4.28	4.49	4.67	4.82	4.96	5.07	5.17	5.27	5.35	5.43	5.50
19	2.96	3.59	3.98	4.25	4.47	4.65	4.79	4.92	5.04	5.14	5.23	5.32	5.39	5.46
20	2.95	3.58	3.96	4.23	4.45	4.62	4.77	4.90	5.01	5.11	5.20	5.28	5.36	5.43
24	2.92	3.53	3.90	4.17	4.37	4.54	4.68	4.81	4.92	5.01	5.10	5.18	5.25	5.32
30	2.89	3.49	3.84	4.10	4.30	4.46	4.60	4.72	4.83	4.92	5.00	5.08	5.15	5.21
40	2.86	3.44	3.79	4.04	4.23	4.39	4.52	4.63	4.74	4.82	4.91	4.89	5.05	5,11
60	2.83	3.40	3.74	3.98	4.16	4.31	4.44	4.55	4.65	4.73	4.81	4.88	4.94	5.00
120	2.80	3.36	3.69	3.92	4.10	4.24	4.36	4.48	4.56	4.64	4.72	4.78	4.84	4.90
∞	2.77	3.31	3.63	3.86	4.03	4.17	4.29	4.39	4.47	4.55	4.62	4.68	4.74	4.80

$P = 0.99 \quad (\alpha = 0.01)$

df \ k	2	3	4	5	6	7	8	9	10	11	12	13	14	15
1	90	135	164	186	202	216	227	237	246	253	260	266	272	277
2	14.0	19.0	22.3	24.7	26.6	28.2	29.5	30.7	31.7	32.6	33.4	34.1	34.8	35.4
3	8.26	10.6	12.2	13.3	14.2	15.0	15.6	16.2	16.7	17.1	17.5	17.9	18.2	18.5
4	6.51	8.12	9.17	9.96	10.6	11.1	11.5	11.9	12.3	12.6	12.8	13.1	13.3	13.5
5	5.70	6.97	7.80	8.42	8.91	9.32	9.67	9.97	10.2	10.5	10.7	10.9	11.1	11.2
6	5.24	6.33	7.03	7.56	7.97	8.32	8.61	8.87	9.10	9.30	9.49	9.65	9.81	9.95
7	4.95	5.92	6.54	7.01	7.37	7.68	7.94	8.17	8.37	8.55	8.71	8.86	9.00	9.12
8	4.74	5.63	6.20	6.63	6.96	7.24	7.47	7.68	7.87	8.03	8.18	8.31	8.44	8.55
9	4.60	5.43	5.96	6.35	6.66	6.91	7.13	7.32	7.49	7.65	7.78	7.91	8.03	8.13
10	4.48	5.27	5.77	6.14	6.43	6.67	6.87	7.05	7.21	7.36	7.48	7.60	7.71	7.81
11	4.39	5.14	5.62	5.97	6.25	6.48	6.67	6.84	6.99	7.13	7.25	7.36	7.46	7.56
12	4.32	5.04	5.50	5.84	6.10	6.32	6.51	6.67	6.81	6.94	7.06	7.17	7.26	7.36
13	4.26	4.96	5.40	5.73	5.98	6.19	6.37	6.53	6.67	6.79	6.90	7.01	7.10	7.19
14	4.21	4.89	5.32	5.63	5.88	6.08	6.26	6.41	6.54	6.66	6.77	6.87	6.96	7.05
15	4.17	4.83	5.25	5.56	5.80	5.99	6.16	6.31	6.44	6.55	6.66	6.76	6.84	6.93
16	4.13	4.78	5.19	5.49	5.72	5.92	6.08	6.22	6.35	6.46	6.56	6.66	6.74	6.82
17	4.10	4.74	5.14	5.43	5.66	5.85	6.01	6.15	6.27	6.38	6.48	6.57	6.66	6.73
18	4.07	4.70	5.09	5.38	5.60	5.79	5.94	6.08	6.20	6.31	6.41	6.50	6.58	6.65
19	4.05	4.67	5.05	5.33	5.55	5.73	5.89	6.02	6.14	6.25	6.34	6.43	6.51	6.58
20	4.02	4.64	5.02	5.29	5.51	5.69	5.84	5.97	6.09	6.19	6.29	6.37	6.45	6.52
24	3.96	4.54	4.91	5.17	5.37	5.54	5.69	5.81	5.92	6.02	6.11	6.19	6.26	6.33
30	3.89	4.45	4.80	5.05	5.24	5.40	5.54	5.65	5.76	5.85	5.93	6 01	6.08	6.14
40	3.82	4.37	4.70	4.93	5.11	5.27	5.39	5.50	5.60	5.69	5.77	5.84	5.90	5.96
60	3.76	4.28	4.60	4.82	4.99	5.13	5.25	5.36	5.45	5.53	5.60	5.67	5.73	5.79
120	3.70	4.20	4.50	4.71	4.87	5.01	5.12	5.21	5.30	5.38	5.44	5.51	5.56	5.61
∞	3.64	4.12	4.40	4.60	4.76	4.88	4.99	5.08	5.16	5.23	5.29	5.35	5.40	5.45

Table A.21 Critical values of U and U'

One-tailed test at $\alpha = 0.05$ or a two-tailed test at $\alpha = 0.10$*

n_2 \ n_1	1	2	3	4	5	6	7	8	9	10	11	12	13	14	15	16	17	18	19	20
1	—	—	—	—	—	—	—	—	—	—	—	—	—	—	—	—	—	—	0	0
																			19	20
2	—	—	—	—	0	0	0	1	1	1	1	2	2	2	3	3	3	4	4	4
					10	12	14	15	17	19	21	22	24	26	27	29	31	32	34	36
3	—	—	0	0	1	2	2	3	3	4	5	5	6	7	7	8	9	9	10	11
			9	12	14	16	19	21	24	26	28	31	33	35	38	40	42	45	47	49
4	—	—	0	1	2	3	4	5	6	7	8	9	10	11	12	14	15	16	17	18
			12	15	18	21	24	27	30	33	36	39	42	45	48	50	53	56	59	62
5	—	0	1	2	4	5	6	8	9	11	12	13	15	16	18	19	20	22	23	25
		10	14	18	21	25	29	32	36	39	43	47	50	54	57	61	65	68	72	75
6	—	0	2	3	5	7	8	10	12	14	16	17	19	21	23	25	26	28	30	32
		12	16	21	25	29	34	38	42	46	50	55	59	63	67	71	76	80	84	88
7	—	0	2	4	6	8	11	13	15	17	19	21	24	26	28	30	33	35	37	39
		14	19	24	29	34	38	43	48	53	58	63	67	72	77	82	86	91	96	101
8	—	1	3	5	8	10	13	15	18	20	23	26	28	31	33	36	39	41	44	47
		15	21	27	32	38	43	49	54	60	65	70	76	81	87	92	97	103	108	113
9	—	1	3	6	9	12	15	18	21	24	27	30	33	36	39	42	45	48	51	54
		17	24	30	36	42	48	54	60	66	72	78	84	90	96	102	108	114	120	126
10	—	1	4	7	11	14	17	20	24	27	31	34	37	41	44	48	51	55	58	62
		19	26	33	39	46	53	60	66	73	79	86	93	99	106	112	119	125	132	138
11	—	1	5	8	12	16	19	23	27	31	34	38	42	46	50	54	57	61	65	69
		21	28	36	43	50	58	65	72	79	87	94	101	108	115	122	130	137	144	151
12	—	2	5	9	13	17	21	26	30	34	38	42	47	51	55	60	64	68	72	77
		22	31	39	47	55	63	70	78	86	94	102	109	117	125	132	140	148	156	163
13	—	2	6	10	15	19	24	28	33	37	42	47	51	56	61	65	70	75	80	84
		24	33	42	50	59	67	76	84	93	101	109	118	126	134	143	151	159	167	176
14	—	2	7	11	16	21	26	31	36	41	46	51	56	61	66	71	77	82	87	92
		26	35	45	54	63	72	81	90	99	108	117	126	135	144	153	161	170	179	188
15	—	3	7	12	18	23	28	33	39	44	50	55	61	66	72	77	83	88	94	100
		27	38	48	57	67	77	87	96	106	115	125	134	144	153	163	172	182	191	200
16	—	3	8	14	19	25	30	36	42	48	54	60	65	71	77	83	89	95	101	107
		29	40	50	61	71	82	92	102	112	122	132	143	153	163	173	183	193	203	213
17	—	3	9	15	20	26	33	39	45	51	57	64	70	77	83	89	96	102	109	115
		31	42	53	65	76	86	97	108	119	130	140	151	161	172	183	193	204	214	225
18	—	4	9	16	22	28	35	41	48	55	61	68	75	82	88	95	102	109	116	123
		32	45	56	68	80	91	103	114	123	137	148	159	170	182	193	204	215	226	237
19	0	4	10	17	23	30	37	44	51	58	65	72	80	87	94	101	109	116	123	130
	19	34	47	59	72	84	96	108	120	132	144	156	167	179	191	203	214	226	238	250
20	0	4	11	18	25	32	39	47	54	62	69	77	84	92	100	107	115	123	130	138
	20	36	49	62	75	88	101	113	126	138	151	163	176	188	200	213	225	237	250	262

SOURCE: From Audrey Haber and Richard P. Runyon, *General Statistics*, Addison-Wesley, Reading, Mass., 1969.

* To be significant to any given n_1 and n_2: Obtained U must be equal to or *less than* the value shown in the table. Obtained U' must be equal to or *greater than* the top/lower value shown in the table. Dashes in the body of the table indicate that no decision is possible at the stated level of significance.

Table A.21 *(continued)*

One-tailed test at $\alpha = 0.025$ or a two-tailed test at $\alpha = 0.05$*

n_1 / n_2	1	2	3	4	5	6	7	8	9	10	11	12	13	14	15	16	17	18	19	20
1	—	—	—	—	—	—	—	—	—	—	—	—	—	—	—	—	—	—	—	—
2	—	—	—	—	—	—	—	0 16	0 18	0 20	0 22	1 23	1 25	1 27	1 29	1 31	2 32	2 34	2 36	2 38
3	—	—	—	0 15	1 17	1 20	2 22	2 25	3 27	3 30	4 32	4 35	5 37	5 40	6 42	6 45	7 47	7 50	8 52
4	—	—	0 16	1 19	2 22	3 25	4 28	4 32	5 35	6 38	7 41	8 44	9 47	10 50	11 53	11 57	12 60	13 63	13 67
5	—	—	0 15	1 19	2 23	3 27	5 30	6 34	7 38	8 42	9 46	11 49	12 53	13 57	14 61	15 65	17 68	18 72	19 76	20 80
6	—	—	1 17	2 22	3 27	5 31	6 36	8 40	10 44	11 49	13 53	14 58	16 62	17 67	19 71	21 75	22 80	24 84	25 89	27 93
7	—	—	1 20	3 25	5 30	6 36	8 41	10 46	12 51	14 56	16 61	18 66	20 71	22 76	24 81	26 86	28 91	30 96	32 101	34 106
8	—	0 16	2 22	4 28	6 34	8 40	10 46	13 51	15 57	17 63	19 69	22 74	24 80	26 86	29 91	31 97	34 102	36 108	38 111	41 119
9	—	0 18	2 25	4 32	7 38	10 44	12 51	15 57	17 64	20 70	23 76	26 82	28 89	31 95	34 101	37 107	39 114	42 120	45 126	48 132
10	—	0 20	3 27	5 35	8 42	11 49	14 56	17 63	20 70	23 77	26 84	29 91	33 97	36 104	39 111	42 118	45 125	48 132	52 138	55 145
11	—	0 22	3 30	6 38	9 46	13 53	16 61	19 69	23 76	26 84	30 91	33 99	37 106	40 114	44 121	47 129	51 136	55 143	58 151	62 158
12	—	1 23	4 32	7 41	11 49	14 58	18 66	22 74	26 82	29 91	33 99	37 107	41 115	45 123	49 131	53 139	57 147	61 155	65 163	69 171
13	—	1 25	4 35	8 44	12 53	16 62	20 71	24 80	28 89	33 97	37 106	41 115	45 124	50 132	54 141	59 149	63 158	67 167	72 175	76 184
14	—	1 27	5 37	9 47	13 51	17 67	22 76	26 86	31 95	36 104	40 114	45 123	50 132	55 141	59 151	64 160	67 171	74 178	78 188	83 197
15	—	1 29	5 40	10 50	14 61	19 71	24 81	29 91	34 101	39 111	44 121	49 131	54 141	59 151	64 161	70 170	75 180	80 190	95 200	90 210
16	—	1 31	6 42	11 53	15 65	21 75	26 86	31 97	37 107	42 118	47 129	53 139	59 149	64 160	70 170	75 181	81 191	86 202	92 212	98 222
17	—	2 32	6 45	11 57	17 68	22 80	28 91	34 102	39 114	45 125	51 136	57 147	63 158	67 171	75 180	81 191	87 202	93 213	99 224	105 235
18	—	2 34	7 47	12 60	18 72	24 84	30 96	36 108	42 120	48 132	55 143	61 155	67 167	74 178	80 190	86 202	93 213	99 225	106 236	112 248
19	—	2 36	7 50	13 63	19 76	25 89	32 101	38 114	45 126	52 138	58 151	65 163	72 175	78 188	85 200	92 212	99 224	106 236	113 248	119 261
20	—	2 38	8 52	13 67	20 80	27 93	34 106	41 119	48 132	55 145	62 158	69 171	76 184	83 197	90 210	98 222	105 235	112 248	119 261	127 273

* To be significant for any given n_1 and n_2: Obtained U must be equal to or *less than* the value shown in the table. Obtained U' must be equal to or *greater than* the top/lower value shown in the table.

Table A.21 (continued)

One-tailed test at $\alpha = 0.01$ or a two-tailed test at $\alpha = 0.02$*

n_2 \ n_1	1	2	3	4	5	6	7	8	9	10	11	12	13	14	15	16	17	18	19	20
1	—	—	—	—	—	—	—	—	—	—	—	—	—	—	—	—	—	—	—	—
2	—	—	—	—	—	—	—	—	—	—	—	0/26	0/28	0/30	0/32	0/34	0/36	0/38	1/37	1/39
3	—	—	—	—	—	—	0/21	0/24	1/26	1/29	1/32	2/34	2/37	2/40	3/42	3/45	4/47	4/50	4/52	5/55
4	—	—	—	0/20	1/23	1/27	2/30	3/33	3/37	4/40	5/43	5/47	6/50	7/53	7/57	8/60	9/63	9/67	10/70	
5	—	—	0/20	1/24	2/28	3/32	4/36	5/40	6/44	7/48	8/52	9/56	10/60	11/64	12/68	13/72	14/76	15/80	16/84	
6	—	—	1/23	2/28	3/33	4/38	6/42	7/47	8/52	9/57	11/61	12/66	13/71	15/75	16/80	18/84	19/89	20/94	22/98	
7	—	0/21	1/27	3/32	4/38	6/43	7/49	9/54	11/59	12/65	14/70	16/75	17/81	19/86	21/91	23/96	24/102	26/107	28/112	
8	—	0/24	2/30	4/36	6/42	7/49	9/55	11/61	13/67	15/73	17/79	20/84	22/90	24/96	26/102	28/108	30/114	32/120	34/126	
9	—	1/26	3/33	5/40	7/47	9/54	11/61	14/67	16/74	18/81	21/87	23/94	26/100	28/107	31/113	33/120	36/126	38/133	40/140	
10	—	1/29	3/37	6/44	8/52	11/59	13/67	16/74	19/81	22/88	24/96	27/103	30/110	33/117	36/124	38/132	41/139	44/146	47/153	
11	—	1/32	4/40	7/48	9/57	12/65	15/73	18/81	22/88	25/96	28/104	31/112	34/120	37/128	41/135	44/143	47/151	50/159	53/167	
12	—	2/34	5/43	8/52	11/61	14/70	17/79	21/87	24/96	28/104	31/113	35/121	38/130	42/138	46/146	49/155	53/163	56/172	60/180	
13	0/26	2/37	5/47	9/56	12/66	16/75	20/84	23/94	27/103	31/112	35/121	39/130	43/139	47/148	51/157	55/166	59/175	63/184	67/193	
14	0/28	2/40	6/50	10/60	13/71	17/81	22/90	26/100	30/110	34/120	38/130	43/139	47/149	51/159	56/168	60/178	65/187	69/197	73/207	
15	0/30	3/42	7/53	11/64	15/75	19/86	24/96	28/107	33/117	37/128	42/138	47/148	51/159	56/169	61/179	66/189	70/200	75/210	80/220	
16	0/32	3/45	7/57	12/68	16/80	21/91	26/102	31/113	36/124	41/135	46/146	51/157	56/168	61/179	66/190	71/201	76/212	82/222	87/233	
17	0/34	4/47	8/60	13/72	18/84	23/96	28/108	33/120	38/132	44/143	49/155	55/166	60/178	66/189	71/201	77/212	82/224	88/234	93/247	
18	0/36	4/50	9/63	14/76	19/89	24/102	30/114	36/126	41/139	47/151	53/163	59/175	65/187	70/200	76/212	82/224	88/236	94/248	100/260	
19	1/37	4/53	9/67	15/80	20/94	26/107	32/120	38/133	44/146	50/159	56/172	63/184	69/197	75/210	82/222	88/235	94/248	101/260	107/273	
20	1/39	5/55	10/70	16/84	22/98	28/112	34/126	40/140	47/153	53/167	60/180	67/193	73/207	80/220	87/233	93/247	100/260	107/273	114/286	

* To be significant for any given n_1 and n_2: Obtained U must be equal to or *less than* the value shown in the table. Obtained U' must be equal to or *greater than* the top/lower value shown in the table.

Table A.21 *(continued)*
One-tailed test at α = 0.005 or a two-tailed test at α = 0.01*

n_2 \ n_1	1	2	3	4	5	6	7	8	9	10	11	12	13	14	15	16	17	18	19	20
1	—	—	—	—	—	—	—	—	—	—	—	—	—	—	—	—	—	—	—	—
2	—	—	—	—	—	—	—	—	—	—	—	—	—	—	—	—	—	—	0 / 38	0 / 40
3	—	—	—	—	—	—	—	0 / 27	0 / 30	0 / 33	1 / 35	1 / 38	1 / 41	2 / 43	2 / 46	2 / 49	2 / 52	2 / 54	3 / 54	3 / 57
4	—	—	—	—	0 / 24	0 / 28	1 / 31	1 / 35	2 / 38	2 / 42	3 / 45	3 / 49	4 / 52	5 / 55	5 / 59	6 / 62	6 / 66	7 / 69	8 / 72
5	—	—	—	0 / 25	1 / 29	1 / 34	2 / 38	3 / 42	4 / 46	5 / 50	6 / 54	7 / 58	7 / 63	8 / 67	9 / 71	10 / 75	11 / 79	12 / 83	13 / 87
6	—	—	0 / 24	1 / 29	2 / 34	3 / 39	4 / 44	5 / 49	6 / 54	7 / 59	9 / 63	10 / 68	11 / 73	12 / 78	13 / 83	15 / 87	16 / 92	17 / 97	18 / 102
7	—	—	0 / 28	1 / 34	3 / 39	4 / 45	6 / 50	7 / 56	9 / 61	10 / 67	12 / 72	13 / 78	15 / 83	16 / 89	18 / 94	19 / 100	21 / 105	22 / 111	24 / 116
8	—	—	1 / 31	2 / 38	4 / 44	6 / 50	7 / 57	9 / 63	11 / 69	13 / 75	15 / 81	17 / 87	18 / 94	20 / 100	22 / 106	24 / 112	26 / 118	28 / 124	30 / 130
9	—	—	0 / 27	1 / 35	3 / 42	5 / 49	7 / 56	9 / 63	11 / 70	13 / 77	16 / 83	18 / 90	20 / 97	22 / 104	24 / 111	27 / 117	29 / 124	31 / 131	33 / 138	36 / 144
10	—	—	0 / 30	2 / 38	4 / 46	6 / 54	9 / 61	11 / 69	13 / 77	16 / 84	18 / 92	21 / 99	24 / 106	26 / 114	29 / 121	31 / 129	34 / 136	37 / 143	39 / 151	42 / 158
11	—	—	0 / 33	2 / 42	5 / 50	7 / 59	10 / 67	13 / 75	16 / 83	18 / 92	21 / 100	24 / 108	27 / 116	30 / 124	33 / 132	36 / 140	39 / 148	42 / 156	45 / 164	48 / 172
12	—	—	1 / 35	3 / 45	6 / 54	9 / 63	12 / 72	15 / 81	18 / 90	21 / 99	24 / 108	27 / 117	31 / 125	34 / 134	37 / 143	41 / 151	44 / 160	47 / 169	51 / 177	54 / 186
13	—	—	1 / 38	3 / 49	7 / 58	10 / 68	13 / 78	17 / 87	20 / 97	24 / 106	27 / 116	31 / 125	34 / 135	38 / 144	42 / 153	45 / 163	49 / 172	53 / 181	56 / 191	60 / 200
14	—	—	1 / 41	4 / 52	7 / 63	11 / 73	15 / 83	18 / 94	22 / 104	26 / 114	30 / 124	34 / 134	38 / 144	42 / 154	46 / 164	50 / 174	54 / 184	58 / 194	63 / 203	67 / 213
15	—	—	2 / 43	5 / 55	8 / 67	12 / 78	16 / 89	20 / 100	24 / 111	29 / 121	33 / 132	37 / 143	42 / 153	46 / 164	51 / 174	55 / 185	60 / 195	64 / 206	69 / 216	73 / 227
16	—	—	2 / 46	5 / 59	9 / 71	13 / 83	18 / 94	22 / 106	27 / 117	31 / 129	36 / 140	41 / 151	45 / 163	50 / 174	55 / 185	60 / 196	65 / 207	70 / 218	74 / 230	79 / 241
17	—	—	2 / 49	6 / 62	10 / 75	15 / 87	19 / 100	24 / 112	29 / 124	34 / 136	39 / 148	44 / 160	49 / 172	54 / 184	60 / 195	65 / 207	70 / 219	75 / 231	81 / 242	86 / 254
18	—	—	2 / 52	6 / 66	11 / 79	16 / 92	21 / 105	26 / 118	31 / 131	37 / 143	42 / 156	47 / 169	53 / 181	58 / 194	64 / 206	70 / 218	75 / 231	81 / 243	87 / 255	92 / 268
19	—	0 / 38	3 / 54	7 / 69	12 / 83	17 / 97	22 / 111	28 / 124	33 / 138	39 / 151	45 / 164	51 / 177	56 / 191	63 / 203	69 / 216	74 / 230	81 / 242	87 / 255	93 / 268	99 / 281
20	—	0 / 40	3 / 57	8 / 72	13 / 87	18 / 102	24 / 116	30 / 130	36 / 144	42 / 158	48 / 172	54 / 186	60 / 200	67 / 213	73 / 227	79 / 241	86 / 254	92 / 268	99 / 281	105 / 295

* To be significant for any given n_1 and n_2: Obtained U must be equal to or *less than* the value shown in the table. Obtained U' must be equal to or *greater than* the top/lower value shown in the table.

A.21 Critical Values of U and U' for the Mann-Whitney Test

This test is used to determine if there is a difference between two populations. Two independent samples, Sample A of size n_1, and Sample B of size n_2, are ranked together. For each B the number of A that have a smaller rank than the B are counted. The sum for all of the B is called U. U' would be the other sum, the sum of all the B that come before each A.

This set of tables is good only if n_1 and n_2 are less than or equal to 20. See Chapter 17, Section 17.4, for n greater than 20.

Locate n_1 along the top and n_2 along the side. At the intersection find the critical values of U and U'.

EXAMPLE 1

For a two-tailed test at $\alpha = 0.05$, with $n_1 = 7$ and $n_2 = 8$ what are the critical limits on U and U'?

At the intersection we find

$$10$$
$$46$$

The top number, 10, stands for the lower limit and the bottom number, 46, stands for the upper limit. Thus, a sum of less than or equal to 10 and a sum of greater than or equal to 46 would be significant.

		n_1	
		7	
n_2	8	10 46	

To use the table for a *one-tailed* test you must decide which of the samples is expected to be the larger. If you classify the expected larger distribution as Sample A and compute the number of A less than each B, you would use only the top values in the table, the smaller numbers. (Thus, for a one tail with a $\alpha = 0.05$ limit and 2 samples of 6, you would consult the first Table A.21 and the critical value of U would be 7.)

A.22 Critical Values of r_{rho} (Spearman Rank Order Correlation Coefficient)

This table gives critically large values of r_{rho}. Sample size n = number of *pairs* of measurements. Critical values are given for one- and two-tailed tests. Each set of measurements is separately ranked from 1 to n. The difference in rank, called d, is calculated for *each* individual. These d are squared and all the d^2 are added up. The correlation coefficient is then found as

$$r_{rho} = 1 - \frac{6 \sum d^2}{n(n_2 - 1)}.$$

Table A.22 Critical values of r_{rho}*

	Level of significance for one-tailed test			
	0.05	0.025	0.01	0.005
	Level of significance for two-tailed test			
n*	0.10	0.05	0.02	0.01
5	0.900	1.000	1.000	—
6	0.829	0.886	0.943	1.000
7	0.714	0.786	0.893	0.929
8	0.643	0.738	0.833	0.881
9	0.600	0.683	0.783	0.833
10	0.564	0.648	0.746	0.794
12	0.506	0.591	0.712	0.777
14	0.456	0.544	0.645	0.715
16	0.425	0.506	0.601	0.665
18	0.399	0.475	0.564	0.625
20	0.377	0.450	0.534	0.591
22	0.359	0.428	0.508	0.562
24	0.343	0.409	0.485	0.537
26	0.329	0.392	0.465	0.515
28	0.317	0.377	0.448	0.496
30	0.306	0.364	0.432	0.478

SOURCE: Adapted from "The Five Percent Significance Levels of Sums of Squares of Rank Differences and a Correction" by E. G. Olds, *Annals of Mathematical Statistics*, Vol. 20 (1949), pp. 117–118, with the kind permission of the author and publishers.
* Number of pairs.

Appendix B
Summation and Subscript Notation

Algebraic Symbols and Notation

In all cases statistics deals with numbers. The numbers themselves are specific or particular measurements of a certain characteristic. However, often we desire to designate or discuss the characteristic without being restricted to an individual measurement of the characteristic. To alleviate this problem we use a letter to represent whatever the numerical value of the measurement might be.

As an example of data we will be considering the number of broken dishes in a restaurant. A possible situation of interest might be the keeping of a record of the total number of dishes broken by various employees during one day's work. For instance, Ann might have broken 2 dishes yesterday, Jill 3 dishes, Frank 1, and so forth. It is quite possible to work with such a type of designation (Ann broke 2 dishes yesterday, and so forth). However, it is cumbersome and does not lend itself easy to generalizations such as discussing "the number of dishes broken by Ann—whatever they might be."

We will now substitute a symbol to represent the number of dishes broken. In this way we will remove any restriction on the particular or exact number. We might decide that we will designate "the number of dishes broken" with the symbol X. Of course, we could use any symbol, as long as we agree upon the meaning of the symbol. Now observe several things:

1. The symbol stands for the measurement of the characteristic (broken dishes).
2. The symbol represents many different possible values. It may be standing for 0, 1, 2, 3, and so forth, depending upon, in this case, the actual number of dishes broken.
3. By stating the letter X by itself we do not restrict ourselves to a particular value and thus we have generalized on the idea (dishes broken).

One difficulty now arises—how to designate or distinguish the different individuals. We might be interested in discussing or separately designating dishes broken by Ann, Jill, Frank, and so forth. We then are to be concerned with the X-values, the measurements, for each of these people, and you should realize that the X- *value for Ann is* 2 (2 dishes broken), the X for Jill is 3 (3 dishes broken), and the X for Frank is 1. But this too is awkward.

Although several possibilities for simplifying those statements exist, probably the easiest and most convenient method will be to rename each individual with a number. We will call Ann 1, Jill 2, and Frank 3. We will then distinguish the X by calling one X_1, the next X_2, the third X_3, and so forth. We read these as X sub 1, X sub 2, and X sub 3, with the numbers being called *subscripts*. Thus 1 represents Ann and X_1 represents the number of dishes she broke. Jill is 2 and X_2 is the number of dishes she broke. In general we observe that each subscript corresponds to a particular individual being measured and each X corresponds to the measurement.

An additional advantage of this method of designation is that we do not have to specify *how many* X there are. We can state that the values being considered are:
$$X_1, X_2, X_3, \ldots$$
where the ... means "and so on."

We can end the list of X's by calling the last value X_n. So we obtain:
$$X_1, X_2, X_3, \ldots, X_n.$$
At any time by just specifying what n is, we also designate the number of X. Thus, if there are 5 employees in our restaurant (David and Sara being the additional two) then $n = 5$ and we have 5 values of X:
$$X_1, X_2, X_3, X_4, X_5.$$

One possible concern should be mentioned at this point. The order of these subscripts or individuals is not rigid but must be agreed upon. We might have scrambled the five people's names and ended up with a different order, perhaps Frank as 1, David as 2, and so forth. Now X_1 stands for the number of dishes Frank has broken. In many cases it won't really matter what order we place the individuals in and we can even juggle them around. However, you should always remember that the subscript implies that the value corresponds to a particular individual and if we desire to learn which individual that is we need some method of locating or finding it.

Often we want to talk about or refer to the measured quantity and recognize that it belongs to a *particular individual, but* we are not concerned about *which* individual. We do not care if the discussion centers about the dishes broken by Frank or Jill, but rather the dishes *someone* broke. Here we are still considering an X-value and in a sense a particular value. The particular one however, will be generalized and called
$$X_i \text{ (read } X \text{ sub } i\text{)}$$
with the assumption that i can mean *any* number from 1 to n.

Thus, I can talk about "the number of dishes broken by *an* employee" with the single symbol X_i, the X meaning some number of dishes and the i declaring that the number of dishes is to correspond to a particular employee, the ith employee. But we do not have to state *who* or *how many*.

Addition

Often we want to perform the operation of addition. Here we want to know the total number of dishes broken yesterday. So we add up the 2 dishes broken by Ann, the 3 by Jill, the 1 by Frank and the 1 and 2 by David and Sara (the last two were not mentioned previously). The total is 9.

If we wanted to make the statement "add the 5 values" without referring to the particular values themselves we could do this with the algebraic symbols as follows:

$$X_1 + X_2 + X_3 + X_4 + X_5 \tag{B.1}$$

and in our example this would be the same as

$$2 + 3 + 1 + 1 + 2 = 9.$$

The advantages of using the algebraic method (using X) should be evident. We do not need to write as much. Juggling the X around does not affect our understanding of the operation. And, most of all, we will be able to generalize one step further. If we do not know how many individuals are to be included in the addition we can end the addition with X_n, yielding:

$$X_1 + X_2 + X_3 + X_4 + \cdots + X_n. \tag{B.2}$$

In this way by merely stating what n is on a particular day we can complete the expression for addition, but we do not *need* to do so in order to discuss the operation. Observe that if yesterday 5 people were working and today 8 people were working, this general equation works. We just state that for yesterday n was 5 and today n is 8.

Summation Notation

In most problems the use of expressions like (B.2) are still too cumbersome so I will introduce another symbol to simplify the writing. The symbol is a Greek letter Σ (sigma), which is to be read as "the sum of" and tells you to take the sum of the quantities to the right of it. Using this symbol we will replace Equation (B.1) by:

$$\sum_{i=1}^{5} X_i. \tag{B.3}$$

The symbol has something on top of it and something on the bottom of it. These things tell you how many and which individuals you are going to be adding. The things we are adding will be X or terms involving X, but in order to condense Equation (B.1) we only want to look at *one of the terms* that sit between the plus signs. The general term in this case is X_i. But which i are to be included? That is where the $i = 1$ and the 5 come in. The number on the bottom of the $\sum$ tells where to start. The number on the top of the $\sum$ tells where to end.

It is also understood that the i increases in increments of 1. Thus,

$$\sum_{i=1}^{5} X_i = X_1 + X_2 + X_3 + X_4 + X_5. \tag{B.4}$$

(tells us where to stop; first increased by 1)

As another example, where T may mean the size of tips received by waiters;

$$\sum_{i=1}^{3} T_i = T_1 + T_2 + T_3. \tag{B.5}$$

The lower limit does *not* have to be 1.

$$\sum_{i=3}^{7} X_i = X_3 + X_4 + X_5 + X_6 + X_7. \tag{B.6}$$

The upper limit may be n, so that Equation B.2 may be simplified as:

$$\sum_{i=1}^{n} X_i = X_1 + X_2 + X_3 + \cdots + X_n. \tag{B.7}$$

Sometimes the numbers on the top and bottom of the $\sum$ are omitted. In these cases it is *understood* that we mean:

$$\sum X_i = X_1 + X_2 + X_3 + \cdots + X_n. \tag{B.8}$$

As I have stated previously, the summation sign refers to *terms* that are to be added. When the terms are more complicated than a single symbol we often need to use parentheses to identify the complete terms. Perhaps the owner disregards one broken dish so he is really concerned with adding the quantities $X - 1$. Each of the terms would be of the form $X_i - 1$, so the first would be $X_1 - 1$, the second $X_2 - 1$, and so forth, and the sum would be:

$$(X_1 - 1) + (X_2 - 1) + (X_3 - 1) + (X_4 - 1) + (X_5 - 1). \tag{B.9}$$

Using the summation notation we could replace this by:

$$\sum_{i=1}^{5} (X_i - 1). \tag{B.10}$$

For the particular values this would be:

$$\sum_{i=1}^{5} (X_i - 1) = (X_1 - 1) + (X_2 - 1) + (X_3 - 1) + (X_4 - 1) + (X_5 - 1) \tag{B.11}$$

$$= (2 - 1) + (3 - 1) + (1 - 1) + (1 - 1) + (2 - 1)$$
$$= 1 \quad\quad + 2 \quad\quad + 0 \quad\quad + 0 \quad\quad + 1 = 4.$$

Let me compare this to something that looks very similar:

$$\sum_{i=1}^{5} X_i - 1. \tag{B.12}$$

All I have done is *remove the parentheses*, but the consequence is that the number 1 is subtracted *only once* and plays no part in the summation. Thus,

$$\sum_{i=1}^{5} X_i - 1 = \underbrace{(X_1 + X_2 + X_3 + X_4 + X_5)}_{\text{this is the } \sum X_i} \underbrace{- 1}_{\substack{\text{this is the} \\ -1 \text{ part}}} \tag{B.13}$$

$$= 2 + 3 + 1 + 1 + 1 + 2 - 1$$

$$= 8.$$

The value here is 8 and was only 4 using Equation (B.11).

Let us look at a few more examples (using the same values of X).

$$\sum_{i=1}^{5} (X_i + 1)^2 = (X_1 + 1)^2 + (X_2 + 1)^2 + (X_3 + 1)^2 + (X_4 + 1)^2$$
$$+ (X_5 + 1)^2 \tag{B.14}$$
$$= (2 + 1)^2 + (3 + 1)^2 + (1 + 1)^2 + (1 + 1)^2$$
$$+ (2 + 1)^2$$
$$= (3)^2 + (4)^2 + (2)^2 + (2)^2 + (3)^2$$
$$= 9 + 16 + 4 + 4 + 9 = 42.$$

Observe that each term we added was of the form $(X + 1)^2$ and the squaring was done *before* the terms were added.

$$\sum_{i=1}^{5} (X_i^2 + 1) = (X_1^2 + 1) + (X_2^2 + 1) + (X_3^2 + 1) + (X_4^2 + 1)$$
$$+ (X_5^2 + 1) \tag{B.15}$$
$$= (2^2 + 1) + (3^2 + 1) + (1^2 + 1) + (1^2 + 1) + (2^2 + 1)$$
$$= (4 + 1) + (9 + 1) + (1 + 1) + (1 + 1) + (4 + 1)$$
$$= 5 + 10 + 2 + 2 + 5 = 24.$$

The only case when parenthesis may be omitted is with multiplication.

$$\sum_{i=1}^{5} 2X_i = 2X_1 + 2X_2 + 2X_3 + 2X_4 + 2X_5 \tag{B.16}$$
$$= 2 \cdot 2 + 2 \cdot 3 + 2 \cdot 1 + 2 \cdot 1 + 2 \cdot 2$$
$$= 4 + 6 + 2 + 2 + 4 = 18.$$

Sometimes different quantities may be involved with the same subscripts. Let X still be dishes broken and let Y be glasses broken. If Ann broke 1 glass, Jill 3, Frank 2, David 2, and Sara 3, then $Y_1 = 1$, $Y_2 = 3$, $Y_3 = 2$, $Y_4 = 2$, and $Y_5 = 3$.

$$\sum_{i=1}^{5}(X_i + Y_i) = (X_1 + Y_1) + (X_2 + Y_2) + (X_3 + Y_3) + (X_4 + Y_4)$$
$$+ (X_5 + Y_5) \quad \text{(B.17)}$$
$$= (2 + 1) + (3 + 3) + (1 + 2) + (1 + 2) + (2 + 3)$$
$$= 3 + 6 + 3 + 3 + 5 = 20.$$

Several important differences between some types of summations frequently encountered in statistics should be pointed out. The following pairs of summations look similar but are *not* similar. I will drop the $i = 1$ and the n (or 5 for the example) from these.

$$(\sum X)^2$$

and

$$\sum X^2.$$

For the example,

$$(\sum X)^2 = (X_1 + X_2 + X_3 + X_4 + X_5)^2 = (9)^2 = 81.$$

The addition is done first, then the squaring. On the other hand,

$$\sum X^2 = X_1^2 + X_2^2 + X_3^2 + X_4^2 + X_5^2$$
$$= 2^2 + 3^2 + 1^2 + 1^2 + 2^2$$
$$= 4 + 9 + 1 + 1 + 4 = 19.$$

Similarly:

$$(\sum X_i + \sum Y_i)^2$$

is not the same as

$$\sum (X_i + Y_i)^2.$$

In our example the first expression is 400 and the second is only 88.

Three Convenient Theorems of Summation

I will not prove the following, but I will demonstrate their validity using some examples.

(1) If C is a constant (say the number 6.4) then

$$\sum_{i=1}^{n} C = n \cdot C. \quad \text{(B.18)}$$

EXAMPLE

$$\sum_{i=1}^{4} 6.4 = C_1 + C_2 + C_3 + C_4$$
$$= 6.4 + 6.4 + 6.4 + 6.4$$
$$= 4(6.4) = n \cdot C.$$

(2) If C is a constant (again say 6.4) then

$$\sum_{i=1}^{n} CX_i = C \sum_{i=1}^{n} X_i. \qquad (B.19)$$

EXAMPLE

$$\sum_{i=1}^{3} 6.4 X_i = 6.4 X_1 + 6.4 X_2 + 6.4 X_3$$
$$= 6.4(X_1 + X_2 + X_3) = 6.4 \left(\sum_{i=1}^{3} X_i \right).$$

Note that this does *not* work for an expression of the type $\sum (2X_i + 1)^2$.

(3)
$$\sum_{i=1}^{n} (X_i + Y_i) = \sum_{i=1}^{n} X_i + \sum_{i=1}^{n} Y_i. \qquad (B.20)$$

This is the only one of the three to use with extra caution.

EXAMPLE

Examine Equation (B.17) and observe how you can drop all of the parentheses around the X and Y to obtain

$$\sum (X_i + Y_i) = X_1 + Y_1 + X_2 + Y_2 + X_3 + Y_3 + X_4 + Y_4 + X_5 + Y_5.$$

Rearranging these we could obtain

$$\sum_{i=1}^{5} (X_i + Y_i) = (X_1 + X_2 + X_3 + X_4 + X_5) + (Y_1 + Y_2 + Y_3 + Y_4 + Y_5)$$
$$= \sum_{i=1}^{5} X_i + \sum_{i=1}^{5} Y_i.$$

Double Subscripts and Double Summations

We have been recording as one measurement the number of dishes broken by an employee on a particular day. We have identified the individuals by using a subscript. Now let us also note the particular day and incorporate this with the X-value. Perhaps the owner decided to start keeping his record on February 1 and plans to observe how many dishes are broken for one week. So he will call February 1 by 1, February 2 by 2, and so forth. We will apply this number to our X-value also as a subscript, so we will have *two* subscripts for each number of broken dishes. Thus, we will call the number of dishes broken by Ann on that first day as X_{11} where the first 1 stands for Ann and the second 1 refers to February 1. Similarly, if Frank broke 2 dishes on February 4th we would state that $X_{34} = 2$ because Frank was employee 3, February 4th is the 4th day, and that particular X-value is equal to 2.

To generalize, we will designate the second item, day, by a different letter, j, and we now have as our general X-value the symbol

$$X_{ij}.$$

A particular individual measurement is designated by specifying both i and j. Summations may now be carried out with *either* of the items. We may either add up all the broken dishes on a given day, take the sum of all the i cases for a given j-value, or add up all the broken dishes by a given employee, take the sum of all the j cases for a given i-value. Thus we would state that "the number of broken dishes on a given day" is

$$\sum_{i=1}^{n} X_{ij}. \tag{B.21}$$

or for our example, since $n = 5$

$$\sum_{i=1}^{5} X_{ij} = X_{1j} + X_{2j} + X_{3j} + X_{4j} + X_{5j}. \tag{B.22}$$

This is a general statement for any day. By stating a day you are interested in you would obtain a more particular case. Thus, February 5th would be

$$X_{15} + X_{25} + X_{35} + X_{45} + X_{55}.$$

(Be careful to read these as X sub one, five; X sub two, five; and so forth, *not* X sub fifteen, X sub twenty-five, and so forth.)

Similarly, the number of dishes broken by someone would be

$$\sum_{j=1}^{m} X_{ij}.$$

We use a different letter for the number of j-values since it does not necessarily have to be the same as n. In fact it is different, it is 7. Since $m = 7$ this becomes:

$$\sum_{j=1}^{7} X_{ij} = X_{i1} + X_{i2} + X_{i3} + X_{i4} + X_{i5} + X_{i6} + X_{i7} \tag{B.23}$$

For a particular individual, say David (or $i = 4$) we would have 'total dishes broken' $= X_{41} + X_{42} + X_{43} + X_{44} + X_{45} + X_{46} + X_{47}$.

Often we desire to discuss the grand total summation. We will add up *all j* for all of the possible i's. This is written as a double sum, or

$$\sum_{j=1}^{m} \sum_{i=1}^{n} X_{ij} = \sum_{j=1}^{m} \left(\sum_{i=1}^{n} X_{ij} \right). \tag{B.24}$$

568   Appendix B

As the right-hand expression implies we do the right summation first and then the left one, although it generally does not matter which actually comes first.

Usually, in order to write down the actual values it is most convenient to create a kind of table that we call an array or a matrix. It has rows and columns. We will consider each row as the i category or the employees. The column will be headed as days and will be the j categories. Table B.1 is a table with all of

Table B.1 An array containing all the numbers of dishes broken by each employee on each day

		Days							Sum of number of dishes employee broke*
	j $\diagdown$ i	Feb 1 1	Feb 2 2	Feb 3 3	Feb 4 4	Feb 5 5	Feb 6 6	Feb 7 7	
Employee	Ann 1	2	1	1	0	1	1	1	7
	Jill 2	3	3	2	1	2	2	3	16
	Frank 3	1	1	1	2	0	1	1	7
	David 4	1	1	3	1	1	2	2	11
	Sara 5	2	2	1	2	3	2	3	15
Dishes broken on each day†		9	8	8	6	7	8	10	56 Total dishes broken‡

$$* \sum_{j=1}^{7} X_{ij} \qquad † \sum_{i=1}^{5} X_{ij} \qquad ‡ \sum_{j=1}^{7} \sum_{i=1}^{5} X_{ij} = 56$$

the numbers of broken dishes for all employees on all of the days. Also included are the different possible sums. Along the bottom are the sums of the dishes broken on the *particular* days. Thus $\sum_{i=1}^{5} X_{ij}$ when $j = 1$ would be $\sum_{i=1}^{5} X_{i1} = 9$, and $\sum_{i=1}^{5} X_{i2} = 8$ would be for the $j = 2$, or the second day, and so forth.

Along the right are the sums of dishes broken by different employees over the entire week. These are symbolized as $\sum_{j=1}^{7} X_{ij}$. For a specific employee, say $i = 2$ (Jill), we have $\sum_{j=1}^{7} X_{2j} = 16$.

The grand sum would be the sum of all broken dishes for all the employees for all the days and could be the double sum

$$\sum_{j=1}^{7} \sum_{i=1}^{5} X_{ij} = 56, \qquad (B.25)$$

which is found by first taking the sum of each column (each day) and then adding up those sums. The way Equation (B.25) is written we would be performing the inside summation first and then the outside summation. Thus, the steps involved are as follows:

$$\sum_{j=1}^{7} \sum_{i=1}^{5} X_{ij} = \sum_{j=1}^{7} (X_{1j} + X_{2j} + X_{3j} + X_{4j} + X_{5j}) \quad (B.26)$$

$$= \overbrace{(X_{11} + X_{21} + X_{31} + X_{41} + X_{51})}^{\text{first column, sum for } j = 1}$$

$$+ \overbrace{(X_{12} + X_{22} + X_{32} + X_{42} + X_{52})}^{\text{2nd column, sum for } j = 2}$$

$$+ \cdots + \overbrace{(X_{17} + X_{27} + X_{37} + X_{47} + X_{57})}^{\text{last column, sum for } j = 7}$$

$$= 9 + 8 + 8 + 6 + 7 + 8 + 10 = 56.$$

More Than Two Subscripts

Sometimes more than two classifications are considered. In many real applications we look at repeats or replications of measurements, and so for each X_{ij} we will perhaps have 2 or 3 separate values. Again to distinguish them apart we designate them by another letter, usually k. Thus, we end up with

$$X_{ijk}$$

and *any* of the many possible summations might be desired. In this case the *overall* sum would be

$$\sum_{i=1}^{n} \sum_{j=1}^{m} \sum_{k=1}^{r} X_{ijk}$$

(where I am calling the possible number of k-values by the letter r).

This notation is of particular use in deriving and understanding the analysis of variance. However, we will not utilize subscript notation beyond two values in this text.

PROBLEMS

For Problems B.1 to B.14 use the following values of X and Y:

$X_1 = 1$, $X_2 = 3$, $X_3 = 4$, $X_4 = 2$, $X_5 = 3$, $X_6 = 1$, $X_7 = 4$, $X_8 = 3$

$Y_1 = 2$, $Y_2 = 4$, $Y_3 = 5$, $Y_4 = 3$, $Y_5 = 7$, $Y_6 = 2$, $Y_7 = 5$, $Y_8 = 5$

Calculate the given summations:

B.1 $\sum_{i=1}^{3} X_i$

B.2 $\sum_{i=3}^{7} X_i$

B.3 $\sum_{i=1}^{4} X_i^2$

B.4 $\sum_{i=3}^{6} (X_i + 1)^2$

B.5 $\sum_{i=5}^{8} X_i + 2$

B.6 $\sum_{i=1}^{5} (X_i + Y_i)^2$

B.7 $\sum_{i=1}^{4} 2X_i$

B.8 $\sum_{i=1}^{8} 2(X_i + 2)$

B.9 $\sum_{i=1}^{4} \left(\frac{X_i}{Y_i}\right)$

B.10 $\sum_{i=5}^{8} X_i Y_i$

B.11 $\sum_{i=1}^{5} X_i Y_i^2$

B.12 $\sum_{i=1}^{4} X_i (X_i + Y_i)^2$

B.13 $\sum_{i=3}^{7} (2X_i + Y_i + 3X_i)$

B.14 $\sum_{i=1}^{5} (2X_i Y_i - 2X_i + X_i^2)$

For Problems B.15 to B.23 use the following values

		\multicolumn{5}{c}{j}
		1	2	3	4	5
i	1	2	2	3	1	4
	2	6	3	5	0	2
	3	1	4	7	3	6
	4	5	3	2	6	3

B.15 What is X_{23}?

B.16 What is X_{35}?

B.17 What is X_{32}?

B.18 What is X_{11}?

Calculate the following sums:

B.19 $\sum_{i=1}^{4} X_{i2}$

B.20 $\sum_{j=1}^{5} X_{3j}$

B.21 $\sum_{i=1}^{3} X_{i3}$

B.22 $\sum_{j=1}^{5} \sum_{i=1}^{4} X_{ij}$

B.23 $\sum_{j=1}^{3} \sum_{i=1}^{4} X_{ij}$

Appendix C
Probability

A comprehensive understanding of statistics generally requires a fair background in the concepts and fundamentals of probability. This section should provide a minimum exposure to the principal topics and rules of probability that are touched on in this text. Those desiring a more complete treatment of probability should refer to the bibliography for other sources.

In order to formally define what probability is we must first examine the term *possibility*. The idea of what is a possibility should be intuitively familiar to you. From the standpoint of probability and statistics we must recognize that we will always be working with some limited collection of possibilities. This collection of possibilities establishes a framework for what might be called an experiment. It is the set of all possible occurrences that might take place. The particular occurrences are not specified, they are random and unpredictable, but they are limited to this list of possibilities.

The most elementary types of situations are those where all of the possibilities are *equally likely*. In tossing a coin the two sides of our coin are the two possibilities and we would generally agree that they have an equal chance of appearing each and every time the coin is flipped. All six sides of a die should have the same chance of turning up, as would each card in a well shuffled deck of cards. When we sample and believe that the sample was *random* we are assuming that every individual in the population that we sampled from had the same chance of being chosen.

If we can further enumerate or list all of the equally likely possibilities we can arrive at a *total* number of possibilities. This then leads us to one definition of probability. (Several other definitions and even other fundamental principles exist. We will discuss one of them shortly.)

> The probability of one of a collection of equally likely possibilities equals $1/(Total\ number\ of\ possibilities)$.

Thus, if there are n possibilities, the probability of one of them randomly being selected is

$$P(A) = 1/n, \qquad (C.1)$$

where we read the $P(A)$ as Probability of A (where A stands for one of the possible items). This is sometimes written as $Prob(A)$ or $Pr(A)$.

The probability of a head would be $\frac{1}{2}$ because there are 2 possibilities. The probability of a 3 on a die would be $\frac{1}{6}$ since there are 6 sides to a die. The probability that I hit you with an eraser if I just randomly toss it into the midst of the 25 students in my class would be $\frac{1}{25}$.

We define the occurrence that we are considering as an *event*. Thus, an event would be the head coming up, a 3 on a die, or you being hit by the eraser. These are all simple events. We can also have more complex events that may involve more than just one of the equally likely occurrences or simple events. We could be interested in the event that an odd number of points turns up on the die, or of the case when I hit someone in the front row. Each of these consists of several simple events. The probability of any of these events is defined as

$$P(A) = \frac{\text{Number of possibilities included in the event } A}{\text{Total number of possibilities}} = \frac{n_A}{n}.$$

where n_A is the number of possibilities associated with the event called A.

This is a more general equation than the previous one. If the event we are considering only consists of 1 possibility then the two equations are identical.

We can now state that

Probability of an odd number of points coming up on a die, is

$$P(A) = \frac{3 \text{ possible ways of an odd number}}{\text{Total of 6 possible numbers}} = \frac{3}{6} = 0.5.$$

If there are 5 students in the front row, then

$$P(\text{hitting a student in the front row}) = \frac{5 \text{ possible ways}}{25 \text{ total students}} = 0.20.$$

A pictoral way of representing these events is to draw what is called a "sample space" and show all the possibilities as dots or areas on this drawing. One such picture is shown in Figure C.1. This is for the possibilities of hitting students in my class. The probability of hitting or picking one student is $\frac{1}{25}$. The probability of a more complex event involves a collection of points and the probabilities associated with each of the points.

Often it is difficult to list all of the possible occurrences. To describe all of the possible bridge hands, or even poker hands, that could be dealt with a deck of cards is an enormous task. If each student had an equally likely chance of picking any seat to sit in, then to find the probability of an empty seat left in the front row would require a listing of *all* the possible ways that seats might be occupied. This too is as mammoth an undertaking. For those occasions that we need to calculate such probabilities there are certain counting formulas to assist us. These are formulas to determine all possible combinations and all permutations (or arrangements) of various types. We will not use them in this text.*

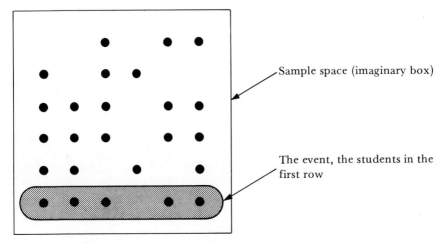

Figure C.1 A sample space for the students in my class. The points have been arranged in accordance with the seating of the students.

Another way of defining probability is empirically. We declare that the probability of a particular event is equivalent to the *relative frequency* that the particular event occurs. This assumes that we take a very large group of trials or we use a long sequence of attempts to arrive at our estimations of the relative frequencies. We will use this definition somewhat interchangeably with the previous one. However, you should realize that the equally likely definition, when applicable, provides a stronger foundation for the theoretical concepts of probability (although we will not need to be concerned about such differences).

Some Rules of Probability

RULE 1 *The probability of a single event, $P(A)$ is always some number from 0 to 1* or

$$0 \leq P(A) \leq 1$$

(*where $\leq$ means less than or equal to*).
In addition,

if $P(A) = 0$, *the event is* impossible;
if $P(A) = 1$, *the event is* certain to occur.

★ The main formulas are: 1. Number of ways N distinct objects can be arranged in a sequence is $N! = 1 \cdot 2 \cdot 3 \cdot 4 \ldots N$ or "N factorial"; 2. Number of ways of arranging r objects from N distinct objects is $P(N,r)$, the permutations of N things taken r at a time $= n!/(N - r)!$; 3. Number of combinations is $C(N,r) = N!/r!(N - r)!$.

The probability of obtaining a 9 on a single die is 0 since a single die does not have a 9 on it. The probability of obtaining more than 0 dots is 1 since there are no sides with no dots on a die.

RULE 2 *The probability that a particular event will not occur is 1 minus the probability that it will occur. This is called the complement of the event and is written as A' (A prime):*

$$P(A') = 1 - P(A). \tag{C.3}$$

Then

$$P(\text{not getting a 3 on the die}) = 1 - P(\text{getting a 3 on a die})$$
$$= 1 - \tfrac{1}{6} = \tfrac{5}{6}$$

$$P(\text{not getting an odd value on a die}) = 1 - P(\text{get odd value})$$
$$= 1 - \tfrac{3}{6} = 1 - \tfrac{1}{2} = \tfrac{1}{2}$$

$$P(\text{not hitting someone in front row}) = 1 - P(\text{hit someone in front row})$$
$$= 1 - \tfrac{5}{25} = \tfrac{20}{25} = \tfrac{4}{5}.$$

Odds: The odds of an event occurring is the ratio of

$$P(A) : P(A'),$$

or

(Probability of the event occurring) to (Probability of the event *not* occurring)

(the : is read as *to*).

It can also be found from the ratio of the number of possibilities of A to the remaining (not A) possibilities, or

$$n_A : n_{A'}.$$

The odds of *getting* a 3 on a die are 1:5 (1 to 5) since there is 1 way of getting a 3 and 5 ways of not getting a 3. This is often reversed. Thus, if the odds are 1:5 for rolling a 3, then the odds are 5:1 for *not* rolling a 3 (or against getting a 3).

The odds of hitting a student in the front row are 5:20 or 1:4. The odds of *not* hitting someone in the front row are 20:5 or 4:1.

Probability Rules for More Complex Events

The areas of statistics which utilize probability often require us to look at somewhat more complex events. For example, samples are not usually single individuals, but consist of the selection of several individuals often with varying probabilities of selection. Two types of problems will be discussed. The first tackles the questions of *either-or* situations while the second looks at *and* cases. These will closely parallel and be related to addition and multiplication.

ADDITION RULE *If two events, A and B are* mutually exclusive *then*

$$P(A \text{ or } B) = P(A) + P(B). \tag{C.4}$$

By *mutually exclusive* we mean that no individual belongs to both A and B. Odd and Even are mutually exclusive, since a number is either odd *or* even, but cannot be both. A student may sit in the front row or the second row, but he cannot be in both locations at the same time. If you use a sample space to represent the possibilities, such as that in Figure C.1, then mutually exclusive events are sets of points or areas that do not overlap.

The probability of hitting a student in the first row (event A) *or* hitting a student in the back row (call this event B) is

$$P(A \text{ or } B) = P(A) + P(B) = \tfrac{5}{25} + \tfrac{3}{25} = \tfrac{8}{25} = 0.32.$$

When the two events are *not* mutually exclusive we arrive at a more general equation. If we just add up the two probabilities we will count those individuals that are in *both* events each time. So we will obtain too high a value for the probability.

Consider the probability of rolling an odd number *or* a number larger than 2 on a single roll of a die. We have already found that the probability of rolling an odd number was $\tfrac{1}{2}$ and the probability of a number larger than a 2 would be $\tfrac{4}{6}$ (since there are 4 values larger than 2). Then using C.4 we would have

$$P(A \text{ or } B) = P(A) + P(B) = \tfrac{3}{6} + \tfrac{4}{6} = \tfrac{7}{6} = 1\tfrac{1}{6},$$

which is impossible. We have counted all of the odd numbers that are also larger than two in both of the probabilities. Three and five were counted twice and we must subtract off this overcount. The general equation is

$$P(A \text{ or } B) = P(A) + P(B) - P(A \text{ and } B). \tag{C.5}$$

For the die this is

$$P(A \text{ or } B) = \tfrac{3}{6} + \tfrac{4}{6} - \tfrac{2}{6} = \tfrac{5}{6}.$$

In Figure C.2 I have indicated which students are males and which are females. I have also outlined the Event "hit a female" as Event B. Now we ask the question what is the probability of hitting someone in the front row or a female. Then the last term of Equation C.5, $P(A \text{ and } B)$, covers the females in the front row, since they are in *both* events and would otherwise be counted twice. Therefore:

$$P(\text{in front row or female}) = \tfrac{5}{25} + \tfrac{12}{25} - \tfrac{2}{25} = \tfrac{15}{25} = 0.60.$$

You should also realize that if you have mutually exclusive events then $P(A \text{ and } B) = 0$ and Equation (C.5) is identical to Equation (C.4). (We will devote more time to the *and* case under the multiplication rule.)

Probability and Independence

Sometimes we will partially reduce a sample space or the collection of individuals and then need to describe a set of probabilities for the reduced sample space. We may be given some information that clarifies or simplifies the distribution of possibilities or we may just be interested in part of the overall collection of possibilities.

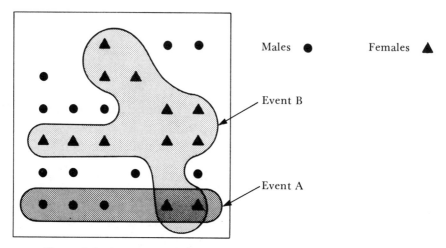

Figure C.2 Sample space for students in my class. I have now indicated which are males and which are females. Two events are illustrated: *A* is students in the front row, *B* is a female.

If I revealed that I could only throw an eraser two rows, then you might ask what is the probability of hitting a female. We state this as "what is the probability of hitting a female *given* that she must be in the first two rows?" We have *reduced* the overall sample space to a *total of only 9 students and 2 females*. Symbolically we write this as

$$P(\text{female}|\text{first 2 rows}) = P(B|C) = \tfrac{2}{9}$$

(where *C* is the event "the first two rows"). The vertical line is to be read as *given*.

As another example, if you knew that the person hit was a female, what is the probability that she was in the first row? The total *n* is now 12 and you are reduced to looking only at the area enclosed by event B. Given that you must remain in this area, there are two in the first row. Therefore

$$P(\text{hit a person in front row}|\text{female}) = \tfrac{2}{12}.$$

From a probability standpoint, if the *given* information or given event does not assist or change the probabilities at all, then we define the two events as independent. Symbolically we say

if
$$P(A|B) = P(A),$$
then *A* and *B* are independent.

If I threw two erasers and the tossing of the first changed nothing, then the two events are independent. The probability of hitting a particular person was $\frac{1}{25}$ and would not be different as a result of knowing that he was or was not hit on the previous toss. The tosses are independent. If, however, getting hit one time meant you learn to duck or you leave the room then there is a change in the probability of being hit that second time.

Multiplication Rule—Joint Probability

We are now able to examine more carefully and more fully the probability associated with an *and* statement. We briefly discussed one situation under the addition rule, but the *and* statement is more general than was implied there and might even be considered as playing a greater role in the field of statistics.

Whenever we discuss a sample we are considering the results of several observations. In order to describe the probabilities for samples we must first be able to define the set of probabilities for all of the *single* observations. But the sample is the result of one observation *and* then another *and* perhaps another, and so forth, so that we amass a set of *and* statements. We are concerned about the probability of my hitting you today *and* tomorrow or not today *and* not tomorrow. We ask about the rolling of a 3 *and* then a 4. I ask what is the probability you will show up for the test *and* receive a 100 on the test.

The formula for the probability of these *joint* events is

$$P(A \text{ and } B) = P(A) P(B|A) = P(B) P(A|B). \tag{C.6}$$

If the two events are *independent* then this reduces to

$$P(A \text{ and } B) = P(A) P(B).$$

Let us look at a few examples. We already described P(female and in front row) in a previous section. Now let us redo that problem using Equation (C.6).

$$P(\text{female } and \text{ in first row}) = P(\text{female}) P(\text{in first row}|\text{female}).$$

Now

$$P(\text{female}) = \tfrac{12}{25}$$

and we just found the P(in first row|female) was $\tfrac{2}{12}$, therefore

$$P(\text{female } and \text{ in the first row}) = P(A) \cdot P(B)$$
$$= \tfrac{12}{25} \cdot \tfrac{2}{12} = \tfrac{2}{25},$$

which agrees with the result we found before.

Conditional events are common if we take a sample and do not replace the individual we sample. Thus, if after taking the first sample we have changed the population or sample space, we also alter the probabilities associated with the second sample. In such cases we must concern ourselves with the conditional probabilities.

More frequently we work with the preferable cases of independent samples and replacement. We essentially do not change the population and repeated samples have the same probabilities associated with them. Thus the probability of joint occurrences is the product of the individual probabilities. For example

$$P(\text{hitting 2 males on 2 tosses}) = P(\text{hit a male on first toss})$$
$$\cdot P(\text{hit a male on second toss})$$
$$= \tfrac{13}{25} \cdot \tfrac{13}{25} = \tfrac{169}{625} = 0.27.$$

It is for this reason that we declare that if the probability of a particular observation is very small, say 0.05, then the probability of it occurring twice is even smaller, namely

$$P(\text{this rare event occurring two times}) = P(\text{event})\, P(\text{event})$$
$$= (0.05)(0.05) = 0.0025.$$

PROBLEMS

The area codes for five widely separated states are listed here. For the problems in this appendix we will only concern ourselves with these 20 area codes.

New York	212, 315, 516, 518, 716, 607, 914
California	209, 916, 717, 213, 408, 707, 415
Florida	305, 904, 813
Kentucky	502, 606
Colorado	303

C.1 Draw a sample space for these area codes.
For Problems C.2 to C.8 determine the probability for randomly picking the appropriate area code(s).
C.2 I will reach California.
C.3 I will reach Colorado.
C.4 The area code starts with a 2.
C.5 The area code is odd.
C.6 The second number is a 0.
C.7 The second number is not a 0.
C.8 The area code starts with a 1.
C.9 What are the odds that the first number is even?
C.10 What are the odds that I reach Kentucky?
C.11 What is the probability of reaching New York or California?
C.12 What is the probability of an area code ending in a 0, 1, 2, or 3?
C.13 Which states have mutually exclusive first digit area codes?

For Problems C.14 to C.23 determine the appropriate probabilities.

C.14 The area code starts with a 5 or is from New York.
C.15 Area code starts with an odd number or is for a state bordering an ocean.

C.16 It is an odd number area code given that it must be from California.
C.17 It is an odd number area code given that it must be from New York.
C.18 It is an odd number area code given that it must be from Colorado.
C.19 Picking a number from New York three times.
C.20 Picking a number from Colorado twice.
C.21 Picking a number from New York and with middle number 1.
C.22 The number starts with a 7 and is from California.
C.23 The first and last numbers are the same.

Bibliography

The accompanying list of references covers a broad range of general and specific texts in both elementary and advanced levels of statistics and probability. I have also provided a guide to these texts according to mathematical level and by topic. This bibliography is by no means exhaustive and merely serves as a point of departure for further work.

GENERAL BOOKS FOR THE LAYMAN. The following are essentially nonmathematical treatments of some areas of statistics and can be read for pleasure.
Huff (20)
Slonim (36)
Scott (34)
Clague (6)

GENERAL STATISTICS BOOKS (BASICALLY NON-PROBABILITY ORIENTED). These should serve as good references for more examples on statistical analysis and other techniques.
Dixon and Massey (10)
Haber and Runyon (16)
Moroney (24)
Roscoe (32)
Snedecor (37)
Walker and Lev (38)
Weinberg and Schumaker (39)

ELEMENTARY STATISTICS BOOKS WITH A PROBABILITY ORIENTATION. These books generally first develop some of the probability theory and then begin the study of statistics.

Alder and Roessler (1)
Freund (13) (the longer of the two books)
Freund (14)
Mendenhall (23)
Mosteller, Rourke, and Thomas (27)
Wilks (42)

MORE ADVANCED STATISTICS BOOKS WITH EMPHASIS ON METHODOLOGY.
Anderson and Bancroft (2)
Bennett and Franklin (3)
Brownlee (5)
Miller and Freund (24)

MORE ADVANCED THEORETICAL OR MATHEMATICAL STATISTICS BOOKS.
Hoel (19)
Hoog and Craig (20)
Mood and Graybill (25)

PROBABILITY THEORY.
Feller (12)
Parzan (29)

TABLES.
Beyer (4)
Hald (17)
Owen (28)
Pearson and Hartly (30)
RAND (32)
Romig (33)

Some Specialized Statistics Books

NONPARAMETRIC STATISTICS.
Siegal (37)

QUALITY CONTROL.
Grant (15)
Juran and Gryna (22)

DESIGN OF EXPERIMENTS.
Cochran and Cox (8)
Davies (9)
Hicks (18)

SAMPLING.
Cochran (7)

REGRESSION.
Draper and Smith (11)

BAYESIAN STATISTICS.
Riaffa (31)
Schmitt (35)

1. Alder, H. L., Roessler, E. B., *Introduction to Probability and Statistics*. W. H. Freeman and Co., San Francisco, Calif., 1968.
2. Anderson, R. L., Bancroft, T. A., *Statistical Theory in Research*. McGraw-Hill Book Co., New York, 1952.
3. Bennett, C. A., Franklin, N. L., *Statistical Analysis in Chemistry and the Chemical Industry*. John Wiley and Sons, Inc., New York, 1954.
4. Beyer, W. H., Ed., *Handbook of Tables for Probability and Statistics*. Chemical Education Rubber Co., Easton, Pa., 1966.
5. Brownlee, K. A., *Statistical Theory and Methodology in Science and Engineering*, 2nd ed. John Wiley and Sons, Inc., New York, 1965.
6. Clague, E., *The Bureau of Labor Statistics*. Frederick A. Praeger, Inc., New York, 1968.
7. Cochran, W. G., *Sampling Techniques*, 2nd ed. John Wiley and Sons, Inc., New York, 1963.
8. Cochran, W. G., Cox, G. M., *Experimental Designs*, 2nd ed. John Wiley and Sons, Inc., New York, 1957.
9. Davies, O. L., *Design and Analysis of Industrial Experiments*, 2nd ed. Hafner Publishing Co., New York, 1963.
10. Dixon, W. J., Massey, F. J., Jr., *Introduction to Statistical Analysis*, 3rd ed. McGraw-Hill, Inc., New York, 1969.
11. Draper, N. R., Smith, H., *Applied Regression Analysis*. John Wiley and Sons, Inc., New York, 1966.
12. Feller, W., *An Introduction to Probability Theory and Its Application*, 3rd ed. John Wiley and Sons, Inc., New York, 1968.
13. Freund, J. E., *Modern Elementary Statistics*, 3rd ed. Prentice-Hall, Englewood Cliffs, N.J., 1967.
14. Freund, J. E., *Statistics, A First Course*. Prentice-Hall Inc., Englewood Cliffs, N.J., 1970.
15. Grant, E. L., *Statistical Quality Control*, 3rd ed. McGraw-Hill, New York, 1964.
16. Haber, A., Runyon, R. P., *General Statistics*. Addison Wesley Publishing Co., Reading, Mass., 1969.
17. Hald, A., *Statistical Tables and Formulas*. John Wiley and Sons, Inc., New York, 1952.
18. Hicks, C. R., *Fundamental Concepts in the Design of Experiments*. Holt, Rinehart, and Winston, Inc., New York, 1964.
19. Hoel, P. G., *Introduction to Mathematical Statistics*, 3rd ed. John Wiley and Sons, Inc., New York, 1962.

20. Hogg, R. V., Craig, A. T., *Introduction to Mathematical Statistics*, 2nd ed. The MacMillan Co., New York, 1965.
21. Huff, D., *How to Lie with Statistics*. W. W. Norton and Co., Inc., New York, 1954.
22. Juran, J. M., Gryna, F. M., Jr., *Quality Planning and Analysis*. McGraw-Hill, New York, 1970.
23. Mendenhall, W., *Introduction to Probability and Statistics*, 2nd ed. Wadsworth Publishing Co., Inc., Belmont, Calif., 1967.
24. Miller, I., Freund, J. E., *Probability and Statistics for Engineers*. Prentice-Hall Inc., Englewood Cliffs, N.J., 1965.
25. Mood, A. M., Graybill, F. A., *Introduction to the Theory of Statistics*, 2nd ed. McGraw-Hill Book Co., New York, 1963.
26. Moroney, N. J., *Facts From Figures*, 3rd ed. Penguin Books, Baltimore, Md., 1956.
27. Mosteller, F., Rourke, R. E. K., Thomas, G. B., Jr., *Probability and Statistics*. Addison Wesley Publishing Co., Reading, Mass., 1961.
28. Owen, D. B., *Handbook of Statistical Tables*. Addison Wesley Publishing Co., Reading, Mass., 1962.
29. Parzen, E., *Modern Probability Theory and Its Applications*. John Wiley and Sons, Inc., New York, 1960.
30. Pearson, E. S., Hartley, H. O., *Biometrika Tables for Statisticians*, 3rd ed. Cambridge University Press, New York, 1966.
31. Raiffa, H., *Decision Analysis*. Addison Wesley Publishing Co., Reading, Mass., 1968.
32. RAND Corporation, *A Million Random Digits with 100,000 Normal Deviates*. The Free Press, Macmillan Corp., New York, 1955.
33. Romig, H. G., *50–100 Binomial Tables*. John Wiley and Sons, Inc., New York, 1953.
34. Roscoe, J. T., *Fundamental Research Statistics for the Behavioral Sciences*. Holt, Rinehart, and Winston, Inc., New York, 1969.
35. Schmitt, S. A., *Measuring Uncertainty: An Elementary Introduction to Bayesian Statistics*. Addison Wesley Publishing Co., Reading, Mass., 1969.
36. Scott, A. H., *Census, U.S.A*. The Seabury Press, New York, 1968.
37. Siegel, S., *Nonparametric Statistics for the Behavioral Sciences*. McGraw-Hill Book Co., Inc., New York, 1956.
38. Slonim, M. J., *Sampling*. Simon and Schuster, New York, 1960.
39. Snedecor, G. W., Cochran, W. G., *Statistical Methods*, 6th ed. Iowa State University Press, Ames, Iowa, 1967.
40. Walker, H. M., Lev, J., *Elementary Statistical Methods*, 3rd ed. Holt, Rinehart, and Winston, Inc., New York, 1969.
41. Weinberg, G. H., Schumaker, J. A., *Statistics: An Intuitive Approach*, 2nd ed. Brooks/Cole Publishing Co., Wadsworth Publishing Co., Inc., Belmont, Calif., 1969.
42. Wilks, S. S., *Elementary Statistical Analysis*. Princeton University Press, Princeton, N.J., 1948.

ANSWERS

Answers to Selected Problems and Portions of Problems

1.1 For example: Has good flavor, aroma, sweetness; also, different grinds, different perking times, water temperature, amount of coffee. Perk coffee for different lengths of time using different amounts of coffee and have several people taste it.

1.3 (a) Amount of reading (how many books), using a poll by librarian. Reading speed, using a test by a teacher. (b) Number, or proportion, of people wearing glasses, survey by an insurance company. Types of frames, survey by an eyeglass manufacturer.

1.5 For example: Long life, if bulbs are in out-of-the-way places; regular in other places. Wattage is associated with cost of operation. In some places the voltage is high and low voltage bulbs will burn out quickly. Amount of light (lumens) would be specific measure of the ability of the bulb to light up a particular area to a particular degree. Each requires a different instrument (clock, wattmeter, voltmeter, light meter).

1.7 (a) SAT of all college students; SAT of all high school graduates; SAT of all high school students (the first is mostly within the second, the second is within the third). SAT of all males; SAT of all males over 20 years old (the last is within the next to last).
(c) Number of divorces in N.Y.; number in Nevada; number of divorces of people earning over $20,000 a year; number who had been divorced before; number of teenage divorces. (All can be different populations).

1.9 (a) Students who read faster also read more. Students who took a reading course read faster than those who did not take the course. Using a test with a national norm is working with a known population. The second hypothesis above is a comparison of two populations.   (b) College students have a greater proportion of eyeglasses than the general public (known population). More males wear glasses than females (two populations).

2.1 (a) (1) a psychologist; (2) measure response time for a particular learning task; (3) all persons; (4) tiredness and illness not related to the experiment, different people might react very differently to the drugs; (5) might conclude that a particular drug helps learning when in fact it does not, and this might lead to a poor recommendation regarding its use; (6) if a person volunteered, determine why he volunteered. Must make sure that there are control persons who are not using the drugs and the person using the drug should not be aware that he is definitely the one with the drug (must select the user randomly).

2.3 Were they all *able* to take the advanced course? Were some of the students in their last semester and would graduate before being able to take the course? The students should have been polled after the course was over.

2.4 Since this is an overall average, the cost is less than for a private college but more than for a commuter going to a state supported school. In fact, it might not represent *any* student.

2.7 (a) It is questionable that the government releases exact figures. Do you define a missile as one capable of being fired, being installed, being repaired, being built?   (c) With every wave the area of a beach changes. Only, if you chose a particular instant, then it would be impossible to measure all of the beach area of so large a region.

3.1 0–23.3%, 1–26.7%, 2–30%, 3–6.7%, 4–6.7%, 6–3.3%

3.2 There are 100 places for every 0.001 from 19.955 to 20.055. This makes it very spread out.

3.3 Categorical; B–1, L–1, D–1, BL–2, BD–6, LD–4, BLD–9

3.6 Number of customers; need at least 135 places on the x-axis. Very unwieldy.

3.7 This becomes reasonable. Histogram depends on intervals selected, for example:

Midpoint	Frequency	Midpoint	Frequency
19.9595	4	20.0195	13
19.9695	8	20.0295	7
19.9795	14	20.0395	5
19.9895	15	20.0495	4
19.9995	14	20.0595	1
20.0095	15		

3.9 For example: Midpoints 1.5, 3.5, 5.5, etc.; limits 0.5–2.5, 2.5–4.5, 4.5–6.5, etc.

3.11 19.994, 20.007, 0.53

3.13 Number of customers up to a particular day; number of days with at least so many customers

3.14 (b) 0.80   (c) 6 or 30%   (d) 21 to 21.75 (actually from 21.195–21.711)

3.15 (a) 25th   (b) 75th, 3   (c) 2   (d) 5   (e) 30   (f) 60   (g) 9   (h) $Q_3 - Q_2$

3.17 (a) Skewed to left; most should do well   (b) Skewed to right; most will do very poorly   (c) Bimodal—2nd graders do poorly, freshmen do well   (d) Slightly skewed to right; 2nd graders do poorly and freshmen may do slightly above poor (perhaps)

4.1 Mean = 29485, median = 29665, mode = none, midrange = 29345

4.3 Must convert to numbers; mean = 2.25, median = mode = midrange = 2. All except the mean are about the same difficulty.

4.4 Mean = 75.3, median = 48.5, mode = none, midrange = 99.5. All are easy. The midrange is probably the easiest. Median.

4.6 (a) Means: diameter, 2.49; height, 4.41; diameter, 21.490. Median: diameter, 2.49; height, 4.41; diameter, 21.426. Midrange: diameter, 21.48; height, 4.40; diameter, 21.328.   (b) Volume calculated from the average diameter and height, using means = 21.47; using medians = 21.47; using midrange = 21.25.

4.8 Mean involves too much addition; median = 20.002, mode = 20.013, midrange = 20.005

4.10 Depends on intervals, using 8 intervals of $500 each with midpoints starting at $27,500. Mean = 29536. Slight difference.

4.11 I used groups of 0.010 with midpoint starting at 19.960, mean = 20.0025; two modes, median = 20.000 and midrange = 20.010.

4.13 (a) From Tables 4.3–4.11, actual = 10   (b) Median = 10.08

4.16 (a) $C = 3.00$, $L = 0.25$ (b) For $X = 1.00$, $X' = -8$; $X = 1.25$, $X' = -7$; etc.   (c) Mean = 3.05, median = 3.00, mode = 3.25, midrange = 3.625

4.18 190.83

4.19 $2.375

4.20 71.4

4.23 Range = 15, variance = 16.73, $s = 4.09$

4.24 Range = 6, $s^2 = 1.97$, $s = 1.40$

4.25 Range = 3, $s^2 = 0.46$, $s = 0.678$

4.26 $s = 51.7$

4.27 $s = 0.85$

4.28 $s = 917$

4.29 Average deviation = 3.17

4.30 Average deviation = 1.07
4.35 Diameter, $s = 0.0295$; height, $s = 0.169$; volume, $s = 0.488$
4.39 (a) $\bar{X}$ (b) 0
4.40 3.4, -2.4, 0, 157, 131.5, 148.5
4.42 19.057, 20.051
4.43 -98.8, 125.4, much smaller set of limits

5.6 (a) 2.574 $E01$ (b) 2.69108 $E05$ (c) 6.873 $E-02$
5.7 (a) 6786500 (b) 2750000000000 (c) 0.0005982
5.8 0, 16.67

6.1 $r = 0.69$
6.2 $r = -0.978$
6.3 $r = 0.64$
6.4 $r = 0.989$
6.6 Yes, at 0.05
6.7 Yes, at 0.01
6.8 Yes, at 0.01
6.10 (a) Yes, at 0.001 (b) yes, at 0.05 (c) no
6.11 $r = 0.218$, not significant
6.12 Doubtful if cause and effect; other factors like differences between typists could influence values. High speed and few errors might be generally associated with better typists.
6.15 Cause and effect (income leads to tax)
6.16 $Y = \text{rating} = -0.55 + 1.01X$, 7.03, 4.5
6.17 $Y = \text{staples} = 27.6 - 1.3X$, 14.43, 1.22
6.18 Errors $= 0.509 + 0.0437S$, $E_{90} = 4.11$
6.22 $r = 0.897$

7.1 (a) There may be changes in the machine over long spans. Different batches of aspirin may cause jumps in concentration. Aging. Control is applicable—same consistancy between tablets should be possible as between bottles. (b) As you become older (time), you should get higher test results because of accumulated learning. Repeating in short interval probably also leads to higher test results. Not truly controllable. (d) Generally, most drivers drift either up or down in speed as they drive. They then abruptly change speed as they notice the speed at which they are traveling. Somewhat controllable.
7.3 $\bar{\bar{X}} = 8.26$, $\bar{R} = 0.46$, $UCL_R = 1.04$; all points in control
$UCL_{\bar{X}} = 8.59$, $LCL_{\bar{X}} = 7.93$; many points out of control (6)
7.4 $\bar{\bar{X}} = 4.99$, $\bar{R} = 0.054$, $UCL_R = 0.12$; in control
$UCL_{\bar{X}} = 5.03$, $LCL_{\bar{X}} = 4.95$; 2 points out of control
7.9 (a) Only certain seasons are busy ones. Except certain peak times.
(c) More advertising on Sunday; little on Saturday; different

7.10 Limits for 0.05 are 4.38 and 5.38. Averages are Brag–4.6; AndNoneAsFine–4.82; X–5.22. No significant differences.

7.11 Limits for 0.05 are 8.008 and 8.404. One row average outside limits.

7.13 North to South are 1% limits, 8.26 and 8.70. All outside limits. East to West are 1% limits, 8.23 and 8.73. 4 out of 5 outside limits.

8.1 (a) Weight: 16 20 21 25 30 31 35 40
    Frequency: 2 1 2 4 1 1 2 2
(b) Weight: 26 31 35 36 40 41 45 46 50 55
    Frequency: 1 4 2 1 2 2 1 2 4 1
(c) Weight: 12 16 20 21 25 30 31 35 40
    Frequency: 1 2 4 2 4 1 1 2 2
(This does not mean that I go home with 12 ounces of fish, etc.)

8.2 (a) Cost: 13 14 15 16 17 18 19
    Frequency: 1 2 3 3 3 2 1
(b) No, 20% of the cases will be $16. (c) Impossible, wrong book, or gypped

8.5 With replacement when we toss the fish back; without replacement when we keep the fish

8.7 (a) If you permit each person who has been asked to be asked again. This can slant the results if you use a small sample. (b) If one of the teachers gives an unusual amount and is permitted to be sampled twice. This may distort the results especially if only a small sample is taken.

8.8 (a) discrete (b) continuous (c) discrete (d) discrete

9.1 (a) For example: The two dies are the same or different; the sum is 7, or not 7. (b) For example: The party I call is there or is not there; I get the correct number or a wrong number.

9.2 (a) Some students come together and if one decides not to come some others may be influenced. (b) Different students may understand different amounts of the subject or have studied more. (c) Some matches may be damper or damaged and have a smaller chance of igniting than the rest of the matches.

9.6 (a) 0.3125 (b) 0.1094 (c) 0.0413 (d) 0.1738
    (e) 0.6477 (f) 0.0708

9.7 (a) 30% (b) 0.1681 (c) 0.0282 (d) none 0.0000
    (1) 0.001   (6) 0.2001
    (2) 0.0014   (7) 0.2668
    (3) 0.0090   (8) 0.2335
    (4) 0.0368   (9) 0.1211
    (5) 0.1029   (10) 0.0282
    (e) 0.0473 (f) 0.0007

Answers

9.8   0.0509, 0.2447
9.9   (a) 0.0625   (b) 0.0000   (c) 0.0207   (d) 0.9793
      (e) 0.0000
9.12 (a) 3.5   (b) 7, 1.45   (c) 7000   (d) Between 6004 and 7916
9.13 (a) 15   (b) 120, between 106 and 134
9.16 160, between 153 and 177
10.1 (a) 0.0793   (b) 0.4920   (c) 0.4292   (d) 0.3256
      (e) 0.6406   (f) 0.0656   (g) 0.0778   (h) 0.4404
      (i) 0.9332   (j) 0.3085
10.2 (a) 0.253   (b) 1.037   (c) -0.319 and $+0.319$
      (d) -0.253
10.3 (a) 0.1915   (b) 0.3830   (c) 0.1586   (d) 0.9053
10.4 53, 85
10.5 0.0918, 16.94
10.6 0.1112, 0.8888
10.7 (a) 4.75%,   (b) 19.71%,   (c) 93.8
10.9 0.0475, 0.4325, 0.3642
10.11 (a) 44%,   (b) 24.1,   (c) 12.4
10.12 0.1379, no
10.13 0.9976
10.14 0.0019

11.1 The person is *not* guilty.
      Alternate: The person is guilty.
      Errors: Decide an innocent person is guilty, decide a guilty person is innocent.
      Consequence: Imprison an innocent person, let guilty person go free. Society generally considers the first as a more serious error.
11.3 You have cancer. You do not have cancer. The second is preferred.
      Errors: Declare no cancer when you do have it (alpha error).
              Declare you have cancer when you do not (beta error).
      Consequence: May miss opportunity to treat it before it spreads.
                   May operate when unnecessary.
11.5 No change in yield. Alternate: There is an increase.
      Not concerned with a decrease, a one-sided, one-directional test.
11.7 No. Hypothesis: Percent germinating is no different than 90%.
      Alternate: Less than 90%, one-direction.
      85% of planted seeds germinate. The statistic would be the number of seeds germinating.
      Errors: Can agree that 90% germinate but actually less germinate (beta error).
              Can say that less than 90% germinate but actually 90% germinate (alpha error).

11.10 Inspect a sample. Yes, he can make an error. Probably not sell them (?) The customer might send them back after he examines them and, or, he might not buy any more.

12.1 Means: 5.00, 5.25, 3.00, 5.50, 6.25
s^2: 2.67, 0.92, 9.00, 5.67, 7.58
12.3 Expect: mean = 4.5, $\sigma = 3.61$, std. error = 1.80
12.4 0.0793
12.5 (a) 336 (b) 0.61 (c) 12.8–15.2 (d) 307.2–364.8
12.6 51.3
12.11 12. A sample of 12 has variance of 0.0571 σ^2.
12.13 (a) 1.812 (b) −1.337 (c) −1.321 to +1.321
 (d) 1.860 (e) −1.310 to +1.310
12.14 (a) 18.3 (b) 1.15 (c) 18.5 (d) 16.5 to 34.4
 (e) From Table A.4, 226
12.15 (a) 5.41 (b) 2.16 (c) 3.21 (d) 2.78 (e) 2.46
 (f) 16.26 (g) 0.023, 6.48

13.3 45.5; 44.1 to 46.9
13.4 80.2 to 119.8
13.6 6.4; 4.7 to 8.1
13.7 36.7 to 59.3
13.8 23.7–30.3. Disagree with company's claim, 19 not in this interval.
13.11 Hypothesis: Texts cost no different than $10.
 Alternate: Texts are different than $10.
 Limits: 8.65 to 11.35.
13.12 Limits: 12,980 and 17,020
13.14 15.21 and 16.79; 0.3015; 0.0384
13.15 225.2 and 284.8; 0.1762; 0.0003
13.16 0.2119
13.17 0.0375, 0.4880, 0.9357, 0.9998, 1.0000

14.1 Correlation coefficients, z values, χ^2 values
14.2 0.001 would be more unusual than even the 0.01 level; 0.10 level is less unusual than any discussed
14.5 (a) Symmetrical (b) Not symmetrical until large df—lower limit tends to be very small and bunched near peak (c) Symmetrical (d) Similar to description of χ^2
14.6 (a) ±1.645 (two-sided) vs. 1.282 (one-sided) (b) 2.56 and 23.2 vs. approximately 21 (c) ±2.845 vs. 2.528
14.7 Runs = 13; 0.01 limit = 13, highly significant
14.8 Runs = 9, not significant
14.9 Runs = 2, highly significant
14.11 Runs = 15, not significant
14.12 $r_{10} = 0.61$, significant at 0.05
14.13 $r_{10} = 0.296$

14.15 r_{21} (two-sided) = 0.267 or 0.364

15.1 $\chi^2 = 3.14$, not significant
15.2 $\chi^2 = 0.960$, not significant
15.3 $\chi^2 = 19.64$; $\chi^2_{0.99} = 11.3$
15.4 $\chi^2 = 8.83$, not significant
15.5 $\chi^2 = 82.66$
15.6 $\chi^2 = 2.56$, not significant
15.7 Depends on the number of categories; $\chi^2 = 2.12$ using 8 = areas
15.8 $\bar{X} = 50.5$; $s = 24.4$; using 5 categories, $\chi^2 = 2.790$
15.10 Using 5 categories, $\chi^2 = 7.84$; significant at 0.025
15.12 $D = 0.03$, 0.05 limit is 0.111, not significant
15.13 $D = 0.17$, not significant
15.14 $D = 0.02$
15.15 $D = 0.30$
15.19 (a) Limit: $z_{0.99} = 2.33$; test value $z = 2.2$, significant at 0.025
    (b) 18.0% to 36.6%   (c) 1080 to 2196
15.20 Very significant, $P = 0.005$
15.22 199
15.23 0.01, z limit $= -1.28$, $z = 1$, not significant
15.25 $z = 20.3$; yes, 5.70 to 23.70 per hour; 95%
15.27 1024.3
15.28 1067.0
15.29 4150

16.1 One-sided only if $\sigma^2 < \sigma^2_{theo}$
    Limit = 0.416, $s^2/\sigma^2 = 0.64$, not significant
16.2 0.05, limit = 2.01, $s^2/\sigma^2 = 1.43$, not significant
16.4 $s^2/\sigma^2 = 6.75$, very significant; $1.11 < \sigma^2 < 7.87$
16.5 $s^2/\sigma^2 = 3.0$, significant
16.6 0.10, limit = 0.652
    $s^2/\sigma^2 = 0.111$, significant
16.8 Limit: $z = \pm 1.96$, $z = 4$, yes
16.9 $z_{0.01} = -2.326$, $z = -3.8$, significant
16.10 $z_{95} = 1.645$, $z = 1.51$, not significant, $-\$23,600$
16.12 $z = \pm 1.96$, $z = 2.44$, significant
16.13 Limit: $t = \pm 2.262$, $t = -2.98$, significant, 44.7 to 49.3
16.14 $t = 2.62$
16.15 $t = 4.53$
16.16 $t = 4.24$
16.19 (a) 62   (b) 27
16.20 16
16.21 49

17.1 $\chi^2 = 2.65$, not significant
17.2 $\chi^2 = 20.25$, not significant

17.3 $\chi^2 = 22.88$
17.4 $\chi^2 = 3.90$
17.5 $z = -2.71$, limit $= \pm 1.645$, significant
17.6 $z = 1.75$, not significant
17.7 $z = -2.12$
17.8 $z = 0.325$
17.10 $U = 24.5$, not significant
17.11 $U = 7.5$, very significant
17.12 $z = 1.2$, not significant
17.13 $z = 2.68$
17.14 $F = 1.17$; $F_{0.05}(9, 9) = 3.18$
17.15 $F = 10$, significant
17.17 $F = 39.0$, significant
17.19 $t = 3.06$, significant
17.20 $z = 1.85$, not significant
17.21 $t = 0.41$, not significant
17.22 $t = 3.87$, significant
17.23 $z = 0.87$, not significant
17.24 $t = 3.44$, significant
17.25 $t = 2.71$, significant
17.27 $t = 1.47$

18.1 4 minus, 3 plus, not significant
18.2 8 plus, 4 minus, not significant
18.3 7 plus, 2 minus, 1 zero, significant at 0.10 level
18.5 One-tail; 10 minus, 1 plus, 1 zero; highly significant
18.6 $T = 9.5$, not significant
18.7 $T = 15.0$, significant
18.8 $T = 9$, significant
18.9 $T = 30.0$, not significant
18.10 $T = 4$, significant
18.11 $t = -0.94$, not significant
18.12 $t = 2.29$, significant
18.13 $t = -1.87$, just significant
18.14 $t = -2.12$, significant
18.15 $t = 1.6$, not significant
18.16 $t = 3.65$, highly significant
18.17 $t = 1.19$

19.1

Source	df	SS	MS	F
Between schools	2	23.17	11.58	9.93
Within schools	9	10.5	1.17	
Total	11	33.67		

Significance at 0.01; $F_{0.99}(2, 9) = 8.02$.

19.2

Source	df	SS	MS	F
Between areas	2	643	321.5	15.51
Within areas	15	311	20.73	
Total	17	954		

Significance at 0.01.

19.3 MS between $= 6.67$; MS within $= 5.92$; $F = 1.12$, not significant

19.4 MS between $= 14.39$; MS within $= 3.17$; $F = 4.54$, significant at 0.05

19.5

Source	df	SS	MS	F
Between industries	3	2.52	0.84	0.75
Within industries	4	4.48	1.12	
Total	7	7.00		

Not significant.

19.6 MS between $= 0.401$; MS within $= 0.155$; $F = 2.59$, not significant

19.7 Max distance $= 2.14$; no difference between nursery and Seseme but both are higher than the control group.

19.8 Max distance $= 6.83$; all are significant differences

19.10 Max distance $= 2.67$

19.13

Source	df	SS	MS	F
Between months	5	121.3	24.27	1.28
Between areas	2	643.0	321.5	16.94
Residual	10	189.7	18.97	
Total	17	954		

Only areas are significant (at 0.01); months are not significant.

19.14

Source	df	SS	MS	F
Between cars	3	11.0	3.67	0.55
Between brands	3	20.0	6.67	1.0
Residual	9	60.0	6.67	
Total	15	91		

Neither is significant.

19.15

Source	df	SS	MS	F
People	2	14.78	7.39	
Drugs	2	28.78	14.39	
Interaction	4	29.22	7.30	18.95
Within	9	3.50	0.39	
Total	17	76.28		

The interaction is significant at 0.01 level, therefore do not test the other F values.

20.1 $C = 0.105$
20.2 $C = 0.53$
20.3 $C = 0.13$
20.5 $C = 0.32$
20.8 $r = -0.969$
20.9 $r = 0.641$
20.10 0.843
20.11 0.956
20.13 0.89
20.15 0.978

20.16

Source	df	SS	MS	F
Regression	1	86.01	86.01	75.31 Significant
Error	14	15.99	1.14	
Total	15	102.00		

20.17

Source	df	SS	MS	F
Regression	1	315.2	315.2	175.5
Error	8	14.4	1.8	
Total	9	329.6		

20.19 Dropping the last category

Source	df	SS	MS	F
Regression	1	50.27	50.27	71.98
Error	9	6.29	0.699	
Total	10	56.56		

20.21

Source	df	SS	MS	F
Regression	1	2514400000	2514400000	447
Error	10	56280000	5628000	
Total	11	2570680000		

20.22 4.189 and 8.871
20.23 11.16 and 17.70
20.24 1.59 and 3.81

B.1   8
B.2   14
B.3   30
B.4   54
B.5   13
B.6   164
B.7   20
B.8   46
B.9   2.72
B.10 58
B.11 317
B.12 530
B.13 92
B.14 133
B.15 5
B.17 4
B.19 12
B.20 21
B.21 15
B.22 68
B.23 43

C.2   0.35
C.3   0.05
C.4   0.15
C.5   0.5
C.6   0.45
C.7   0.55
C.8   0
C.9   2:3
C.10 1:9
C.11 0.70
C.12 0.25
C.14 0.40

C.15 0.95
C.16 0.43
C.17 0.71
C.19 0.043
C.21 0.30
C.23 0.25

Index

A_2, 198, 337
Abcissa, 54
Abnormal variation, 204
Acceptance of hypothesis, 282–87, 336–37. *See also* Hypothesis
Acceptance region, 334, 351
Acceptance sampling, 293
Accumulator (calculator), 133
Accuracy
   in calculating means, 97
   with grouping, 62, 65, 67
   rounding off, 65
Addition, 562
Algol, 137
Alpha error, 285–91
   and beta error 342–43
   and hypothesis 334–35
Alternate hypothesis, 285–91, 334, 336–41
   and beta error, 336–41
Analysis of means, 206
   interpretation, 206–208
   limits, 207
   when σ is not known, 209
Analysis of variance, (ANOVA), 443–64
   ANOVA table
      one-way, 451–52
      two-way, 458
   basic principle, 445–48
   between samples variance, 447–48
   classifications, kinds of, 463
   comparison of variances, reasons, 446–48
   degrees of freedom, 451, 453
   distribution of means and samples, 446–48
   estimating variances, 463–64

F test, 447, 455
   graphing 456, 458–60
   interaction, 444, 458–61
   kinds of ANOVA, 461–64
   mean squares, 455
   one-way ANOVA, 449–51
   overall variance, 447
   randomization, 464
   for regression, 475–76
   sums of squares, 450–52, 457
      one-way, 450–52
      for regression, 472, 475–76
      relation among sums of squares, 452
      two-way, 457
   table for presenting data
      one-way, 449
      two-way, 459
   test for significant differences among means, 455–57
   two-way ANOVA, 457–61
   within treatments variance, 446
ANOVA. *see* Analysis of variance
Approximation to binomial distribution, 248
   normal approximation 273–75
Area
   and histogram, 59–60
   and median, 90
Attribute data. *See* Categorical data
Average, 88
   choosing between different kinds, 101–105, 310–15
   effect of shape of distribution, 104
   effect of wild values, 103
   mean, 91–101, 102–103.*See also* Mean
   median, 89–90, 102–103

midrange, 102
mode, 89, 102
Average deviation, 126
Axis, 53–54

b, (regression), 164. *See also* Regression
Basic (computer language), 142
Bell shaped distribution. *See* Normal distribution
Best linear regression line, 163–64
Beta error, 285–91
  and alpha error, 342–43
  and hypothesis, 335–41
Bias
  in estimating mean, 301
  in estimating standard deviation, 302
  in estimating variance, 301–302
Bimodal distribution, 75–76
  from two distributions 75–76
Binomial distribution, 236–48
  approximation to, 248
    normal approximation, 273–75
  assumptions, 238
  effect of changing n and p, 244
  formula, 241
  mean, 246–47
  what p is, 239–40
  what q is, 240–41
  shape, 244
  standard deviation, 246–47
  use of table, 242–43
  what X is, 236–38
Binomial formula, 241
Binomial tests
  proportions, 375–78
    differences in, 410–13
  runs test, 354–58
  sign test, 432–35

C, contingency coefficient, 469
Calculators, 129, 131–35
  accumulator, 133
  memory, 133
  programmable, 135
Categorical data. *See also* Binomial tests, Chi-square tests
  chi-square tests, 366–73, 404–10
  percent data, 29
  scale (measurement), 40
  tests, 365–78
Cause and effect, 161–62
Central limit theorem, 307–308
  and confidence intervals, 331
Central value, 88. *See also* Average
Chebyshev's inequality, 127
Chi-square distribution, 318–20
  degrees of freedom, 319
  limits, 319, 320
Chi-square tests
  general one way test, 366–69

  degrees of freedom, 368
  test for independence, 404–10
    calculating X^2, 407–408, 409–10
    contingency coefficient, 469
    degrees of freedom, 410
    finding theoretical values, 405–407, 408–10
  test of normality, 370–73
    degrees of freedom, 370–73
  two-way test, 404–10
CI. *See* confidence interval
Coding
  in calculating means, 97–101
  in calculating variances, 112–15
Comparisons
  average, 43. *See also* Means tests
  between two samples, 17–18
  direction, 44
  of independent samples, 403
  of means in analysis of variance, 455–57
  nonmathematical, 4
  shape of distribution, 43
  spread, 43
  standard given, 44
Complement, 574
Computer, 136
Confidence intervals, 17, 328–33
  effect of changing probability level, 331
  and hypothesis testing, 334–35
  for means
    variance known, 390
    variance unknown, 394
  for proportions
    difference in proportions, 412
    large samples, 377
    small samples, 376
  for regression line, 477–80
  and size of samples, 332
  for variance, 386
Continuous measurements, 228–29, 259–61
Control, 189–92
Control charts, 193–202. *See also* Control limits
  analysis of means, 204
  assumptions, 196
  and beta error, 338–41
  means chart, 198
  range chart, 196
Control limits. *See also* Control chart
  and beta error, 338–41
  for means chart, 198
    derivation, $337n$
  odds for exceeding, 197, 199
  for range chart, 196
Correlation, 20, 155–60. *See also* Correlation coefficient
  compared to regression, 167

contingency coefficient, 469
effect of combining distributions, 171
extreme values of correlation, 155, 156
linear, 155–60
multiple, 480–82
nonlinear, 171
nonparametric tests, 469–71
scatter diagram, 156
Spearman rank order test, 470–71
Correlation coefficient. *See also* Correlation
Pearson product moment, r, 156–61
effect of sample size, 160
format for grouped data, 179
formula, 158
interpretation, 156, 159, 169–73
significance of, 160–61
Spearman rank order, r_{rho}, 470–71
Cumulative frequency diagram, 67–71
modified, 70–71
Cumulative percent, 72
finding the median using, 90
Cumulative probability for normal distributions, 263
Cycles, 204

D (Kolmogorov-Smirnov test), 373
d_2, 197
D_3, D_4, 196
Data
definition, 7
gathering, 33
questions to ask, 30, 35
reasonableness, 34
relevance, 31
reliability, 30–31
reporting, 32–34
source, 7, 13 (population), 32
Decile, 84
Decision. *See also* Hypothesis
based on a sample, 221–23
consequences, 283–90
errors, 282
hypothesis, 281
Degrees of freedom
analysis of variance
one-way, 450–52
two-way, 458
chi-square (X^2)
distribution, 319
one-way test, 368
test of normality, 370, 372
two-way test, 410
correlation coefficient, 160
F distribution, 320
t distribution, 316–17
Descriptions, 51
picture, 5
statistical, 10–12
word, 3
Descriptive statistics, 11–12

Design of experiments, 19, 444
analysis of means, 206–209
analysis of variance, 445–62
df. See Degrees of freedom
Differences
in analysis of variance, 455–57
in means. *See also* Analysis of means
independent samples, 419–22
related samples, 437–38
in proportions, 410–13
assumptions, 411
in related samples, 431–38
paired t test, 437–38
sign test, 433–35
Wilcoxen matched pairs test, 435–37
Tukey's test for multiple differences, 455–57
Direction of test, 44, 350–51
Discrete measurements, 228–29
Distribution
bell shaped. *See* Normal distribution
binomial, 236–48
chi-square, 316, 318–20
chi-square over df, 385
combining several, 75–78, 198, 275–77
comparing to a theoretical distribution, 365–74
continuous, 228–29
cumulative
frequency diagrams, 67–71
Kolmogorov-Smirnov, 373–74
discrete, 228–29
F, 316. 320–22
J-shaped, 79
model of, 78
normal, 78–79, 253–76
probability, 219–21
of population, 219–21
rectangular, 79
of runs, 354–58
sampling from any, 229–31
sampling, 297
of sample means, 300–301, 304–308
of sample variances, 301–302
shape, 73–76
skewed, 74
symmetrical, 74
as a spread of values, 11
t, 316–18
theoretical, 366–67
need for, 235
U-shaped, 79
variances, ratio of, 417
Dot chart, 57
Draft lottery, 37–38

E notation, 148
Efficiency, 310
between parametric and nonparametric tests, 353

related to sample size, 312
Empirical rule, 119
  and binomial distribution, 247
Equally likely sampling, 218–19
  and random number tables, 229–31
Error
  alpha, 285–91, 334–35
  beta, 285–91, 336–41
  extreme value, 358
  in measurements, 38–9
  related to estimation, 12
  relation of alpha and beta, 342–43
Estimation. *See also* Confidence intervals, Hypothesis
  best, 327
  bias in, 301
  efficiency of, 310
  interval, 328
  of mean, 300–301, 304–308
  point, 327
  of standard deviation, 302
    using $\bar{R}$, 197
  of variance, 301–302
  variation in, 12–13
Estimator
  of mean, 300
  selecting between, 310–15
  of standard deviation, 197, 302
  unbiased, 301
  of variance, 301–302, 314
Event, 572
  mutually exclusive, 575
Exactness. *See also* error
  in measuring, 7
  precision, 8
Experiment, 33
Experimental design. *See* Design of experiments
Extreme value test, 358–61

F distribution, 320–22
  degrees of freedom, 320–22
  limits, 320–22
  shape, 320
F tests
  and analysis of variance, 447, 455
  differences in variances, 417–19
    assumptions, 417
    one- versus two-tail tests, 419
  in regression analysis of variance, 475–76
Fit to a line, 153, 163
Floating point, 148
Fluctuations, abnormal, 204
Fortan, 130
Frequency polygon, 57
Frequency. *See also* Histogram
  actual, 58
  Cumulative, 67–69
  percent, 59

  relative, 59
  scale, 55–57

Gaussian distribution. *See* Normal distribution
Graphs, 53–54
  and analysis of variance, 456, 458–60
  control charts, 192
  correlation, 155
    reason for plotting, 169, 171
  interaction, 459–60
  pairs of measurements, 153
  regression, 162–67. *See also* Regression
    deviations, 472
    fit, 472–77
    nonlinear, 483–86
  scatter diagrams, 156
  univariate, 53–54
Grouping
  histogram, 60
    unequal categories, 63–65
  means, calculating, 93
  variance, calculating, 106–107
Grouping (nonrandom variation), 204

Histogram, 54
  area, 59
  categories, 60, 65
    intervals, 60–63, 65
    limits, 60–63, 65
    midpoints, 60–63, 65
    unequal, 63–64
  frequency scale, 55
  grouping, 60–67
    accuracy with, 62
  limits on scale, 58
  rules for constructing, 58
  shape, 73
  smooting, 73
"Hold everything constant" technique, 444
Hypothesis, 14–16, 281–91
  alternate, 285–91, 334
  in analysis of variance, 455
  compared to confidence intervals, 334–35
  decisions concerning, 281–90
    acceptance, 282–87, 336–37
    consequences, 283–87, 337–38
    rejection, 282–87, 336–37
  error, 282–91
    alpha, 285–91, 334–35
    beta, 285–91, 334–35
    how related, 336, 342–43
  null, 291–92
  testing, 333–35

Independence
  probability, 576–78
  statistical, 39, 361
Independent samples, tests with, 403–404

Index of determination, 474–75
  use in selecting curve types, 484–86
Inference statistics, 12–13. *See also*
    Estimation, Hypothesis
Interval estimates, 328. *See also*
    Confidence intervals
Instability, 185, 197
Instantaneous time sequence, 185
Interaction, 444, 458–61
Interval
  category of histogram, 60–63, 65
  unequal, 63
Interpolation, in normal distribution
    tables, 269–70

J-shaped distribution, 79

Language, computer, 130, 136
$LCL_{\bar{R}}$, 196
$LCL_{\bar{X}}$, 198
Least squares. *See* Regression
Limit of category of histogram, 60–63, 65
Library program, 136
Line chart, 57
Linear correlation, 156
Linear equations, 164
Lottery, draft, 37–38

Mann-Whitney U test, 413–17
  calculating U, 415
  large samples, 416
  ranking data, 414
  reason for using, 413–14
  ties, 416
Many sample cases, 19, 443–44
  reason for special techniques, 443–44
Matched pairs tests, 431–38
  paired t test, 437–38
  sign test, 433–35
  Wilcoxen, 435–37
Matrix in multiple correlation, 481
Mean. *See also* Average, Means tests
  of binomial distribution, 246–47
  central limit theorem, 307
  coding, 97–101
  compared to other averages, 310–14
  effect of dividing by a constant, 100–101
  effect of subtracting a constant, 97–99
  error from grouping, 97
  estimation of, 300–301
    unbiased, 310
  general formula, 91–92
  for grouped data, 93–97
  variance of sample means, 305–308
  weighted mean, 125
Means tests
  difference in, 419–22
    choosing correct test (table), 419–20
    related samples, 437–38
  t test, 393–96
  variance known, 390
  variance unknown, 394
  z test, 389–93
Means chart, 198. *See also* Control chart,
    Control limits
Mean deviation, 126
Measurement
  accuracy, 8, 65
  adding errors, 121
  scales
    attribute, 40
    categorical, 40
    interval, 41
    ordinal, 40
    ratio, 41–42
Measurement data
  continuous or discrete, 228–29
  as sampling with replacement, 228
Measuring devices, relevance of, 31–32
Median, 89–90
  given cumulative frequency, 90
  distribution of sample, 310–14
  efficiency, 310–14
  for grouped data, 124
Memory (calculator), 133
Midpoint
  for calculating means, 93
  category of histogram, 60–63, 65
Midrange, 88
  distribution of samples, 310–13
  efficiency, 310–13
Mode, 74, 89
Model of a distribution, 78, 353
Multiple correlation, 480–82
Multiple regression, 482–83

Nonlinear relationships, 173–76, 483–86
Nonlinear regression, 171, 173–76,
    483–86
Nonparametric tests, 352–53. *See also*
    individual listings
  binomial, 374–78
  chi-square
    one-way, 366–73
    two-way (independence), 404–10
  correlation, 469–71
  contingency coefficient, 469
  extreme value, 358–61
  Kolmogorov-Smirnov, 373–74
  Mann-Whitney U, 413–17
  proportions, 374–78
    differences in, 410–13
  runs, 354–58
  sign test, 433–35
  Spearman rank order correlation,
    470–71
  Wilcoxen matched pairs, 435–36
Normal distribution
  area, 259–61
  approximation to binomial

distribution, 273–75
central limit theorem, 307–309
description, 78
determining if a distribution is normal, 362
    chi-square test, 370–73
effect of changing means and standard deviations, 254–58
empirical rule, 119
formula, 258–59
probability of individual value, 258–59, 261
interrelation, 275–76
height at a point, 258–59
ordinates, 259
standardizing, 262–63
tables
    finding X values, 271–73
    finding z values, 265–71
    for area in tail, 266–67
    for 0 to X, 265, 269
    given an area, 269–71
    interpolation, 269–70
    reason for, 262–24
    two limits, 267–69
    up to X, 265–66
    use of, 265–73
Normal probability paper, 79–81
Normality, tests for
    chi-square, 370–73
    Kolmogorov-Smirnov, 381
    reason to test, 362
Null hypothesis, 291–92

Odds, 12, 574
Ordered data
    extreme value test, 358–61
    Kolmogorov-Smirnov test, 373–74
    Mann-Whitney U test, 413–17
    scale, 40
    Spearman rank order correlation, 470–71
    Wilcoxen matched pairs, 435–36
Ordinate, 54
Operating characteristic curve, 338, 341

Paired t test, 437–38
    calculating the statistic, 437
    comparison to other tests, 438–39
Parameter, 39, 253, 353
    of normal distribution, 253
Parametric test. *See also individual listings*
    means
        difference in independent samples, 419–22
        difference in related samples, 437–38
        variance known, 388–93
        variance unknown, 393–96

variance, 385–88
    difference in, 417–19
Percent
    change, 31–32
    relative, 59
Percentile, 71
Population, 13, 296
    binomial distribution, standard deviation of, 244
    confidence intervals for, 328–29
    distribution, 219–21
    effect of replacement, 224–28
    finite versus infinite, 225–26
    relation of samples from, 15–16
    selection of, 13–14
    standard deviation, 107
    variance, 107
    variance of samples from, 307
Power curve, 338, 341
Power of a statistical test
    comparison among tests, 438–39
    most powerful test, 439n
Prediction. *See also* Regression
    linear, 163–68
    range of, 169
Probability, 12, 217–24, 571–78
    addition rule, 575
    binomial distribution, 238–40
    comparison with statistics, 217–24
    complement, 574
    counting formulas, 573n
    definitions, 219, 571, 573
    equally likely cases, 571
    event, 572
    independence, 576–78
    joint, 577–78
    multiplication rule, 577–78
    possibility, 571
    relative frequency, 573
    rules, 573–75, 577
    sample space, 572–73
    underlying principle of, 217–21
Probability distribution, 219–21, 235
    effect of replacement, 224–27
    sampling from, 219–21
Probability level, 350. *See also* Significance level
Probability paper, normal, 79
Process, 185
    control of, 189–92. *See also* Control chart
Program (computer), 130
    library program, 136
Programmable calculator, 135
Proportions. *See also* Binomial distribution
    difference in, 410–13
    sample size for estimating, 383
    tests for, 375–78

q (Tukey's test), 456
Quality control, 27, 192–203. *See also* Control, Control charts
Quartile, 84

r (Pearson product moment correlation coefficient), 156–61. *See also* Correlation coefficient.
r_{rho} (Spearman rank order correlation coefficient), 470–71
$\bar{R}$, 196
r^2, interpretation, 473–75
Random numbers, equally likely sampling, 229–31
Random number table, 229–31
Random sampling, 37, 218–19
  in analysis of variance, 464
Randomness
  distribution of sample means, 380
  nonrandom variation, 203–204
  runs test for, 354–58
Range, 105
  efficiency of, 314
Range chart, 196. *See also* Control chart
Range of prediction in regression, 168, 172
Rectangular distribution, 79
Regression, 20, 162–66
  best line, 163
  calculating b, 164
  cause and effect, 162
  compared to correlation, 167
  effect of combining curves, 171–72
  how good is the fit, 471–77
  index of determination, 474–75
  linear regression equation, 163–64
  nonlinear, 171, 173–76, 483–86
  predicting X from Y, 166
  r^2, interpretation of, 473–75
  range of prediction, 168, 172
  residuals, 472–73
  sums of squares, 472–77
Rejection of hypothesis, 282–87, 336–37. *See also* Hypothesis
Rejection region, 334, 351
Related samples
  test, 431–38
  comparison among tests, 438–39
Relationship, 152–55. *See also* Correlation, Regression, Nonlinear relationship
  cause and effect, 161
  nonlinear, 171, 173–76, 483–86
  prediction, 152, 161, 163
Relative cumulative frequency, 69
Relative efficiency, 312
Relative frequency, 59, 67
Relative percent, 59
Replacement, 224–28
  and random number table, 229–30
  and sample size, 227
Residuals, 472
Rounding off, 65
Runs test (above and below the median), 354–58

Sample
  standard deviation, 107
  variance, 107
  variation between, 193–94
Sample size, 36, 396–97
  alpha and beta error, 342–43
  and binomial distribution, 244
  and binomial test, 375
  and confidence intervals, 332
  and control limits, 196–98
  effect on correlation coefficient, 160
  efficiency, 312–14
  for estimating means, determining, 400
  for estimating proportions, determining, 383
  randomness, 37
  and replacement, 227
  and significant difference, 396–97
Sample space, 572–73
Sampling, 295–97
  from any distribution, 229
  distribution, 297–302, 305–310, 315–16. *See also* t, Chi-square, and F distributions
  effect of sample size, 227
  equally likely, 219–20
  from finite population, 225–28
  from a normal distribution, 305–10
  random, 218–19, 229–31
  replacement, 224–28
  stratified, 37n
Scales. *See also* Measurements, Histogram
  in correlation and regression, 168
    effect of size of scale, 168
    nonlinear, 173–76
    truncated, 172
  in histograms, 55, 58, 60–65
  measurement, 40–41
Scatter diagram, 156. *See also* Correlation
Sequence of data, 185
Shifts in data, 204
Sigma, 91, 562
Sigmas, 117–21
  empirical rule, 119
  Tchebychev's rule, 127
Sign test, 433–35
  comparison to other tests, 438–39
  effect of sample size, 433–34
  limits, 434
Significance level. *See also* Statistical tests
  and confidence interval, 328–31
  of correlation coefficient, 160, 169
  one- and two-tail tests, 350–52
  and sample size, 396–97

Simulation, 229–31
Skewed distribution, 74
  binomial, when skewed, 244
  chi-square, 319
  comparing averages from, 104
  s and s^2, 302
  transformation of, 302–303
Source
  of data. *See* Population
  stability of, 190–92. *See also* Control
Spearman rank order correlation
  coefficient, 470–71
Spread. *See also individual listings*
  average deviation, 126
  range, 105
  standard deviation, 106, 116–21
  variance, 106–15
SS. *See* Sums of squares
Stability, 361. *See also* Control charts
Standard deviation, 106–21
  for binomial distribution, 246–47
  definition, 106
  estimation of, 302
    from average range, 197
  interpretation of, 116–21
  need for in describing a normal
    distribution, 254–58
  Tchebychev's rule, 127
Standard error
  of mean, 307, 310, 331
  for regression estimate, 477–80
Standardized normal distribution, 262–63
Statistic, 350
  chi-square, 318–20
  D, 373
  definition, 350
  F, 320–22
  nonparametric versus parametric,
    352–53
  r, 156–61
  r_{rho}, 470–71
  r_{xx}, 360–61
  t, 316–18
  T, 435
  test, 350
  u, 354–56
  U, 413–17
  z, 263
Statistics
  comparison with probability, 217–24
  purpose, xiii
  underlying principle of, 217, 221–24
Statistical test. *See also individual listings*
  analysis of variance, 443–64
  binomial, 375–78
  categorical data, 365–78
  chi-square
    normal distribution, 370–73
    one-way, 366–69
    two-way, 404–10

  chi-square over degrees of freedom,
    385–88
  correlation coefficient, significance
    of, 160
  direction of, 350
  extreme value, 358–61
  independence, 404–10
  Kolmogorov-Smirnov, 373–74
  Mann-Whitney U, 413–17
  means, 389–96, 419–22, 437–38
  one- versus two-sided, 350
  paired t, 437–38
  proportions, 374–78, 410–13
  shape of distribution, 366–74
  sign test, 433–35
  spread, 385–88
  t test, 393–96
  two independent samples, 403–22.
    See also Chi-square, Proportions,
    Mann-Whitney, Variance, Means
    list of tests, 404
  two related samples, 431–39. *See also*
    Sign test, Wilcoxen, paired t test
    comparison among different tests,
    438–39
  variance, 385–88, 417–19
  Wilcoxen matched pairs, 435–36
  z test, 389–93
Student's t-distribution. *See* t-distribution
Storage (calculator), 133
Subscripts, 91, 561
  double, 566–69
Summation notation, 91–92, 562–69
  double summation, 566–69
  multiple, 569
  theorems on, 565–66
Sums of squares, 450
  in analysis of variance, 450–52
  in regression, 472–77
Symmetrical distribution, 74
  comparing averages from, 104
Symbols, algebraic, 560–61
  X-values, 560–61

t-distribution, 316–18
  degrees of freedom, 316–17
  and normal distribution, 317
  shape, 317
  and significance, 318
  table, use of, 317–18
t test, 393–96
  difference in mean for independent
    sample, 419–22
  for mean, 393–96
  and multiple comparisons, 444
  paired t test, 437–38
  with related samples, 437–38
T (Wilcoxen matched pairs test), 435
Tables, use of, 491. *See also individual listings*
  and inside back cover

Tail of distribution, 350–51
  normal, 267
Tally sheet, 54–56
Tchebychev's rule, 127
Test. *See* Statistical test
Test-retest, 432
Time dependency, 185, 361
Time sharing, 130
  examples, 137–47
Transformation, 28, 483
  and normal distribution, 263
  square and square roots, 302–303
  standardizing, 263
Tree diagram, 23
Trend, 204
Type I error. *See* Alpha error
Type II error. *See* Beta error

u (runs test), 354–58
U (Mann-Whitney test), 413–17
U-shaped distribution, 79
UCI$_{\bar{R}}$, 196
UCI$_{\bar{X}}$, 198
Unbiased estimator, 301, 310
Univariate data, 42
Univariate test, 353–54
Units
  sigmas, 116
  standard deviations, 116–21
Unusual values, extreme value test, 358–61
  nonrandom variation, 203–206

Variability. *see* Variation
Variance, 106–15
  of binomial distribution, 246–47
  calculating
    for a calculator, 107–108
    coded, 112
    by the definition, 106, 107
    ungrouped, 108–109
  and chi-square, 318
  definition, 106
  effect of dividing by a constant, 113–15
  effect of subtracting a constant, 112
  estimation of, 301
  and F distribution, 320
  for grouped data, 106–107
  interpretation, 121
  of population, 106
  why s^2, 302
  of sample, 106
  table formats for calculating
    for calculators, 110
    for coded data, 113
    by the definition, 109
  tests, 386–88, 417–19
Variation
  between samples, 305–10
    control charts, 193–95
  and correlation, 164–73
  due to time, 188
  in measurements, 8–9
  nonrandom, 203–206
  within a sample, 26

Wild values, 35
  effect on averages, 103
  extreme values test, 358–61
Weighted mean, 125
Wilcoxen matched pairs test, 435–37
  calculating T, 435–36
  comparison to other tests, 438–39

$\bar{X}$, X-bar, 91
$\bar{\bar{X}}$, 198

Z_{α}, 207
z values, 263
z test, 389–93
  difference in means, independent samples, 419–22
    estimating sample size, 400